Topics in Applied Physics   Volume 59

Topics in Applied Physics Founded by Helmut K. V. Lotsch

Tunable Lasers

Edited by
L. F. Mollenauer, J. C. White,
and C. R. Pollock

With Contributions by
K. Cheng   M. H. R. Hutchinson   T. Jaeger   C. Lin
L. F. Mollenauer   C. R. Pollock   M. J. Rosker
C. L. Tang   C. R. Vidal   J. C. Walling   G. Wang
J. C. White

Second Updated Edition
With 230 Figures

Springer-Verlag
Berlin Heidelberg GmbH

Dr. *Linn F. Mollenauer*

AT&T Bell Laboratories, Crawfords Corner Road,
Holmdel, NJ 07733, USA

Dr. *Jonathan C. White*

Mallinckrodt Institute of Radiology, Washington University Medical Center,
510 South Kingshighway Boulevard, St. Louis, MO 63110, USA

Professor *Clifford R. Pollock,* Ph.D.

Cornell University, School of Electrical Engineering, Phillips Hall,
Ithaca, NY 14853-5401, USA

ISBN 978-3-540-55571-1   ISBN 978-3-540-47165-3 (eBook)
DOI 10.1007/978-3-540-47165-3

Library of Congress Cataloging-in-Publication Data. Tunable Lasers / edited by L. F. Mollenauer and J. C. White; with contributions by K. Cheng ... [et al.]. – 2nd ed / C. R. Pollock. p. cm. – (Topics in applied physics; v. 59) Includes bibliographical references and index. ISBN 978-3-540-55571-1 (New York) ISBN 978-3-540-55571-1 (Berlin) 1. Tunable lasers. I. Mollenhauer, L.F. (Linn Frederick), 1937- . II. White, J.C. (Jonathan Curtis), 1952- . III. Pollock, C.R. (Clifford R.) IV. Series. TA1706.T86   1992   621.3'6–dc20   92-19149

© Springer-Verlag Berlin Heidelberg 1992
Originally published by Springer-Verlag Berlin Heidelberg New York in 1992

Typesetting: K + V Fotosatz, W-6124 Beerfelden
Production Editor: P. Treiber

54/3140-5 4 3 2 1 0 – Printed on acid-free paper

Preface to the Second Edition

There has been much activity in the general area of tunable lasers since the first edition of this book appeared 5 years ago. In this second edition, I have tried to compile a listing of significant developments and references in the nine topics contained in the first edition. There are, without doubt, some important papers or references which were overlooked. To the authors of such papers, and to the readers who will be short-changed, I extend my apologies. Every effort was made to get information from leaders in the fields on the topics covered in this broad text. I would especially like to thank Prof. Frank Tittel of Rice University for contributions to Chap. 2, Professor C. R. Vidal for his aid in updating Chap. 4, Prof. Chung Tang for updating Chap. 5, and Dr. Peter Wiggley of CREOL for his help on fiber Raman lasers.

As in the first edition, dye lasers are not included here. They are covered in detail in the books edited by F. P. Schäfer (*Dye Lasers*, 3rd edn., Topics Appl. Phys., Vol. 1), F. J. Duarte (*High-Power Dye Lasers*, Springer Ser. Opt. Sci., Vol. 65) and M. Stuke (*Dye Lasers: 25 Years*, Topics Appl. Phys., Vol. 70). The reader should also be aware of the classic text by W. Koechner on *Solid-State Laser Engineering* (Springer Ser. Opt. Sci., Vol. 1), which recently appeared in a third edition. Several topics are discussed only briefly in Chap. 10 due to the existence of this treatise.

I would like to thank Dr. Helmut Lotsch of Springer-Verlag for requesting that I serve as an editor of this second edition, and for his suggestions on topical coverage depth.

Ithaca, NY
January 1992

C. R. Pollock

Preface to the First Edition

Ever since the invention of the laser itself, the spectroscopist has dreamed of lasers that could be tuned continuously over whatever set of resonances he wished to study. Two developments of the mid-1960s — the optical parametric oscillator and the dye laser — were the first to begin to fulfill that dream. The cw dye laser, with its ability to produce extremely narrow linewidths, was particularly successful and revitalized the study of atomic physics.

Other, complementary developments soon followed. These included the excimer, color center, and high pressure gas lasers, as well as Raman shifting and four-wave mixing techniques for further extending the tuning ranges of such primary tunable laser sources. By the end of the 1970s, continuously or quasi-continuously tunable coherent sources were thus available for the visible and the near infrared, and a good part of the ultraviolet and the far infrared.

Despite the existence of a number of excellent treatises on individual technologies, to the best of our knowledge, no one has yet attempted to survey the entire field of tunable lasers in a single volume. The purpose of this book is to fill that void. It is particularly aimed at those who are not necessarily laser experts, but who may wish to discover quickly and with a minimum of effort the best technology to satisfy a particular problem, and what the possibilities and limitations of that technology are.

The introductory chapter describes the basic principles and common features of tunable lasers at an elementary level. The remaining chapters then describe particular laser types (or related techniques), beginning with the shortest wavelengths (excimer lasers and four-wave mixing), continuing with the visible and near infrared (the optical parametric oscillator, color center, fiber Raman, and paramagnetic ion lasers), and finally considering the longest wavelengths, with tunable high pressure IR lasers. At the request of the publishers, dye lasers are not included here, as there already exist several detailed treatises devoted exclusively to them. Also missing is a chapter on tunable lead salt diode lasers, an omission which we admit is somewhat arbitrary.

We would like to thank Dr. Helmut Lotsch of Springer-Verlag for suggesting this project several years ago and for his patience in awaiting the results. An equal note of thanks to each of the contributing authors, especially to those who supplied contributions on short order.

Holmdel and Stanford,
January 1987

L. F. Mollenauer
J. C. White

Contents

Contributors

Cheng, Kevin
  E. I. DuPont de Nemours & Co., Experimental Station,
  Central Research and Development Department, P. O. Box 80306,
  Wilmington, DE 19880-0306, USA

Hutchinson, M. Henry R.
  Imperial College of Science, Technology and Medicine,
  The Blackett Laboratory, Prince Consort Road,
  London SW7 2BZ, UK

Jaeger, Tycho
  Norsk Elektro Optikk A/S, P. O. Box 17, N-2001 Lellestrom, Norway

Lin, Chinlon
  Bell Communications Research, 331 Newman Springs Road,
  Red Bank, NJ 07701, USA

Mollenauer, Linn F.
  AT & T Bell Laboratories, Crawfords Corner Road,
  Holmdel, NJ 07733, USA

Pollock, Clifford R.
  School of Electrical Engineering, Cornell University,
  Phillips Hall, Ithaca, NY 14853-5401, USA

Rosker, Mark J.
  Rockwell International Science Center, 1049 Camino Dos Rios,
  Thousand Oaks, CA 91360, USA

Tang, Chung L.
  School of Electrical Engineering, Cornell University,
  Ithaca, NY 14853, USA

Vidal, Carl R.

Max-Planck-Institut für Extraterrestrische Physik und Astrophysik, Giessenbachstrasse, W-8046 Garching, Fed. Rep. of Germany

Walling, John C.

5833 Lomond Drive, Santiago, CA 92120, USA

Wang, Gunnar

Norwegian Defence Research Establishment, P.O. Box 25, N-2007 Kjeller, Norway

White, Jonathan C.

Mallinckrodt Institute of Radiology, Washington University Medical Center, 510 South Kingshighway Boulevard, St. Louis, MO 63110, USA

1. General Principles and Some Common Features

Linn F. Mollenauer and Jonathan C. White

With 4 Figures

This chapter is intended to serve as a general introduction to tunable lasers. Basic principles of the optical parametric oscillator and of those lasers based on vibronically broadened transitions are discussed. Also briefly discussed are such common matters as optical gain, laser efficiency, laser cavities, frequency selection and linewidth, and the production of ultrashort pulses by modelocking.

1.1 Tunable Lasers and Nonlinear Spectroscopy

By making possible the study and exploitation of nonlinear phenomena, lasers have brought about a revolution in optics and spectroscopy. Before the laser, spectroscopy was confined to the study of linear susceptibilities and their resonances, with resolution and sensitivity severely limited by inhomogeneous line broadening and the feeble intensity of filtered white light. Now, however, various "hole burning" and "tagging" techniques [1.1] (based on the laser's ability to selectively manipulate populations) allow for frequency resolution limited only by transition lifetimes. Other nonlinear techniques [1.1] such as two-photon absorption or CARS (coherent anti-Stokes Raman spectroscopy), make accessible transitions that are either forbidden or are extremely weak in linear spectroscopy. Furthermore, such nonlinear phenomena are of great intrinsic interest, and yield qualitatively new information about the material system under study. At the opposite extreme, of broad frequency but narrow temporal resolution, model-locked lasers allow for the study of transient phenomena well into the femtosecond regime [1.2].

Of necessity, the first experiments in nonlinear optics and laser spectroscopy were performed with lasers of "fixed" frequency. (The frequencies were potentially tunable, but over a span of at most a few wave numbers.) Thus, those experiments were largely limited to the study of the laser medium iteself or to the measurement of nonresonant susceptibilities.

It was clear, however, that the benefits of laser spectroscopy could be fully realized only with lasers that could be continuously tuned through particular resonances of the system of interest. Thus it became the spectroscopist's dream to have lasers of "broad" tunability. Two developments of the mid-sixties, the optical parametric oscillator [1.3] (see [Ref. 1.4, Chap. 3]), shortly followed by the dye laser [1.5,6] promised to fulfill that dream. The cw dye laser, with its ability to produce extremely narrow linewidths, was particularly successful, and became the paradigm for a large class of tunable lasers based on "vibronically broadened" transitions [1.7].

Following these early discoveries, great progress has been made in the generation of tunable, coherent light, and techniques to span the region from the vacuum ultraviolet to the far infrared are now available. High-energy excimer lasers based on bound-free molecular transitions have permitted tunable laser operation over much of the ultraviolet spectrum. Nonlinear frequency shifting techniques allow increased spectral coverage with just a few tunable, primary laser sources. Two techniques of frequency conversion have been particularly successful, namely, stimulated Raman processes and parametric sum or difference frequency generation, such as four-wave mixing [Ref. 1.4, Chap. 5]. These techniques are now routinely employed in the physics and chemistry laboratory.

In general, the principles of the various techniques for the generation of tunable coherent light are most efficiently described in the individual chapters. Nevertheless, in the following, we shall sketch the basic operating principles underlying the two most fundamental classes of tunable laser, viz., the optical parametric oscillator and those lasers using vibronically broadened transitions.

1.2 The Optical Parametric Oscillator

It is perhaps not surprising that the first broadly tunable laser evolved from the understanding of nonlinear optics made possible by the laser itself. In an optical parametric oscillator (OPO), a suitable nonlinear crystal is placed in an optical cavity and pumped by another laser, such that two new frequencies, ω_S (the "signal" frequency) and ω_I (the "idler" frequency) are produced [Ref. 1.4, Chap. 3]. As both energy and momentum must be conserved, one has

$$\omega_P = \omega_S + \omega_I \tag{1.1}$$

and

$$n_P \omega_P = n_S \omega_S + n_I \omega_I \tag{1.2}$$

$$(k_P = k_S + k_I)$$

as the waves are colinear. Thus, the desired output (the "signal" frequency) results from difference frequency generation between the pump and idler frequencies. Nevertheless, unlike ordinary difference generation between two laser beams, in the OPO, only the pump is supplied externally, while the idler is an internal byproduct of the nonlinear process.

From (1.1,2), one can easily show that

$$\omega_S = \omega_P \frac{n_P - n_I}{n_S - n_I} \ . \tag{1.3}$$

Equation (1.3) can be satisfied only in anisotropic crystals; however, in such crystals, for a given $n(\omega)$, (1.3) has just two solutions. Tunability is thus obtained by varying $n(\omega)$, by adjusting either the angle between the crystal axes and the light beams, or the crystal temperature. The overall tuning range can be large, and frequency ratios of $3:1$ or more have been obtained in practice.

In principle, the cavity of an OPO can be resonant for both the signal and idler frequencies at once. Nevertheless, in such a doubly resonant cavity, the need to satisfy the cavity resonance condition for two frequencies simultaneously, in addition to satisfying (1.1), results in poor output frequency stability of the OPO. That is, it can be shown [1.8] that a change in cavity length δl will result in a shift in output frequencies

$$\delta \omega \approx \frac{n_I}{n_I - n_S} \omega_P \frac{\delta l}{l} \ . \tag{1.4}$$

Thus, since the quantity $n_I/(n_I - n_S)$ is typically of order 100, the frequency instability of a doubly resonant OPO will be several orders of magnitude larger than in an ordinary laser. Thus, the OPO usually is constructed with a singly resonant cavity, i.e., the cavity mirrors are made to have low reflectivity at the idler frequency.

On the other hand, the singly resonant cavity requires a much greater pump power for attainment of laser threshold, such that in practice, the OPO must be pumped with a Q-switched laser. The accompanying short pulse duration (typically < 10 ns) then places its own limitations on frequency definition of the OPO. Nevertheless, recent developments, such as the use of nonlinear crystals as guiding fibers, may soon make possible the creation of a cw OPO; such an OPO would be expected to be capable of precise frequency definition.

1.3 Tunable Lasers Based on Vibronically Broadened Bands

A large and important class of tunable lasers is based on the vibronically broadened transitions that can occur in certain gain media, such as organic dyes (usually in liquid solvents), color centers, and certain transition metal ions in crystalline hosts. When such a medium is placed in a tuned cavity and pumped above laser threshold, stimulated emission can be made to occur at any desired frequency within the emission band. Furthermore, as will be shown, the vibronically broadened emission bands are *homogeneously* broadened, meaning that each dye molecule, color center, or ion of the gain medium is equally capable of contributing its energy at the desired frequency. Thus, when the laser is operated well above threshold, the broad, spontaneous emission is largely suppressed, and most of the emission is concentrated at one frequency, that of the selected laser mode. Such spectral condensation is a common feature of lasers based on homogeneously broadened bands.

In the following, the vibronic broadening mechanism will be described in some detail. For convenience, let us refer to the dye molecule, color center, or other ionic or atomic system used to provide gain as the "optical gain center" or simply "center". For the gain centers of interest here, the immediately surrounding atoms or ions of the host provide much, and often nearly all, of the potential experienced by the center. Thus the center is intimately coupled to the surrounding medium and its phonons, and in particular to certain localized phonon modes. It is this strong coupling that makes the optical absorption and emission appear mostly as broad bands. As will be shown, the band widths are typically several times greater than the characteristic phonon energy, and hence can be as great as $\approx 1000\,\mathrm{cm}^{-1}$.

The major features of vibronically broadened bands can be obtained from the simple *configuration coordinate model* [1.9, 10]. In that model, only one predominant mode, usually a breathing mode of the surrounding atoms or ions, is considered. Let Q be the associated generalized coordinate. It is assumed that the overall wave function can be written as

$$\Psi_{k,n}(Q,r) = \Phi_k(Q,r)\chi_n(Q) \tag{1.5}$$

where Φ_k and r are the electronic wave function and position, respectively, and where χ_n is a simple harmonic oscillator eigenfunction corresponding to vibrations of the generalized coordinate.

The configuration coordinate diagram of Fig. 1.1 shows the ground and first optically excited states of a center, where the two parabolas represent the associated vibronic potentials. For the ground state, the system is char-

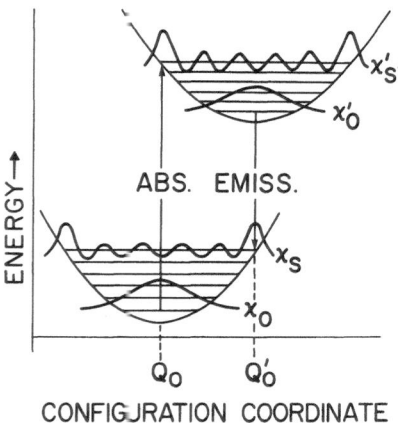

Fig. 1.1. Configuration coordinate diagram, showing ground and optically excited states of a center with vibronically broadened absorption and emission bands (see text)

acterized by effective ionic mass M, vibrational frequency ω, and an average coordinate value Q_0; for the excited state, the corresponding quantities are designated with primes.

The probability of an absorption transition involving the particular vibronic states χ_n and χ'_m is proportional to the square of the quantity

$$\langle \Psi_2 | eR | \Psi_1 \rangle \langle \chi'_m | \chi_n \rangle \tag{1.6}$$

where the first term is the usual electric dipole matrix element. For most gain centers of interest here, $\delta Q \equiv Q'_0 - Q_0$ is large enough that the overlap integrals take on their greatest values for a range of $m > n$. Thus the absorption tends to involve the creation of several phonons; also the "no-phonon" line (corresponding to transitions for which $m = n$) is weak and often undetectable. The Huang-Rys S parameter, defined (for the excited state) as

$$S' \equiv \tfrac{1}{2} M' \omega'^2 (\delta Q)^2 / h \omega' \tag{1.7}$$

represents the mean number of phonons generated in an absorption. (A similar, but not necessarily identical S applies to the ground state.)

For low temperatures, absorption takes place primarily from the lowest vibronic state (χ_0). The relative sizes of the overlap integrals $\langle \chi_0 | \chi'_m \rangle$ tend to reflect the Gaussian shape of the χ_0 function, since for $m \gg 0$ ($m \sim s'$) the χ_m tend to be strongly peaked near one overlapping point (the neighborhood of the limit of classical motion) and oscillate rapidly elsewhere. As the temperature is increased, higher vibronic levels of the ground state become populated and contribute to the absorption. This causes the band to broaden further, its shape to become less Gaussian, and its peak to shift position.

Strictly speaking, according to our simple model, the absorption should appear as a sequence of sharp bands, each representing the creation of a specific number of phonons; however, in reality, many secondary phonon modes of differing frequencies are involved, and the pattern fills out to form a smooth and usually featureless band. Note that for a given photon energy within the band, each center has equal probability of absorption. Thus, the band broadening is indeed *homogeneous*, with all the positive implications for efficient laser operation discussed above.

Following optical excitation, the highly excited local mode rapidly dissipates its energy through coupling to phonon modes of the bulk host, and the system relaxes (again, for low temperature) to the vibrational state χ'_0. Substantial changes are also often involved in the electronic part of the wavefunction Ψ_2. The system is then said to be in the relaxed-excited state (RES), and in the relaxed configuration. Direct measurements on both dyes and color centers have shown that except for extremely low temperatures ($T \lesssim 30 \, \text{K}$) the characteristic time for such relaxation is in the subpicosecond range. Thus, in the practical temperature range for laser operation, the relaxation is so fast that significant luminescent emission takes place only from the RES. Finally, following emission, a second rapid relaxation returns the system to the vibronic ground state (χ_0) of the normal configuration.

The shape of the emission band will be determined similarly to the absorption. It should also be clear from Fig. 1.1 that the mean photon energy will be lower for emission than for absorption; this reduction is known as the Stokes shift. For laser action, a center with Stokes shifted bands in effect constitutes a so-called "four-level" system. For such a system, the populations of the χ_0' and χ_s levels of the emitting system are always inverted for any finite rate of optical pumping, thus making it easier to obtain optical gain.

1.4 Optical Gain

For a given rate of excitation, the optical gain of a laser medium is maximum when the rate of stimulated emission is zero or negligibly small. Nevertheless, as laser action builds up, the increasing rate of stimulated emission reduces the excited state population; equilibrium is achieved when the gain is thus lowered to a value just large enough to overcome cavity losses. The former, maximum gain is known as the small signal gain. The magnitude of small signal gain is of central importance to laser performance. With high gain, greater cavity loss can be tolerated, high efficiency is more easily attained, cavity alignment is less critical, and attainable tuning ranges become broader. High gain is also useful in the production of short pulses by mode locking.

The net gain of a uniformly excited medium is $G \equiv I_{out}/I_{in} = \exp(\alpha z)$ where z is the gain path length. The net gain coefficient α can be written as $\alpha = \alpha_G - \alpha_L$ where α_G is the coefficient calculated for the inverted luminescence levels alone, and where α_L is the coefficient of absorption loss. Of course, only when $\alpha_G > \alpha_L$ can there be a net gain. Even when the unexcited gain medium is perfectly transparent in the region of the luminescence band, the optically excited centers themselves may absorb there; this is the phenomenon of *self-absorption*. Thus, not all materials that luminesce efficiently are potential laser-gain media. In practice, self-absorption has been known to limit the tuning range possible with certain dyes and transition metal ions to a fraction of that expected from the width and shape of the luminescence band.

For a four-level system, with a Gaussian luminescence band of full width at the half-power points $\delta\nu$, the gain cross section σ_0 at the band peak can be calculated from the well-known formula

$$\sigma_0 = \frac{\lambda_0^2 \eta}{8\pi n^2 \tau_l} \frac{1}{1.07\,\delta\nu} \tag{1.8}$$

where λ_0 is the wavelength at band center, n is the host index, η is the quantum efficiency of luminescence, and τ_l is the measured luminescence decay time.

(The quantity τ_1/η is the true radiative decay time.) In the absence of self-absorption, the gain coefficient α_0 at the band peak is then simply computed as $\alpha_0 = \sigma_0 N'$, where N' is the population density in the relaxed excited state.

For operation below or just at the threshold of laser action, N' and hence the gain, is a simple function of the pump intensity alone. Since the pump rate out of the ground state is equal to the photon absorption rate, one has $Nu = \beta I/E_p$, where u is the pump rate, β is the absorption coefficient at the pump wavelength, E_p is the pump photon energy, and I is the pump beam intensity. Thus, $N' = Nu\,\tau_1 = \beta(I/E_p)\,\tau_1$. Combining the above equations, one obtains for the *small-signal* gain coefficient [1.11]:

$$\alpha_0 = \frac{1}{8\pi} \frac{\lambda_0^2}{n^2} \frac{\eta}{(1.07\,\delta\nu)} \frac{\beta I}{E_p} \, . \tag{1.9}$$

The large-signal gain coefficient can be obtained by multiplying (1.6) by the ratio of spontaneous to total (stimulated plus spontaneous) emission rates.

1.5 Laser Efficiency

The great variety of pumping and cavity arrangements used in tunable lasers mitigate against an all-encompassing discussion of efficiency. Thus, in general, specialized treatment will be required for each particular class of laser. Nevertheless, it is possible to write down an approximate expression for the efficiency of optically pumped lasers, and in particular to show from that expression the close connection between efficiency and gain.

Even for optically pumped lasers, exact theoretical treatment is difficult; the gain coefficient, pump rate, and emission rate are all interdependent, and all vary significantly over the region of interaction in the gain medium. Therefore, let us consider a simplified model in which those variable quantities have been replaced by suitable averages. One can then easily derive the following relation [1.11]

$$\eta = f \frac{h\nu_1}{h\nu_p} \frac{T}{L+T} \left(1 + \frac{\ln[(1-L)(1-T)]}{2\alpha l} \right) , \tag{1.10}$$

where f is the effective fraction of pump photons absorbed by the gain medium, $h\nu_1$ and $h\nu_p$ are the laser and pump photon energies respectively, $(1-L)$ is an effective end mirror reflectivity, where L represents the effects of total internal cavity loss, $(1-T)$ is the output mirror reflectivity, and $2\alpha l$ is the ln of the double-pass, small-signal gain, i.e., (the ln of) the gain that would exist if there were no laser action.

The first three terms in (1.10) are obvious in derivation. That is, only those pump photons actually absorbed in the gain region are used and η must be degraded by the ratio of photon energies. The third term simply reflects division of the stimulated emission into two parts: the useful fraction, transmitted by the output coupler, and that part dissipated internally.

The origin of the final term, which represents the ratio of stimulated to total emission, is less obvious. Note that for the laser operating at threshold, $(1 - L)(1 - T)\exp(2\,\alpha l) \cong 1$, and hence the final term becomes nearly zero, as expected. On the other hand, for operation far above threshold, $2\,\alpha l$ is considerably greater than $-\ln[(1 - L)(1 - T)]$, and the final term should be a fair fraction of unity. Nevertheless, $2\,\alpha l$ cannot increase indefinitely with increasing pump power, but is limited by the density of laser-active centers and by the gain cross section σ.

The fraction f is often difficult to evaluate accurately, especially in transversely pumped lasers, where it is often small. Nevertheless, in a coaxially pumped laser with a large absorption coefficient at the pump frequency, f can approach unity. Also, if the gain is great enough, efficient operation can be obtained with T large enough to swamp out internal cavity losses, i.e., where the term $T/(L + T)$ is close to unity. Thus, in a coaxially pumped laser with high small signal gain, it is possible for the overall efficiency to approach the ratio of output to pump photon energies. This ideal figure is seldom realized in practice however, and efficiencies at band center are more often less than half that value.

1.6 Optical Cavities for Tunable Lasers

The efficiency and frequency definition of tunable lasers are strongly influenced by the spatial nature of the cavity mode. Therefore, most tunable lasers use the simplest possible mode, known as the "fundamental" or "Gaussian" mode, because of its simple Gaussian intensity profile; precautions are usually taken to assure that the laser will operate in that spatial mode to the exclusion of all others. One primary advantage of the fundamental mode is that it allows for production of the minimum focal spot size.

1.6.1 Properties of the Gaussian Mode

Figure 1.2 illustrates the principal features of the Gaussian mode. The distance at which field amplitudes are $1/e$ times that on axis is called w; at the beam waist, $w = w_0$ and $z = 0$. The mode is characterized by any two of the three basic parameters λ (wavelength), w_0 and b, the *confocal parameter,* defined as the distance between the points where $w = \sqrt{2}\,w_0$. The parameters are related to each other as follows

$$b\lambda = 2\,\pi n w_0^2 \qquad\qquad (1.11)$$

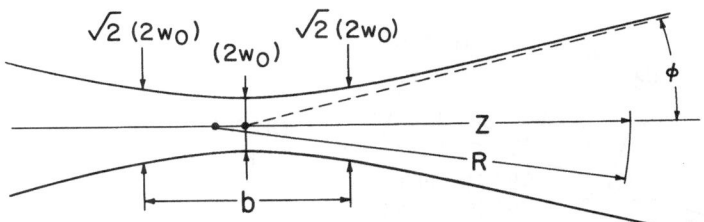

Fig. 1.2. Gaussian mode in the region of its beam waist (see text)

where n is the refractive index. It can be shown [1.12] that $w(z)$ expands as

$$w(z) = w_0\sqrt{1+(2z/b)^2}\qquad(1.12)$$

and that the radius of curvature of the wavefront, $R(z)$, is given by the expression

$$R(z) = z\left[1+\left(\frac{b}{2z}\right)^2\right].\qquad(1.13)$$

As required by symmetry, $R(0) = \infty$, but note that R quickly descends to a minimum, $R(b/2) = b$, from whence it begins to approach the asymptote $R(z) = z$. The behavior of R is of great importance in cavity design.

The domain $|z| \gg b/2$ is known as the far-field region. In that region the fundamental mode behaves like a point source of geometrical optics; i.e., $R(z) \cong z$, as already noted, and $w \cong (2w_0/b)z$. From the latter approximation, one can obtain a far field (half) angle ϕ, where

$$\phi = 2w_0/b = \lambda/\pi w_0 .\qquad(1.14)$$

1.6.2 The Astigmatically Compensated Cavity

In cw dye and color center lasers, where the necessary gain can be achieved over a path of a few millimeters or less, the gain medium is located at a tightly focused beam waist whose diameter is typically $\sim 10-20\ \mu\text{m}$. This small spot size allows the pump beam to be similarly focused, such that maximum intensity and gain are obtained for a given pump power. In this way, the pump power required for laser threshold is minimized. Also, as already noted in Sect. 1.5, such tightly focused, coaxially-pumped cavities tend to be efficient, since practically all the incident pump power can be absorbed in a volume of the amplifying medium that is coincident with that swept out by the laser mode itself.

The most commonly used arrangement for such a tightly focused cavity is shown in Fig. 1.3. Note that the cavity has two legs, the upper one in the figure characterized by a short confocal parameter (b_1, $\cong$ gain medium thick-

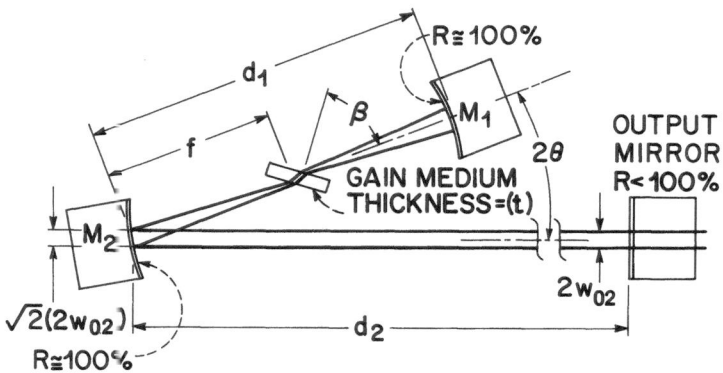

Fig. 1.3. Folded, astigmatically compensated cavity (see text)

ness, t) and by a tight beam waist w_{01} at the gain medium, while the corresponding quantities (b_2, w_{02}) for the lower leg are usually much larger. The angle between the beam and the normal to the surfaces of the gain medium β is made equal to Brewster's angle, such that reflection losses at the surfaces (for a mode whose electric field is in the plane of the paper) can be avoided. (The alternative, the use of a dye cell or color center crystal with antireflection coated surfaces is either awkward or impractical.)

Rotation of the gain medium surfaces away from the normal creates its own problem: a severe astigmatism that must be compensated if the cavity is to have low loss and a stable mode. In the design of Fig. 1.3, this compensation is provided by the opposing astigmatism induced by mirror M_2, and can be made exact through adjustment of the reflection angle 2θ.

The mode is said to be stable as long as w_2 (at M_2) is of finite size. In Fig. 1.3, let δ be such that

$$d_1 = r_1 + f + \delta \tag{1.15}$$

where r_1 is the radius of curvature of M_1. In the following, it will be assumed that $r_1 \gg t_1$, such that the beam waist (in the upper leg) occurs at a distance $\sim r_1$ from M_1. It has been shown formally [1.13], that a stable mode will then be obtained (in a compensated cavity) as long as δ lies within the range

$$0 \leq \delta \leq f^2/d_2 \equiv 2S \tag{1.16}$$

where $f(=r_2/2)$ is the focal length of M_2. The quantity $2S$ is known as the *stability range*.

There is a simple but insightful way to derive (1.16). Clearly, one limit will be obtained when $b_2 \gg d_2$; then $R \sim \infty$ everywhere in the lower leg, and the beam waist in the upper leg will then be formed a distance $\sim f$ from M_2, thereby making $\delta \cong 0$. The other limit is obtained for $b_2 \ll d_2$; since M_2 is then in the far field region of both legs of the cavity, the two beam waists (at M_0

and at the crystal) can be treated as object and image of a thin lens, respectively. One can then quickly calculate that the upper beam waist will be formed at a distance $f + f^2/d_2$ from M_2; thus $2S = f^2/d_2$.

The mirror spacing d_1 is usually adjusted such that the cavity is in the middle of its stability range. For such adjustment, one then has the following important relations:

$$b_1 = 2S \;, \tag{1.17a}$$

$$b_2 = 2d_2 \;, \tag{1.17b}$$

$$w_{01} = f(\lambda/2\pi d_2)^{1/2} \;, \tag{1.17c}$$

$$w_{02} = (d_2\lambda/\pi)^{1/2} \;. \tag{1.17d}$$

Note that the beam waist size w_{01} given by (1.17c) represents the maximum for that quantity, as d_1 is varied.

The astigmatic compensation can be computed on the following basis. First, let z be the propagation direction just outside the gain medium, and let yz be the plane of Fig. 1.3. Then the focal lengths for sagittal (xz) ray bundles and tangential (yz) ray bundles differ by

$$f_x - f_y = f/\cos\theta - f\cos\theta \;. \tag{1.18a}$$

Similarly, the sagittal and tangential rays must travel effective distances differing by

$$d_x - d_y = t\sqrt{n^2+1}/n^2 - t\sqrt{n^2+1}/n^4 \tag{1.18b}$$

through the gain medium, where t and n are the gain medium thickness and index of refraction, respectively. Thus, by equating (1.18a and b), one obtains the condition for astigmatic compensation [1.13]:

$$f\sin\theta\tan\theta = t(n^2-1)\sqrt{n^2+1}/n^4 \;. \tag{1.18c}$$

Note, however, that perfect compensation is not required. That is, failure to satisfy (1.18c) reduces the stability range, but not necessarily to zero.

It is also useful to know the cross sectional area of the beam waist inside the amplifying medium. In general, the behavior of the cross section is complicated [1.13]. Nevertheless, for the case $S \gg t$, the area is

$$A \cong \pi n w_{01}^2 \tag{1.19}$$

since from simple geometrical optics one can show that the beam waist has elliptical cross section, with minor radius w_{01} and major radius $n w_{01}$.

1.7 Frequency Selection in Tunable Lasers

To meet the demands of high-resolution spectroscopy with cw lasers, it is usually necessary to operate the laser in only one longitudinal mode at a time.

In such "single-frequency" operation, the line-widths can be a few kHz or less, and long-term frequency definition can be to within one MHz or better through locking of the laser frequency to a carefully controlled reference cavity.

The degree of tuning-element selectivity necessary for the achievement of single-frequency output depends on how the laser is operated. For example, in a cw ring laser, lasing in the most favored longitudinal (traveling wave) mode reduces the gain for all other modes to values below threshold; thus, in principle, single-frequency operation should be achieved no matter how poor the tuning element resolution. With a low resolution element, however, minor perturbations can change the identity of the most favored mode, and the laser frequency will dither about over several modes. Thus, even with a ring laser, a certain minimal degree of selectivity is required to guarantee true single-frequency operation.

Nevertheless, a great many cw lasers use a standing-wave cavity, as, for example, the one shown in Fig. 1.3. In a standing-wave cavity, there is the additional problem of *spatial hole burning* [1.14, 15]. That is, excess excited state population builds up in the region of the standing-wave nulls of the lasing mode, thereby allowing large enough gain for simultaneous operation on a second longitudinal mode. Later in this section, we shall derive an exact formula for the achievement of single-frequency operation in the face of spatial hole burning.

1.7.1 Tuning Elements

To aid in the choice of tuning elements, whether for the achievement of single-frequency operation or for other applications, the following is a brief description of the most commonly used elements and their selectivities.

a) Prisms

A simple prism, most often of fused silica, represents the least expensive and simplest tuning element. With an apex angle such that both employed surfaces are at Brewster's angle to the beam, it is virtually lossless. (Nevertheless, for operation in the 2.7 μm region of the infrared, even infrasil exhibits significant absorption, and OH^- free materials such as CaF_2 or sapphire must be used. For wavelengths greater than ~3 μm, the only transmissive materials are the alkali halides.)

In the cavity of Fig. 1.3, the beam waist at the gain medium has a fixed location (at the center of curvature of M_1) and behaves as an aperture whose diameter is the diffraction-limited spot size. Thus, when a prism is inserted in to the long arm of the cavity, it behaves much as in an ordinary spectrometer, and its resolving power [1.16] is

$$\delta \tilde{v} = \frac{1}{l_p} \left(\lambda \frac{dn}{d\lambda} \right)^{-1} , \qquad (1.20)$$

where l_p is the difference in the extreme paths through the prism. For a fused silica, Brewster's angle prism, $l_p \sim 2a$ where a is the beam diameter and the dimensionless quantity $\lambda(dn/d\lambda) \sim 2.5 \times 10^{-2}$ in the middle of the visible; for a beam of 1 mm dia., (1.20) then yields ~ 200 cm^{-1} for $\delta\tilde{\nu}$. From this estimate it can be seen that the prism constitutes a low-resolution element. It is particularly useful in mode-locked lasers, where high selectivity is often not desirable.

b) Diffraction Gratings

When higher selectivity is needed, a diffraction grating can be used. Gratings can be used in many ways, but the most common is as a first-order retro-reflector. If the grove spacing is fine enough with respect to the wavelengths involved, the incident energy (except for absorption loss) will be divided entirely between first- and zeroth-order reflections, and the latter can be used for output coupling. Furthermore, when attached to a sine-bar drive, such a grating retroreflector allows for precise and linear wavelength readout, and hence is easily adapted for wavelength scanning.

The relative power response to an incident Gaussian beam of a grating in retroreflection (in any order) is

$$R(\phi) = \exp(-2\phi^2) \; , \qquad (1.21)$$

where the phase angle $\phi = \pi l_g \delta\tilde{\nu}$, l_g is the depth of grating illuminated between the $\exp(-2)$ power points, and $\delta\tilde{\nu}$ is the detuning in cm^{-1}. From (1.21), the full width at half-maximum response of the grating is $\sim 0.375/l_g$ cm^{-1}. In a cavity with typical parameters, even without the use of beam expanding optics, l_g can be several millimeters. The corresponding half-power bandwidth of the diffraction grating is then only a few cm^{-1}.

Gratings are particularly suitable for use in the infrared, where they can have high efficiency. For example, in the near infrared, and with blaze angle chosen correctly, some gratings have exhibited first-order retroreflection of $\gtrsim 95\%$. Furthermore, those high reflectivities were nearly constant over a wide wavelength range [1.17].

Where extreme efficiency is not required, gratings can used at grazing incidence for increased selectivity. Such an arrangement is often used, for example, in pulsed dye lasers.

c) Birefringence Plates

A tuning element of more or less intermediate selectivity can be made from one or more birefringence plates oriented at Brewster's angle to the beam [1.18]. It is usual for the optic axis to be in the plane of each plate, and for the axes to be parallel to each other when more than one plate is involved. Tuning is by rotation of the plates about their normal axes. The phase lag ϕ between waves linearly polarized along the fast and slow axes, is

$$\phi = 2\pi \frac{(n_0 - n_e)t}{\lambda \sin\beta}(1 - \cos^2\beta \sin^2\alpha) \qquad (1.22)$$

where n_0 and n_e are the ordinary and extraordinary indices, respectively, t is the plate thickness, λ the wavelength, β Brewster's angle, and α is the angle between the fast axis of the birefringence plate and the s (high loss) polarization of the Brewster surfaces. In general, a linearly polarized mode is made elliptically polarized by the plates, and substantial reflection loss results. For those wavelengths for which ϕ is an integral multiple of 2π, however, the linear mode is unaffected, and there is no loss. From (1.22) one has immediately that the lasing wavelength varies with plate rotation angle (α) as

$$\lambda = (\lambda_p/m)(1 - \cos^2\beta \sin^2\alpha) \;, \qquad (1.23)$$

where $\lambda_p = (n_0 - n_e)t/\sin\beta$, and m is an integer. For crystalline quartz, $\cos^2\beta \cong 0.31$, such that a practical tuning range somewhat less than 30% wide is possible with a given set of plates.

The frequency response of a birefringence tuner is a complicated function of the number of plates, the number of additional Brewster surfaces in the

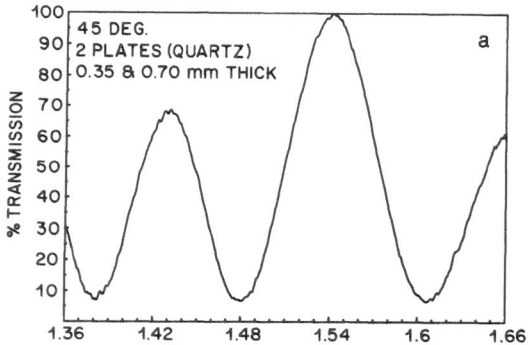

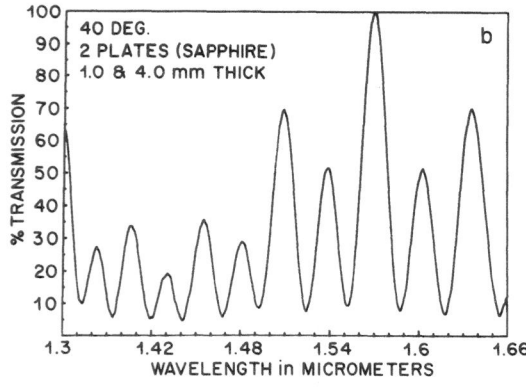

Fig. 1.4a,b. The measured double-pass response of two birefringence plate tuners for α at or near the angle of maximum selectivity (45°). Note that the quantity $n_0 - n_e \approx 0.008$ for both quartz and sapphire

cavity, and whether the resonator is a ring or standing-wave cavity [1.18]. Nevertheless, the effective bandpass is always inversely proportional to the quantity $(n_o - n_e) t$, where t refers to the thickest plate. The selectivity is also maximum for $\alpha \sim 45°$. Figure 1.4 shows the measured double-pass response of typical two plate birefringence tuners.

The limited tuning range cited above may be a disadvantage for certain applications. Since the tuner does not depend on spatial dispersion, however, the resonance wavelength is insensitive to changes in the laser beam direction; other advantages include nearly zero insertion loss and the ability to handle high intensities. Birefringence plates of appropriate thickness are often used to control the wavelength and to help shape the pulses in mode-locked lasers.

d) Etalons

A pair of partially transmissive plane-parallel mirrors, or etalon, oriented more or less normally to the beam, is often used to increase overall selectivity. The etalon usually takes the form of a solid plate (which is then tuned by tilting), or has an air gap that is varied by a piezoelectric transducer.

The transmission of an etalon whose mirrors have negligible absorption and whose normals are at angle θ with respect to the beam is [1.16]

$$T = \left(1 + \frac{4R_m}{(1 - R_m)^2} \sin^2(\phi/2)\right)^{-1} , \qquad (1.24)$$

where $\phi = 2\pi(n2d/\lambda)\cos\theta$, R_m is the mirror reflectivity, and n is the index of refraction of the material filling the etalon. Thus, the (100%) transmission maxima of (1.24) are periodic with spacing $\delta\tilde{\nu} = 1/(n2d)$; $\delta\tilde{\nu}$ is the "free spectral range". From (1.24) it can be seen that even for R_m as low as 50%, the etalon response is ~ 9 times narrower than its free spectral range.

1.7.2 Single-Frequency Operation in the Presence of Spatial Hole Burning

Consider a laser operating on just one mode, whose standing wave in the gain medium is represented by the function $\sin^2(4\pi nz/\lambda)$, where n is the index of refraction, z measures distance in the propagation direction, and λ is the vacuum wavelength. It can be shown [1.19] that the excited state population N_2 has the following spatial distribution

$$N_2(z) = \frac{u_p N_0}{u_p + w_l} [1 + a\sin^2(4\pi nz/\lambda)]^{-1} , \qquad (1.25)$$

where $a = w_{max}/(u_p + w_l)$, and where, in turn, w_{max}, u_p, and w_l are the rates of stimulated emission at the standing wave peaks (this is, of course, the maximum rate), optical pumping, and spontaneous emission ($w_l = 1/\tau_r$), respectively, and where N_0 is the total center density. It can also be shown

[1.19] that the gain available to a potentially competing mode, phase shifted by angle ϕ from the lasing mode, is simply

$$g(\phi) = g_t [1 + (G-1) \sin^2 \phi] \tag{1.26}$$

where the quantity $G = (a+1)^{1/2}$ is the maximum relative gain available to a competing mode, and g_t is the gain at threshold.

Let us define N_{min} as the excited-state population density at the nulls of the spatial distribution (1.25). G is always less than $(N_0/N_{min})^{1/2}$, *a quantity independent of the pump power*. (In practice, G is rarely much greater than ~ 10.)

Now consider the cavity shown in Fig. 1.3; let D represent the distance between the closest mirror (M_1) and the gain medium. Since all modes must have the same phase at M_1, the two potentially competing modes must be frequency shifted from each other by an amount

$$\Delta v = \frac{\phi}{2\pi} \frac{c}{D} . \tag{1.27}$$

The quantity

$$\Delta v_0 = \frac{c}{4D} \tag{1.28}$$

is known as the "spatial hole burning frequency", but it represents the spacing between simultaneously lasing modes only in the special case where the net response of all tuning elements is broad compared to Δv_0 itself.

To achieve single-frequency operation, the response profile of the tuning element(s), when multiplied by (1.26), must always yield less than unity except, of course, at the desired frequency of operation. The necessary selectivity is usually provided by one or more intracavity etalons, (usually used with a wide band tuning element such as a diffraction grating or set of birefringence plates). Scanning of the laser frequency over even a limited range then requires simultaneous adjustment of the laser cavity length and the etalon or etalons; this adjustment is often made with piezoelectric transducers. For tuning over greater ranges, the wide band tuning element must be adjusted as well.

Single-frequency operation can also be achieved with the use of a diffraction grating alone, provided the inequality

$$g_t (G-1) \leqslant \left(\frac{l_g}{2D} \right)^2 \tag{1.29}$$

is satisfied, where l_g is the depth of the illuminated spot on the grating. In this way, much of the awkwardness associated with the tracking of multiple tuning elements can be avoided. An example [1.19] of such single-knob tuning is discussed in Chap. 6.

1.8 Mode Locking

The large homogeneous bandwidths of certain tunable laser materials, especially dyes and color centers, are ideally suited to the production of ultra-short (picosecond and femtosecond) pulses [1.20]. The corresponding lasers have found as much or more use in such pulse production as in high-resolution spectroscopy. Therefore, although a full treatment of the vast subject of ultrashort pulse production is beyond the scope of this book, it is only appropriate that we at least sketch the basic principle by which such pulses are produced, i.e., mode locking.

The term mode locking refers the phase locking of many hundreds or thousands of adjacent, simultaneously oscillating cavity modes, such that they sum to form a stream of short pulses. (For example, consider a pulse 0.1 ps wide generated in a mode-locked laser with 100 MHz mode spacing; since such a pulse has a bandwidth at half intensity points of ~3000 GHz, more than 30000 modes must be locked together to generate the pulse.) Mode locking is initiated and sustained through violent gain (or loss) modulation at the frequency ($c/2L$) of the cavity mode spacing; such modulation serves to excite many side bands with that same frequency spacing. The side bands then provide, through stimulated emission, for the required mutual phase locking of the modes.

The modulation may be provided by insertion of an externally excited modulator or a saturable absorber into the cavity (loss modulation), or by pumping of the gain medium by the pulse train from a second mode-locked laser. In the latter technique, known as synchronous pumping, mode locking is achieved by adjusting the length of the pumped laser's cavity such that the time $2L/c$ corresponds exactly to the period between pulses of the mode-locked pump laser. Effects of an externally excited modulator are usually too mild for the production of the shortest pulses, and thus synchronous pumping, either by itself or combined with a saturable absorber or other nonlinear device, is most commonly used. (In Chap. 6, we will describe the combination of synchronous pumping and pulse compressive effects in an optical fiber (the "soliton laser") to generate femtosecond pulses in the infrared.)

The principal requirement for the achievement of mode locking by synchronous pumping is for a large enough gain cross section, such that the gain change ΔG brought about by a single pump pulse, is a large fraction of the net gain required to bring the laser above threshold. Thus, one has

$$\ln \Delta G \approx N\sigma/A \ , \tag{1.30}$$

where N is the number of absorbed pump photons pulse, σ is the gain cross section, and A is the cross sectional area of the beam at the crystal. When practical values for N and A are inserted into (1.30), σ is required to be $\sim 10^{-17}$ cm^2 or greater. This criterion is easily met by organic dyes and by most color centers.

References

1.1   M. Levinson: *Introduction to Nonlinear Laser Spectroscopy* (Academic, New York 1982)
1.2   For example, D. H. Auston, K. B. Eisenthal (eds): *Ultrafast Phenomena IV,* Springer Ser. Chem. Phys. Vol. 38 (Springer, Berlin, Heidelberg 1984)
1.3   J. A. Giordmaine, R. C. Miller: Phys. Rev. Lett. **14**, 973 (1965)
1.4   Y.-R. Shen (ed.): *Nonlinear Infrared Generation,* Topics Appl. Phys., Vol. 16 (Springer, Berlin, Heidelberg 1977)
1.5   F. P. Schäfer, W. Schmidt: Z. Naturforschg. **19a**, 1019 (1964)
1.6   P. P. Sorokin, J. R. Lancard: IBM J. Res. Develop. **10**, 162 (1966)
1.7   F. R. Schäfer (ed.): *Dye Lasers,* 2nd ed., Topics Appl. Phys., Vol. 1 (Springer, Berlin, Heidelberg 1977)
1.8   Y. R. Shen: *The Principles of Nonlinear Optics* (Wiley, New York 1984)
1.9   G. F. Imbusch, R. Kopelman: In *Laser Spectroscopy of Solids,* 2nd ed., ed. by W. M. Yen, P. M. Selzer, Topics Appl. Phys., Vol. 49 (Springer, Berlin, Heidelberg 1986) Chap. 1
1.10 J. J. Markham: "F-Centers in Alkali Halides", in *Solid State Physics* ed. by F. Seitz, D. Turnbull (Academic, New York 1966) Supp. 8
1.11 L. F. Mollenauer: "Color Center Lasers", in *Laser Handbook,* Vol. 4, ed. by M. L. Stitch, M. Bass (North-Holland, Amsterdam 1985) Chap. 2
1.12 H. W. Kogelnik, T. Li: Appl. Opt. **5**, 1550 (1966)
1.13 H. W. Kogelnik, E. P. Ippen, A. Dienes, C. V. Shank: IEEE J. Quantum Electron. **QE-8**, 373 (1972)
1.14 C. L. Tang, H. Statz, G. Demars: J. Appl. Phys. **34**, 2289 (1963)
1.15 C. T. Pike: Opt. Commun. **10**, 14 (1974)
1.16 F. A. Jenkins, H. E. White: *Fundamentals of Optics,* 3rd ed. (McGraw-Hill, New York 1957)
1.17 K. R. German: Appl. Opt. **18**, 2348 (1979)
1.18 A. Bloom: J. Opt. Soc. Am. **64**, 447 (1974)
1.19 N. D. Vieira, L. F. Mollenauer: IEEE J. Quantum Electron. **QE-21**, 195 (1985)
1.20 S. L. Shapiro (ed.): *Ultrashort Light Pulses,* 2nd ed., Topics Appl. Phys., Vol. 18 (Springer, Berlin, Heidelberg 1984)

2. Excimer Lasers

M. Henry R. Hutchinson

With 24 Figures

Excimer lasers are increasingly being employed as high-power sources of tunable laser light in the uv and vuv spectral regions. This chapter describes the electronic structures, basic kinetic models, various pumping configurations, and operating parameters of some of the more widely utilized excimer systems. Particular attention is given to discussion of the rare-gas dimer and rare-gas halide excimer lasers. The possibilities for new lasers using triatomic and rare-gas oxide excimer complexes are considered.

2.1 Introduction

Tunable lasers operating at wavelengths in the visible and infrared regions of the spectrum have been available for many years. However, the last ten years have seen the development of a new type of gas laser which can produce pulses of radiation in the ultraviolet and vacuum ultraviolet, a region of the spectrum which had hitherto been distinguished by a marked lack of powerful, efficient lasers. This development has been due to the discovery and study of molecules of the rare gases such as Xe_2, KrF, and XeCl. That the rare gases should form molecules may be surprising since they are well known to be chemically inert. However, as will be discussed in more detail later, bound electronically excited molecules of the rare gases can be formed from excited rare-gas atoms, even though the interaction between the constituent atoms (when in their electronic ground states) is almost entirely repulsive. These molecules are called "excimers", the term referring to a molecule which is bound in an excited electronic state, but which is dissociated or dissociative in the electronic ground state. The potential energy diagram for such a diatomic molecule AB is shown schematically in Fig. 2.1.

Excimers may be formed by the interaction between two atoms or molecules, one of which is electronically excited, e.g.,

$$A + B^* \rightarrow AB^* \ .$$

The bound molecule AB^* may then decay radiatively to the ground state and dissociate

$$AB^* \rightarrow A + B + h\nu \ .$$

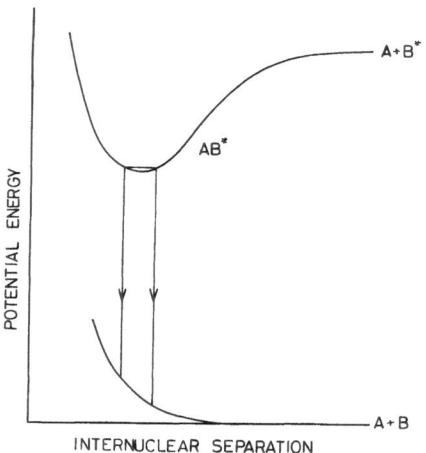

POTENTIAL ENERGY

INTERNUCLEAR SEPARATION

A + B*

AB*

A + B

Fig. 2.1. Schematic energy level diagram of an excimer AB

The ground state of the excimer may be repulsive, as shown in Fig. 2.1 or be sufficiently weakly bound so as to be unstable at normal temperatures. Excimers with strongly repulsive ground states emit radiation which is characterized by a broad continuum which appears to the long-wavelength side of the resonance lines of the excited atom (B*) from which it is formed. On the other hand, excimers with weakly bound, dissociative ground states, produce emission spectra displaying vibrational and rotational structure, characteristic of a conventional bound-bound molecular transition. Broadband excimer emission from discharges in atomic vapours of the group II B elements was observed as long ago as 1901 by *Hartley* and *Ramage* [2.1] and later *Wood* [2.2], and in 1927 *Lord Rayleigh* [2.3] attributed the emission from mercury discharges to a transition between an excited state of the Hg_2^* dimer and the repulsive ground state.

Population inversion and laser action can be created in excimers since the lifetime of the bound excited electronic state is generally much longer than that of the dissociative molecular ground state. The potential of excimers as laser media was first pointed out by *Houtermans* [2.4] in 1969, although the first demonstration of laser action in an excimer system, liquid xenon (Xe_2) was not made until 1970 [2.5]. Laser action was subsequently demonstrated in the rare-gas excimers Xe_2 ($\lambda = 172$ nm) [2.6], Kr_2 ($\lambda = 146$ nm) [2.7], and Ar_2 ($\lambda = 126$ nm) [2.8] by irradiation of the pure noble gas at very high pressures ($p > 10$ atm) with high-current, relativistic electron – beams, and this led to the development of pulsed, tunable, high-power lasers operating in the VUV region of the spectrum. These are discussed in Sect. 2.2.

The possibility of laser action utilizing heteronuclear excimers (sometimes referred to as "exciplexes") was first demonstrated with XeF in 1964 [2.9], but the real potential of these systems was not recognised until some ten years later. The rare-gas halide excimers are important as efficient, powerful sources in the ultraviolet region of the spectrum and have been the object of intense research and development in recent years. The most important rare-gas halide lasers operate on the *B-X* transitions of the diatomic excimers; and of the ten systems from which excimer emission has been observed, laser action has been produced on seven, with wavelengths between 175 nm (ArCl) and 351 nm (XeF). Lasing has also been produced on the weaker, broadband *C-A* transition of XeF ($\lambda = 490$ nm) and in the triatomic excimers Kr_2F ($\lambda = 435$ nm) and Xe_2Cl ($\lambda = 520$ nm). The *C-A* transition and triatomic lasers are discussed in Sect. 2.3.1.

The rare gases also form very weakly bound excimers with oxygen which have emission bands lying close to the atomic resonance lines of oxygen. Lasers operating on the oxides of Xe, Kr, and Ar have been developed and these are discussed briefly in Sect. 2.4. The properties of excimers and excimer lasers have been reviewed by several authors [2.10 – 13].

2.2 Rare-Gas Excimers

The broad VUV continua produced by the rare-gas excimers have been known for many years. In 1930, *Hopfield* [2.14] observed the continuum in the VUV emission of high-pressure helium and attributed it to $He_2^*(^1\Sigma_u^+) - He_2(^1X_g)$ emission. Broad-band excimer emission from all the rare gases has been studied by *Tanaka* and co-workers [2.15] and has been used for many years as a convenient continuum source for vacuum ultraviolet spectroscopy. The approximate wavelengths of the excimer emission spectra of all the rare gas and rare-gas halide excimers are indicated in Fig. 2.2.

2.2.1 Electronic Structure and Spectroscopy

The atomic ground states of Ne, Ar, Kr, and Xe have the electronic configuration $s^2p^6(^1S_0)$. For a pair of these atoms in contact, the twelve p electrons are distributed into the four valence orbitals σ_g, π_u, π_g, and σ_u. The removal of one electron to form the positive molecular ion results in four possible states, $^2\Sigma_u^+$, $^2\Pi_g$, $^2\Pi_u$, and $^2\Sigma_u^+$, depending in which of the four orbitals the vacancy occurs. If the vacancy occurs in the strongly anti-bonding σ_u orbital or the weakly-bonding π_g orbital, the resulting $^2\Sigma_u^+$ and $^2\Pi_g^+$ states are strongly and weakly bound, respectively; otherwise the states are repulsive. In the heavier gas atoms, spin-orbit splitting in the ^2P ground state of the atomic ion gives rise to further splitting of the two $^2\Pi$ states of the molecular ionic states. The excited states of the diatomic excimer are Rydberg in character, four of which are bound; $^1\Sigma_u^+$, $^3\Sigma_u^+$, $^1\Sigma_g^+$, and $^3\Sigma_g^+$.

Spectroscopic [2.16, 17] and theoretical [2.18 – 20] studies of the helium excimer have enabled detailed potential curves for He_2^* to be drawn [2.21]. Less detailed information is available for the heavier excimers, but calcula-

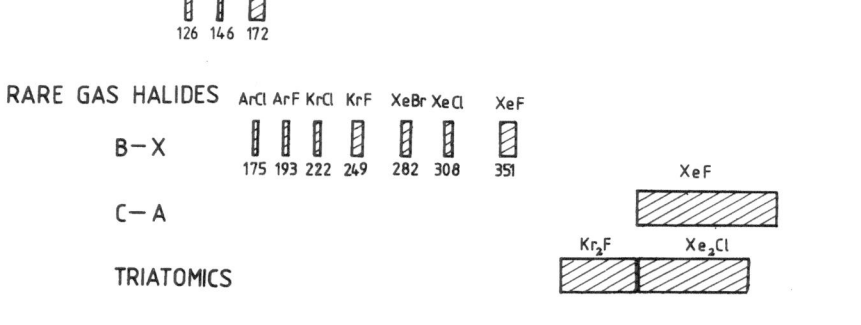

Fig. 2.2. Wavelengths of the rare gas and rare-gas halide lasers

tions of the potential curves for Xe_2^* [2.22], Kr_2^* [2.23], and Ar_2^* [2.24] have been made. The curves for Ar_2^* are shown in Fig. 2.3 and are typical of the heavier noble gases. These potential energy curves indicate the existence of strongly bound ($E_b \sim 1$ eV) $^1\Sigma_u^+ (0_u^+)$ and $^3\Sigma_u^+ (0_u^-, 1u)$ states which correlate with the atomic $^3P_1$ and $^3P_2$ states, respectively. They also show weakly bound $^1\Sigma_g^+ (0_g^+)$ and $^3\Sigma_g^+ (0_g^-, 1g)$ states which correlate whith the $^3P_1$ and $^3P_2$ levels, but with a pronounced potential maximum at interatomic distances of $R \sim 3.5$ Å. A small van der Waals minimum in the ground-state potential is also indicated.

The $^3P_0$ and $^3P_2$ states are metastable because of the $\Delta J = 1$ selection rule. At high pressures, resonance trapping of radiation from the $^3P_1$ and $^1P_1$ resonance states means that all four states may be regarded as metastable. At relatively low pressures, two-body collision processes populate vibrational excimer levels close to the dissociation limit and give rise to broadening to the long-wavelength side of the $^3P_1$ atomic resonance line. This is referred to as the "first continuum". At higher pressures (>100 Torr), excimers are formed by three-body collisions, as follows:

$$Xe^* (^3P_{1,2}) + Xe + Xe \rightarrow Xe_2^* (^{1,3}\Sigma_u^+) + Xe \ .$$

By absorbing energy, the third body enables lower vibrational levels to be populated and emission occurs over a broad continuum, as shown in Fig. 2.4. As the pressure is further increased, the increased rate of collisionally induced vibrational relaxation [2.25] establishes a thermal distribution of population in the vibrational states. The emission spectrum becomes that of a Boltzmann-averaged set of states, and is thus reduced in width at normal temperatures [2.26].

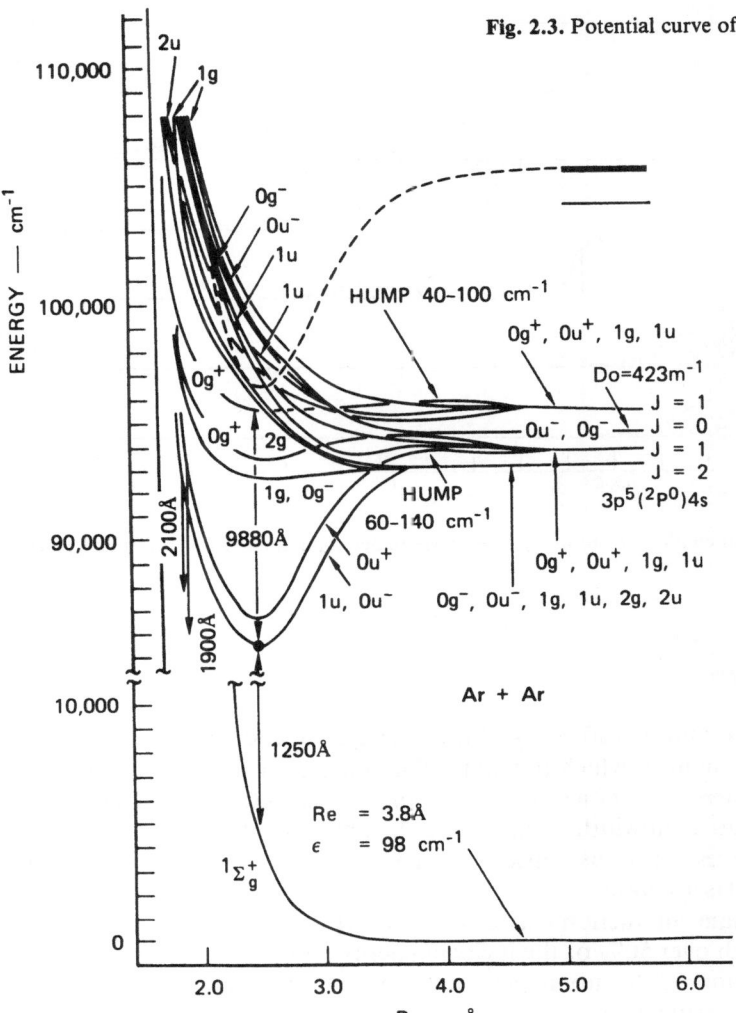

Fig. 2.3. Potential curve of the Ar₂ excimer

2.2.2 Kinetic Models

For a transition with a homogeneously broadened line of spectral width $\Delta\lambda$ and wavelength λ the stimulated emission cross section σ_s is given by

$$\sigma_s \simeq \frac{\lambda^4}{8\pi\tau c\Delta\lambda} \ , \tag{2.1}$$

where τ is the spontaneous decay time and c is the speed of light. For a four-level laser operating close to threshold, the pumping power per unit volume P required to produce a gain coefficient of unity per unit length is given by

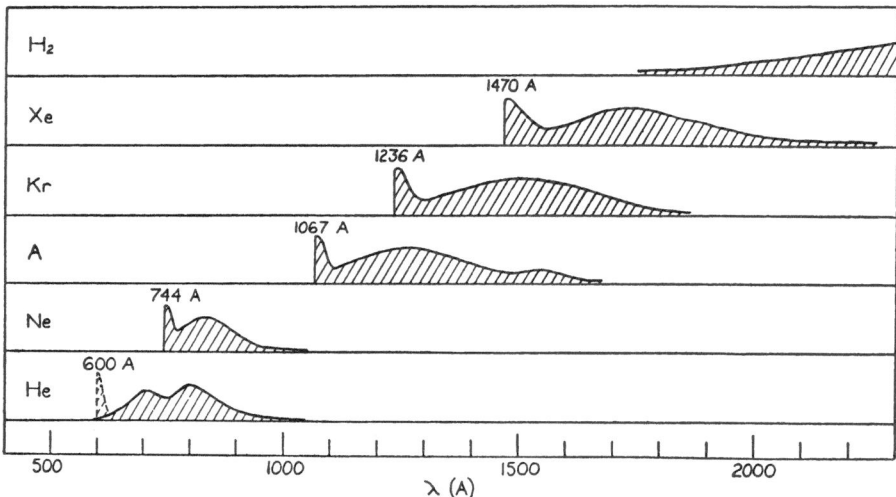

Fig. 2.4. Emission continua from rare-gas excimers under conditions of low pressure [2.15]

$$P = \frac{8 \pi h c^2 \Delta \lambda}{\lambda^5 \eta \phi} \, , \tag{2.2}$$

where ϕ is the quantum efficiency of the laser system and η is the efficiency of the kinetic mechanism which populates the upper laser level. Emission from rare-gas excimers is characterised by short wavelengths ($\lambda < 200$ nm) and relatively broad bandwidths ($\Delta \lambda \sim 15$ nm). Consequently, very high-power pumping sources, such as pulsed electron – beams, are required for the generation of laser action.

The dominant interaction between high-energy electrons and a rare gas is ionization, with over 50% of the incident energy absorbed by this mechanism [2.27]. Excitation of the atoms plays a relatively minor role, as a significant fraction of the incident energy is used in heating the gas. Thus, if excimers are formed from ions and excited atoms with unit efficiency, the overall efficiency of converting the energy of the high-energy electrons into excimer radiation would be approximately 50%. This is confirmed experimentally at gas pressures of ~1 atm and low-excitation densities, and illustrates the efficient channeling of energy into the excimer states [2.28].

The very large number of processes and atomic and molecular species in electron-beam excited noble gases might appear to preclude any attempt to construct a detailed model for the production of population inversion in noble-gas excimer lasers. However, by first identifying all the possible processes which may take place and then eliminating those which are non-essential in describing the experimental results, considerable progress has been made in describing the main characteristics of excimer emission [2.29 – 34].

Table 2.1. Reactions for rare-gas excimer production

	Reaction	Rate coefficient or cross section	Reference
(R1)	$Xe + e_p^- \rightarrow Xe^+ + e_s^- + e_p^-$		
(R2)	$Xe + e_p^- \rightarrow Xe^*(Xe^{**}) + e_p^-$		
(R3)	$Xe^* + e^- \rightleftharpoons Xe^+ + e^- + e^-$	$\sim 10^{-15}$ cm^2	[2.35]
(R4)	$Xe^+ + Xe + Xe \rightarrow Xe_2^+ + Xe$	$1.8 - 3.6 \times 10^{-31}$ cm^6 s^{-1}	[2.36–38]
(R5)	$Xe_2^+ + e^- \rightarrow Xe^{**} + Xe$	1.4×10^{-6} cm^3 s^{-1}	[2.39–41]
(R6)	$Xe_2^+ + Xe + Xe \rightarrow Xe_3^+ + Xe$	9×10^{-32} cm^6 s^{-1}	[2.34]
(R7)	$Xe_3^+ + e^- \rightarrow Xe^{**} + Xe + Xe$	$\sim 9.5 \times 10^{-5}$ cm^3 s^{-1} (estimate)	[2.34]
(R8)	$Xe^*(^3P_1) + e^- \rightleftharpoons Xe^*(^3P_2) + e^-$		
(R9)	$Xe^*(^3P_1) + Xe + Xe \rightarrow Xe_2^*(^1\Sigma_u^+) + Xe$	1.7×10^{-32} cm^6 s^{-1}	[2.42]
	$Xe^*(^3P_2) + Xe + Xe \rightarrow Xe_2^*(^3\Sigma_u^+) + Xe$	7.0×10^{-32} cm^6 s^{-1}	[2.43]
(R10)	$Xe_2^*(^1\Sigma_u^+) + e^- \left.\right\} \rightleftharpoons Xe_2^*(^3\Sigma_u^+) + e^- \left.\right\}$	2.3×10^{-7} cm^3 s^{-1}	[2.34]
(R11)	$Xe \left.\right\}$ $Xe \left.\right\}$	$1.2 - 4.3 \times 10^{-13}$ cm^3 s^{-1}	[2.34,44]
(R12)	$Xe_2^* + Xe_2^* \rightarrow Xe_2^+ + Xe + Xe + e^-$	3.5×10^{-10} cm^3 s^{-1} 8×10^{-11} cm^3 s^{-1}	[2.45] [2.46]
(R13)	$Xe_2^* + e^- \rightarrow Xe_2^+ + 2e^-$		[2.35]
(R14)	$Xe_2^*(^1\Sigma_u^+) \rightarrow 2Xe + h\nu$	1.8×10^8 s^{-1}	[2.47]
	$Xe_2^*(^3\Sigma_u^+) \rightarrow 2Xe + h\nu$	1.0×10^7 s^{-1}	[2.47]
(R15)	$Xe_2^* + h\nu \rightarrow Xe_2^+ + e^-$	2×10^{-18} cm^2	[2.48]
(R16)	$Xe_2^*(^1\Sigma_u^+) + h\nu \rightarrow 2Xe + 2h\nu$	1.5×10^{-17} cm^2	[2.34]
	$Xe_2^*(^3\Sigma_u^+) + h\nu \rightarrow 2Xe + 2h\nu$	1.0×10^{-18} cm^2	[2.34]

The principal reactions relevant to the kinetics of the xenon excimer laser are shown in Table 2.1.

Reactions (R1,2) (Table 2.1) represent the ionization and excitation of ground-state atoms by fast primary or secondary electrons. Reaction (R3) describes the ionization of excited atoms by electrons. At high pressures, the atomic ions form molecular ions by three-body association (R4) and this is followed by rapid (≤ 1 ns) dissociative recombination (R5) or, as suggested by *Werner* et al. [2.34], by the formation of triatomic ions (R6) which then undergo rapid dissociative recombination (R7). The exact spectroscopic description of the reaction products Xe** is unspecified but is chosen to represent the behavior of levels lying above the $6p$ atomic state and below the minimum of the potential well of Xe. Electron collisions (R8) tend to equalize the populations of the $6s(^3P_1, {}^3P_2)$ states, and the rare-gas excimers are formed from the $6s$ atomic states by three-body association described by (R9).

In addition to the singlet Xe_2^* [$^1\Sigma_u^+ (0_u^+)$] state, there exists a triplet state [$^3\Sigma_u^+ (0_u^-, 1u)$], lying approximately 0.1 eV below the singlet. (See the similar energy level diagram of Ar, Fig. 2.3.) Exchange of population between these states, as shown by (R 10, 11), is possible by electron and heavy-atom collisions. Losses due to Penning ionization and electronic ionization are described by (R 12, 13). The measured lifetimes of Xe_2^* [$^1\Sigma_u^+$] and Xe_2^* [$^3\Sigma_u^+$] are 5.5 and 96 ns, respectively [2.47], and the corresponding stimulated emission gain cross sections, as calculated from (2.1), are $\sim 1.5 \times 10^{-17}$ and 1.0×10^{-18} cm^2. But since the Xe_2^* states are more than halfway to the continuum, photons emitted by Xe_2^* have sufficient energy to ionize other Xe_2^* excimers [2.15]. Thus, the effective gain cross section is reduced by self-absorption, such that net gain is possible only on the singlet-state transition.

2.2.3 Pumping Methods

Although the pumping of rare-gas excimer lasers by electron-beam sustained discharges has been studied, the only method that has been successful so far is the irradiation of the high-pressure gas by intense beams of relativistic electrons. Unfortunately, electron-beam generators capable of producing the short-duration (50 – 100 ns), high-current (10 – 100 kA) pulses of 1 MeV electrons are bulky, thus restricting the availability of this type of laser. These generators consist of a high-voltage source, such as a Marx bank or a pulse transformer, a pulse forming line of relatively low impedance (to produce a voltage pulse with a fast risetime), and a vacuum diode in which high-energy electrons are produced by field emission from a cold cathode. The anode must be sufficiently thin (< 50 μm) to permit penetration by electrons of energy greater than ~ 200 keV, and can be made from titanium, aluminium, or aluminised dielectric.

The most common geometrical configurations of electron-beam diodes are shown in Fig. 2.5. When a biplanar diode is used with high-pressure gases, the gas is pumped in a direction transverse to the optical axis of the laser to accommodate the short electron penetration depths (Fig. 2.5a). Scattering of electrons by the entrance foil (which separates the diode vacuum from the laser gas) causes the energy to be deposited nonuniformly, with the greatest excitation density occurring close to the entrance foil. More uniform de-position, and therefore more efficient use of the pumping energy, may be obtained using a coaxial geometry where the gas is contained in a thin-walled metal tube (Fig. 2.5b). The tube acts as the anode and is concentric with a cylindrical cathode. The longitudinal geometry shown in Fig. 2.5c is suitable for pumping gases with long electron penetration depths, and can be used with the rare-gas halide lasers discussed in Sect. 2.3.

a) The Xe_2^* Excimer Laser

The first indication of gain in an excimer system was the observation by *Basov* et al. of spectral narrowing in liquid xenon pumped by a high-current relativistic electron beam [2.5]. Since then, considerable progress has been made in the development of gaseous xenon excimer lasers. The concentrated

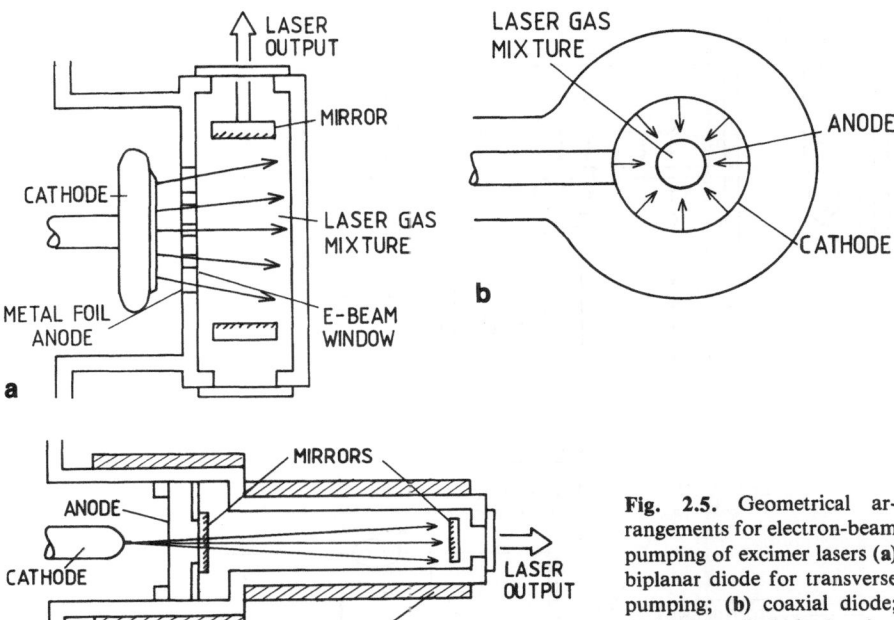

Fig. 2.5. Geometrical arrangements for electron-beam pumping of excimer lasers (**a**) biplanar diode for transverse pumping; (**b**) coaxial diode; (**c**) biplanar diode for longitudinal pumping

effort on xenon, rather than krypton and argon, has been due in part to difficulties in fabricating short-wavelength mirrors with low losses and high reflectivity, that are also capable of withstanding the high intensities produced by these lasers. Multilayer dielectric coatings with high-damage thresholds are commercially available for $\lambda \sim 172$ nm, but at shorter-wavelengths, Al/MgF$_2$ mirrors, with much lower damage thresholds, must be used.

The bound-free transitions of the rare-gas excimers at high pressures are effectively homogeneously broadened, with linewidths of ~ 2000 cm^{-1}. This enables the lasers to be tuned with relatively high efficiency. A schematic diagram of a coaxial, tunable xenon laser is shown in Fig. 2.6. Purified xenon at a pressure of $7 - 12$ bar is pumped by short (~ 50 ns) current pulses of ~ 1 kA and voltages of $0.5 - 0.75$ MV. A circuit diagram of the power supply is shown in Fig. 2.7. This system can generate 1 GW electrical pulses (750 kV, 50 J, 50 ns) at a repetition rate of 10 Hz, and makes use of a pulse transformer to generate the high voltage, rather than a more conventional Marx bank. The xenon gas is cooled by continuous circulation through a water-cooled heat exchanger. Nevertheless, the optical homogeneity of the gas is adversely affected at high temperatures, and in this particular laser, the repetition rate is limited to ~ 1 Hz. The maximum laser output power of ~ 2 MW is limited by the damage threshold of the laser mirrors.

In principle, the xenon laser may be tuned using prisms, gratings, or etalons. In practice, however, the intracavity power is sufficiently high to

28   *M. H. R. Hutchinson*

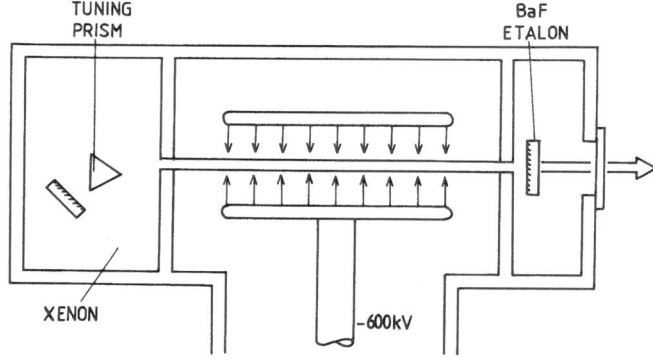

Fig. 2.6. Schematic
diagram of a coaxial,
tunable xenon laser

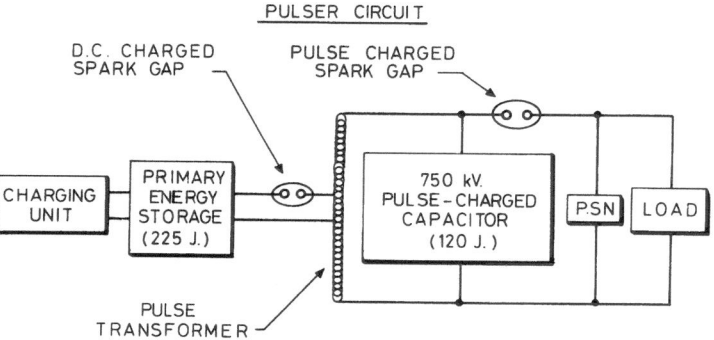

Fig. 2.7. Circuit diagram of a 750 keV pulse generator for a
rare-gas excimer laser operating at a high repetition rate

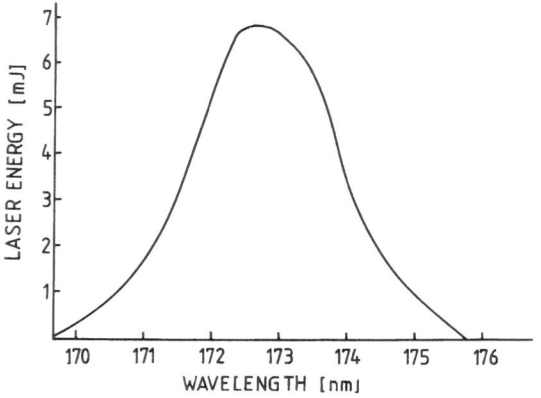

Fig. 2.8. Tuning range of the
xenon excimer laser

damage gratings or etalons so that prisms are preferred. Moreover, the dispersion of fused quartz at 172 nm ($\sim 3 \times 10^{-3}$ nm^{-1}) is so large that the angular dispersion of a single prism can exceed that of a 1200 mm^{-1} grating used in auto-collimation. The tuning range of the laser is shown in Fig. 2.8. Output energies of up to 8 mJ and linewidths of 0.25 nm can be produced. By using very powerful electron beams (3 MeV, 100 kA) [2.49], (850 keV, 50 kA) [2.50] in a transverse pumping configuration, output powers of 400 MW have been obtained with intrinsic efficiencies of up to 2%. At high-pumping rates, laser action is observed to terminate before the pumping pulse, and the longest pulses (200 ns) have been produced using low-gain systems operating at atmospheric pressure [2.51].

b) The Ar$_2^*$ Excimer Laser

Argon excimers emit broad bandwidth radiation centered around 126 nm, and using techniques similar to those employed for pumping the xenon laser, tunable laser outputs can be produced. The Ar$_2^*$ laser has attracted interest because of the apparent possibility of tuning to the Lyman-α ($\lambda = 121.6$ nm) transition of atomic hydrogen to provide a diagnostic for magnetically confined fusion plasmas. Although gain at this wavelength has been measured [2.52], attempts to tune the laser to the Lyman-transition have been frustrated by (1) the lack of low-loss, damage resistant windows, mirrors, and gratings; (2) the high gain of the laser at line center ($\lambda = 126$ nm); (3) absorption in the gas at the Lyman-wavelength due to the production of hydrogen atoms in the laser gas when the electron beam is fired [2.52, 53]. The output power of the laser has so far been limited to ~ 50 kW because of damage to the Al/MgF$_2$ resonator mirrors.

2.3 Rare-Gas Halide Excimers

Rare-gas halides were recognized as suitable laser media following extensive study of the reactions of the metastable states of the rare gases [2.54]. The electronic configuration of an excited rare-gas atom is very similar to that of an alkali metal, i.e., a single s electron orbiting a core of unit positive charge. Thus there is a strong similarity between the ionization potentials and polarizabilities of the metastable states ($^3P_{0,2}$) of Ne, Ar, Kr, and Xe and the ground states of Na, K, Rb, and Cs, respectively. In particular, the excited rare gases form very strong ionic bonds by charge transfer to electronegative atoms such as the halogens, forming excimers which radiate in the ultraviolet and vacuum ultraviolet. By taking advantage of the similarity between rare-gas halides and alkali halides, *Ewing* and *Brau* [2.55] predicted the emission wavelengths of many molecules. Detailed spectroscopic [2.56 – 60] analyses and *ab initio* calculations [2.61 – 63] of the relevant potentials have been made, so that the characteristics of these excimers are quite well understood.

2.3.1 Electronic Structure and Kinetics

That some excimers are formed and radiate with high efficiencies, while others radiate less strongly or not at all, can be understood by considering the mechanism by which rare-gas halides are formed [2.64]. Excited rare-gas atoms (A*) have relatively low ionization potentials (4 – 5 eV), and can interact with electronegative molecules (RX) acting as halogen (X) donors by a charge transfer or "harpooning" mechanism, as, for example:

$$A^* + RX \rightarrow A^+ + RX^- \ .$$

As shown in Fig. 2.9, this charge transfer may take place at relatively large atom-molecule separations (0.5 – 1 nm), where the covalent (A*, RX) and ionic (A$^+$, RX$^-$) potential curves cross. The donor ion RX$^-$ may then dissociate in the field of the rare-gas ion to form the ionic excimer (A$^+$X$^-$)* in a vibrationally excited state, e.g.,

$$A^+ + RX^- \rightarrow (A^+RX^-) \rightarrow (A^+X^-)^* + R \ .$$

For this reaction to lead to the formation of an excimer, the dissociation energy $D(A-X)^*$ of the excimer to A* and X must be greater than the dissociation energy $D(R-X)$ of the donor molecule, i.e., the reaction A* + RX → AX* + R must be exothermic. The excimer is therefore formed in a range of vibrationally excited states, where the highest vibrational energy is $D(AX)^* - D(R-X)$, i.e., the maximum total energy of the excimer is given by the energy of the initial reactants, $E(A^*) - D(R-X)$.

The generalized potential curves of an excimer AX* are shown in Fig. 2.10. If, as in the case of ArBr, the covalent potentials (A + X*) cross the ionic excimer potentials close to their minima, the probability of predissociation (A$^+$X$^-$)* → A + X* is very high, and for this reason no ArBr excimer emission is observed. In ArCl, potential crossings occur at much higher energies. Although predissociation leading to emission from atomic chlorine does occur when the excimer is formed from Ar and Cl$_2$, excimer emission and weak laser action is observed [2.65].

The ionic B states correlate with the separated ion pair A$^+$ (^2P) + X$^-$ (^1S) which has a total energy of I.P. (A) – E.A. (X), where I.P. (A) is the ionization potential of A and E.A. (X) is the electron affinity of X. The smaller the halogen ion, the smaller will be the equilibrium length of the ionic A-X bond, and so the greater will be the ionic dissociation energy. Since all the halogen atoms have similar electron affinities (~3 eV), the smallest ions, the fluorides, will have the lowest B states. Hence the B-X emission wavelengths of the halides of a given rare gas decrease monotonically with increasing atomic number of the halogen. The ionization potentials of the rare gases increase with decreasing atomic number and therefore B-X emission wavelengths of a given halide of different rare gases decreases monotonically with decreasing atomic number of the rare gas. The lighter rare gases, He

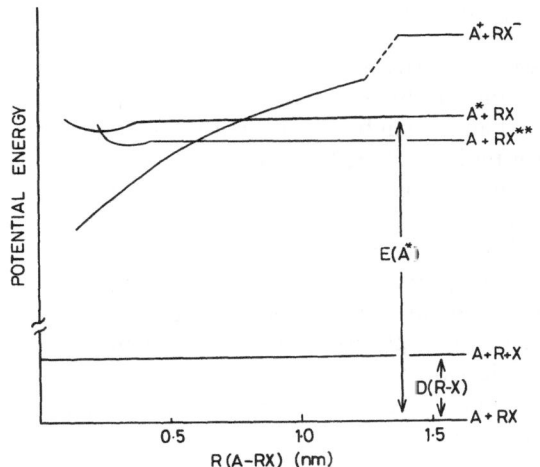

Fig. 2.9. Potential energy curves describing the interaction of a rare gas atom (A) with a halogen donor molecule (RX)

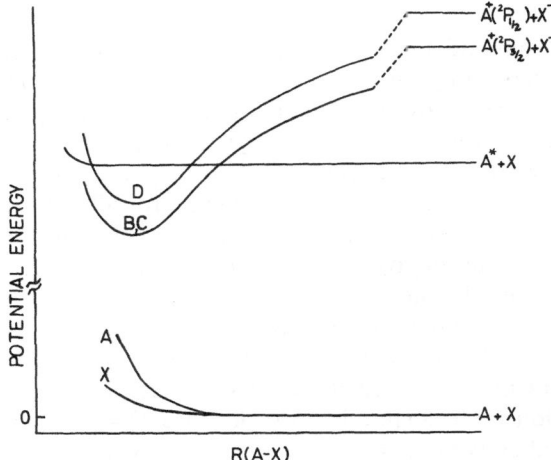

Fig. 2.10. Generalised potential curves for the rare-gas halide excimer AX*

and Ne, would form excimers with ionic potential minima well above the covalent potentials correlating with A + X*. Under those circumstances the formation of stable excimers is unlikely, and the only neon-halide excimer emission to be observed is from NeF.

Instead of producing charge exchange reactions, collisions between excited rare gas atoms and halogen donors can continue along the (A*, RX) covalent potential. Excitation transfer

$$A^* + RX \rightarrow A + RX^{**}$$

or Penning ionization

$$A^* + RX \rightarrow A + RX^+ + e^- \qquad [\text{if } E(A^*) > \text{I.P. (RX)}]$$

at shorter ranges may take place if the covalent potential has a sufficiently large minimum and suitable crossings occur. However, in spite of these other possible reactions, many rare-gas halides can be formed with branching ratios of 0.5−1.0 [2.66]. The high formation efficiencies and high reaction rates associated with the long-range interactions make rare-gas halide molecules very useful as high-power laser systems.

As will be discussed later, high-power lasers are often pumped by electron beams in a manner similar to that employed for the rare-gas excimer lasers. Under those circumstances most of the energy is deposited in the gas in the creation of ions, where both positive rare-gas ions and negative halogen ions are formed by rapid dissociative electron attachment, e.g., $F_2 + e^- \rightarrow F^- + F$. The rare-gas-halogen ion pairs may then form ionic excimers directly, provided predissociation does not occur.

a) Spectroscopy

The interaction of a rare-gas atom and a halogen atom gives rise to two states designated $1\,{}^2\Sigma^+$ and $1\,{}^2\Pi$, depending on the orientation of the singly occupied $2p$ halogen orbital [2.62,63]. However, due to spin-orbit splitting of ground state of the halogen atoms, the ${}^2\Pi$ state is split, giving rise to $X(1/2)$, $A(1/2)$, and $A(3/2)$ states. Nevertheless, the $A(1/2)$ and $A(3/2)$ states which lie above the $X(1/2)$ state may be regarded as being degenerate for all but the heavier halogens. Similarly, the charge transfer interaction between the excited rare-gas atom and halogen atom gives rise to two ionic states, $2\,{}^2\Sigma^+$ and $2\,{}^2\Pi$, which would be almost degenerate, in the absence of spin-orbit coupling. However, spin-orbit effects are very pronounced in the rare-gas ions where, for example, the $Kr^+\,({}^2P)$ ground state is split $({}^2P_{1/2}, {}^2P_{3/2})$ by 0.666 eV (the analogous splitting in F is only 0.050 eV). Thus the ionic ${}^2\Pi$ state which correlates with the ${}^2P$ ion ground state is also split into $\Omega = 1/2$ and $\Omega = 3/2$ components. The higher lying $\Omega = 1/2$ state is referred to as the $D(1/2)$ state, and the lower $\Omega = 3/2$ state, which is almost degenerate with $B(1/2)$ (derived from the ${}^2\Sigma^+$ state), is referred to as the $C(3/2)$ state. The normal designation of the spin-orbit corrected potentials is to label them A, B, C, etc., in order of increasing energy. Since the B and C states are nearly degenerate, their designation in terms of the axial angular momentum Ω may be reversed [2.61].

Radiative transitions may occur on the $D(1/2) \rightarrow X(1/2)$, $B(1/2) \rightarrow X(1/2)$ and $C(3/2) \rightarrow A(3/2)$ transitions. In all the rare-gas halides, the A states are strongly repulsive, whereas the $X(1/2)$ state is at most weakly repulsive at internuclear separations at which transitions from the upper manifolds take place. The B-X bands all show pronounced structure, which is expected for the fairly flat potentials of the lower states. However, the $X(1/2)$ states of XeF and XeCl are bound by 1065 cm^{-1} [2.67] and 255 cm^{-1} [2.68], respectively; so that for these molecules, transitions terminating in the $X(1/2)$ states show normal bound-bound vibrational structure. At high pressures, the emission bandwidths for $B \to X$ transitions are typically 2 nm. Weaker bands due to $C \to A$ emission are observed at longer wavelengths in each molecule [2.64, 69]. Because they terminate on the purely repulsive A state, these $C \to A$ transitions have much greater bandwidth (~ 70 nm for XeF) than the $B \to X$ emission. The $D \to X$ emission has been observed for most of the rare-gas halides [2.69 – 72] and vibrational and rotational analyses for the $D \to X$ emission from XeF has been carried out [2.60]. Since the D and B potentials are approximately parallel and separated by the atomic $^2P_{1/2} - {}^2P_{3/2}$ splitting, the D-X bands are similar to the B-X bands, but blue-shifted by an energy comparable to the ionic spin-orbit splitting. However, there is evidence [2.69] that the D state is strongly quenched, and at high pressures the emission is very weak.

The B-X and C-A transitions are the most important laser transitions, and an accurate determination of radiative lifetimes is therefore significant. The measured lifetimes of the B-X transitions in KrF and XeF are 9 ns [2.72 – 73] and 14.25 – 18.8 ns [2.74 – 76], respectively, and compare reasonably well with the calculated values of 6.5 ns [2.61] and 12 ns [2.63]. The lifetimes of many of the other excimers have also been calculated [2.63, 77] and are shown with the emission wavelengths in Table 2.2. The calculated lifetimes for the C-A transitions are much longer (~ 120 ns) and, in the case of XeF, agree reasonably well with the measured lifetime of 93 ns [2.78].

The stimulated emission cross sections can be calculated using (2.1) from a knowledge of the emission wavelength, bandwidth, and radiative decay times. The cross sections for the B-X bands mostly lie within the range $2 - 5 \times 10^{-16}$ cm^2 [2.50, 69, 79, 80], whereas the C-A transitions, with longer lifetimes and broader bandwidths [2.63], have much smaller cross sections, e.g., $\sigma[\text{XeF}(C\text{-}A)] = 5 \times 10^{-18}$ cm^2 [2.81].

Table 2.2. Emission wavelengths [nm] and lifetimes [ns] of the B-X and C-A transitions of the rare-gas halides. Laser action has been observed on those transitions whose wavelengths are in italics

		F		Cl		Br		I	
		λ [nm]	τ [ns]	λ [nm]	τ [ns]	λ [nm]	τ [ns]	λ [nm]	τ [ns]
Xe	B-X	*351*	(12 – 19)	*308*	(11)	*282*	(12)	253	(12)
	C-A	490	(93, 113)	350	(120)	302	(120)	263	(110)
Kr	B-X	*249*	(6.5 – 9)	*222*		206			
	C-A	275	(63)						
Ar	B-X	*193*	(4.2)	*175*					
	C-A	203	(48)	199					
Ne	B-X	108	(2.6)						
	C-A	117	(38)						

The first observation of laser action on the *B-X* transitions of the noble-gas halide excimers was made in 1975 by *Searles* and *Hart* [2.82], who constructed an electron-beam pumped XeBr* laser oscillator. Following the observation of gain on the *C-A* band of XeF* [2.81], *Bishel* et al. [2.83] observed laser oscillation by photolysis of XeF, using Xe_2^* ($\lambda = 172$ nm) fluorescence as an excitation source. The complete list of rare-gas halides is shown in Table 2.2. Many of them have been made to lase by pumping with electron beams, electron-beam controlled discharges, or self-sustained discharges preionized by uv or x-radiation.

b) Kinetics of Rare-Gas Halides

The *B-X* Transitions

All the *B-X* band rare-gas halide lasers employ electronically excited mixtures of the rare gas and halogen donor required to form the excimer and another rare gas as buffer. Nevertheless, the detailed kinetic route by which the excimer is formed depends upon the excimer, the donor molecule, the excitation conditions, and partial pressures of the constituent gases. Several authors [2.84 – 88] have considered in detail the formation kinetics and loss processes in electron-beam and discharge pumped KrF and XeF lasers, using models which take into account up to one hundred different reactions. To illustrate the main features, however, only the kinetics of KrF in mixtures of argon, krypton, and fluorine are discussed here in detail.

Krypton Fluoride. The principal reactions in the formation of KrF mixtures in electron-beam excited Ar:Kr:F mixtures and their rate coefficients are shown in Table 2.3. A simplified flow diagram is shown in Fig. 2.11.

As has already been noted in the consideration of the rare-gas excimer lasers (Sect. 2.2.2), irradiation of rare gases by high-energy electrons produces ions and excited atoms whith very high efficiency. Typical gas mixtures contain 90% Ar, 5 – 10% Kr, and 0.2 – 0.5% F_2 at total pressures of 1 – 3 atm. Hence most of the electron energy is used in the production of Ar^+ and Ar* reactions (R17), (R18), and the slower electrons undergo dissociative attachment with F_2 to form F^- ions (R19). There are therefore two distinct channels by which the excimers are formed: an ionic channel with Ar^+ as the precursor, and a neutral or covalent channel via the rare-gas metastables.

At total pressures less than one atmosphere, the Ar^+ and F^- ions rapidly combine to form ArF* (R20). At higher pressures, however, the production of molecular ions Ar_2^+ by three-body association is dominant (R21). The molecular ion may then form ArF* by recombination with F^- (R22), with a reaction rate which is expected to be similar to that of the atomic ion [2.87] (R21). Alternatively, a charge exchange reaction (R23) with Kr can produce Kr^+. The KrF* excimer is then created by an exchange reaction between ArF* and Kr (R24) or by recombination of Kr^+ and F^- (R25). If the partial

Table 2.3. Reactions for the production of KrF

	Reaction	Rate coefficient	Reference
(R17)	$Ar + e_p^- \rightarrow Ar^+ + e_s^- + e_p^-$		
(R18)	$Ar + e_p^- \rightarrow Ar^* + e_p^-$		
(R19)	$F_2 + e_s^- \rightarrow F + F^-$	$\sim 2.3 \times 10^{-9}$ cm^3 s^{-1}	[2.89]
(R20)	$Ar^+ + F^- + (Ar) \rightarrow ArF^* + (Ar)$	$\sim 2 - 3 \times 10^{-6}$ cm^3 s^{-1}	[2.87]
(R21)	$Ar^+ + Ar + Ar \rightarrow Ar_2^+ + Ar$	$1.9 - 3 \times 10^{-31}$ cm^6 s^{-1}	[2.85, 90 – 92]
(R22)	$Ar_2^+ + F^- (+ Ar) \rightarrow ArF^* + Ar(+ Ar)$	$\sim 10^{-6}$ cm^3 s^{-1}	[2.93]
(R23)	$Ar_2^+ + Kr \rightarrow Kr^+ + 2 Ar$	7.5×10^{-10} cm^3 s^{-1}	[2.94]
(R24)	$ArF^* + Kr \rightarrow KrF^* + Ar$	1.6×10^{-9} cm^3 s^{-1}	[2.95]
(R25)	$Kr^+ + F^- (+ Ar) \rightarrow KrF^*(+ Ar)$	$1 - 2 \times 10^{-6}$ cm^3 s^{-1}	[2.87]
(R26)	$Kr^+ + Kr + Ar \rightarrow Kr_2^+ + Ar$	2.4×10^{-31} cm^6 s^{-1}	[2.84]
(R27)	$Kr_2^+ + F^- (+ Ar) \rightarrow KrF^*(+ Ar)$	$\sim 10^{-6}$ cm^3 s^{-1}	[2.93]
(R28)	$Ar^* + F_2 \rightarrow ArF^* + F$	7.8×10^{-10} cm^3 s^{-1}	[2.96]
(R29)	$Kr^* + F_2 \rightarrow KrF^* + F$	7.2×10^{-10} cm^3 s^{-1}	[2.66]

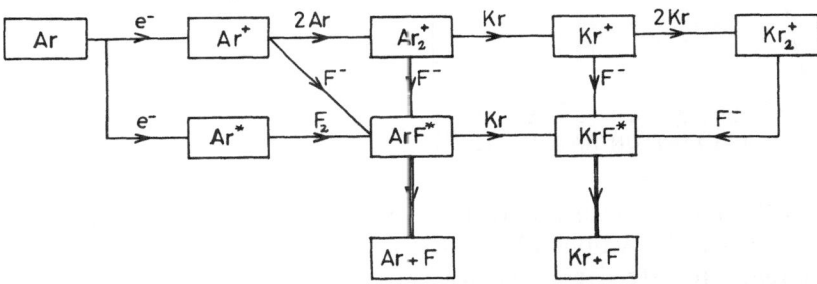

Fig. 2.11. Simplified flow diagram representing the kinetics of KrF in Ar:Kr:F$_2$ mixtures

pressure of krypton is relatively high, Kr$^+$ ions may be formed from Kr by three-body association (R26) and KrF* may then be formed by recombination between Kr$_2^+$ and F$^-$ (R27).

In electron-beam excited gas mixtures, only a relatively small fraction (15 – 20%) of the total energy is used in the creation of excited atoms. In a stable electrical discharge, however, most of the energy deposited in the gas should go towards excitation rather than ionization. Thus the metastable rare-gas atoms would be expected to be the principal precursors of the excimers. The metastable argon atoms (principally $^3P_2$) [2.97] react with F$_2$ by a "harpooning" reaction to form ArF* (R28). The ArF* may then form KrF* by an exchange reaction with Kr (R24). However, if the proportion of krypton in the gas mixture is sufficiently high ($> 10\%$), Kr* may be formed in preference to Ar* in the discharge (the excited states of krypton lie lower than those of argon), and the harpooning or charge transfer reaction with F will produce KrF* (R29). Halogen donor species other than halogen dimer molecules, as, for example, NF$_3$, N$_2$F$_4$, SF$_6$, HCl, CCl$_3$, and several chlorinated hydrocarbons, have been used with varying effect on the

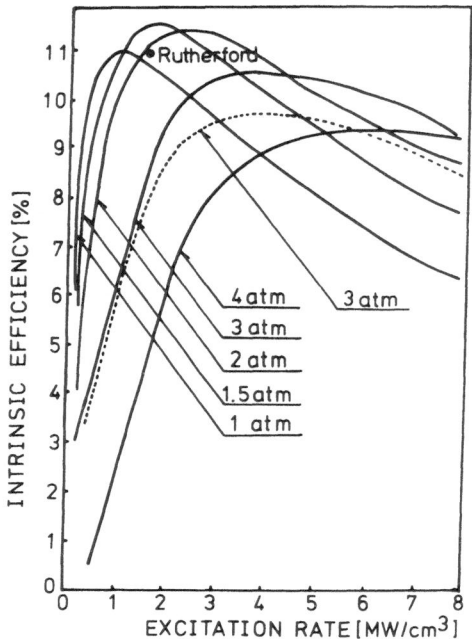

Fig. 2.12. Calculated intrinsic efficiency of electron-beam pumped KrF laser as a function of excitation rate [2.86]

efficiency. Helium and neon are frequently used as buffer gases. Nevertheless, there exists a set of reactions which, over a wide range of pressures and gas composition, can effectively channel the energy deposited in the gas by relativistic or nonrelativistic electrons into the B states, and possibly the C states, of the rare-gas excimers.

The density of excimers which can be produced is reduced, however, due to quenching by electrons and the atomic and molecular gases. At high pumping rates the number density of electrons is large. Electron quenching of KrF* and ArF* then becomes important, giving rise to reduced intrinsic efficiency of KrF* lasers, as shown in Fig. 2.12. The two-body quenching rate coefficient of 2×10^{-7} cm^3 s^{-1} indicates that this process becomes important for electron number densities $\gtrsim 5 \times 10^{14}$ cm^{-3}. KrF* may be quenched by Kr in two-body $(8.6 - 1.6 \times 10^{-12}$ cm^3 s^{-1} [2.98, 99]) or three-body $(2.9 - 9.7 \times 10^{-31}$ cm^6 s^{-1} [2.98 – 101] processes, forming the triatomic excimer KrF*, which radiates in a broad band centred at 420 nm. The three-body process is

$$KrF^* + Kr + [Kr,Ar] \rightarrow Kr_2F^* + [Kr,Ar] \ .$$

Thus, in spite of the expected increase in efficiency from direct production of Kr* at increased krypton pressure, the partial pressure of krypton must be kept relatively low ($\leq 10\%$) to minimize the effects of quenching. The two body quenching of KrF* by Kr is thought to result from collisional mixing of the $B(1/2)$ and $C(3/2)$ states [2.98], i.e.,

$$KrF^*(B) + Kr \rightleftharpoons KrF^*(C) + Kr .$$

Two-body quenching with F_2 is important only at low pressures. Although the quenching rate coefficients for argon are smaller than for krypton, quenching by argon is the most important mechanism, because it is the majority gas. Two-body $(1.8 \times 10^{-12} \, cm^3 \, s^{-1})$ [2.98] and three-body $(7-11 \times 10^{-32} \, cm^6 \, s^{-1})$ [2.98-100] processes are important, the latter forming the triatomic excimer $ArKrF^*$, which subsequently undergoes an exchange reaction with krypton to form Kr_2F^*. In a typical mixture of $Ar:Kr:F_2$ = 95% : 4.8% : 0.2% at a total pressure of 2100 Torr, quenching by two-body collisions with argon accounts for ~17% of the total quenching rate, three-body collisions with argon accounts for 66%, F_2 for 11%, and all krypton collisions for 6% [2.98]. Thus the lifetime of the $KrF^*(B)$ state is reduced from 7 to 1.5 ns. This loss in fluorescence efficiency is less serious in a laser where, if the gain is saturated, the total radiative emission rate is increased by stimulated emission.

The efficiency of some excimer lasers is further reduced by photoabsorption, both by the halogen donor molecules and by short-lived species created during excitation of the gas mixture. "Self-absorption" from the B state of the excimer is of primary importance, since this effect, as for the rare-gas excimer lasers, will reduce the gain cross section. The absence of significant self-absorption in KrF^* is indicated by the close agreement between the measured gain cross section and the calculated stimulated emission cross section [2.102]. Absorption in $Ar:Kr:F_2$ mixtures at the $KrF^*(B\text{-}X)$ emission wavelength (249 nm) is significant, however, and is due to $F_2 (\sigma = 1.3 \times 10^{-19} \, cm^2)$ [2.103], $F^- (\sigma = 5.6 \times 10^{-18} \, cm^2)$ [2.104], $Ar_2^+ (\sigma = 1.3 \times 10^{-17} \, cm^2)$ [2.105], $Kr_2^+ (\sigma = 3.1 \times 10^{-19} \, cm^2)$ [2.106], $Ar^*(4p)$ and $Kr^*(5p)(\sigma = 6 \times 10^{-18} \, cm^2)$ [2.107]. The cross sections for Ar_2F and Kr_2F are expected to be similar to those for Ar_2^+ and Kr_2^+. Due to the long lifetime of $Kr_2F(\sim 150 \, ns)$ [2.101, 108-110], relatively large densities $(\sim 10^{15} \, cm^{-3})$ can build up and give rise to significant absorption. Measurements of absorption [2.102, 88] in electron-beam pumped mixtures indicate that Kr_2F is the most strongly absorbing species. Some loss can be avoided by replacing F_2 by NF_3, which does not absorb at KrF^* (or XeF^*) emission wavelengths, but NF_3 is less efficient in forming KrF^*, because of a lower branching ratio with Kr^* and charge transfer collisions with Kr^+ [2.111]. Nevertheless, the reaction of NF_3 with Xe^* has a branching ratio of unity, and is used to advantage in XeF lasers. Similarly, Cl_2 can be replaced by HCl in $XeCl^*$ lasers. The absorption cross sections of many of the absorbing species in all the noble-gas halides are shown in Fig. 2.13. Since the largest absorption cross sections are ~10× smaller than the stimulated emission cross sections of the excimers, the gain of a laser will saturate at a lower-power density than that required to saturate the absorption. The nonsaturable absorption limits the length of gain medium that can be used effectively to 1-2 m, and it limits the optical extraction efficiency typically to less than 50% [2.88].

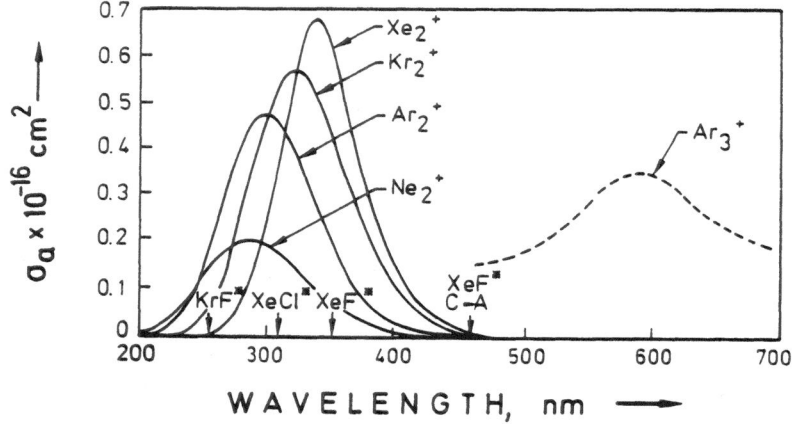

Fig. 2.13. Calculated absorption cross sections of species present in rare-gas halide mixtures [2.112]

Xenon Chloride. Some differences between the kinetics of KrF and XeCl are worth noting. The XeCl laser is usually operated in mixtures of Ne : Xe : HCl, although it has been shown that the neutral channel is ineffective, since reactions between metastable xenon atoms and HCl do not produce excited XeCl excimers [2.113]. Formation via the ionic channel would be efficient if the formation rate of Xe^+ and Cl^- were sufficiently rapid; Xe^+ can be formed rapidly enough by the charge transfer process

$$Ne_2^+ + Ne + Xe \rightarrow Xe^+ + 3\,Ne$$

with a rate coefficient of $3.9 \times 10^{-30}\,cm^6 s^{-1}$ [2.114]. However, the formation rate of XeCl* is limited by the rate of formation of Cl^- by dissociative attachment of HCl. The cross section for this process is very small ($\sim 10^{-17}\,cm^2$) for the $v = 0$ state but, at the threshold electron energy, it is 38 times greater for $v = 1$ and 880 times greater for $v = 2$ [2.115]. Cl^- can therefore be formed rapidly by vibrational excitation of HCl followed by dissociative attachment. When argon is used as the diluent gas, Xe^+ ions may be formed by the process

$$Ar^+ + 2\,Ar \rightarrow Ar_2^+ + Ar$$

$$Ar_2^+ + Xe \rightarrow Xe^+ + 2\,Ar \ ,$$

Cl_2 can be used as a Cl donor, but it has a small electron attachment cross section [2.116]. It can undergo a harpooning reaction with Xe*, however, and the neutral channel can in this case contribute to XeCl* production.

Xenon Fluoride. The XeF excimer is different from the other rare gas halide excimers in that the C state lies significantly lower ($\sim 700\,cm^{-1}$) than the

B state. It will be shown in the following subsection that the *C* state can be populated by collisional relaxation from the *B* state, and that at high pressures, lasing involves the *C-A*, rather than the *B-X*, transition.

Argon Fluoride. The kinetics of ArF production in Ar: F_2 mixtures mimic the corresponding reactions of the KrF production scheme. Nevertheless, the excimer has a higher quantum efficiency than KrF, and suffers less absorption from the molecular ions (Ar_2^+), as shown in Fig. 2.13. When neon is added as a buffer gas, Ne^+ is formed. Nevertheless the ionic channel is inefficient because of the reaction

$$Ne^+ + F^- \rightarrow Ne^* + F$$

and the neutral channel is therefore dominant.

C-A Transitions

The XeCl and XeF excimers are distinguished not only by having weakly-bound ground states, but also by the fact that the *C* states lie below the *B* state by approximately $230 \, cm^{-1}$ and $700 \, cm^{-1}$, respectively. An energy level diagram for XeCl is shown in Fig. 2.14. For XeF in thermal equilibrium at room temperature, approximately 95% of the combined population of the *B* and *C* states is in the *C* state. Transitions from *C-X* are forbidden but, as noted in Sect. 2.3.1, transitions to the strongly repulsive *A* state produce a broad continuum centered at 490 nm. The relatively long radiative lifetime ($\tau \sim 100$ ns) of the *C* state and the large bandwidth ($\Delta\lambda \sim 70$ nm) give rise to a relatively small stimulated emission cross section of $\sim 10^{-17} \, cm^2$, see (2.1). As a result, very intense pumping is required. The resultant high density of

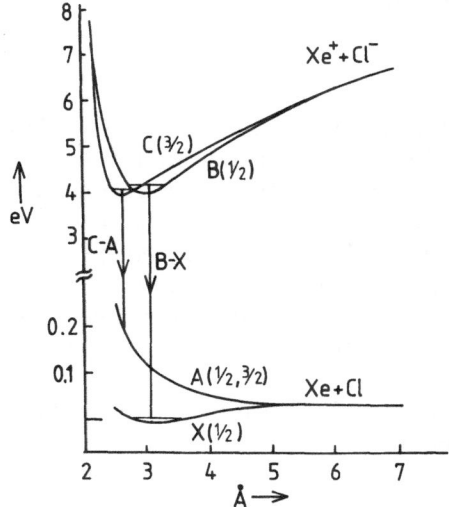

Fig. 2.14. Schematic energy level diagram of the XeCl excimer

electrons tends to thermalize the populations of the B and C states at a high electron temperature, thereby reducing the C-state population. Other excited and ionic species, principally Ar^*, Xe^*, Ar_2^*, and Ar_3^+, with absorption bands in the blue-green region of the spectrum, produce losses. Thus, the overall efficiency of C-A transition lasers is expected to be much lower than that of B-X transition lasers.

In high-pressure gas mixtures containing Xe, and either F_2 or NF_3 as the fluorine donor, $XeF(C$-$A)$ transition lasers have been pumped by electron beams [2.117, 118] and by discharges [2.119, 120], but output powers have been very low. Because of absorption by the excited species produced during the pumping of the gas mixtures, laser oscillation cannot occur until after the pumping has been terminated. The problem of electron thermalization and quenching can be overcome, but at the price of increased complexity of the pumping system [2.121]. That is, the $XeF(C)$ state is populated by photolysis of XeF_2 with fluorescence from Xe_2^* produced by electron-beam pumping of xenon at high pressures.

Most studies of the $XeF(C$-$A)$ laser have used NF_3 as the fluorine donor because of its high rate coefficient for dissociative attachment [2.122]. However, the rate coefficient for quenching Xe^*, Xe_2^*, and similar species by the harpooning reaction is relatively small [2.123]. Thus, although XeF^* is produced rapidly by an ionic channel, the concentrations of excited species which absorb the laser radiation remain high. On the other hand, F_2 has an electron attachment coefficient which is much smaller than that of NF_3 for electron energies $\gtrsim 1$ eV, and a quenching rate coefficient for Xe^*, etc., which is typically $\sim 10 \times$ larger. Consequently, the use of mixtures of F_2 and NF_3 have enabled the performance of the $XeF(C$-$A)$ laser to be improved. Intrinsic efficiencies of approximately 0.1%, and energy densities of 0.1 J/l have been produced in electron-beam pumped $Ar : Xe : NF_3 : F_2$ systems, an increase of almost $100 \times$ over previous mixtures [2.124].

Triatomic Excimers

The diatomic rare-gas halide excimers are distinguished by the fact that they provide, at several wavelengths in the ultraviolet, laser transitions with high gain and consequently high laser output powers. Nevertheless, at internuclear separations in the vicinity of the laser transitions, the lower level is only weakly repulsive, or as in the case of XeF and XeCl, is actually bound. Thus the bandwidths of the transitions are relatively narrow ($\Delta\lambda \lesssim 2$ nm), and the spectral range over which they may be tuned is therefore very limited. This restricts their more widespread application, for example, in atomic and molecular spectroscopy.

As already seen in Sect. 2.3.1.6, the rare-gas halides are quenched at high pressures by a three-body reaction giving rise to a triatomic molecule:

$$RX^* + 2R \rightarrow R_2X^* + R \ .$$

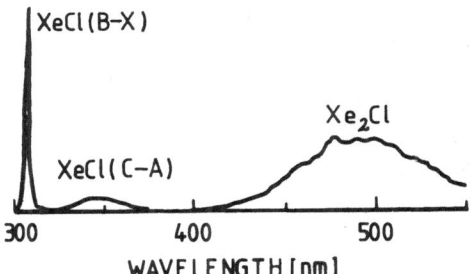

The triatomic molecule is an excimer and decays radiatively to a dissociative ground state. The emission is a broad, unstructured continuum, red-shifted by 1.5 to 2.2 eV from both the diatomic (*B-X*) and (*C-A*) excimer bands, as shown for the case of Xe$_2$Cl in Fig. 2.15. At the relatively high pressure of 8 atm, most of the emission lies in the broad band between 450 and 550 nm, which is attributed to Xe$_2$Cl. Similar fluorescence bands are produced by Ar$_2$Cl, Ar$_2$F, Kr$_2$F, Xe$_2$Br, and Xe$_2$F; the wavelengths and bandwidths of which are shown in Fig. 2.2. Nevertheless, laser action has been produced only in Kr$_2$F and Xe$_2$Cl. Emissions from several mixed trimers (ArXeF*) have also been observed.

As noted in Sect. 2.2.1, the positive diatomic molecular ion A_2^+ is stable with a binding energy of ~1 eV. Consequently, an ionic complex $A_2^+X^-$ may be formed. The strongly repulsive nature of the ground state in the region of the optical transition accounts for the very broad continuum. Theoretical calculations of the ionic complex structure indicate an isosceles triangular geometry with A-A and A-X bond lengths very similar to those of the isolated A_2^- and AX* diatomic molecules.

The fluorescence decay times of these triatomic excimers is relatively long ($\tau \sim 200$ ns). The long decay times, combined with the broad bandwidth of the transitions, give rise to stimulated emission cross sections of approximately 5×10^{-18} cm^2 (a cross section ~100× less than that of the corresponding diatomic excimers). Systems with low stimulated emission cross sections are much more vulnerable to losses arising from excited-state absorption. Calculations of the cross section for the $8\,^2\Gamma - 4\,^2\Gamma$ transition in Xe$_2$Cl indicate that it is large enough to significantly reduce gain on the $4\,^2\Gamma - 2\,^2\Gamma$ trimer band near $\lambda = 438$ nm. It has also been suggested that excited-state trimer absorption may be a significant loss mechanism in the dimer-excimer lasers operating on the *B-X* and *C-A* bands.

Pumping by electron beams has so far been the only successful method for producing laser action in Xe$_2$Cl [2.126] and Kr$_2$F [2.127, 68]. As already shown in Sect. 2.2.2, irradiation of rare gas by electrons produces relatively large transient concentrations of rare-gas molecular ions (Ar$_2^+$, Kr$_2^+$, Xe$_2^+$) (absorption bands shown in Fig. 2.13), and rare-gas excimers (Ar$_2^*$, Kr$_2^*$, Xe$_2^*$) with lifetimes of $10-20$ ns. As in the case of the XeF(*C-A*) laser, these species act as absorbers, with the net effect of producing optical loss while the laser is being pumped. Therefore, laser action is restricted to the after-glow when the short-lived absorbers have decayed. Nevertheless, long-lived rare-gas atomic metastables, for example, Xe* and Kr*, have been shown to survive for several hundred nanoseconds, and to give rise to absorption which further reduces the laser efficiency and tuning range.

2.3.2 Excitation Methods

Rare-gas halide lasers may be pumped by high-current, relativistic electron beams or by rapid electrical discharges in which the gas is either preionised by uv or x-radiation or ionised by electron-beam injection during the discharge pulse.

a) Electron-Beam Pumping

The techniques of electron-beam pumping rare-gas halides are similar to those employed for the noble-gas excimers, but are different in two respects. First, the diatomic rare-gas halides have narrow bandwidths, longer wavelengths, and similar lifetimes. The stimulated emission cross sections are therefore much larger, and hence the required pump power density is much less. Second, because the kinetics favor operation at relatively low pressures ($1-3$ atm rather than $10-60$ atm for the rare-gas excimers), the stopping power of the gas is much less. Even though lower voltage may be used, the range of the electrons may exceed the transverse dimensions of the gas cell. Thus, much of the incident energy is lost to the walls of the cell. This loss may be overcome in principle by the use of an axial pumping geometry in which the electron beam is injected into the laser resonator and guided along the optical axis by an externally applied magnetic field, as shown in Fig. 2.5 c. Since the dominant excitation mechanism in an electron-beam pumped laser is ionization of the dilutent gas, the ionic formation channel is favored. The quantum efficiency is given by $h\nu/W$, where $h\nu$ is the photon energy and W is the energy required to produce an ion-electron pair. For the KrF laser ($h\nu \sim 5.0$ eV) in an argon-based gas mixture ($W \sim 20$ eV), the quantum efficiency is 25%, although other losses which have been discussed earlier will reduce this figure substantially. Nevertheless, intrinsic efficiencies as high as 15% have been reported [2.129] in a short-pulse (~ 150 ns) coaxially pumped KrF laser, and efficiencies of 10% have been obtained for pulses of $0.5-1$ µs duration using much longer gain lengths ($1-2$ m) and lower current densities [2.88].

A KrF laser pumped by electron beams arranged at 90° intervals around the circumference of a cylindrical laser volume is shown in Fig. 2.16. This system has an output energy of 250 J in a pulse of 50 ns duration, produced with an intrinsic efficiency of up to 11% (see Fig. 2.12). For this system, a capacitor, with water as the dielectric medium, is charged from a Marx bank

Fig. 2.16. An electron-beam pumped KrF laser of 250 J output energy [courtesy of the Rutherford Appleton Laboratory, England]

and discharged into four separate pulse-forming lines which shape the high-voltage pulse to be applied to the electron guns. When losses in the electron-beam generator (particularly in the pulse-forming system and in the injection of high-energy electrons into the gas) are included, overall efficiencies are reduced. Nevertheless, values of several percent are possible. Electron-beam pumping is generally used in large, high-peak power systems at low repetition rates.

b) Self-Sustained Discharges

The most convenient method of pumping rare-gas halide excimer lasers is by a self-sustained discharge. In the discharge, the sudden application of a high electric field in excess of the breakdown threshold causes the initial low level of ionization in the gas to grow exponentially due to an electron avalanche process. The growth stops when the plasma impedance becomes significantly smaller than the output impedance of the driving electrical circuit. Nevertheless, the avalanche from a single electron can grow to such an extent that its space-charge field becomes comparable with the magnitude of the applied field. Electrons from weaker avalanches can then be drawn into stronger ones, thus further increasing the charge density.

   If only a limited number of avalanches are produced, the first one to close the gap between the electrodes will produce a conducting filament, and an arc will develop. The time taken for this to occur under typical excimer laser conditions is a few tens of nanoseconds. If the initial electron number density in the gas is sufficiently high, however, diffusion of the electrons will produce some overlap of neighbouring avalanche heads and smooth out the developing space charge heads. In this way, a relatively homogeneous discharge can be sustained, provided the impedance of the driving circuit remains matched to the plasma impedance.

The simplest form of self-sustained discharge-pumped excimer laser uses low-inductance discharge circuits to supply a short high-voltage pulse to the laser gas. The gas is preionised by uv radiation from an array of synchronised spark gaps placed close to the discharge volume. The circuits can be of various designs, such as simple capacitor dumping circuits, or *LC* inversion networks. For efficient pumping, however, the energy must be supplied to the gas in a time comparable with that of the collapse of the plasma impedance. Hence the inductance of the circuit must be kept to a minimum, and the duration of the output pulse is limited to ~10 ns. A simple circuit for a synchronously-pumped oscillator-amplifier system is shown in Fig. 2.17. There the uv preionisation is produced by spark gaps in series with the discharge circuit [2.129].

The problem of the discharge instability to arc formation can be alleviated to some degree by the provision of distributed resistive ballasting of the discharge electrodes. Then, if a localized region of high-current density should develop, the voltage in that region of the electrode would decrease, and growth of the developing arc would be suppressed. By this means pulses of up to 250 ns duration have been produced, but with relatively low pulse-energies of 10 mJ [2.67]. Even in this case, however, the discharge eventually becomes unstable. It has been suggested [2.130] that this instability may be due to a

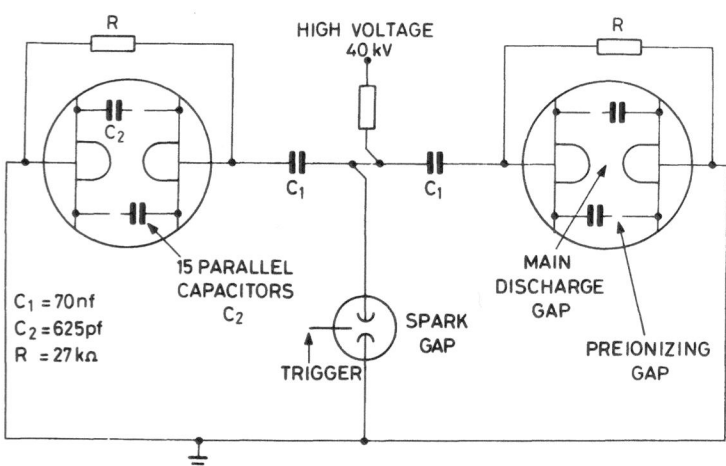

Fig. 2.17. Circuit diagram of a UV-preionised discharge oscillator pumped amplifier excimer laser system [2.129]

local depletion of the electronegative HCl molecules and a consequent local increase in current. This in turn further depletes the HCl, and a streamer develops.

Ultraviolet preionized, discharge-pumped excimer lasers are now widely available, and output pulse energies up to 5 J [2.131], or repetition rates of 100 Hz can be obtained. The output-beam divergence and tuning ranges of these systems is discussed in Sect. 2.3.3.

Self-sustained, discharge-pumped excimer lasers can be scaled to higher-output powers by increasing both the apertures and the output-pulse duration. Such scalling requires, however, a means of producing sufficiently uniform high densities of preionization throughout the discharge volume. It also requires a driving circuit such as a low impedance ($\sim 1 \, \Omega$) pulse forming network which can provide a relatively continuous impedance match with the discharge plasma over the time of interest. Because of their very long penetration depths in gases at atmospheric pressures, x-rays provide a useful method for preionization of large volumes [2.132]. Consequently, x-ray preionized excimer lasers have been developed in recent years, and output-pulse energies of several Joules and durations of up to 250 ns [2.133] have been produced. It seems likely that pulse energies of 100 J or higher could be produced by this method, but, as in the case of the uv preionized lasers, the pulse duration may be limited by the local depletion of the halogen donor molecules (e.g., HCl and F_2), which are regenerated relatively slowly.

c) Electron-Beam Sustained Discharges

As noted in the previous section, self-sustained electrical discharges in high-pressure gases can be unstable unless they are operated under well-defined conditions. The dominant mechanism for ionization of noble-gas atoms at high values of E/N is a two-step process in which the atoms are first excited by electron collisions and then ionized. If the discharge is run at values of E/N low enough that the rate of ionization by electron collisions is less than the rate of electron loss by electron attachment, then the spatial distribution of the discharge is stable. For a steady state, the difference in the two rates must be made up from an external ionization source which sustains the discharge, normally an electron beam [2.134, 135]. A typical arrangement is shown in Fig. 2.18. The efficacy of the discharge is measured by the increase in laser output power over that obtained with only electron-beam pumping, and this is limited by the stability requirement. Nevertheless, the pumping power supplied to the gas might well be increased by factors of $5 - 10$ by this technique [2.136].

Pumping by electron-beam sustained discharges offers several other potential advantages over pure electron-beam pumping. Since the main precursors of the excimer are metastable atoms rather ions, the quantum efficiency is given by the ratio of the photon energy to the atomic excitation energy. In KrF, $E(Kr^*) = 10$ eV, $h\nu = 5$ eV, and the quantum efficiency is

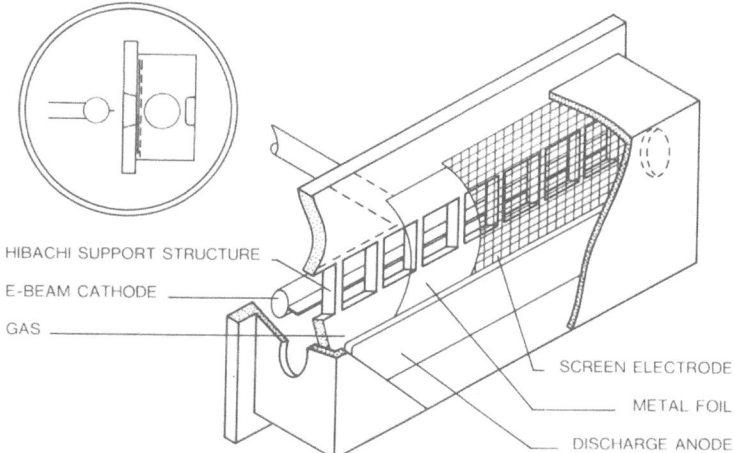

Fig. 2.18. Schematic diagram of an electron-beam sustained discharge-pumped excimer laser

~0.5, approximately twice that of an electron-beam pumped system. In a sustained discharge, the impedance of the plasma is well defined, although very low ($<1\ \Omega$), and electrical coupling from the discharge circuit to the plasma should be relatively efficient. Since most of the pump power is supplied to the gas from the discharge circuit rather than from the more complex electron beam, these systems could ultimately have higher efficiencies, especially for lasers with high-output energies and pulses of long duration ($\sim 1\ \mu s$). Nevertheless, for laser oscillators with high pulse powers and for amplifiers, the high pump powers of electron beams make direct pumping more suitable.

2.3.3 Characteristics of Rare-Gas Halide Excimer Lasers

a) Frequency Control

With the exception of XeF and XeCl, the rare-gas halide excimers have excited (B) states with large binding energies (~ 5 eV) and relatively flat ground states. They thus give rise to emission spectra of relatively narrow spectral width ($\Delta\lambda \sim 2$ nm). The ability to extract energy in a narrow bandwidth with a tuned laser depends upon the distribution of energy among the vibrational levels of the excimer and the rate of vibrational relaxation within the manifold. Calculations of the frequency-dependent A-coefficient for the lowest-lying vibrational states shows that the population in the $v = 0-5$ states can all be extracted with similar efficiency at the same wavelength, but higher-lying states are inaccessible. In the case of KrF, the vibrational relaxation rate is a function of v and has a minimum at $v = 14$ for a gas pressure of 1 atm [2.137]. This minimum rate of de-excitation determines the effective vibrational relaxation time, which is of the order of 200 ps under laser conditions, in agreement with experimental estimates [2.138]. While this finite relaxation time may restrict the accessibility of the stored energy when the laser is used to amplify short, narrow-bandwidth pulses [2.139], it is sufficiently rapid to enable a laser oscillator with output pulse durations of >10 ns to be tuned without a substantial intrinsic loss in power.

The lasers can be frequency tuned over a narrow range using conventional intracavity tuning elements such as prisms, gratings, and etalons. The output energy of the laser decreases significantly under these circumstances, however, because of increased cavity losses. This is particularly true in the case of the uv-preionized, discharge-pumped lasers for which the gain is present typically for only 30 – 40 ns, and the desired radiation has insufficient time to build up and saturate the gain of the medium. Furthermore, intracavity apertures are often required to discriminate against higher-order modes propagating at large angles to the optical axis, and this restricts the mode volume of the laser to a small fraction of the total which is excited. As a result, amplification of the output of the tuned laser is often necessary.

Some narrowing of the linewidth of a uv-preionized discharge can be produced relatively easily without substantial loss of power by using quartz prisms. By this method output pulses of up to 30 mJ have been obtained in KrF and ArF lasers with linewidths of ~0.1 nm ($\sim 10\,\mathrm{cm}^{-1}$) and tuning ranges of ~2 nm [2.140,141]. A substantial decrease in linewidth is possible by inserting a small aperture into the cavity, as shown in Fig. 2.19a, and by using beam-expansion prisms to reduce the diffraction-limited beam divergence and hence increase the resolving power of the dispersive prisms. Linewidths of $0.4\,\mathrm{cm}^{-1}$ can be obtained by this method [2.142]. Further reduction in linewidth can be achieved by increasing the output mirror reflectivity from 10 to 50% to increase the number of cavity transits, and by using a solid etalon with a free spectral range of $1.0\,\mathrm{cm}^{-1}$ and a finesse of approximately 10 (Fig. 2.19b). Linewidths of $<0.1\,\mathrm{cm}^{-1}$ have been produced in this way [2.143].

Etalons can be used alone to produce frequency narrowing of excimer lasers. Using a single etalon of $\sim 50\,\mathrm{cm}^{-1}$ and a finesse of approximately 10, a linewidth of $3-4\,\mathrm{cm}^{-1}$ and a pulse energy of 1 mJ can be produced with a simple UV preionized discharge KrF laser [2.144]. Further spectral narrowing can be achieved using several etalons of different free spectral ranges within the laser cavity [2.145]. A system with three etalons is shown in Fig. 2.20. A master oscillator operates in a stable cavity, containing 1 mm and 0.1 mm etalons and a mode-limiting aperture, and produces an output of a few microjoules with a bandwidth of 3 GHz ($0.1\,\mathrm{cm}^{-1}$). This output is injected into a second, slave oscillator containing a $5\times$ beam-expanding telescope, a 1 cm etalon, and a adjustable aperture. A 10 ns output pulse of 500 MHz ($0.02\,\mathrm{cm}^{-1}$) bandwidth is produced with energy of a few microjoules [2.146].

Multiple etalons or combinations of etalons and prisms can provide very narrow linewidths, but continuous tuning over the full bandwidth of the laser

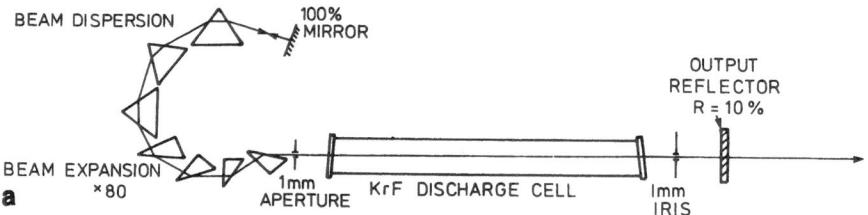

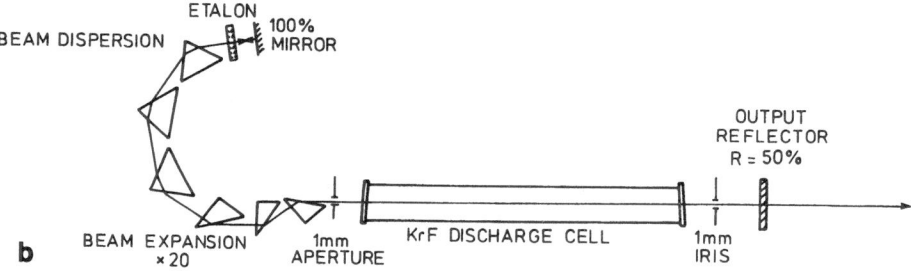

Fig. 2.19. A tunable discharge-pumped KrF laser with (**a**) beam expansion and dispersive prisms; (**b**) with an additional etalon

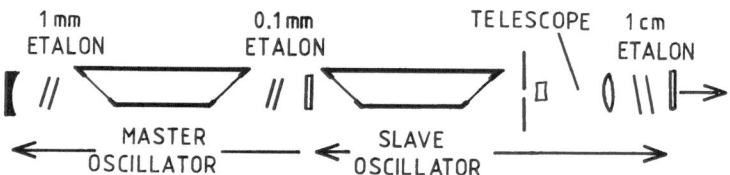

Fig. 2.20. A frequency narrowed KrF laser system using three etalons

transition is complicated and can be done reliably only under computer control. Diffraction gratings used close to grazing incidence [2.129] provide a simple method of providing very narrow linewidths, and only one cavity element needs to be adjusted to provide continuous spectral tuning. This technique is shown in Fig. 2.21. Using in third order a 1200 lines/nm grating blazed at 7500 A, linewidths of <0.3 cm^{-1}, tunable over the spectral range $248.05 < \lambda < 248.6$ nm, have been obtained for KrF. Using the same tuning technique, linewidths of 0.5 cm^{-1} have been produced from an ArF laser [2.147]. The spectrum of the instrument-limited laser output, as well as the amplified spontaneous emission, is shown in Fig. 2.22. The modulation in the spectrum of the spontaneous emission is due to absorption by atmospheric oxygen.

The output power of narrow-bandwidth excimer lasers is usually much too low for most applications, and further amplification is generally necessary. The most convenient method is to use a synchronously-pumped, uv-preionized discharge amplifier which can operate either as a single- or multiple-

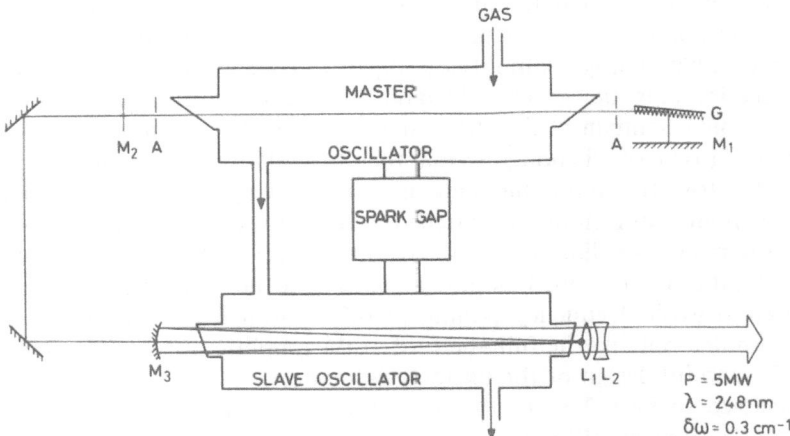

Fig. 2.21. A tunable KrF laser system employing a grazing incidence grating

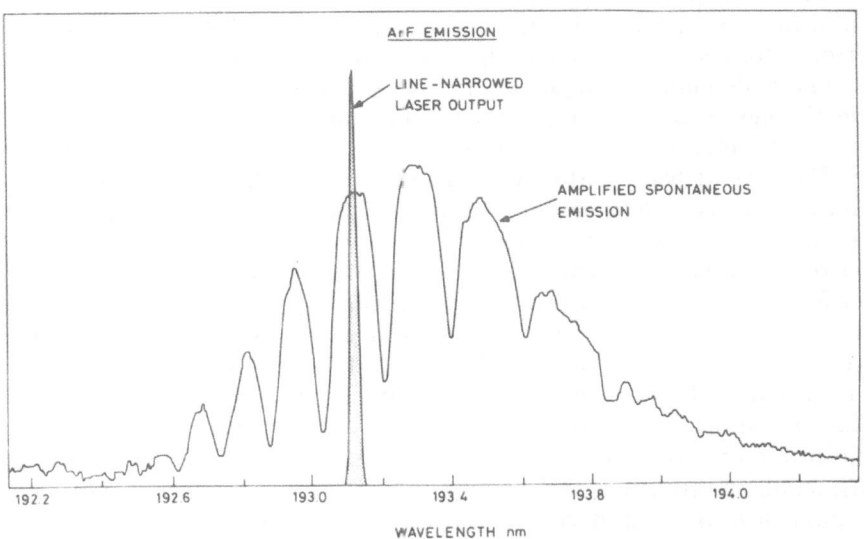

Fig. 2.22. Line narrowed output of a grating-tuned ArF laser oscillator and ArF amplified spontaneous emission (Intensity scale factors for the two are different)

pass amplifier, or an injection-locked oscillator as shown in Fig. 2.21. For a diffraction-limited input beam, input powers of ~ 1 mW are sufficient to lock $\sim 90\%$ of the output of the injection-locked oscillator to the input frequency, and output powers of 5 MW can be produced [2.129].

The injection-locked oscillators may incorporate either stable or unstable resonators, but the latter has the advantages that much of the energy stored in the amplifier may be extracted, and low-divergence, near diffraction-limited beams may be obtained. In designing the unstable resonator for optimum performance with a high-gain amplifier pumped with a short excitation pulse, the magnification must be chosen to be large enough for a diffraction-limited mode to become dominant in the short time available. The gain of the laser medium must also be small enough so that the threshold for laser oscillation is not reached before this mode has had time to build up. Furthermore, the mode of the input beam should be matched to that of the unstable resonator. When very narrow laser linewidths are required, the length of the unstable resonator should also be tuned simultaneously with the wavelengths of the input pulse, to avoid frequency pulling [2.148]. Nevertheless, beams with divergence limited only by the full aperture of the amplifier discharge can be produced by careful design of the unstable resonator.

As discussed in Sect. 2.3.2b, the discharge in uv-preionized systems is unstable, and in some amplifier designs the optical homogeneity of the excited gas is very poor. This gives rise to a deterioration in the beam quality when a single- or double-pass configuration or a stable resonator is used with the amplifier. This problem can be overcome by replacing a conventional mirror with a phase-conjugate mirror, as shown in Fig. 2.23. The phase-conjugate mirror is formed simply by focusing the amplified narrow-bandwidth radiation in a Brillouin-active liquid such as hexane. Provided the intensity is sufficiently high, a counter-propagating beam which is the phase-conjugate of the input beam, is then produced by stimulated-Brillouin scattering. Although the efficiency of back-scattering is significantly less than unity, diffraction-limited beams can be produced after a double-pass of the amplifier in both KrF and ArF laser systems [2.149].

Excimer laser systems with near diffraction-limited beams and Fourier-transform-limited bandwidths have been created by using a continuous-wave frequency-narrowed dye laser as the primary source of radiation. The required excimer wavelength must then be produced by frequency conversion, as for example, by third harmonic generation. (The harmonic generation is usually preceeded by amplification at the dye laser frequency.) By subsequent amplification in consecutive ArF excimer amplifiers of high optical quality, (diffraction and transform-limited) pulses of up to 450 mJ energy and ~ 7 ns duration have been produced [2.150]. Diffraction-limited ouput pulses of 60 mJ energy and 150 MHz bandwidth have been similarly produced in a KrF system, where the injected 248 nm radiation was produced by second harmonic generation in a cooled ADP crystal [2.151].

An alternative approach to the generation of tunable uv radiation for amplification in excimer lasers is by frequency doubling of a pulsed dye laser followed by anti-Stokes stimulated Raman scattering in a high-pressure gas. By frequency doubling a dye laser pulse of 60 mJ energy, frequency shifting to the fourth anti-Stokes line of H_2 and amplifying in two ArF amplifiers, 125 mJ pulses of ~ 1.6 cm^{-1} bandwidth and low divergence have been produced [2.152]. This method of generating very short wavelengths is more

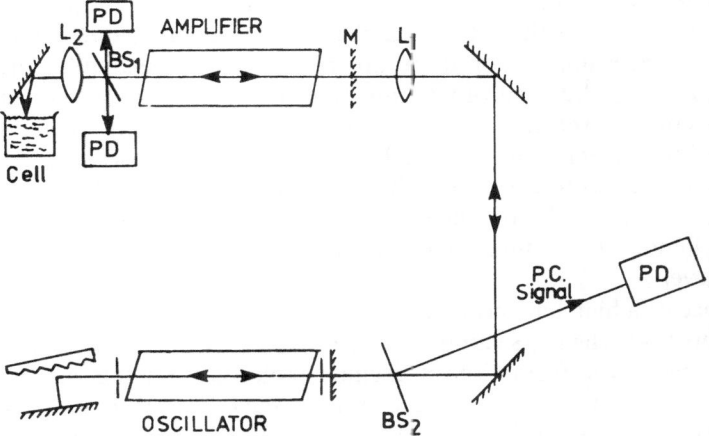

Fig. 2.23. Oscillator-amplifier system with a phase-conjugate mirror to improve beam quality

efficient than third harmonic generation, but the linewidth which can be produced is limited by the linewidth of the Raman medium.

An advantage of producing tunable uv radiation by nonlinear optical methods prior to amplification is that it enables ultrashort pulses to be produced. Advantage can therefore be taken of the relatively wide bandwidth of most rare-gas halide excimers (~ 100 cm^{-1}) which enables pulses as short as 0.1 ps to be amplified. By replacing the narrow-bandwidth dye laser by a mode-locked laser, output pulse energies of 40 mJ have been produced in pulses of 10 ps duration [2.153]. Intensities of $>10^{17}$ W cm^{-2} may be produced in the focused beam. Laser systems of this type have been used to study a wide range of nonlinear processes in atoms and molecules [2.154 – 156].

2.4 Rare-Gas Oxide Excimer Lasers

The need for high-power lasers for laser plasma research has stimulated a search for media which can store large amounts of energy to be extracted in a short pulse of radiation. For an amplifier to be efficient, however, the gain must be saturated. Therefore, for an efficient storage laser the stimulated emission cross section must be large enough so that the saturation energy density ($h\nu/\sigma_s$) does not exceed the damage threshold of the associated optics (~ 5 J cm^{-2}) and yet be small enough to avoid parasitic oscillation in the excited medium. Cross sections of $\sim 10^{-19}$ cm^2 offer a reasonable compromise.

The group VI-A elements (oxygen, sulfur, and selenium) have been suggested [2.157] as suitable laser systems. The p^4 ground electron configuration of these elements give rise to a ground $^3P_{0,1,2}$ state. The first excited electronic level is a $^1D_2$ state, and the next higher electronic level is a $^1S_0$ state. Pure electronic dipole transitions between these levels are forbidden, and radiative lifetimes exceeding 1 s or more are typical. As an example, the Doppler-broadened $^1S_0 - {}^1D_2$ "auroral" transition in atomic oxygen has an emission cross section on the order of 10^{-19} cm^2, and is therefore a potential candidate for a high-power laser system. In addition, buffer gases may be added to the laser medium to act as selective collisional quenchers of the $^1D_2$ level. In this manner it may be possible to construct a laser on the $^1S_0 - {}^1D_2$ transition that will not self-terminate due to an accumulation of population in the lower laser level.

In the presence of a high background pressure of a noble gas, the emission lines of the group VI-A elements can be considerably broadened and slightly shifted. This is due to the formation of excimer molecules. Of the various combinations possible, the rare-gas − oxide systems have been studied most extensively. The oxygen $^1S_0$ and $^1D_2$ states interact with xenon and krypton to form weakly-bound excimers (with a binding energy of about 500 cm^{-1} for XeO [2.158]): although, as shown in Fig. 2.24, the ArO potentials are repulsive [2.159]. Because of the weak binding, the atomic and excimer populations are in equilibrium, as for example in the XeO system,

$$O(^1S) + Xe + Xe \rightleftharpoons XeO(2\,^1\varSigma^+) + Xe \ .$$

As the pressure of xenon is increased, the equilibrium shifts in favor of the excimer, which has a radiative lifetime shorter than that of the atoms. Thus,

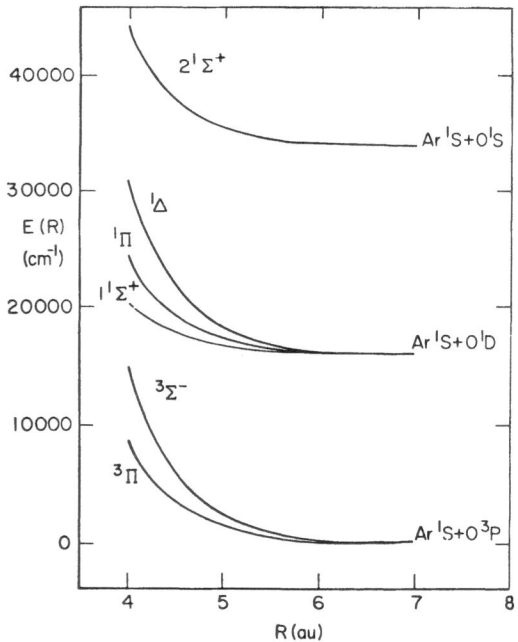

Fig. 2.24. Potential curves for ArO [2.159]

the overall lifetime is decreased, and the optical cross section is increased. By irradiating mixtures of the noble gas and trace amounts of oxygen with high-current electron beams, output powers of 100 kW have been obtained on the "auroral" bands of XeO, KrO [2.159, 160], and ArO [2.161], but with disappointingly low efficiencies (<0.1%), due in part to quenching by electrons.

2.5 Conclusion

The development of excimer lasers in recent years has been extremely rapid. Output pulse powers of several GW, mean powers of 25 W, and repetition rates of >1 kHz have been produced. Lasers have been developed whose spectral, spatial, and temporal characteristics are limited only by fundamental physical principles, and these lasers are now opening up new areas of non-linear optics. Excimer lasers, especially rare-gas halide lasers, will become increasingly important as a primary source of intense, tunable uv laser radiation.

References

2.1   W. N. Hartley, H. Ramage: R. Dublin Soc. Trans. 7, 339 (1901)
2.2   R. W. Wood: Phil. Mag. 18, 240 (1909)
2.3   Lord Rayleigh: Proc. Roy. Soc. London A 116, 702 (1927)
2.4   F. G. Houtermans: Helv. Phys. Acta 33, 933 (1960)
2.5   N. G. Basov, V. A. Danilychev, Yu. M. Popov, D. D. Khodkevich: J.E.T.P. Lett. 12, 329 (1970)
2.6   H. A. Koehler, L. J. Ferderber, D. L. Redhead, P. J. Ebert: Appl. Phys. Lett. 21, 198 (1972)
2.7   P. W. Hoff, J. C. Swingle, C. K. Rhodes: Appl. Phys. Lett. 23, 245 (1973)
2.8   W. M. Hughes, J. Shannon, R. Hunter: Appl. Phys. Lett. 24, 488 (1974)
2.9   L. A. Kuznetsova, Y. Y. Kugyarkov, V. A. Shpanskii, V. M. Shutoretskii: Vestn. Mosk. Univ. Ser II Khim 19 (1964)
2.10 C. A. Brau: In Excimer Lasers, 2nd. ed., ed. C. K. Rhodes, Topics Appl. Phys., Vol. 30 (Springer Berlin, Heidelberg 1984) Chap. 4
2.11 J. J. Ewing: In Laser Handbook, Vol. 3 (North-Holland, Amsterdam, 1979)
2.12 M. J. Shaw: Progr. Quantum Electron. 6, 3 (1979)
2.13 M. H. R. Hutchinson: Appl. Phys. 21, 95 (1980)
2.14 J. J. Hopfield: Phys. Rev. 35, 1133 (1930); Astrophys. J. 72, 133 (1930)
2.15 Y. Tanaka, A. S. Jursa, F. J. LeBlanc: J. Opt. Soc. Am. 48, 304 (1958)
2.16 M. L. Ginter: J. Chem. Phys. 42, 561 (1965)
2.17 Y. Tanaka, K. Yoshino: J. Chem. Phys. 50, 3087 (1969)
2.18 R. S. Mulliken: J. Am. Chem. Soc. 91, 4615 (1969)
2.19 B. Liu: Phys. Rev. Lett. 27, 1251 (1971)
2.20 S. L. Guberman, W. A. Goddard: Chem. Phys. Lett. 14, 460 (1972)
2.21 M. L. Ginter, R. Battino: J. Chem. Phys. 52, 4469 (1970)
2.22 R. S. Mulliken: J. Chem. Phys. 32, 5170 (1970)
2.23 T. L. Barr, D. Dee, F. R. Gilmore: J. Quant. Spect. Rad. Trans. 15, 625 (1975)

2.24   R. C. Michaelson, A. L. Smith: J. Chem. Phys. **61**, 2566 (1974)

2.25   E. H. Fink, F. J. Comes: Chem. Phys. Lett. **30**, 267 (1975)

2.26   F. H. Mies: Mol. Phys. **26**, 1233 (1973)

2.27   R. L. Platzman: Intern. J. Appl. Rad. Isotopes **10**, 116 (1961)

2.28   H. Powell, J. Murray: Lawrence Livermore Laboratory Annual Report, UCRL-50021-75, 502 (1975)

2.29   C. W. Werner, E. V. George, P. W. Hoff, C. K. Rhodes: Appl. Phys. Lett. **25**, 235 (1974)

2.30   E. V. George, C. K. Rhodes: Appl. Phys. Lett. **23**, 139 (1973)

2.31   D. C. Lorents: Physica **82 C**, 19 (1976)

2.32   G. R. Fournier: Opt. Commun. **13**, 385 (1975)

2.33   M. H. R. Hutchinson: Inst. Phys. Conf. Ser. **29**, 109 (1976)

2.34   C. W. Werner, E. V. George, P. W. Hoff, C. K. Rhodes: IEEE J. QE-**13**, 769 (1977)

2.35   D. Ton-That, M. R. Flannery: Phys. Rev. A**15**, 517 (1977)

2.36   A. K. Bhattacharya: Appl. Phys. Lett. **17**, 521 (1970)

2.37   A. P. Vitols, H. J. Oskam: Phys. Rev. A**8**, 1860 (1973)

2.38   D. Smith, A. G. Dean, I. C. Plumb: J. Phys. B**5**, 2134 (1972)

2.39   J. J. Lennon, M. C. Sexton: J. Electronics and Control **1**, 123 (1959)

2.40   H. J. Oskam, V. R. Mittlestadt: Phys. Rev. **132**, 1445 (1963)

2.41   J. N. Bardsley, M. A. Biondi: Adv. At. Mol. Phys. **6**, 2 (1970)

2.42   C. J. Freeman, M. J. McEwan, R. F. Claridge, L. F. Phillips: Chem. Phys. Lett. **10**, 530 (1971)

2.43   P. R. Tumpson, J. M. Anderson: Can. J. Phys. **48**, 1817 (1970)

2.44   J. W. Keto, R. E. Gleason, Jr., T. D. Bonfield, G. W. Walters, F. K. Soley: Chem. Phys. Lett. **42**, 125 (1976)

2.45   A. Johnson, J. B. Gerardo: J. Chem. Phys. **59**, 1738 (1973)

2.46   E. Zamir, C. W. Werner, W. P. Lapatovich, E. V. George: Appl. Phys. Lett. **27**, 56 (1975)

2.47   J. W. Keto, R. E. Gleason, G. K. Walters: Phys. Rev. Lett. **33**, 1375 (1974)

2.48   O. Cheshnovsky, A. Gedanken, B. Raz, J. Jortner: Chem. Phys. Lett. **22**, 23 (1973)

2.49   E. R. Ault, R. S. Bradford, M. L. Blaumik: (unpublished)

2.50   R. O. Hunter, J. Shannon, W. Hughes: Maxwell Laboratories Report MLR-378 (1975)

2.51   C. E. Turner: Appl. Phys. Lett. **31**, 659 (1977)

2.52   M. H. R. Hutchinson, M. R. O. Jones: (unpublished)

2.53   W. G. Wrobel, H. Rohr, K. H. Steuer: Appl. Phys. Lett. **36**, 113 (1980)

2.54   For a review see M. F. Golde: Gas Kinetics and Energy Transfer **2**, 123 (1977)

2.55   J. J. Ewing, C. A. Brau: Phys. Rev. A**12**, 129 (1975)

2.56   C. A. Brau, J. J. Ewing: J. Chem. Phys. **63**, 4640 (1976)

2.57   J. Tellinghuisen, J. M. Hoffman, G. C. Tisone, A. K. Hays: J. Chem. Phys. **64**, 2484 (1976)

2.58   J. Tellinghuisen, A. K. Hays, J. M. Hoffman, G. C. Tisone: J. Chem. Phys. **65**, 4473 (1976)

2.59   J. Tellinghuisen, P. C. Tellinghuisen, G. C. Tisone, J. M. Hoffman, A. K. Hays: J. Chem. Phys. **68**, 5177 (1978)

2.60   P. C. Tellinghuisen, J. Tellinghuisen, J. A. Coxon, J. E. Velagos, D. W. Setser: J. Chem. Phys. **68**, 5187 (1978)

2.61   T. H. Dunning, Jr., P. J. Hay: Appl. Phys. Lett. **28**, 649 (1976)

2.62   P. J. Hay, T. H. Dunning Jr.: J. Chem. Phys. **66**, 1306 (1977)

2.63   P. J. Hay, T. H. Dunning Jr.: J. Chem. Phys. **69**, 2209 (1978)

2.64   L. A. Gundel, D. W. Setser, M. A. A. Clyrre, J. A. Coxon, W. Nip: J. Chem. Phys. **64**, 4390 (1976)

2.65   R. W. Waynant: Appl. Phys. Lett. **30**, 234 (1977)

2.66   J. E. Velazeo, J. H. Kolts, D. W. Setser: J. Chem. Phys. **65**, 3468 (1976)

2.67   D. C. Hogan, R. Bruzzese, A. J. Kearsley, C. E. Webb: J. Phys. D**14**, L157 (1981)

2.68   N. G. Basov, V. S. Zuev, A. V. Kanev, L. D. Mikheev, D. B. Shavrovskii: Sov. J. Quantum. Electron. **10**, 1561 (1980)

2.69   C. A. Brau, J. J. Ewing: J. Chem. Phys. **63**, 4640 (1975)

2.70 J. R. Murray, H. T. Powell: Appl. Phys. Lett. **29**, 252 (1976)
2.71 R. Shuker: Appl. Phys. Lett. **29**, 285 (1976)
2.72 G. P. Quigley, W. M. Hughes: Appl. Phys. Lett. **32**, 649 (1978)
2.73 R. Burnham, S. K. Searles: J. Chem. Phys. **67**, 5967 (1977)
2.74 R. Burnham, N. W. Harris: J. Chem. Phys. **66**, 2742 (1977)
2.75 J. G. Eden, S. K. Searles: Appl. Phys. Lett. **30**, 287 (1977)
2.76 J. G. Eden, R. W. Waynant: Opt. Lett. **2**, 13 (1978)
2.77 T. H. Dunning Jr., P. J. Hay: J. Chem. Phys. **69**, 134 (1979)
2.78 R. W. Waynant, J. G. Eden: IEEE J. QE-15, 61 (1978)
2.79 M. Rokni, J. H. Jacob, J. A. Mangaro, R. Brochu: Appl. Phys. Lett. **30**, 458 (1977)
2.80 Yu. A. Kudryartsev, N. P. Kuzmina: Appl. Phys. **13**, 107 (1977)
2.81 R. M. Hill, P. L. Trevor, D. L. Huestis, D. C. Lorents: Appl. Phys. Lett. **34**, 137 (1979)
2.82 S. K. Searles, G. A. Hart: Appl. Phys. Lett. **27**, 243 (1975)
2.83 W. K. Bischel, H. H. Nakano, D. J. Eckstrom, R. M. Hill, D. L. Huestis, D. C. Lorents: Appl. Phys. Lett. **34**, 565 (1979)
2.84 W. B. Lacina, D. B. Cohn: Appl. Phys. Lett. **32**, 106 (1978)
2.85 T. H. Johnson, A. M. Hunter: J. Appl. Phys. **51**, 2406 (1980)
2.86 F. Kannari, M. Obara, T. Fujioka: J. Appl. Phys. **53**, 135 (1982)
2.87 M. R. Flannery, T. P. Yang: Appl. Phys. Lett. **32**, 327 (1978)
2.88 M. Rokni, J. A. Mangan, J. H. Jacob, J. C. Hsia: IEEE J. QE-14, 464 (1978)
2.89 Hao-Lin Chen, R. E. Center, D. W. Trainor, W. I. Fyfe: Appl. Phys. Lett. **30**, 99 (1977)
2.90 W.-C. F. Liu, D. C. Conway: J. Chem. Phys. **60**, 784 (1974)
2.91 D. Smith, P. R. Cromey: J. Phys. **B1**, 638 (1968)
2.92 J. C. Cronin, M. C. Sexton: J. Phys. **D1**, 863 (1969)
2.93 M. R. Flannery, T. P. Yang: Appl. Phys. Lett. **32**, 356 (1978)
2.94 D. K. Bohme, N. G. Adams, M. Moselman, D. B. Dunkin, E. E. Ferguson: J. Chem. Phys. **52**, 5094 (1970)
2.95 M. Rokni, J. H. Jacob, J. A. Mangano, R. Brochu: Appl. Phys. Lett. **31**, 79 (1977)
2.96 L. G. Piper, J. E. Velazes, D. W. Setser: J. Chem. Phys. **59**, 3323 (1973)
2.97 C.-H. Chen, M. G. Payne: IEEE J. QE-15, 149 (1979)
2.98 J. G. Eden, R. W. Waynant, S. K. Searles, R. Burnham: Appl. Phys. Lett. **32**, 733 (1978)
2.99 V. H. Shui: Appl. Phys. Lett. **31**, 50 (1977)
2.100 M. Rokni, J. H. Jacob, J. A. Mangano: Phys. Rev. A**16**, 2216 (1977)
2.101 G. P. Quigley, W. M. Hughes: Appl. Phys. Lett. **32**, 649 (1978)
2.102 A. M. Hawryluk, J. A. Mangano, J. H. Jacob: Appl. Phys. Lett. **31**, 164 (1977)
2.103 R. K. Steunenberg, R. C. Vogel: J. Am. Chem. Soc. **78**, 901 (1956)
2.104 A. Mandl: Phys. Rev. A**3**, 251 (1971)
2.105 W. J. Stevens, M. Gardner, A. Karo: J. Chem. Phys. **67**, 2860 (1977)
2.106 W. R. Wadt, D. C. Cartwright, J. S. Cohen: Appl. Phys. Lett. **31**, 672 (1977)
2.107 H. A. Hyman: Appl. Phys. Lett. **31**, 14 (1977)
2.108 W. R. Wadt, P. J. Hay: J. Chem. Phys. **68**, 3850 (1978)
2.109 J. F. Eden, R. S. F. Chang, L. J. Palumbo: J. QE-15, 1146 (1979)
2.110 A. Luches, V. Nassisi, A. Perrone, M. R. Perrone: Opt. Commun. A**4**, 109 (1982)
2.111 M. J. Shaw, J. D. C. Jones: Appl. Phys. **14**, 393 (1977)
2.112 H. H. Michels, R. H. Hobbs, L. A. Wright: J. Chem. Phys. **71**, 5053 (1979)
2.113 J. H. Kolts, J. E. Velazco, D. W. Setser: J. Chem. Phys. **71**, 1247 (1979)
2.114 a C. B. Collins, F. W. Lee: J. Chem. Phys. **72**, 5381 (1980)
2.114 b D. Kligler, H. H. Nakano, D. L. Huestis, W. K. Bishel, R. M. Hill, C. K. Rhodes: Appl. Phys. Lett. **3**, 39 (1978)
2.115 M. Allan, S. F. Wong: J. Chem. Phys.
2.116 W. V. Kurepa, D. S. Belic: J. Phys. B**11**, 3719 (1978)
2.117 W. E. Ernst, F. K. Tittel: Appl. Phys. Lett. **35**, 36 (1979)
2.118 J. D. Campbell, C. H. Fisher, R. E. Center: Appl. Phys. Lett. **37**, 348 (1980)
2.119 R. Burnham: Appl. Phys. Lett. **35**, 48 (1979)

2.120 C. H. Fisher, R. E. Center, G. J. Mullaney, J. P. McDaniel: Appl. Phys. Lett. **35**, 26 (1979)
2.121 D. J. Eckstrom, H. C. Walker: IEEE J. QE-**18**, 176 (1982)
2.122 P. J. Chantry: In *Applied Atomic Collision Physics,* Vol. 3 Gas Lasers (Academic, New York 1982)
2.123 J. E. Velazco, J. H. Kolts, D. W. Setsir: J. Chem. Phys. **69**, 4357 (1978)
2.124 W. L. Nighan, Y. Nachshon, F. K. Tittel, W. L. Wilson: Appl. Phys. Lett. **42**, 1006 (1983)
2.125 D. L. Huestes, G. Marowsky, F. K. Tittel: In *Excimer Lasers,* 2nd ed., ed. by Ch. K. Rhodes, Topics Appl. Phys., Vol. 30 (Springer, Berlin, Heidelberg 1984) Chap. 6
2.126 F. K. Tittel, W. L. Wilson, R. E. Stickel, G. Marowsky, W. E. Ernst: Appl. Phys. Lett. **36**, 405 (1980)
2.127 F. K. Tittel, G. Marowsky, M. C. Smayling, W. L. Wilson: Appl. Phys. Lett. **37**, 862 (1980)
2.128 M. L. Bhaumik, R. S. Bradford, Jr., E. R. Ault: Appl. Phys. Lett. **28**, 23 (1976)
2.129 R. G. Caro, M. C. Gower, C. E. Webb: J. Phys. D **15**, 767 (1982)
2.130 J. Coutts, C. E. Webb: J. Appl. Phys. **59**, 704 (1986)
2.131 R. S. Taylor, P. B. Corkum, S. Watanabe, K. Leopold, A. J. Alcock: IEEE J. QE-**19**, 416 (1983)
2.132 S. C. Lin, J. I. Leratter: Appl. Phys. Lett. **34**, 505 (1979)
2.133 M. R. Osborne, P. W. Smith, M. H. R. Hutchinson: Opt. Commun. **52**, 415 (1985)
2.134 J. A. Mangano, J. H. Jacob: Appl. Phys. Lett. **27**, 495 (1975)
2.135 M. Casey, P. W. Smith, M. H. R. Hutchinson: Rev. Sci. Instrum. **54**, 458 (1983)
2.136 J. D. Daugherty, J. A. Mangano, J. H. Jacob: Appl. Phys. Lett. **28**, 581 (1976)
2.137 Lawrence Livermore National Laboratory: Laser Laboratory Annual Report – UCRL-50021-80 (1980)
2.138 K. T. V. Grattan, M. H. R. Hutchinson, K. J. Thomas: Opt. Commun. **39**, 303 (1981)
2.139 J. H. Jacob, D. W. Trainor, M. Rokni, J. C. Hsia: Appl. Phys. Lett. **37**, 522 (1980)
2.140 T. R. Loree, K. B. Butterfield, D. L. Barker: Appl. Phys. Lett. **32**, 171 (1978)
2.141 J. Boker, J. Zavelovich, C. K. Rhodes: Phys. Rev. A **21**, 1453 (1980)
2.142 Lambda Physik GmbH
2.143 J. N. Ross: (private communication)
2.144 J. Goldhar, J. R. Murray: Opt. Lett. **1**, 199 (1977)
2.145 J. Goldhar, W. R. Rapoport, J. R. Murray: IEEE J. QE-**16**, 235 (1980)
2.146 J. Goldhar, M. W. Taylor, J. R. Murray: IEEE J. QE-**20**, 772 (1984)
2.147 M. C. Gower: Rutherford Appleton Laboratory Annual Report RL-83-043 (1983)
2.148 I. J. Bigio, M. Slalkine: Opt. Lett. **6**, 36 (1981)
2.149 M. C. Gower, R. G. Caro: Rutherford Appleton Laboratory Annual Report RL-82-039 (1982)
2.150 H. Egger, T. Srinivasan, K. Hohla, H. Scheingraber, C. R. Vidal, H. Pummer, C. K. Rhodes: Appl. Phys. Lett. **39**, 37 (1981)
2.151 R. T. Hawkins, H. Egger, J. Boker, C. K. Rhodes: Appl. Phys. Lett. **36**, 391 (1980)
2.152 J. C. White, J. Bokor, R. R. Freeman, D. Henderson: Opt. Lett. **6**, 293 (1981)
2.153 H. Egger, T. S. Luk, K. Boyer, D. F. Muller, H. Pummer, T. Srinivasan, C. K. Rhodes: Appl. Phys. Lett. **41**, 1032 (1982)
2.154 T. S. Luk, H. Pummer, K. Boyer, M. Shahidi, H. Egger, C. K. Rhodes: Phys. Rev. Lett. **51**, 110 (1983)
2.155 H. Pummer, H. Egger, T. S. Luk, T. Srinivasan, C. K. Rhodes: Phys. Rev. A **28**, 795 (1983)
2.156 T. Srinivasan, H. Egger, H. Pummer, C. K. Rhodes: IEEE J. QE-**19**, 1270 (1983)
2.157 R. C. Sze, P. B. Scott: Appl. Phys. Lett. **33**, 419 (1978)
2.158 C. D. Cooper, M. D. Lichtenstein: Phys. Rev. **109**, 2026 (1958)
2.159 P. S. Julienne, M. Krauss, W. J. Stevens: Chem. Phys. Lett. **38**, 174 (1976)
2.160 H. T. Powell, J. R. Murray, C. K. Rhodes: Appl. Phys. Lett. **25**, 730 (1974)
2.161 H. T. Powell, J. R. Murray, C. K. Rhodes: IEEE J. QE-**11**, 27 D (1975)
2.162 H. Powell: 8th Winter Colloquium on High Power Visible Lasers, Park City, Utah (1977)

3. Four-Wave Frequency Mixing in Gases

Carl R. Vidal

With 14 Figures

Coherent light can be generated in the vacuum ultraviolet by third-harmonic generation and sum-frequency mixing in gaseous nonlinear media. Conversion efficiencies of up to several percent allow the generation of kW and even MW peak powers in nanosecond pulses, and linewidths as low as 0.05 cm^{-1} are known. Such four-wave mixing extends the range of high-resolution laser spectroscopy, and is useful in plasma diagnostics and laser chemistry. In this chapter, theoretical treatment is followed by detailed description of experimental techniques and a survey of results.

3.1 Background

In recent years methods of nonlinear optics have been successfully used to extend the tuning range of lasers into other spectral regions. In the visible and the infrared spectral regions this has generally been done by sum- or difference-frequency mixing in suitable nonlinear crystals. In the vacuum ultraviolet (vuv) spectral region, however, most solids become opaque and can no longer be employed, because the generated wave of interest is absorbed and cannot escape the nonlinear medium. It is for this reason that sum frequency mixing in the vuv region is generally carried out in gaseous media using four wave parametric processes.

This chapter focuses on the resonant and nonresonant third-harmonic generation and sum-frequency mixing in gaseous nonlinear media which in recent years have been investigated rather extensively [3.1 – 3]. Four-wave frequency mixing in gases has become increasingly interesting as a powerful technique for extending high-resolution laser spectroscopy into the vuv, where many atoms, and a number of the most important small molecules such as H_2, CO, NO and others, have some of their most prominent absorption features. Several applications of this kind have already been reported [3.4, 5].

Harmonic generation in gases was first reported by *New* and *Ward* [3.6, 7] and by *Rado* [3.8]. The conversion efficiencies in these early experiments were extremely low and the results were therefore of little practical interest until *Harris* and coworkers [3.9 – 11] showed that the conversion efficiency in a gaseous system can be raised by several orders of magnitude using a two-component system. Similar to an idea which was originally demonstrated by

Bey et al. [3.12] in mixtures of liquids, Harris and coworkers showed that phase matching in gases can be achieved by a suitable mixture in which one component has been selected according to its nonlinear susceptibility, whereas the other component has to provide the appropriate overall refractive index for phase matching of the gas mixture. A further significant improvement was established by *Bloom* et al. [3.13], *Hodgson* et al. [3.14], and *Leung* et al. [3.15], who demonstrated that the conversion efficiency may be raised by exploiting a two-photon resonant enhancement of the nonlinear susceptibility. The latter effect was first discussed by *Maker* and *Terhune* [3.16] for third-order nonlinear processes in solids and liquids.

With gaseous systems, conversion efficiencies up to a few percent have so far been reported. Peak powers as high as 1 MW have been obtained for nonresonant harmonic generation using fixed-frequency high-power solid-state lasers [3.17, 18]. In order to indicate the technical development, a few further important achievements should be mentioned. Powers of the order of 10 W from nonresonant sum-frequency mixing of dye lasers were reported by *Hilbig* and *Wallenstein* [3.19, 20]. Peak powers of about 400 W at the wavelength of Lyman alpha have been obtained by *Langer* et al. [3.21] from nonresonant harmonic generation in phase-matched krypton argon mixtures, using a high-power dye laser. By using two-photon resonant sum frequency mixing of dye lasers, peak powers of the order of 10 W have been achieved by *Wallace* and *Zdasiuk* [3.22] and by *Junginger* et al. [3.23] in magnesium. Higher peak powers have been obtained by *Freeman* et al. [3.24] and *Mahon* and *Tomkins* [3.25] in mercury. *Scheingraber* and *Vidal* [3.26, 27] reported powers as high as 500 W in a strontium-xenon mixture using a tunable, excimer laser pumped dye laser. With a repetition rate of $100 \, s^{-1}$, this corresponds to a time-averaged power of 0.5 mW. Most recently, *Hilbig* and *Wallenstein* [3.28] have achieved powers as high as $0.5 - 3 \, kW$ by two-photon resonant sum and difference frequency mixing in mercury covering selected spectral regions between 109 and 196 nm.

Even higher peak powers, of typically 10 MW, have been achieved by combining the methods of nonlinear optics with high-power excimer lasers as amplifiers [3.29, 30]. Most recently, peak powers as high as 1 GW (pulse duration: 10 ps) have been demonstrated in the vuv where in the latter system a synchronously pumped, mode-locked dye laser was used in the initial nonlinear system [3.31, 32]. A significant advantage of gaseous nonlinear media over nonlinear crystals is that the gaseous media are not destroyed by laser-induced electric breakdown.

All of the preceding systems have used pulsed lasers to generate fundamental waves of sufficiently high input power. The first cw system was reported by *Freeman* et al. [3.33], who achieved a two-photon resonant enhancement of the nonlinear susceptibility by applying large magnetic fields to a strontium-xenon mixture. By means of a superconducting solenoid, they tuned the nonlinear medium into resonance with the incident cw beams [3.34] and obtained 10^7 photons/s (10^{-11} W, linewidth: 6 GHz). More recently, *Timmer-*

mann and *Wallenstein* [3.35] succeeded in making a cw system with 1.2×10^5 photons/s (2.8×10^{-13} W, linewidth: $\simeq 1$ MHz) in a magnesium-krypton mixture, without resorting to a magnetic field. Both cw systems have so far achieved a conversion efficiency of $10^{-12} - 10^{-11}$. This corresponds to cw output powers of $10^{-13} - 10^{-11}$ W.

It is the purpose of this chapter to outline the basic principles of four-wave frequency mixing in gases and to review the important aspects for designing nonlinear media of high conversion efficiency.

In general, a quantitative analysis of the electric fields interacting with a nonlinear medium requires solutions which obey the Schrödinger equation as well as Maxwell's equations [3.36, 37]. The Schrödinger equation describes the response of the medium to a local electric field and defines the material properties of the nonlinear medium, for example, in terms of the linear and nonlinear susceptibilities. Maxwell's equations, on the other hand, describe the response of the electromagnetic fields to the local properties of the medium. According to these general considerations, we start with a discussion of the nonlinear susceptibilities and their frequency-dependent properties, and present the selection rules associated with them. We then proceed to the fundamental equations of nonlinear optics as derived from Maxwell's equations. The general properties of a gaseous two-component system are then discussed in the small-signal limit for different kinds of incident beams, and considering also the linewidth dependence. With this information we present the general requirements for and the technical realization of suitable nonlinear media, which we follow by a description of experiments done in the small signal limit. Finally, we discuss the onset of saturation and the behavior in the high intensity saturation regime, where additional competing nonlinear processes have to be considered, and which also affect the optimum conditions in practical applications.

3.2 Nonlinear Susceptibilities

If matter is exposed to an electric field, the bound electrons of the atoms or molecules are displaced from their equilibrium positions under the influence of the applied electric field, giving rise to a net polarization of the medium. For small electric fields E, the polarization P of any material is given by the well-known linear relationship

$$P^{L} = N\chi^{(1)}E , \qquad (3.1)$$

where N is the number density of atoms or molecules and where the proportionality constant $\chi^{(1)}$ is the complex linear susceptibility

$$\chi^{(1)} = \bar{\chi}^{(1)} + i\tilde{\chi}^{(1)} = \frac{1}{\hbar} \sum_a \frac{|\mu_{ag}|^2}{(\Omega_{ag} - \omega)} , \qquad (3.2)$$

where μ_{ag} is the dipole matrix element for the transition between state $|g\rangle$ and state $|a\rangle$ and where

$$\Omega_{ag} = \omega_{ag} - i\Gamma_{ag} \tag{3.3}$$

is the corresponding complex transition frequency containing the damping constant Γ_{ag}. The real part of the linear susceptibility defines the refractive index n according to

$$1 + 4\pi N\tilde{\chi}^{(1)} = \varepsilon_r = n^2 , \tag{3.4}$$

whereas the imaginary part gives the absorption coefficient κ according to

$$4\pi N\tilde{\chi}^{(1)} = \varepsilon_i = \frac{cn\kappa}{\omega} . \tag{3.5}$$

For large electric fields the linear relation (3.1) eventually starts to break down, and higher-order nonlinear terms in the electric field have to be included. They may be incorporated by an additional nonlinear correction which can be given as a series expansion

$$P^{NL} = \sum_{n=2}^{\infty} P^{(n)} . \tag{3.6}$$

In order to specify the different series members $P^{(n)}$ we define the total electric field amplitude by its Fourier components according to

$$E(r, t) = \tfrac{1}{2} \sum_j e_j \hat{E}(r, \omega_j) \exp(i k_j r - i\omega_j t) + \text{c.c.} \tag{3.7}$$

where e_j is the unit polarization vector of wave j. Similarly, the total polarization is expressed by its Fourier components according to

$$P(r, t) = \tfrac{1}{2} \sum_j e_j P(r, \omega_j) \exp(-i\omega_j t) + \text{c.c.} . \tag{3.8}$$

In this manner the nth order terms $P^{(n)}$ in (3.6) are given by [3.37]

$$P_{\alpha_s}^{(n)}(r, \omega_s) = \frac{n! N}{2^{n-1}} \sum_{\alpha_1 \dots \alpha_n} \chi_{\alpha_s \alpha_1 \dots \alpha_n}^{(n)} (-\omega_s; \omega_1 \dots \omega_n)$$
$$\times E_{\alpha_1}(r, \omega_1) \dots E_{\alpha_n}(r, \omega_n) , \tag{3.9}$$

where N is the number density of atoms or molecules in the medium. The factor 2^{1-n} originates from the factor $1/2$ in the definitions of (3.7, 8), and the factor $n!$ accounts for the intrinsic permutation symmetry of the nonlinear susceptibilities if all frequencies $\omega_1, \omega_2 \dots \omega_s$ are nonzero and different. The indices $\alpha_1, \alpha_2 \dots \alpha_s$ are cartesian subscripts (x, y or z). Eq. (3.9) defines the nonlinear optical susceptibility tensors $\chi^{(n)}$, which are the key quantities of nonlinear optics, characterizing the nonlinear response of the medium to the simultaneous interaction with the different incident waves. For a summary of a large variety of nonlinear processes and their corresponding nonlinear susceptibilities, the reader is referred, for example, to the very useful Tables 2.1 – 3 of Ref. 3.37.

The nonlinear susceptibility tensors can be determined from a quantum mechanical time-dependent perturbation treatment of the nonlinear medium where one is interested in the expectation values of the field induced polarization P. Those values may be expressed in terms of the diagonal and off-diagonal elements of the density matrix [3.38]

$$\langle P(t) \rangle = N_t \text{Tr}\{\varrho(t)\mu\} = N_t \sum_{mn} \varrho_{mn}(t)\mu_{mn} \, , \qquad (3.10)$$

where N_t is the total number of interacting particles and μ is the dipole moment operator. The variation of the density matrix in time is determined by the von Neumann equation [3.38]

$$i\hbar\dot{\varrho}_{mn} = [H, \varrho]_{mn} \, . \qquad (3.11)$$

The Hamiltonian

$$H = H^0 + H' \qquad (3.12)$$

may be separated into its unperturbed part H^0, which gives the energy eigenstates of the unperturbed system, and the interaction Hamiltonian

$$H' = -\mu E(t) \, . \qquad (3.13)$$

As long as the electric field amplitudes E are sufficiently small, the density matrix elements may be expanded into a series of increasing order in the perturbation [3.38]. This was first done by *Armstrong* et al. in their classical paper [3.39] and by *Orr* and *Ward* [3.40]. It was later extended by *Puell* and *Vidal* [3.41] to systems in which the transient nature of the electric field amplitudes has to be considered explicitly. Different situations were treated in which the pulse duration of the incident beam is long or short compared to the phase and energy relaxation times, respectively.

According to (3.9) the nonlinear susceptibility $\chi^{(n)}$ is a $(n+1)$-rank tensor which in almost all cases of practical interest, such as birefringent nonlinear crystals, collapses to just a few significant numbers. For isotropic gaseous nonlinear media one is left with only one significant number. In general, the nth order nonlinear susceptibility is given by the following expression [3.36, 37]

$$\chi^{(n)}(-\omega_s; \omega_1 \ldots \omega_n)$$

$$= \frac{\mathscr{S}_T}{n! \, \hbar^n} \sum_{gb_1 \ldots b_n} \varrho(g) \frac{\langle g|e_s\mu|b_1\rangle \langle b_1|e_1\mu|b_2\rangle \ldots \langle b_n|e_n\mu|g\rangle}{(\Omega_{b_1 g} - \omega_1 - \ldots - \omega_n)(\Omega_{b_2 g} - \omega_2 - \ldots - \omega_n) \ldots (\Omega_{b_n g} - \omega_n)} \, . \qquad (3.14)$$

The summation has to be carried out over all atomic or molecular states $b_1, b_2 \ldots b_n$, and the permutation operator $\mathscr{S}_T$ requires that the sum has to be taken over all possible permutations of the $n + 1$ interacting waves specified by the individual e_k and ω_k. The factor $\varrho(g)$ takes care of the weighting over the unperturbed equilibrium distribution of initial states. For a single, thermally populated atomic ground state, $\varrho(g)$ can be dropped. For molecular systems in general, however, this cannot be done.

The most important nonlinear susceptibilities for consideration at present are the different $\chi^{(3)}$ terms characterizing a four-wave interaction with the nonlinear medium, which is usually the lowest-order nonlinearity of gaseous media. Similar to nonlinear crystals with a center of symmetry, such as calcite, the even-order terms $\chi^{(2)}, \chi^{(4)} \ldots$ generally vanish for isotropic gaseous nonlinear media, because the nonlinear polarization must be unaffected by reflection of the coordinate system at its origin. This symmetry, however, may be removed by applying, for example, an external electric or magnetic field. For the case of an electric field, this was shown in solids [3.42, 43] as well as in atomic and in molecular gases [3.44 – 50].

Among the different third-order nonlinear processes associated with $\chi^{(3)}$ we shall be interested in the specific case of four-wave mixing in which a sum frequency signal at

$$\omega_s = \omega_1 + \omega_2 + \omega_3 \tag{3.15}$$

is generated and which is illustrated in Fig. 3.1. For practical applications, however, one is more interested in situations where either two or all three of the incident beams have identical frequencies. The first case is usually realized in two-photon resonant sum frequency mixing where, according to Fig. 3.1, $\omega_1 = \omega_2 = \omega_{bg}/2$. The latter case represents third harmonic generation with $\omega_1 = \omega_2 = \omega_3 = \omega_s/3$.

According to (3.14), the nonlinear susceptibility associated with the process defined by (3.15) can readily be given. For explaining the relevant material parameters of four-wave mixing, however, it suffices to present only the simplest possible result of the third-harmonic generation where the three incident waves are identical. This limit already reveals all the important aspects, such as the selection rules, and the different kinds of resonant enhancement of the nonlinear susceptibility. If a nonlinear medium is exposed to an incident wave of angular frequency ω we have to consider the following three different kinds of third-order nonlinear susceptibilities for which, among all the possible permutations, only the dominant terms of interest are given

$$\chi^{(3)}(-3\omega; \omega\omega\omega) = \chi_T^{(3)}(3\omega)$$

$$= \hbar^{-3} \sum_{abc} \frac{\langle g | e_s \mu | a \rangle \langle a | e_1 \mu | b \rangle \langle b | e_2 \mu | c \rangle \langle c | e_3 \mu | g \rangle}{(\Omega_{ag} - \omega)(\Omega_{bg} - 2\omega)(\Omega_{cg} - 3\omega)}$$

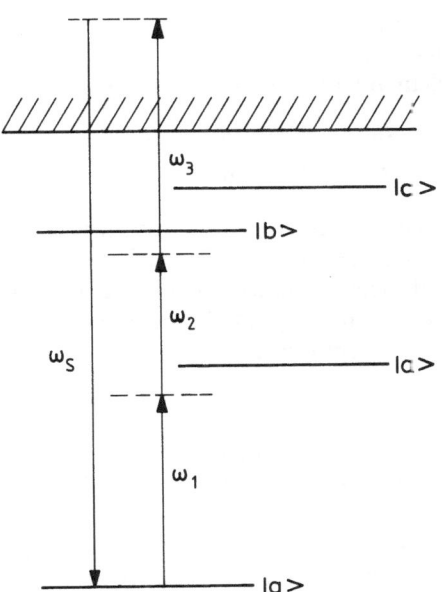

Fig. 3.1. Schematic diagram of sum-frequency mixing in a nonresonant nonlinear medium

$$\chi^{(3)}(-\omega;\omega-\omega\omega) = \chi_{SA}^{(3)}(\omega)$$

$$= \hbar^{-3} \sum_{abc} \frac{\langle g|e_s\mu|a\rangle\langle a|e_1\mu|b\rangle\langle b|e_2\mu|c\rangle\langle c|e_3\mu|g\rangle}{(\Omega_{ag}-\omega)(\Omega_{bg}-2\omega)(\Omega_{cg}-\omega)}$$

$$\chi^{(3)}(-\omega;\omega 3\omega-3\omega) = \chi_{SB}^{(3)}(\omega,3\omega)$$

$$= \hbar^{-3} \sum_{abc} \frac{\langle g|e_s\mu|a\rangle\langle a|e_1\mu|b\rangle\langle b|e_2\mu|c\rangle\langle c|e_3\mu|g\rangle}{(\Omega_{ag}-3\omega)(\Omega_{bg}-2\omega)(\Omega_{cg}-3\omega)} .$$

(3.16)

These nonlinear susceptibilities have the following physical significance. The first expression for $\chi_T^{(3)}(3\omega)$ specifies a polarization at 3ω which gives rise to the desired third harmonic signal. The second expression for $\chi_{SA}^{(3)}(\omega)$ causes a polarization at ω, and can be viewed at as a field-dependent correction to the linear susceptibility. Its real part is responsible for a field-dependent change of the refractive index, whereas the imaginary part entails a two-photon absorption. The third expression for $\chi_{SB}^{(3)}(\omega,3\omega)$ in (3.16) specifies a polarization at the frequency ω which originates from a simultaneous interaction of the fundamental wave and its third-harmonic wave with the nonlinear medium. The real part is therefore a change of the refractive index at the fundamental frequency due to the harmonic wave, and the imaginary part characterizes a Raman-type gain or loss of the fundamental wave under the influence of the harmonic wave.

All three expressions in (3.16) are almost identical, and differ only with respect to the resonant denominators. All numerators contain a product of four matrix elements of $e_k\mu$, where e_k is the unit vector specifying the polarization of wave k, and μ is the dipole-moment operator. By means of the Wigner-Eckart theorem the individual matrix elements may be written as [3.51]

$$\langle \alpha J_a m_a | e_k \mu | \beta J_b m_b \rangle = (-1)^{J_a - m_a} \begin{pmatrix} J_a & 1 & J_b \\ -m_a & \Delta m_k & m_b \end{pmatrix} \langle \alpha J_a \| \mu \| \beta J_b \rangle \tag{3.17}$$

where $\langle \alpha J_a \| \mu \| \beta J_b \rangle$ is the reduced matrix element. The quantity Δm_k inside the $3j$-symbol is equal to $+1(-1)$ for right-(left-) hand circularly polarized radiation and equal to 0 for linearly polarized radiation. The selection rule $-m_a + \Delta m_k + m_b = 0$ associated with the $3j$-symbol [3.51] immediately yields the following condition for the nonlinear susceptibilities in (3.16)

$$\Delta m_s + \Delta m_1 + \Delta m_2 + \Delta m_3 = 0 \ . \tag{3.18}$$

The latter relation implies conservation of angular momentum. As a practical and important example, note that two left-hand circularly polarized waves and a right-hand circularly polarized wave result in a right-hand circularly polarized wave. On the other hand, three right-hand or three left-hand circularly polarized waves do not result in any sum frequency wave at all.

The selection rules also imply that the states $|a\rangle$ and $|c\rangle$ have the opposite parity with respect to the states $|g\rangle$ and $|b\rangle$. This has a very important consequence for optimizing the nonlinear susceptibility of interest, $\chi_T^{(3)}$. For enhancing the nonlinear susceptibility it is obvious that the resonant denominators in (3.16) should be minimized. Since Ω_{ag} and Ω_{cg} are electric dipole allowed transitions and Ω_{bg} is not, it is first of all advantageous to minimize the two-photon resonant factor $(\Omega_{bg} - 2\omega)$ as shown by *Bloom* et al. [3.13], *Hodgson* et al. [3.14], and *Leung* et al. [3.15]. On the other hand, in minimizing the factors containing the dipole allowed transitions one has to be careful, because minimizing the one-photon resonance may lead to a strong absorption of the fundamental wave, whereas minimizing the three-photon resonance, for example, by means of a suitable auto-ionizing transition [3.14] leads to a strong absorption of the harmonic wave. In both cases the achievable harmonic wave will be reduced considerably. As shown below, the electric dipole allowed transitions should be optimized only to the extent that the corresponding optical depth is smaller or comparable to unity [3.52]. Already at this stage it is worth noting that the conversion efficiency for the four-wave mixing process due to the $\chi_T^{(3)}$ term is ultimately limited by the competing $\chi_{SA}^{(3)}(\omega)$ terms. Because of the great similarity between the different third-order nonlinear susceptibilities it is eventually very difficult to optimize one term with respect to the other, since any change in frequencies or in the nonlinear medium will always affect all nonlinear susceptibilities in (3.16) simultaneously.

The preceding discussion was carried out for the simplest case, that of third-harmonic generation. For the general four-wave mixing case there is essentially no significant difference with respect to the physics. The relations are much more cumbersome to work with, however, because the various multiples of the fundamental wave now have to be replaced by all possible permutations containing $\omega_1, \omega_2, \omega_3$, and ω_s.

3.3 Fundamental Equations of Nonlinear Optics

The interaction of electromagnetic waves with a scattering medium is governed by Maxwell's equations

$$\nabla \cdot D = 4\pi\varrho \qquad \nabla \cdot B = 0 \qquad\qquad (3.19)$$

$$\nabla \times E = -\frac{1}{c}\frac{\partial B}{\partial t} \qquad \nabla \times H = \frac{4\pi}{c}j + \frac{1}{c}\frac{\partial D}{\partial t}$$

using the common notation [3.53]. The electric displacement D is given by

$$D = E + 4\pi P = \varepsilon E \qquad\qquad (3.20)$$

and the magnetic induction B by

$$B = H + 4\pi M = \mu H \ . \qquad\qquad (3.21)$$

For the purposes of nonlinear optics, one generally makes the following simplifying assumptions:

1) magnetization $M = 0$: $H = B \rightarrow \mu = 1$,
2) source-free medium: $\varrho = 0$,
3) currentless medium: $j = 0$.

The latter two assumptions, of course, do not hold if the nonlinear medium is, for example, a plasma [3.54]. Otherwise, one is left with the simplified Maxwell equations

$$\nabla \times E = -\frac{1}{c}\frac{\partial B}{\partial t} \qquad\qquad (3.22)$$

$$\nabla \times B = \frac{1}{c}\frac{\partial E}{\partial t} + \frac{4\pi}{c}\frac{\partial P}{\partial t}$$

resulting in the well-known wave equation

$$\Delta E - \frac{1}{c^2}\frac{\partial^2 E}{\partial t^2} = \frac{4\pi}{c^2}\frac{\partial^2 P}{\partial t^2} \ . \qquad\qquad (3.23)$$

For the subsequent considerations it is useful to split the polarization P in (3.23) into its linear and nonlinear contributions, which are defined in (3.1 and 6), respectively,

$$P = P^L + P^{NL} . \tag{3.24}$$

In this manner, (3.23) goes over to the driven wave equation

$$\Delta E - \frac{n^2}{c^2} \frac{\partial^2 E}{\partial t^2} - i \frac{\varepsilon_i}{c^2} \frac{\partial^2 E}{\partial t^2} = \frac{4\pi}{c^2} \frac{\partial^2 P^{NL}}{\partial t^2} . \tag{3.25}$$

Note that according to (3.4), the refractive index n is defined to be a real quantity.

At this stage one has to specify the Fourier components of the electric field amplitudes, which generally are defined in terms of Gaussian beams. To simplify matters, most of the subsequent results will be derived within the plane-wave approximation $\hat{E}(r, \omega) = \hat{E}(z, \omega)$, and Gaussian beams will only be considered if necessary, as for example in the case of the phase matching of focused beams (Sect. 3.4.3). In order to simplify the notation, the Fourier component of the electric field amplitude $\hat{E}(r, \omega_j)$ in (3.7) will be replaced in all subsequent equations by $\hat{E}_j(r)$, where the index j no longer denotes a cartesian subscript as in (3.9), but the jth wave with the frequency ω_j.

At this point the wave equation is generally simplified by making the so-called slow amplitude approximation

$$\frac{\partial \hat{E}_j}{\partial t} \ll \omega \hat{E}_j ; \quad \frac{\partial \hat{E}_j}{\partial z} \ll k \hat{E}_j \tag{3.26}$$

which implies that the electric field amplitudes change sufficiently slowly within times of the order of ω^{-1} and over a length of typically k^{-1}. With this approximation, we finally arrive at the fundamental equations of nonlinear optics

$$\frac{d\hat{E}_j}{dz} = i \frac{2\pi\omega_j}{cn_j} P_j^{NL} \exp(-ik_j z) - \frac{\kappa_j}{2} \hat{E}_j , \tag{3.27}$$

where we have used for the total derivative of the electric field amplitude the relation

$$\frac{d\hat{E}_j}{dz} = \frac{\partial \hat{E}_j}{\partial z} + \frac{n_j}{c} \frac{\partial \hat{E}_j}{\partial t} . \tag{3.28}$$

In (3.27), $\hat{E}_j$ is a field envelope function at the frequency ω_j, which according to the definition of (3.7) is slowly varying in space and time. Within the plane-wave approximation, the spatial dependence of the electric field amplitude in a nonlinear medium is therefore given by

$$E_j = \hat{E}_j \exp(ik_j z) \tag{3.29}$$

and in the limit $P^{NL} = 0$, (3.27) leads to the well-known absorption relation of linear optics

$$E_j = E_{j0} \exp[(-\kappa_j/2 + ik_j)z] \ . \tag{3.30}$$

Since the electric field amplitude E_j is related to the intensity Φ_j according to [3.53]

$$\Phi_j = \frac{n_j c}{8\pi} |E_j|^2 \ , \tag{3.31}$$

one obtains the well-known Beer's law

$$\Phi_j = \Phi_{j0} \exp(-\kappa_j z) \ . \tag{3.32}$$

The Fourier components of the electric field amplitude and of the nonlinear polarization are $\hat{E}_j$ and P_j, respectively. Hence the fundamental equations of nonlinear optics are a coupled set of differential equations for the different $\hat{E}_j$, which are coupled through the nonlinear polarization P_j. These nonlinear polarizations may be viewed at as a phased array of electric dipoles which are capable of radiating at the various frequency combinations of the electric driving fields.

3.4 The Small Signal Limit

In discussing the fundamental equations of nonlinear optics for the problem of four-wave mixing in gases, it is useful to distinguish essentially three regimes which differ with respect to the electric field amplitudes to be considered, viz., (1) the small signal limit, (2) onset of saturation, and (3) high-intensity saturation.

With growing electric field amplitudes, an increasing number of nonlinear polarizations are required for a quantitative description of four-wave mixing in gases.

3.4.1 Plane Waves

In the small signal limit, the only nonlinear polarization to be considered in the fundamental equations of nonlinear optics (3.27) is given, according to (3.9), by

$$P_s^{(3)}(\omega_s) = \tfrac{3}{2} N \chi_T^{(3)}(-\omega_s; \omega_1 \omega_2 \omega_3) E_1 E_2 E_3 \ . \tag{3.33}$$

Since all other nonlinear polarizations are assumed to be negligible, the electric field amplitudes of the incident waves E_1, E_2, and E_3 are given by

(3.30) for collinear incident beams within the plane-wave approximation. Hence, inserting (3.33) into (3.27) and using (3.30), the electric field amplitude $\hat{E}_s$ is defined by

$$\frac{d\hat{E}_s}{dz} = i\frac{3\pi\omega_s}{cn_s}N\chi_T^{(3)}E_{10}E_{20}E_{30}\exp\left[\left(-\frac{\kappa_1+\kappa_2+\kappa_3}{2}-i\Delta k\right)z\right]-\frac{\kappa_s}{2}\hat{E}_s \tag{3.34}$$

where

$$\Delta k = k_s - k_1 - k_2 - k_3 \ . \tag{3.35}$$

The wave vector of the jth wave is k_j which is related to the refractive index n_j of (3.4) according to

$$k_j = \omega_j n_j/c \ . \tag{3.36}$$

We now introduce the optical depth τ_j

$$\tau_j = \kappa_j L = \sigma_j^{(1)}(\omega_j)NL \ , \tag{3.37}$$

where L is the length of the nonlinear medium and $\sigma_j^{(1)}(\omega_j)$ is the one-photon absorption cross section at the frequency ω_j. Equation (3.34) can then be integrated, with the result

$$\hat{E}_s(L) = i\frac{3\pi\omega_s}{cn_s}NL\chi_T^{(3)}E_{10}E_{20}E_{30}\exp(-\tau_s/2)\left[\frac{\tau_s-\tau_i}{2}-i\Delta kL\right]^{-1}$$

$$\times\left[\exp\left(\frac{\tau_s-\tau_i}{2}-i\Delta kL\right)-1\right] \tag{3.38}$$

where τ_i defines the total optical depth for all the incident waves

$$\tau_i = \tau_1 + \tau_2 + \tau_3 \ . \tag{3.39}$$

Replacing the electric field amplitude E_j by the intensity Φ_j according to (3.31), one finally obtains

$$\frac{\Phi_s}{n_s} = \left[\frac{24\pi^2\omega_s}{c^2n_s}NL\chi_T^{(3)}(-\omega_s;\omega_1\omega_2\omega_3)\right]^2\frac{\Phi_{10}\Phi_{20}\Phi_{30}}{n_1n_2n_3}F(\Delta kL,\tau_i,\tau_s) \tag{3.40}$$

where F is the phase-matching factor given by

$$F(\Delta kL,\tau_i,\tau_s) = \frac{\exp(-\tau_i)+\exp(-\tau_s)-2\exp[-(\tau_i+\tau_s)/2]\cos(\Delta kL)}{[(\tau_s-\tau_i)/2]^2+(\Delta kL)^2} \ . \tag{3.41}$$

This result was first reported by *Bey* and coworkers [3.55, 12].

3.4.2 Phase Matching

According to (3.41), the phase-matching factor F depends on three variables: $(\Delta k \cdot L)$, τ_i, and τ_s which one tries to minimize in order for F to approach the maximum value of unity. For discussion of the phase-matching factor F it is useful to extract from (3.41) a few limiting cases. For negligible absorption cross sections, the phase-matching factor has the form

$$F(\Delta k L, \tau_i = 0, \tau_s = 0) \simeq \left(\frac{\sin(\Delta k L/2)}{\Delta k L/2} \right)^2 \tag{3.42}$$

which is already well known from lower-order processes in nonlinear crystals [3.56]. For an optically thick system with $\tau_s \gg 1$ or $\tau_i \gg 1$, the minima of the preceding distribution fill up and one obtains a Lorentzian type profile for the phase-matching factor

$$F(\Delta k L, \tau_i \gg 1, \tau_s \ll 1) = F(\Delta k L, \tau_i \ll 1, \tau_s \gg 1)$$

$$= \frac{1}{[(\tau_s - \tau_i)/2]^2 + (\Delta k L)^2} \; . \tag{3.43}$$

For practical applications, it is generally not very difficult to keep the incident waves away from resonance with any electric-dipole allowed transition of the nonlinear medium, such that $\tau_1 \ll 1$, $\tau_2 \ll 1$, $\tau_3 \ll 1$, and hence $\tau_i \ll 1$. The cross section $\sigma_s^{(1)}(\omega_s)$, however, is frequently associated with a photoionization or auto-ionization, and has therefore to be taken into account. In Fig. 3.2 the phase-matching factor F is therefore plotted for $\tau_i = 0$ and for different values of τ_s as a parameter, showing also the limiting curves of (3.42) (upper curve) and of (3.43) (bottom curves).

Finally, for a phase-matched system with $\Delta k \cdot L = 0$, (3.41) becomes

$$F(\Delta k L = 0, \tau_i, \tau_s) = \left[\frac{\exp(-\tau_i/2) - \exp(-\tau_s/2)}{(\tau_i/2) - (\tau_s/2)} \right]^2 \; . \tag{3.44}$$

In the limit of small optical depths this expression may be approximated by

$$F(\Delta k L = 0, \tau_i \ll 1, \tau_s \ll 1) \simeq 1 - \frac{\tau_s + \tau_i}{2} \; . \tag{3.45}$$

In the case of large values of the optical depth, and for $\tau_s \gg \tau_i$ as in most practical situations, the intensity Φ_s at the sum frequency becomes

$$\frac{\Phi_s}{n_s} = \left(\frac{48 \pi^2 \omega_s}{c^2 n_s} \frac{\chi_T^{(3)}}{\sigma_s^{(1)}} \right)^2 \frac{\Phi_{10} \Phi_{20} \Phi_{30}}{n_1 n_2 n_3} \; . \tag{3.46}$$

It is important to note that the latter expression is independent of the column density NL of the nonlinear medium, and depends only on the ratio of the

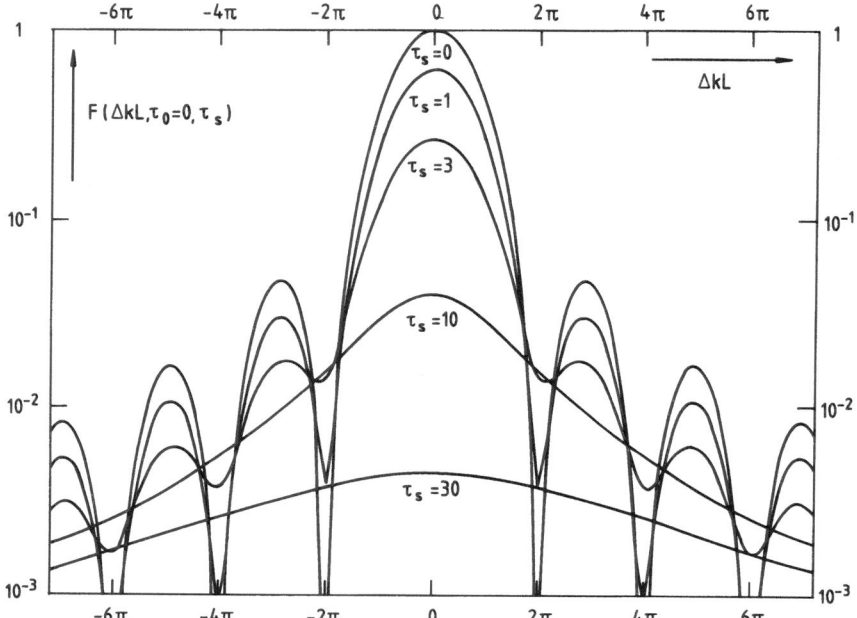

Fig. 3.2. The phase-matching factor F as a function of the mismatch ΔkL for different values of the optical depth τ_s. The optical depths for the incident waves are assumed to be negligible ($\tau_i = 0$)

nonlinear susceptibility $\chi_T^{(3)}$ and of the absorption cross section $\sigma_s^{(1)}(\omega_s)$. In the case of large optical depths, an equilibrium is reached inside the nonlinear medium in which the harmonic intensity is balanced by the power absorbed.

Equation (3.46) shows clearly the limitations which are imposed on a nonlinear medium by large optical depths. As mentioned already in Sect. 3.2, it is for this reason that a resonant enhancement of the nonlinear susceptibility at the frequencies of the incident waves and of the generated sum frequency wave are only useful for optical depths which are smaller or comparable to unity. Any further increase of the optical depths does not further improve the conversion efficiency.

The situation is quite different if four-wave frequency mixing is used as a method for investigating resonances of the nonlinear medium at the sum frequency [3.1]. In this case it is important to provide an optically thin system where, for example, the Fano-type profiles of the auto-ionizing resonances [3.57, 58] and their intensity dependence [3.59] are solely determined by the $\chi_T^{(3)}$ term in (3.40), and are not distorted by the absorption cross sections contained in the phase-matching factor of (3.41).

Besides a proper choice of the optical depths, the mismatch Δk defined in (3.35) is the most important quantity for optimizing the phase-matching factor F of (3.41). In order to achieve $\Delta k = 0$, one has to fulfill the following condition

$$\omega_1 n_1 + \omega_2 n_2 + \omega_3 n_3 = \omega_s n_s \qquad (3.47)$$

where (3.36) was used. For the limiting case of the third harmonic generation, this immediately leads to

$$n_1 = n_s \qquad (3.48)$$

implying that the fundamental and the harmonic wave have to travel with the same phase velocity through the nonlinear medium. In general, (3.47) is not fulfilled. However, as shown by *Bey* et al. [3.12] in liquids and by *Harris* and coworkers [3.9–11] in gaseous media, (3.47) can be satisfied in mixtures containing two components *a* and *b* for a particular density ratio

$$\frac{N_a}{N_b} = \frac{\omega_s \bar{\chi}_b^{(1)}(\omega_s) - \sum\limits_{j=1}^{3} \omega_j \bar{\chi}_b^{(1)}(\omega_j)}{\sum\limits_{j=1}^{3} \omega_j \bar{\chi}_a^{(1)}(\omega_j) - \omega_s \bar{\chi}_a^{(1)}(\omega_s)} . \qquad (3.49)$$

This condition imposes very stringent boundary conditions on the nonlinear medium because (3.49) has to be fulfilled over the entire length, requiring a sufficiently good homogeneity of the nonlinear medium.

Bjorklund and coworkers [3.60,61] have pointed out that phase matching can also be achieved in a one-component system. This is accomplished by properly adjusting the frequency of the incident laser beams such that the following condition, which is independent of the density, is fulfilled:

$$\omega_s \bar{\chi}^{(1)}(\omega_s) = \sum\limits_{j=1}^{3} \omega_j \bar{\chi}^{(1)}(\omega_j) . \qquad (3.50)$$

Hence a homogeneous medium is not required. For practical applications, however, a one-component system appears to be less attractive. For efficient two-photon resonant sum frequency mixing, one generally uses a gaseous two-component system in conjunction with a tunable and a fixed-frequency laser. In most cases the nonlinear medium can be well enough phase matched over the entire tuning range. For the system proposed by *Bjorklund* and coworkers [3.60,61], one needs a one-component system with three tunable lasers, which require a sophisticated (computer-controlled) nonlinear tuning technique. The individual lasers have to be simultaneously tuned in such a manner that they (1) give the desired sum frequency, (2) stay on the two-photon resonance and (3) obey the phase-matching relation (3.50). In addition, one should stay away from any strong one-photon absorption.

3.4.3 Gaussian Beams

In order to raise the conversion efficiency of a nonlinear medium, one frequently focuses the incident laser beams to increase the intensity. As will be shown below, however, focusing of the incident beams is only advantageous as long as higher-order nonlinearities do not reduce the conversion efficiency and cause additional perturbing parametric processes to occur (Sect. 3.8).

For focused beams the plane-wave approximation is no longer valid and the incident beams have to be described by a Gaussian intensity distribution [3.62]. Assuming a TEM$_{00}$-mode, which among all possible modes provides the highest intensity for a given input power, one has

$$E_j = E_j(r,z)\frac{b}{b+i2z}\exp\left(\frac{-k_jr^2}{b+i2z}\right)\exp(ik_jz) \tag{3.51}$$

with

$$b = 2k_jR_j^2 \tag{3.52}$$

where b is the confocal parameter and R_j is the 1/e radius of the intensity distribution in the focal plane.

In order to illustrate more clearly the effects of focusing on the phase-matching curve, absorption effects will be neglected for the moment. Absorption effects tend to smear out the oscillatory structure of the phase-matching curve, as shown for the case of plane waves in Fig. 3.2, and they further obscure some of the focusing effects. For a collinear arrangement of the incident beams with identical confocal parameters, the output beam is also Gaussian with the same confocal parameter. This is, of course, only true as long as one stays within the small signal limit. Since one is mainly interested in situations where the phase-matching condition is already closely satisfied, one has to a good approximation $\Delta k \ll k_j$. As a result, the radial dependence from the exponential factor in (3.51) can be neglected for the evaluation of the fundamental equations of nonlinear optics (3.27). With the preceding asumptions, one finally obtains a result which can be given in a form identical to (3.40). The phase-matching factor F is different however, and may be approximated by

$$F(\Delta kL, b/L) = \frac{1}{L^2}\left(\int_{-L/2}^{+L/2}\frac{e^{-i\Delta kz}}{(1+i2z/b)^2}dz\right)^2. \tag{3.53}$$

This phase-matching integral was first given by *Ward* and *New* [3.7] and has been reanalyzed again by *Bjorklund* [3.63] and *Puell* et al. [3.18]. For the case of plane waves with $b\to\infty$, (3.53) becomes (3.42). Another very useful analytical result can be obtained for $L \gg b$. In this case of tight focusing the limits of the integral can be extended to infinity, and one obtains

$$F(\Delta kb, b/L \ll 1) = \begin{cases} 0 & \text{for } \Delta kb \geqslant 0 , \\ \dfrac{\pi^2}{4}\left(\dfrac{b}{L}\right)^2(\Delta kb)^2 e^{\Delta kb} & \text{for } \Delta kb < 0 . \end{cases} \tag{3.54}$$

The latter function goes through a maximum for $\Delta k b = -2$, where the phase-matching factor takes on the value

$$F(\Delta k b = -2, b/L \ll 1) = \left(\frac{\pi}{e}\frac{b}{L}\right)^2 . \tag{3.55}$$

By going through a similar derivation, *Bjorklund* [3.63] also investigated the optimum conditions for the difference frequency mixing process and for tightly-focused beams. In summary, the optimum conditions for tightly-focused beams are obtained in general with

$$\Delta k b = \begin{cases} -2 & \text{for } \omega_s = \omega_1 + \omega_2 + \omega_3 , \\ 0 & \text{for } \omega_s = \omega_1 + \omega_2 - \omega_3 , \\ +2 & \text{for } \omega_s = \omega_1 - \omega_2 - \omega_3 . \end{cases} \tag{3.56}$$

It should be pointed out that there are different definitions in the literature for the sign of Δk. Here we have adopted the most frequently used convention, that of [3.11,63,64], whereas [3.7,18] use the opposite sign.

The preceding results are particularly important for practical applications. According to (3.54), the dispersion of the medium has to be negative, $\Delta k < 0$, for sum frequency mixing and tightly-focused beams. This constraint has to be modified for the case of frequency difference mixing and, of course, also does not hold for parallel or weakly-focused beams.

The conversion efficiency of the nonlinear medium has so far been expressed as a function of the intensity of the incident waves and of the sum frequency wave, as defined by (3.40). For most applications, however, the more relevant quantity is the power or energy conversion efficiency. Since the power of a particular wave is proportional to $R_j^2 \Phi_j \sim b_j \Phi_j$, (3.55) leads immediately to the important result that for tight focusing, $b/L \ll 1$, and the power conversion efficiency becomes independent of the confocal parameter b_j and the particular focusing associated with it. Hence, the higher intensity is offset by a shorter effective length of the nonlinear medium. For a given length of nonlinear medium any excessive focusing will therefore no longer raise the power conversion efficiency, but will only lead to additional perturbing nonlinear polarizations as a consequence of the increased intensity. As a rule of thumb, the optimum power conversion efficiency in a phase-matched system is given typically for $b \simeq L$.

In Fig. 3.3 several phase-matching curves are shown, including the limit of (3.54) with $b/L \ll 1$ (solid curve) for the case of no absorption losses. With increasing values of b/L, the oscillatory structure becomes increasingly pronounced. One eventually approaches the case of plane waves with the phase-matching curve of (3.42), which is the upper curve of Fig. 3.2.

So far we have considered only the simplest possible case of a Gaussian single-mode beam with a TEM_{00}-mode according to (3.51). In many applica-

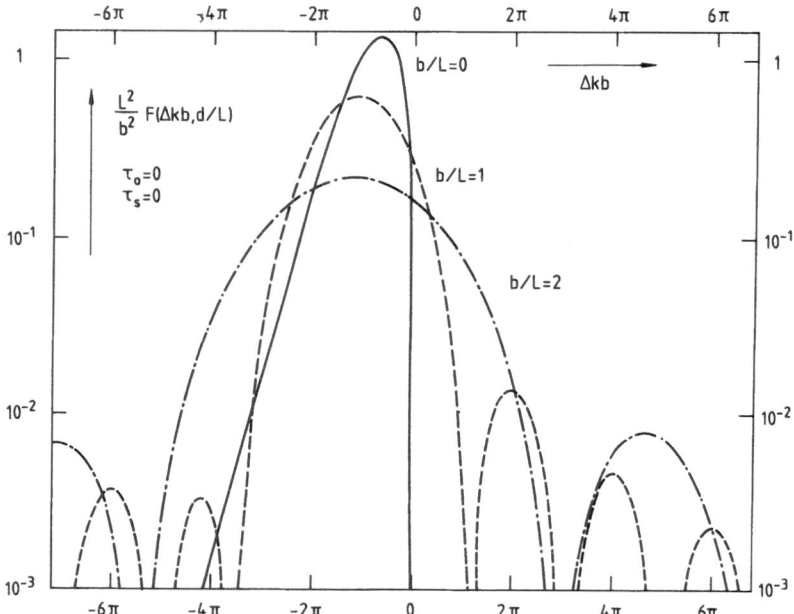

Fig. 3.3. The phase-matching factor $(L/b)^2 F(\Delta k b, b/L)$ as a function of $\Delta k b$ for different values of b/L. All optical absorption cross sections are assumed to be negligible

tions, however, one actually works with a multimode beam. In this case one has a very extensive superposition of contributions originating from the different individual modes. Aş shown by *Mui Yiu* et al. [3.64], the phase-matching curves differ significantly for different laser modes and for finite values of b/L. As a result, the single-mode oscillatory structure is washed out almost completely for a multimode beam, and one obtains an asymmetric phase-matching curve with no pronounced structure. In summary, the phase-matching condition depends essentially on four different parameters:

1) the k-vector mismatch Δk,
2) the absorption losses τ_j,
3) the mode structure of the incident beams, and
4) the focusing condition b/L.

In the limit of plane waves $L/b = 0$, the multimode structure no longer influences the phase-matching curve. Consequently, the phase-matching curve can be used as an extremely sensitive indicator of the plane-wave limit for a multimode beam, if the absorption losses are sufficiently small to yield an oscillatory structure of the phase-matching curve.

3.4.4 Linewidth Dependence

Another important aspect associated with the multimode structure of the incident laser beams is the frequency spectrum of the incident beams. We have considered only monochromatic waves in the preceding derivations of (3.40). Besides the frequency spectrum of the incident and the generated waves, one has to account also for the homogeneous and inhomogeneous linewidth of the nonlinear medium. This is particularly important for the case of two-photon resonant sum frequency mixing ($\omega_1 = \omega_2 = \omega_{bg}/2$), where (3.40) becomes

$$\frac{\Phi_s}{n_s} = \left[\frac{24\,\pi^2\,\omega_s}{c^2 n_s} NL \chi_T^{(3)}(-\omega_s; \omega_1 \omega_1 \omega_2) \right]^2 \frac{\Phi_{10}^2\,\Phi_{20}}{n_1^2 n_2} F(\Delta kL, \tau_i, \tau_s) \quad (3.57)$$

and which will be discussed in the following. In this case it is useful to decompose the nonlinear susceptibility into a resonant and a nonresonant factor

$$\chi_T^{(3)}(-\omega_s; \omega_1 \omega_1 \omega_2) = (\Omega_{bg} - 2\,\omega_1)^{-1} \chi_{NR}^{(3)}(\omega_s) \;, \quad (3.58)$$

where the homogeneous linewidth Γ_{bg} is contained in the complex transition frequency Ω_{bg} according to (3.3). For a Maxwellian velocity distribution, the number of particles with a shifted resonance frequency ω_D is given by a Gaussian distribution

$$N(\omega_D) = \frac{N}{\sqrt{2\,\pi\,\gamma_D}} \exp\left(-\frac{(\omega_D - \omega_{bg})^2}{2\,\gamma_D^2} \right) \quad (3.59)$$

where

$$\gamma_D = \frac{\omega_{bg}}{c} \sqrt{\frac{kT}{M}} \;. \quad (3.60)$$

The laser spectrum is assumed to consist of independent modes with a mode spacing which is small compared with any one of the linewidths involved. In many cases the envelope of the mode intensities can be given to a good approximation by a Gaussian distribution

$$\Phi_1(\bar{\omega}) = \frac{2\,\Phi_1}{\sqrt{\pi}\,\delta} \exp\left[-\frac{4}{\delta^2}(\bar{\omega} - \omega_1)^2 \right] \;, \quad (3.61)$$

where δ is a measure of the laser linewidth. In order to evaluate the frequency distribution at the sum frequency, all possible combinations of laser modes ω_a and ω_b have to be considered. Of these, in particular those contributions will dominate where the sum frequency $\omega_a + \omega_b$ falls within the linewidth of the two-photon resonant transition. It is important to realize that this may also include frequency combinations $\omega_a + \omega_b$, where neither of the two frequencies is resonant. Assuming that Φ_2 is nonresonant, the square of the laser intensity Φ_1^2 in (3.57) has to be replaced therefore by the auto-convolution of (3.61)

$$\Phi_1^2 G(\omega) = \sqrt{\frac{2}{\pi}} \frac{\Phi_1^2}{\delta} \exp\left[-\frac{2}{\delta^2}(\omega - 2\omega_1)^2\right] \tag{3.62}$$

with a subsequent integration over ω. Inserting (3.58 – 62) into (3.57), one finally obtains

$$\Phi_s = \frac{C}{2\pi\gamma_D^2} \int_{-\infty}^{+\infty} d\omega\, G(\omega) \left|\int_{-\infty}^{+\infty} d\omega_D \frac{\exp[-(\omega_D - \omega_{bg})^2/2\gamma_D^2]}{\omega_D - \omega - i\Gamma_{bg}}\right|^2 \tag{3.63}$$

where C is given by

$$C = \left(\frac{24\pi^2\omega_s}{c^2 n_s} NL\chi_{NR}^{(3)}\right)^2 \frac{\Phi_{10}^2 \Phi_{20}}{n_1^2 n_2} F(\Delta kL, \tau_i, \tau_s) \ . \tag{3.64}$$

Equation (3.63) was first given by *Stappaerts* et al. [3.65]. Since (3.63) cannot be integrated in a closed form, they determined asymptotic expressions. Later *Leubner* et al. [3.66] presented results with a greatly extended range of validity which cover most situations of practical interest. As a key step Leubner et al. brought (3.63) into the more convenient form

$$\Phi_s = 4C\,\mathrm{Re}\left\{\int_0^\infty d\eta \exp\left[-\left(\frac{\gamma_D^2}{4} + \frac{\delta^2}{8}\right)\eta^2 + i\Delta\omega\,\eta\right]\right.$$

$$\left. \times \int_\eta^\infty d\xi \exp\left(-\frac{\gamma_D^2\xi^2}{4} - \Gamma_{bg}\xi\right)\right\} \ . \tag{3.65}$$

The latter equation has the virtue that it is amenable to standard asymptotic techniques because all parameters appear in the exponent of exponentials. Furthermore, the integral can be written as a product of two factors, each depending on a single variable. In this manner, one can make successive application of standard asymptotic techniques appropriate for one-dimensional integrals containing one or more large parameters. On the basis of (3.65), [3.66] gives three series expansions for different limits of the linewidth parameters and the frequency detuning.

For practical applications a few limits are of particular interest. In high-resolution spectroscopy, for example, the laser linewidth is generally small compared to any other linewidth of the nonlinear medium. Another case of interest is important for systems which are designed for large conversion efficiencies. For such a situation, the laser linewidth has to be comparable with the Doppler width in order to optimize the number of particles participating in the nonlinear process. Another case, which could be taken directly from (3.57, 58), is that of nonresonant frequency mixing, where the frequency detuning is larger than any one of the linewidths.

A more complicated situation is encountered if line-narrowing effects have to be considered, as done by *Dick* and *Hochstrasser* [3.67] for three-wave mixing processes. In this case, the preceding convolution with the Gaussian-Doppler profile is generally no longer valid because of the statistical correlation between the velocity changing and the quantum-state changing collisions [3.68, 69]. It has also been shown by *Steel* et al. [3.70] that four-wave mixing in gases can be used as a technique of Doppler-free spectroscopy.

3.5 The Nonlinear Medium

3.5.1 General Requirements

For the efficient generation of coherent vuv radiation by means of four-wave mixing, a suitable nonlinear medium has to meet three major requirements:

1) The nonlinear susceptibility $\chi_T^{(3)}(-\omega_s; \omega_1 \omega_2 \omega_3)$ has to be large. This may be enhanced by exploiting a suitable two-photon resonance.
2) In order to exploit large column densities and to achieve large conversion efficiencies, it is necessary to maintain proper phase matching.
3) The optical depths for the incident waves and for the sum-frequency wave have to be small enough such that $\tau_j < 1$.

As discussed below, additional and more subtle requirements become important at large input intensities where one approaches the onset of saturation.

Because of the third requirement gaseous media are generally used for four-wave mixing in the vuv. As already explained, the first two requirements are generally met by means of a gaseous two-component system, where one component has been selected to provide a large enough nonlinear suscepti-bility, whereas the second component is responsible for achieving the phase-matching condition. Since the linear and nonlinear susceptibilities are determined by the transition moments and by the position of the atomic or molecular energy levels, the performance of a particular gaseous nonlinear medium depends strongly on the frequency of the incident waves. Over a wide range of frequencies the linear and nonlinear properties are dominated by the properties of the first few resonance transitions. For this reason, in the near vuv ($\lambda < 200$ nm), one typically uses metal-vapor inert-gas mixtures, whereas in the distant vuv ($\lambda < 100$ nm), also labeled xuv, it is frequently more appropriate to employ inert-gas mixtures. In order to meet the very stringent homogeneity requirements imposed by the phase-matching condition, the metal-vapor inert-gas mixtures are generally realized in devices based on the principle of the heat pipe oven [3.71 – 74]. The following subsection gives a brief description of such a device.

3.5.2 The Heat-Pipe Oven

A heat-pipe oven consists of a tube and a mesh structure covering the tube's inner surface which acts as a wick. In addition, it has a heating element and a heat sink, and contains a working material whose vapor is confined by an inert gas. The heater evaporates the working material of interest inside the central portion of the tube. The vapor flows out of the center section towards the ends of the tube and condenses in the outer, slightly colder parts of the heat-pipe oven. The condensed liquid then returns through the mesh back to the heater section because of capillary forces acting inside the mesh structure, as in the wick of a candle. The vapor zone eventually extends over a length of the heat pipe oven for which the power supplied by the heater is balanced by the losses due to radiation and due to heat conduction through the walls of the tubing. Because of the continuous evaporation and condensation of the working material, a large amount of heat is transferred through the heat pipe oven. This causes a very large heat conductivity which exceeds that of the most conductive solids by several orders of magnitude. Hence it causes an extremely homogeneous temperature and density distribution over the length of the vapor column. A stable equilibrium is finally reached in which the central vapor column is confined by inert gas whose pressure determines the vapor pressure inside the heat-pipe oven. The inert gas has the additional virtue of protecting the windows at the end of the tube from deposition of the vapors and corrosion associated with it.

A concentric heat-pipe oven was developed [3.72] which is able to provide extremely homogeneous metal vapor inert gas mixtures as required for efficient four-wave mixing. The schematic arrangement of the concentric heat-pipe oven is shown in the lower half of Fig. 3.4. An inner heat-pipe oven, containing the metal-vapor inert-gas mixture is surrounded by an outer heat-pipe oven. The outer oven, acting as an isothermal heater, imposes an extremely homogeneous temperature distribution on the inner heat-pipe oven. The temperature is determined by the confining inert-gas pressure P_o of the outer heat pipe oven, according to the vapor pressure curve of the working material. Frequently, the inner and outer heat-pipe oven are operated with the same working material. In this case, if the inert-gas pressure P_i of the inner heat pipe is larger than P_o, then the total pressure of the outer heat pipe oven is equal to the partial pressure of the working material in the inner heat-pipe oven. Consequently, the inner heat-pipe oven is filled with a metal-vapor inert-gas mixture, where the partial pressure of the metal vapor is given by P_o and the partial pressure of the inert gas by $(P_i - P_o)$ requiring $P_i > P_o$. Compared with cells containing inert-gas mixtures, it is one of the major advantages of a concentric heat-pipe oven that the partial pressures can be easily and independently adjusted. This aspect is particularly valuable for meeting the phase-matching condition.

For technical reasons, it can sometimes be advantageous to operate the outer heat-pipe oven with a working material which has a higher vapor

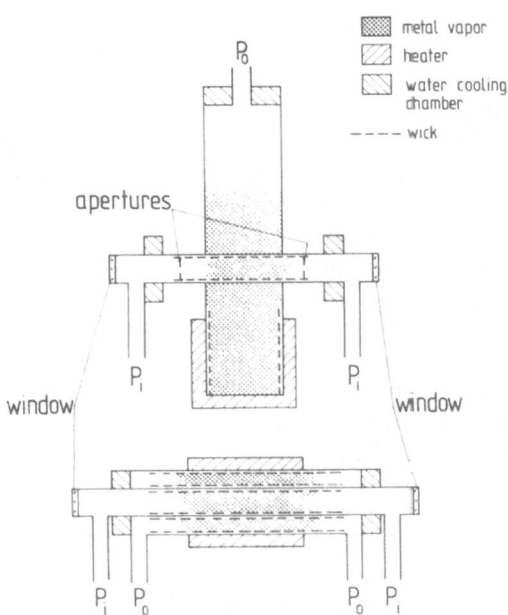

metal vapor
heater
water cooling chamber
----- wick

apertures

P_0

P_i P_i

window   window

P_i P_0 P_0 P_i

Fig. 3.4. Schematic arrangement of the concentric heat-pipe oven (lower half) which provides a homogeneous metal-vapor inert-gas mixture of adjustable length depending on the heater power supplied. The upper-half shows a modification of the concentric heat pipe oven which provides in addition a well-defined column density which is independent of the heater power supplied

pressure for a particular desired temperature than the working material of the inner heat-pipe oven. This is generally recommended if the metal vapor acting as the nonlinear medium has to be operated at rather low pressures for maintaining a small enough optical depth. As an example, Sr-Xe mixtures have frequently been generated in a concentric heat-pipe oven where the outer heat-pipe system was operated with sodium. The larger vapor pressure in the outer heat-pipe oven leads to smaller flow velocities for maintaining a particular power balance [3.74]. This results in turn in greater stability and homogeneity of the heat-pipe oven, as indicated by smaller short-term pressure fluctuations. In this case the partial pressure of the metal vapor inside the inner heat-pipe oven is, of course, no longer given by the total pressure P_0 of the outer heat-pipe oven, but it can be easily calibrated as a function of P_0.

The heat-pipe oven imposes further technical constraints on the particular choice of a nonlinear medium. For stable operation of the heat-pipe oven, the metal vapor should have a vapor pressure of a few Torr in a temperature range below 1200 °C, where high-temperature resistant stainless steel can be used for manufacturing the heat-pipe oven.

In the concentric heat-pipe oven shown in the lower half of Fig. 3.4, the length of the metal-vapor inert-gas mixture depends on the heater power. Small changes in the power balance will lead to corresponding changes of the total column density inside the heat-pipe oven. For maintaining the phase-matching condition in situations with a pronounced oscillatory structure, it may therefore be necessary to regulate the heater power. Furthermore, for obtaining a stable equilibrium, a long warm-up time is generally required. In order to circumvent these problems, a modified concentric heat-pipe oven was developed [3.73] which combines the virtue of the accurately defined partial pressures with a well-defined column density. This has been achieved with the system shown in the upper half of Fig. 3.4, where the length of the vapor column is defined by the diameter of the vertical heat-pipe oven. Any changes in the power balance will now move the transition region between the metal vapor and the confining inert gas inside the vertical heat-pipe oven up and down, but it will not affect the mechanically defined length of the heating zone imposed on the inner heat-pipe oven. In addition, a stable operation of the modified concentric heat-pipe system is achieved long before the vapor column in the vertical heat-pipe oven has reached its final equilibrium position. In practical applications, the modified concentric heat-pipe oven has so far provided the most stable and well-defined metal-vapor inert-gas mixtures. It can also be operated as a sealed-off system.

In order to achieve phase matching in metal-vapor inert-gas mixtures, one may sometimes need rather high inert-gas pressures, which result in a large homogeneous linewidth of the nonlinear medium. According to Sect. 3.4.4 this may lead to an excessive reduction of the conversion efficiency, and in particular for situations with a small laser linewidth. It may therefore be advantageous to use mixtures of metal vapors for phase matching, as shown by *Wynne* et al. [3.75] in the infrared using a Na-K system, and by *Bloom* et al. [3.76] in the ultraviolet spectral region using a Na-Mg system. For a nonlinear medium of this kind a different type of concentric heat-pipe oven can be used which generates mixtures of saturated and unsaturated metal vapors [3.77], where the pressure ratio can be adjusted again. Nevertheless, the latter system is significantly more difficult to operate than a system containing a metal-vapor inert-gas mixture, and has therefore rarely been used.

3.5.3 Other Systems

Concentric heat pipe ovens are easy to operate down to the lithium-fluoride cutoff around 104 nm. Beyond this range no useful window material exists until one can use, for example, thin metal foils. Since in the distant vuv inert-gas mixtures begin to become useful, the windows can be replaced by differential pumping [3.78 – 83] which requires, however, rather large amounts of gas and a powerful pumping system. A variant on that method which requires smaller amounts of gas was demonstrated by *Bonin* and *McIlrath* [3.84,85], who used a rotating disk valve as a fast shutter, opening the optical system only during the duration of the laser pulse. As a further method *Lucatorto* et al. used a glass capillary array [3.86,87], which greatly reduces the gas conductance, but introduces diffraction losses depending on the size of the capillaries. A more elegant technique was demonstrated by *Kung* [3.88], *Marinero* et al. [3.89,90], and by *Bokor* et al. [3.91], who used a pulsed supersonic jet [3.92] of xenon for frequency mixing. The great advantage of pulsed jets is that the nonlinear medium is only "turned on" around the duration of the laser pulse. Hence, the gas consumption is very small and simple vacuum systems can be used. A disadvantage is that because of the large density gradients in the vicinity of the orifice, pulsed jets are difficult to phase match and only small column densities can be used.

Another experimental difficulty in working with pure inert-gas systems is illustrated by experiments in Xe-Ar mixtures of *Kung* et al. [3.93], where it was shown that because of mixing problems, it may be difficult to provide a nonlinear medium homogeneous enough to obtain a meaningful phase-matching ratio. It should be stressed that a concentric heat pipe oven operates in dynamic equilibrium, where the homogeneity of the metal-vapor inert-gas mixture is permanently maintained by the continuous evaporation and condensation of the metal vapor.

In going to the extreme ultraviolet, where eventually inert gases will also cease to be useful, a gas of ions will become advantageous as a nonlinear medium, because ions can have ionization potentials well above those of the inert gases. Ions would also allow a resonant enhancement of the higher-order nonlinear susceptibilities (Sect. 3.8.4). In this context a new technique may become interesting, which promises to provide ions of different ionization stages with sufficient density and an adequate homogeneity. As shown by *Lucatorto* and *McIlrath* [3.94 – 96] and by *Skinner* [3.97], almost complete ionization of dense atomic vapors as generated, for example, inside a heat-pipe oven, can be achieved by irradiating the atomic resonance line with a high-power laser. In this manner ions are produced in a tightly-confined region with a very good homogeneity, as required for phase matching. The broad applicability of this method has been demonstrated for Li [3.95], Na [3.94], Ca, Sr [3.97], and Ba [3.96, 97].

The first harmonic generation using metal vapor ions was reported by *Sorokin* et al. [3.98] for calcium ions, which were generated by two-photon resonant multiphoton ionization, and gave rise to radiation at 127.8 nm. Also, resonantly photoionized magnesium was demonstrated recently by *Lebedev* et al. [3.99] for generating vuv radiation at 123.6 nm. They also observed the influence of magnesium ions diffusing out of the focal region in reducing the conversion efficiency of the ionized nonlinear medium. In this context, the fundamental equations of nonlinear optics (3.27) may have to be modified if the ionized gas acting as the nonlinear medium is no longer sourcefree, $\varrho \neq 0$, and no longer currentless, $j \neq 0$, as assumed above.

In addition to atomic systems, molecular gases such as H_2 [3.48, 79 – 81], CO [3.79, 80, 100 – 103], NO [3.104, 105], N_2 [3.103], and I_2 [3.106 – 109] have also been used successfully as nonlinear media. In these molecules the four-wave mixing may be resonantly enhanced by an intermediate electronic state. Nevertheless, in the vuv, photodissociation of the molecular gas may destroy the nonlinear medium. The conversion efficiencies which have so far been achieved are rather small because the energy levels and the corresponding transition moments are generally spread out over a rather large range. Under those circumstances, only very few levels with a rather small transition moment will efficiently support a resonant enhancement of the nonlinear susceptibility. This situation is quite different from the infrared spectral region, where molecular systems are rather attractive. *Kildal* and coworkers [3.110 – 112] have demonstrated two-photon resonant harmonic generation of CO_2 laser radiation in CO using rovibronic levels of the CO molecular ground state. Systems with a potential conversion efficiency of a few percent were suggested by *Kildal* and *Brueck* [3.112] in CO and by *She* and *Billman* [3.113], and by *Pan* et al. [3.114] in H_2.

3.6 Experiments in the Small-Signal Limit

In this section experiments are presented which illustrate behavior in the small-signal limit. Almost all experiments have been done with pulsed high-power laser systems, including mode-locked systems for ultrashort pulses. High-power solid-state lasers have mostly been employed in nonresonant harmonic generation. For two-photon resonant sum frequency mixing, high-power dye lasers have been used, pumped by flash lamps, frequency-doubled Nd lasers, nitrogen lasers, or excimer lasers.

Quantitative experimental analysis of four-wave mixing in gases requires first of all a well-defined profile of the incident beams. In general, it is advantageous to start with measurements of the phase-matching curve, which reveal the complex linear susceptibilities and verify the properties of the incident beams. In the following steps the nonlinear susceptibility $\chi_T^{(3)}$ can be obtained from a measurement of the conversion efficiency. For the measurement of the higher-order nonlinearities, which become important beyond the range of validity of the small signal limit, a detailed analysis of the field dependence is required, as shown below.

The discussion therefore starts with the phase-matching curves which have been measured most accurately using metal-vapor inert-gas mixtures. For a quantitative analysis it is important to note that, contrary to nonlinear crystals or gases confined by windows, a concentric heat-pipe oven does not provide a rectangular density profile of the nonlinear medium. The partial pressure of the metal vapor inside a concentric heat-pipe oven is constant over almost the entire length of the vapor column, except for a short transition region between the vapor column and the confining inert gas. The length of the transition zone depends on the temperature gradient imposed on the inner heat-pipe oven and the diffusion of the metal vapor into the inert gas. For calculating the phase-matching curves taken with a concentric heat-pipe oven, one has to account for these density gradients in the transition region. They have been approximated by *Puell* and coworkers [3.18, 52], by an empirical relation of the kind

$$N(z) = \tfrac{1}{2}N(0)\{1 + \tanh\left[\alpha(|z| - |L/2|)\right]\} \, , \tag{3.66}$$

where α specifies the steepness of the density gradient. In this case (3.34) can no longer be integrated in closed form, but has to be integrated numerically.

Figure 3.5 shows a phase-matching curve for two-photon resonant third harmonic generation in an optically thick Sr-Xe mixture, using a flash-lamp pumped dye laser [3.52]. The measurements were done with a confocal parameter of the beam $b = 550$ cm, a length $L = 30$ cm, and a strontium pressure $P_{Sr} = 8.5$ Torr. With a measured absorption cross section of $\sigma^{(1)} = 3.5 \times 10^{-18}$ cm^2 at a wavelength of 191.9 nm, the optical depth of the system was $\tau_s = 7.5$. Extrapolating a density gradient $\alpha = 0.6$ cm^{-1} from earlier Rb-Xe measurements [3.18], a phase-matching curve (solid line in Fig. 3.5) was calculated by *Scheingraber* et al. [3.52] which contains no free adjustable parameters and is in excellent agreement with the measurements. Compared with the corresponding phase-matching curve for a rectangular density distribution, as given by (3.43), the phase-matching curve in Fig. 3.5 shows a pronounced asymmetry which originates from the density gradient of the transition zones, and which is typical for all phase-matching curves taken with a heat pipe oven. The origin of this asymmetry can be easily understood. For a xenon pressure well below the optimum phase-matching value there is always a range within the transition zones where the phase-matching condition is met. These parts contribute significantly to the total harmonic output. For a xenon pressure well above the optimum value, phase matching is achieved nowhere within the metal-vapor inert-gas mixture. Around the maximum of the phase-matching curve, however, the density gradients lead only to a minor reduction of the conversion efficiency.

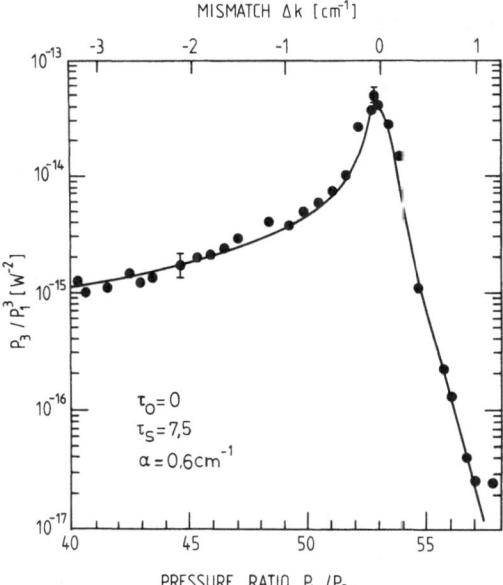

Fig. 3.5. Phase-matching curve for two-photon resonant third-harmonic generation in a Sr-Xe mixture using a parallel beam [3.52]. The *solid line* has been calculated for an optical depth $\tau_s = 7.5$ and a density gradient $\alpha = 0.6$ cm^{-1}

Phase-matching curves in optically thin systems were taken by *Puell* et al. [3.18] in a Rb-Xe mixture using nonresonant third-harmonic generation, and by *Junginger* et al. [3.23] in a Mg-Kr mixture using two-photon resonant third-harmonic generation. Figure 3.6 shows a photoelectric scan of the harmonic intensity as a function of the krypton pressure for a fixed magnesium pressure $P_{Mg} = 14$ Torr. A nitrogen laser pumped dye laser was used. The phase-matching curve shows the oscillatory structure which is typical for optically thin systems and plane waves. A comparison with theory is shown in Fig. 3.7, where only the maxima and minima of the measured phase matching curve are indicated by the dots. Again the asymmetry due to the density gradients can clearly be seen. Excellent agreement with theory was achieved by *Junginger* et al. [3.23] only after accounting for the changes of the homogeneous linewidth Γ_{bg} of the nonlinear medium as a function of the krypton pressure, which was varied in Fig. 3.7 over a wide range, from 42 to 210 Torr. The modulation depth of the phase-matching curve gave a cross section $\sigma^{(1)} = 4.5 \times 10^{-19}$ cm^2 at a wavelength of 143 nm. For the specific example of Fig. 3.7 the column density has to be controlled to within at least 3% in order to stabilize the conversion efficiency and to allow an accurate measurement of the phase-matching factor for a given pressure ratio. For experiments of this kind the modified concentric heat-pipe oven with its well-defined column density is particularly valuable, and the stability requirements are otherwise rather difficult to satisfy.

It should be noted that phase-matching curves for nonresonant harmonic generation using focused beams were also taken by *Young* et al. [3.10] in a Rb-Xe system and by *Kung* et al. [3.115] in a Cd-Ar system. Both experiments have shown good qualitative agreement with theoretical phase-matching curves. However, for a detailed quantitative comparison with theory, the important parameters of the nonlinear medium and of the incident beam were not known accurately enough. The phase-matching curves for the Rb-Xe mixture were taken with a Pyrex cell filled with rubidium and xenon, and for the Cd-Ar mixture with a normal open-ended heat-pipe oven, which for the case of metal-vapor inert-gas mixtures does not operate in the stable heat-pipe mode. Both experiments were done as a function of temperature for a given inert-gas pressure. Consequently, a comparison with theory is not very reliable because of the poorly defined vapor pressures, and because of the temperature-dependent changes in the column density, absorption rates, and the conversion efficiency. Discrepancies between theory and experiment can also be seen in work of *Ferguson* and *Arthurs* [3.116] for Ca-Xe mixtures and of *Taylor* [3.117] for Na-Xe mixtures, which again are most likely due to poorly defined experimental parameters and inhomogeneities of the nonlinear medium.

Tomkins et al. [3.118] have pointed out the effect of optical aberrations as another source of errors on four-wave mixing experiments. The aberrations can distort the phase-matching curve given by (3.54), an effect which has in some cases been erroneously attributed to the presence of higher-order spatial modes.

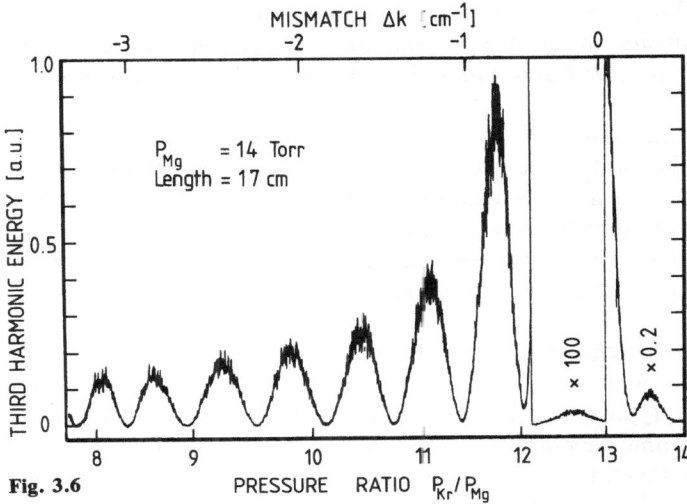

Fig. 3.6

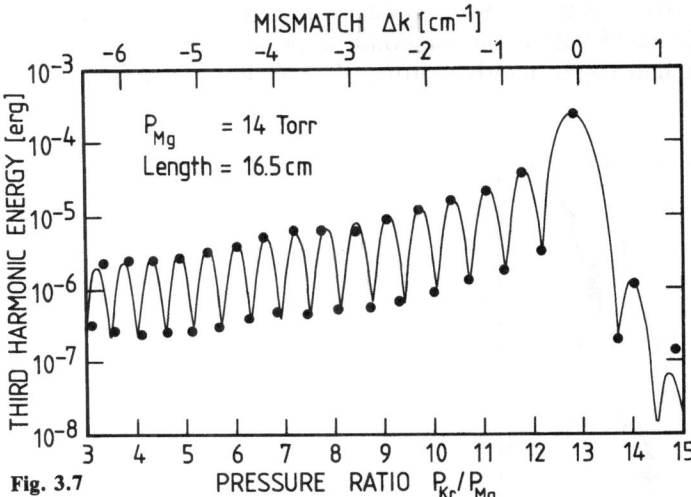

Fig. 3.7

Fig. 3.6. Photoelectric scan of a phase-matching curve for two-photon resonant third-harmonic generation in a Mg-Kr system [3.23]. For a fixed magnesium pressure of 14 Torr, the harmonic intensity is measured as a function of the krypton pressure

Fig. 3.7. Phase-matching curve of a Mg-Kr system. The *dots* give the measured maxima and minima of Fig. 3.6. The *solid line* represents the theoretical curve for $P_{Mg} = 14$ Torr, $L = 16.5$ cm, $\sigma^{(1)} = 4.5 \times 10^{-19}$ cm^2, $\alpha = 0.63$ cm^{-1}, and accounts for the pressure dependence of the homogeneous linewidth Γ_{bg} of the $3s^2$ ^{1}S-3s3d ^{1}D two-photon resonance at 431 nm

As a very useful byproduct, phase-matching curves have also been used for accurate measurements of electric dipole matrix elements according to (3.2, 49), as shown by *Puell* and *Vidal* [3.119] and by *Wynne* and *Beigang* [3.120].

The phase-matching curves define the complex linear susceptibility as well as the profile and the mode structure of the incident beams. With knowledge of the latter quantities, a measurement of the conversion efficiency yields the third-order nonlinear susceptibility $\chi_T^{(3)}$ of the active medium. The susceptibility can differ rather widely from one situation to the other, depending on the particular two-photon resonance employed and the wavelengths of the incident beams.

Nevertheless, measurements of a given resonance, if made carefully, can be entirely consistent with one another and with the models. As an example, Fig. 3.8 shows the typical third-power law as obtained for two-photon resonant third-harmonic generation (at $\Delta k = 0$) in a Sr-Xe system [3.52]. The measurements were taken under conditions which are identical with those at $\Delta k = 0$ of Fig. 3.5. In Fig. 3.8 the solid line is calculated for $\tau_s = 7.5$, whereas the dashed line corresponds to $\tau_s = 0$. The latter calculations were done with a nonlinear susceptibility evaluated according to (3.16). For this purpose the electric-dipole matrix elements were either taken from measurements [3.121, 122] or calculated within the Coulomb approximation [3.123, 124], where some values had to be slightly readjusted according to the preceding

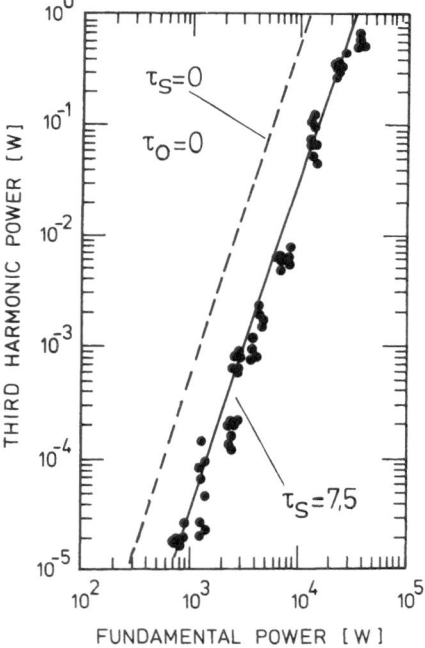

Fig. 3.8. Third-harmonic power versus fundamental power as measured for two-photon resonant third harmonic generation in a Sr-Xe mixture exploiting the $5s^2\,^1S$-$5s5d\,^1D$ two-photon resonance at 575.8 nm. Theoretical curves are shown for $\tau_s = 0$ ($- - -$) and for $\tau_s = 7.5$ (———), indicating the importance of absorption losses

results of the phase-matching curve [3.119]. For the calculation of the nonlinear susceptibilities, frequently an extended set of matrix elements is required. Knowledge of the signs, which may give rise to severe cancellations [3.11,125], must also be included. In those cases where the nonlinear susceptibility is taken directly from the measurements, Fig. 3.8 clearly shows that a knowledge of the phase-matching curve is a prerequisite.

Within the small signal limit, where the only nonlinear polarization to be considered is given by (3.33), measurements of the kind shown in Figs. 3.5 – 8 completely characterize the properties of a nonlinear medium. Higher-order nonlinearities, which are responsible for the onset of saturation, have to be extracted from a field-dependent study to be discussed in the following sections.

3.7 Onset of Saturation

3.7.1 General Considerations

In experiments of the kind shown in Fig. 3.8, the harmonic intensity follows a third-power law over a wide range of input intensities. Eventually, however, the harmonic intensity starts to level off, long before energy conservation leads to a depletion of the fundamental wave. This onset of saturation can be explained by additional nonlinear polarizations which give rise to field-dependent changes of the refractive index, and which are responsible for a destruction of the phase-matching condition. For gaseous media, this was first shown by *Puell* et al. in nonresonant systems [3.18], as well as in two-photon resonant systems [3.126].

In order to illustrate the physics which lead to the onset of saturation and to avoid cumbersome equations, the following discussion will be restricted mainly to the case of resonant and nonresonant third-harmonic generation. An extension to the general case of frequency mixing is then straightforward and reveals no new physics. For the case of third harmonic generation, we give expressions for the nonlinear polarizations at the frequencies ω and $\omega_s = 3\,\omega$, where for the moment, only terms up to third order in the electric field amplitudes are included:

$$P_1^{NL}(\omega) = \frac{N}{4}\,[3\,\chi_T^{(3)}(\omega)E_sE_1^*E_1^* + 3\,\chi_{SA}^{(3)}(\omega)E_1\,|E_1|^2$$

$$+\,6\,\chi_{SB}^{(3)}(\omega,3\,\omega)E_1\,|E_s|^2]\ , \tag{3.67}$$

$$P_s^{NL}(3\,\omega) = \frac{N}{4}\,[\chi_T^{(3)}(3\,\omega)E_1E_1E_1 + 3\,\chi_{SA}^{(3)}(3\,\omega)E_s\,|E_s|^2$$

$$+\,6\,\chi_{SB}^{(3)}(3\,\omega,\omega)E_s\,|E_1|^2]\ . \tag{3.68}$$

In these two equations, the different contributions for which the third-order nonlinear susceptibilities have already been given in (3.16), are distinguished by their physical significance. The first term of (3.68), containing the nonlinear susceptibility $\chi_T^{(3)}(3\,\omega)$, is responsible for the actual third-harmonic generation. It corresponds to the nonlinear polarization which was given in (3.33) for the general case of four-wave mixing, and which has so far been the only term considered in the small signal limit. The first term of (3.67) containing $\chi_T^{(3)}(\omega)$ describes the inverse process by which third-harmonic radiation is transferred back to the fundamental wave. The second term of (3.67,68) containing $\chi_{SA}^{(3)}$ represents an intensity-dependent change of the linear susceptibility at ω and $3\,\omega$, respectively. The real part is responsible for an intensity-dependent change of the refractive index, whereas the imaginary part gives rise to a two-photon absorption. The real part of the last term of (3.67) [or (3.68)] describes the change of the refractive index at the fundamental (harmonic) frequency due to the harmonic (fundamental) intensity. The imaginary part of the last terms in (3.67,68) can be interpreted as a nonlinear polarization which gives rise to Raman-type gain or loss.

Inserting the nonlinear polarizations of (3.67,68) into the fundamental equations of nonlinear optics (3.27) and assuming a rectangular density profile of the nonlinear medium, the electric field amplitudes of the fundamental and the harmonic wave, E_1 and E_s, can be expressed in terms of elliptical integrals [3.39, 127]. Since the resulting expressions are not easy to digest, in the following we will consider the simplest case of physical interest where $E_s \ll E_1$.

3.7.2 Nonresonant Case

In the limit where the electric field amplitudes of the harmonic wave are small enough, all nonlinear polarizations in (3.67,68) containing E_s can be neglected. Consequently, one is left only with the first term of (3.68) containing $\chi_T^{(3)}$, and with the second term of (3.67) containing $\chi_{SA}^{(3)}(\omega)$. With (3.27) the electric field amplitude of the fundamental wave is then defined by the differential equation

$$\frac{d\hat{E}_1}{dz} = i\,\frac{3\,\pi\,\omega}{2\,c\,n_1}\,N\chi_{SA}^{(3)}(\omega)\,\hat{E}_1^3 - \frac{\kappa_1}{2}\,\hat{E}_1 \qquad (3.69)$$

which can be solved immediately to yield

$$\hat{E}_1^2 = \frac{E_{10}^2 \exp(-\kappa_1 z)}{i(3\,\pi\,\omega/c\,n_1)[\chi_{SA}^{(3)}(\omega)/\sigma^{(1)}(\omega)]E_{10}^2[\exp(-\kappa_1 z)-1]+1} \,. \qquad (3.70)$$

This result has to be inserted into the differential equation specifying the electric field amplitude of the harmonic wave. To simplify the further derivation let us assume a small optical depth of the fundamental wave, $\tau_1 \ll 1$, and a small enough intensity of the incident wave such that the condition

$$\frac{3\,\pi\omega}{c\,n_1}\,[\chi^{(3)}_{SA}(\omega)/\sigma^{(1)}(\omega)]\,E^2_{10} \ll 1 \tag{3.71}$$

holds. Then the harmonic intensity can be cast into a form similar to (3.40):

$$\frac{\Phi_s(\omega_s = 3\,\omega)}{n_s} = \left(\frac{4\,\pi^2\omega_s}{c^2 n_s}\,NL\,\chi^{(3)}_T(-\omega_s;\omega\omega\omega)\right)^2 \frac{\Phi^3_{10}}{n^3_1}\,F(\Delta kL,\tau_1,\tau_s) \tag{3.72}$$

where the phase-matching factor $F(\Delta k \cdot L, \tau_1, \tau_s)$ is identical with that given in (3.41), if the wave-vector mismatch Δk of (3.35) is now replaced by the field-dependent mismatch

$$\Delta k(\Phi_{10}) = \Delta k + \omega \left(\frac{6\,\pi}{c\,n_1}\right)^2 N \bar{\chi}^{(3)}_{SA}(\omega)\,\Phi_{10} \; . \tag{3.73}$$

In this manner the phase-matching condition of the small signal limit $\Delta k = 0$ has to be replaced by the intensity-dependent condition $\Delta k(\Phi_{10}) = 0$. The field-dependent correction term of (3.73) contains the real part of the nonlinear susceptibility $\chi^{(3)}_{SA}(\omega)$. It is important that the phase-matching condition can now no longer be satisfied for all electric field amplitudes of the incident wave. The imaginary part of the nonlinear susceptibility $\chi^{(3)}_{SA}(\omega)$ modifies the one-photon absorption by a contribution due to the two-photon absorption. That contribution is given by the cross section

$$\sigma^{(2)}(\omega) = \omega \left(\frac{6\,\pi}{c\,n_1}\right)^2 \tilde{\chi}^{(3)}_{SA}(\omega)\,\Phi_{10} \tag{3.74}$$

which can be neglected off resonance, and which will become important, as shown below, in the two-photon resonant case.

The preceding relations illustrate the importance of additional nonlinear polarizations for the onset of saturation. For comparison with experiments, one generally has to go back to a more accurate numerical integration of the fundamental equations of nonlinear optics using, for example, a Runge-Kutta method. In this manner one can account for all the additional nonlinear polarizations which have been neglected in the preceding simplified treatment. Since all experiments studying the onset of saturation have been carried out with pulsed high-power lasers for which the electric field amplitudes change in space and time, an integration over all field-dependent contributions has to be performed. These experimental difficulties allow for optimum phase matching only for a small fraction of the incident laser pulse. Furthermore, realistic calculations for experiments carried out with a heat pipe oven have to take into account the density distribution along the optical axis according to (3.66).

Figure 3.9 shows an example of a nonresonant third-harmonic experiment which was carried out in a Rb-Xe system using a mode-locked high-power

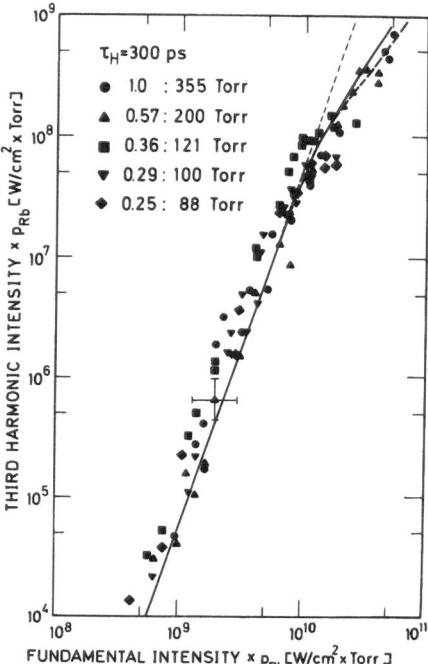

Fig. 3.9. Third-harmonic generation in a Rb-Xe system using a mode-locked Nd : glass laser with a pulse duration of 300 ps. The third-harmonic intensity times Rb vapor pressure is plotted versus the input intensity times Rb vapor pressure. Theoretical curves are given for the small signal limit (*dashed line*), for the third-order calculation (*broken line*) and for the fifth-order calculation (*solid line*)

Nd : glass laser with a pulse duration of 300 ps [3.18]. In order to reconcile measurements taken at different partial pressures of the nonlinear medium, $\Phi_s P_{Rb}$ is plotted versus $\Phi_1 P_{Rb}$. With the pressure normalized intensities, the conversion efficiency in the small signal limit becomes independent of the density N, as can be seen from (3.40). The calculations displayed in Fig. 3.9 are in good agreement with the experiments, and clearly show the onset of saturation at elevated input intensities, as indicated by the deviation from the simple third-power law (dashed line). The broken line in Fig. 3.9 has been calculated from consideration of the nonlinear polarizations up to third order according to (3.67,68). Only a minor modification (solid line) is obtained from extending the calculations up to the fifth order. This shows that, in this case, the onset of saturation originates primarily from the third-order terms in the nonlinear polarization which give rise to intensity-dependent changes of the refractive index. Similar results have recently been obtained by *Ganeev* et al. [3.128] for third-harmonic generation in Xe-Ar mixtures.

Based on the model defined by (3.67,68), *Puell* and *Vidal* [3.127] have investigated the optimum conditions for nonresonant third-harmonic generation. In their theoretical analysis they introduced a dimensionless nonlinear parameter, the inverse of which may be viewed as a figure of merit given by

$$f_M = \left| \frac{12 \sqrt{(n_3/n_1)} \chi_T^{(3)}(3\,\omega)}{(n_3/n_1) \chi_{SA}^{(3)}(\omega) + (n_1/n_3) \chi_{SA}^{(3)}(3\,\omega) - 4\chi_{SB}^{(3)}(\omega, 3\,\omega)} \right| . \qquad (3.75)$$

The figure of merit is essentially the ratio of the third-harmonic coefficient and the different Kerr constants which are responsible for the field-dependent changes of the refractive index. It was shown that energy conversion efficiencies in excess of 10% can be achieved only for $f_M > 1$. Because of possible cancellations in the denominator, the figure of merit depends in a rather intricate way on the energy eigenvalues and the transition moments of the nonlinear medium, and on the frequency of the incident wave. The results of these investigations were then applied to the nonresonant third-harmonic generation of the iodine laser radiation [3.129].

3.7.3 Two-Photon Resonant Case

Saturation on a two-photon transition was first reported by *Wang* and *Davis* [3.130] in thallium, by *Held* et al. [3.131] and by *Ward* and *Smith* [3.132] in cesium, and theoretically predicted by *Chang* and *Stehle* [3.133,134] considering the ac Stark effect and power broadening. In going from the nonresonant to the two-photon resonant case, the theoretical model has to be further refined. The most important extension of the model is required by the two-photon absorption, which has a negligible influence in the nonresonant case, and for which the cross section was given in (3.74). In this case, the population densities associated with the different quantum states of the nonlinear medium have to be determined from a general set of rate equations

$$dN_i/dt = \sum_q R_F(q \to i) - \sum_p R_D(i \to p) \ , \tag{3.76}$$

where $R_F(q \to i)$ is the filling rate to be summed over all initial states $|q\rangle$ and $R_D(i \to p)$ the depletion rate to be summed over all final states $|p\rangle$. The population densities N_i must always obey the boundary condition

$$\sum_i N_i = N_0 \ , \tag{3.77}$$

where N_0 is the total number density of the nonlinear medium. In most practical situations, only radiative rates have to be considered in (3.76) since collisional rates have a minor influence on the population densities within the pulse duration of typical lasers. Using the notation of Fig. 3.1, the most important filling rate is due to the two-photon absorption rate given by

$$R_{TPA}(g \to b) = \left(N_g - \frac{g_g}{g_b} N_b\right) \sigma^{(2)}(\omega_{bg}) \frac{\Phi_1}{2\hbar\omega_{bg}} \ . \tag{3.78}$$

It causes a laser-induced transfer of population density from the ground state $|g\rangle$ to the two-photon resonant state $|b\rangle$, and builds up a population inversion of the state $|b\rangle$ with respect to the lower-lying excited states such as state $|a\rangle$. In general, the one-photon filling and depletion rates are given by

$$R_F(q \to i) = A(q \to i) + \left(N_q - \frac{g_q}{g_i} N_i \right) \sigma^{(1)}(\omega_{qi}) \frac{\Phi_1}{\hbar \omega_{qi}} \,, \qquad (3.79)$$

$$R_D(i \to p) = A(i \to p) + \left(\frac{g_p}{g_i} N_i - N_p \right) \sigma^{(1)}(\omega_{ip}) \frac{\Phi_1}{\hbar \omega_{ip}} \,, \qquad (3.80)$$

where A is the Einstein A-coefficient for spontaneous emission. For situations with no significant population inversion, the total depletion rate of state $|i\rangle$ is immediately given by $R_D(i \to p) = N_i/\tau_i$, where τ_i is the radiative lifetime of state $|i\rangle$. For situations with a population inversion, the stimulated emission will quickly equilibrate the population densities of the two states involved, such that $N_a/N_b = g_a/g_b$ where g_i is the statistical weight of state $|i\rangle$. The latter process generally occurs within times which are short compared with the pulse duration. In this manner the population density N_b of the two-photon resonant state $|b\rangle$ is effectively locked to the population density N_a of state $|a\rangle$. State $|b\rangle$ decays, therefore, with the time constant of state $|a\rangle$, which is determined by the large transition moment of the resonance line $|a\rangle \to |g\rangle$, and possibly modified by reabsorption processes.

For input intensities covering the onset of saturation, it is generally sufficient to consider only the levels $|g\rangle$, $|b\rangle$, and $|a\rangle$, where state $|a\rangle$ may stand for several intermediate states by giving it the appropriate increased statistical weight. As a result of the two-photon absorption, the population densities of all levels will vary in space and time as a function of the applied electric field amplitudes. The population densities then enter into the fundamental equations of nonlinear optics (3.27) which have to be integrated. For this purpose the complex linear and nonlinear susceptibilities are calculated from a superposition of contributions resulting from all those terms which have a significant population density at some point in space and time

$$N_0 \chi^{(n)} = \sum_i N_i \chi_i^{(n)} \,. \qquad (3.81)$$

As a further consequence of the two-photon absorption, the intensity of the fundamental wave is considerably reduced inside the nonlinear medium. For achieving the highest conversion efficiency, it is in general more advantageous to stay in an intensity regime just below the onset of saturation.

With the preceding model, *Puell* et al. [3.126] have carried out calculations which are in excellent agreement with the results of two-photon resonant third-harmonic generation in a Sr-Xe system using flash-lamp pumped dye lasers. Fig. 3.10 shows two phase-matching curves which have been taken for input intensities well below and well above the onset of saturation. For the lower curve, excellent agreement is obtained using a small signal model. For the upper curve, however, the small signal model (dotted line) disagrees considerably, and excellent agreement is obtained only with a model (solid line) taking into account the two-photon absorption. Keeping the pressure

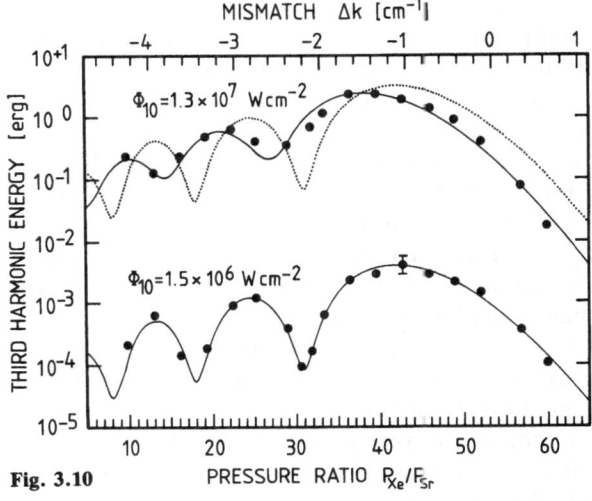

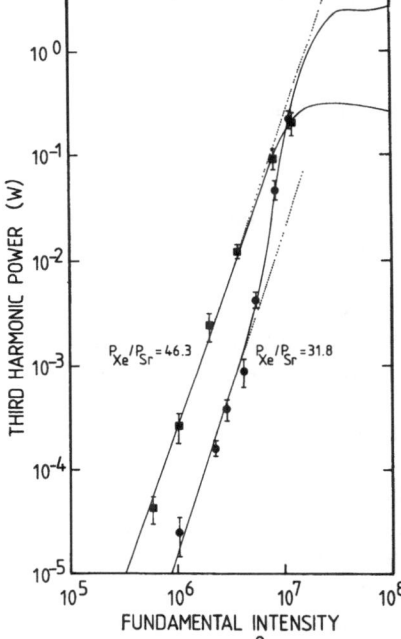

Fig. 3.10. Phase-matching curves of a Sr-Xe system for the $5s^2\ {}^1S\text{-}5s5d\ {}^1D$ two-photon resonance at 575.8 nm, taken with a parallel multimode beam below and above the threshold for the onset of saturation. In the *upper curve*, the lines represent calculations including (———) or neglecting ($\cdots\cdots$) saturation effects

Fig. 3.11. Third-harmonic power versus fundamental intensity for a Sr-Xe system taken at two different pressure ratios P_{Xe}/P_{Sr}. The calculations are carried out including (———) and neglecting ($\cdots\cdots$) saturation effects due to the two-photon absorption

ratio P_{Xe}/P_{Sr} constant, one obtains the intensity dependence shown in Fig. 3.11. The calculations are carried out including (solid lines) and neglecting (dotted lines) the two-photon absorption. It is interesting to note that for the pressure ratio of $P_{Xe}/P_{Sr} = 31.8$, the initial mismatch is actually improved by the field-dependent changes of the refractive index, and the third harmonic intensity increases more rapidly than a cubic law for a limited range of the input intensity. For both curves in Fig. 3.11 the harmonic intensity eventually levels off, as expected for the onset of saturation.

To summarize the results for the onset of saturation, we note that the limitations in the conversion efficiency are always due to intensity-dependent changes of the refractive index. This is true for the nonresonant as well as the two-photon resonant case. For the nonresonant case, it is the real part of $\chi_{SA}^{(3)}(\omega)$ which causes the intensity-dependent change of the refractive index. For the two-photon resonant situation, it is the imaginary part of $\chi_{SA}^{(3)}(\omega)$ which gives rise to a two-photon absorption, and which modifies the refractive index of the nonlinear medium according to (3.81) because of the modified population densities inside the nonlinear medium.

The preceding results clearly demonstrate that for a nonlinear medium of large conversion efficiency, the third requirement of Sect. 3.5.1 ($\tau_j < 1$) has to be extended also to the two-photon absorption. In addition to requiring a small optical depth for the incident and generated waves, it is important to have a sufficiently small two-photon absorption cross section. The latter requirement was implied by *Junginger* et al. [3.23] in their definition of a so-called low-loss medium.

3.7.4 The ac Stark Effect and Multiphoton Ionization

So far we have discussed those nonlinear polarizations which are responsible for destroying the phase-matching condition, and hence for causing the onset of saturation. Nevertheless, before discussing the high-intensity regime (to be described in the following section), other nonlinearities may have to be considered. One of these processes is the optical Stark effect [3.135], which cannot be taken into account by a perturbation approach of the type giving rise to the linear and nonlinear susceptibilities, but which requires a treatment to all orders. The important aspect for present consideration is that the ac Stark effect causes level shifts of the two-photon resonance [3.136], and therefore affects the resonant enhancement of the frequency mixing. These effects can sometimes be favorable because the ac Stark effect tends to smear out the two-photon resonance and thus delays its saturation.

A second nonlinear process not contained in (3.67, 68), is multiphoton ionization. This process can influence the physics in a number of ways. First, it will reduce the intensity of the incident wave due to absorption. It can also modify the population densities of the nonlinear medium. In this manner it changes the linear and nonlinear susceptibilities of the medium according to (3.81), causing secondary effects associated with the modified susceptibilities [3.137]. As a further consequence, multiphoton ionization will generate photoelectrons and ions which can induce Stark broadening. It has even been suspected that the local electric fields due to the charged particles are also responsible for the second-harmonic generation which should not occur for spherically symmetric systems (Sect. 3.2), but which has actually been observed [3.138 – 141]. *Bethune* et al. [3.142] have tried to explain the second-harmonic generation in another way, by resonant quadrupole transitions. The preceding explanations have been disputed by *Freeman* et al. [3.143], who favor an explanation in which the spherical symmetry is removed by a spatial density gradient of ground-state atoms due to multiphoton ionization. However, *Scheingraber* and *Vidal* [3.26] did not see the faintest trace of a second-harmonic signal in strontium, although at large input intensities the density gradients were so large that they even gave rise to self-defocusing (Sect. 3.8.2).

Recent experiments in strontium, magnesium, and cadmium, performed by *Scheingraber* and *Vidal* [3.144], have shown a very strong signal very close to the position of the second harmonic signal. A detailed analysis showed that this signal is due to a redistribution of the population densities to be discussed below in Sect. 3.8.3. It originates from a very efficient two-photon absorption process, in which the excited p-levels of the resonance lines are populated via stimulated emission from the two-photon resonant level. The latter process is particularly efficient for transitions in the infrared. As a result, strong stimulated emission is observed on the corresponding resonance lines. One of these lines, originating from an excited level near the two-photon resonant level, can be erroneously mistaken as a second harmonic if the fine structure splitting is rather small. A similar situation has recently been investigated by *Malcuit* et al. [3.145], who observed a competition between amplified spontaneous emission from the 3d-level in sodium and resonantly enhanced four-wave mixing. The two competing processes were discriminated by two-photon resonant and nonresonant experiments measuring the outcoming radiation in forward and backward directions and also by using counter-propagating waves.

Similar to frequency mixing, multiphoton ionization will be enhanced by a two-photon resonance, which again is subject to the ac Stark effect [3.146 – 149]. Using the example of Fig. 3.1 where three incident photons are sufficient to ionize the nonlinear medium, the overall multiphoton ionization can actually occur through different pathways:

1) two-photon resonant three-photon absorption,
2) two-photon absorption to the intermediate state $|b\rangle$ and subsequent ionization from state $|b\rangle$ by means of an extra photon, and
3) one-photon absorption due to the sum frequency signal.

Processes (1) and (3) depend on the instantaneous electric field amplitude applied, whereas process (2) depends on the population density accumulated in state $|b\rangle$. Furthermore, process (3) is sensitive to the phase matching of the sum-frequency signal, whereas processes (1) and (2) are not. Because of the latter criterion, process (3) was shown to dominate in a Sr-Xe system which was optimized with respect to the conversion efficiency by using a large column density [3.126].

In trying to explain experiments in cesium by *Ward* and *Smith* [3.132], who exploited an accidental coincidence of the ruby laser with a two-photon resonance in cesium, *Georges* et al. [3.150,151] investigated the observed saturation in the conversion coefficiency. Contrary to *Puell* et al. [3.126], they found that the dominant process limiting the conversion efficiency is due to two photon resonant, three-photon ionization, and to the accompanying destruction of the phase matching. The difference in the results appears to be due to the fact that Georges et al. considered a low-density medium and hence rather large electric field amplitudes. This situation is different from the high-density system discussed in [3.126], where the nonlinear medium was optimized with respect to the conversion efficiency, and where the saturation occurs earlier for smaller electric field amplitudes. Georges et al. also did not fully account for the stimulated emission from the two-photon resonant state $|b\rangle$, a process which heavily competes with the multiphoton ionization because it reduces the probability for the process (2) mentioned above.

In this context it is interesting to mention a series of experiments carried out in xenon [3.152–156], where it was observed that the resonantly enhanced multiphoton ionization gradually decreased with increasing pressure, while the third-harmonic generation showed the opposite behavior. Similar results were also obtained by *Normand* et al. [3.157] in mercury. At first sight rather puzzling, this result was explained by *Jackson* and coworkers [3.158–160] as an interference between two different, but coherent pathways to the resonant intermediate level. In this case a three-photon excitation driven by the electric field of the fundamental wave competes with a one-photon excitation driven by the electric field of the harmonic wave. Similar explanations using Bloch type equations were given by *Payne* and *Garrett* [3.161–163], and by *Poirier* [3.164].

3.8 High Intensity Saturation

In the high-intensity saturation regime, one is still primarily dealing with the same nonlinear polarizations which have already been discussed for the onset of saturation. But they are now strongly enhanced and cause additional effects. Furthermore, a variety of higher-order nonlinearities become important which may be viewed in part as field-dependent corrections of the third-order nonlinearities, whereas other nonlinearities give rise to a new class of higher-order parametric processes. All of these higher-order nonlinearities, which will be the subject of this section, show up most clearly in two-photon resonant situations where they are resonantly enhanced.

3.8.1 Conversion Profiles

For unraveling the different nonlinear processes in the high-intensity saturation regime, the key information is contained in the frequency dependence of

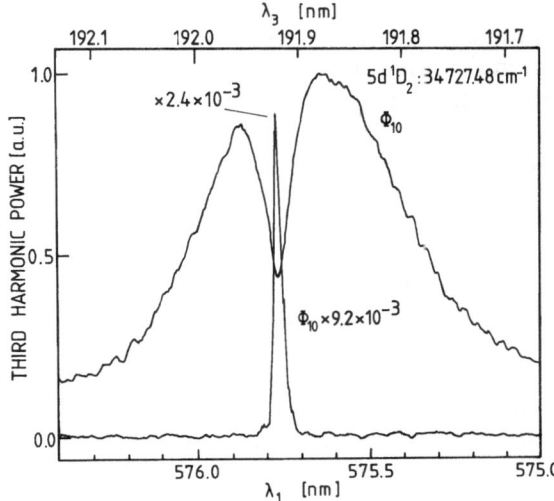

Fig. 3.12. Third-harmonic intensity as a function of wavelength around the $5s^2\,{}^1S$-$5s5d\,{}^1D$ two-photon resonance in strontium at 575.8 nm. The intensity profile showing a pronounced central dip was taken for a pulse intensity of 5×10^9 W/cm^2 in the beam waist. The narrow profile was obtained from lowering the input intensity by a factor of 9.2×10^{-3} and raising the sensitivity by a factor of 420. $P_{Sr} = 3.5$ Torr

the conversion efficiency around a particular two-photon resonance. This frequency dependence, called the conversion profile, shows very characteristic features as a function of the input intensity. Figure 3.12 shows a plot of the third-harmonic power as a function of the wavelength around the $5s^2$ ${}^1S$-$5s5d\,{}^1D$ two-photon resonance in strontium for two different input intensities, as measured by means of an excimer laser pumped dye laser [3.26]. Similar conversion profiles have been obtained for a number of other two-photon resonances. The example of Fig. 3.12 is particularly instructive because it is the same two-photon resonance described by the measurements of Figs. 3.5, 8, 10, 11. Hence all the linear and third-order nonlinear susceptibilities are well defined, so that modifications due to higher-order non-linearities can be clearly distinguished. Figure 3.12 shows, for the highest input intensity, a very pronounced power broadening and a sharp dip, which gives rise to a minimum of the conversion efficiency right on the two-photon resonance. These features were first predicted in [3.165], and were also observed on different transitions in mercury [3.24, 25, 28] and cadmium [3.166]. Further theoretical studies on this subject were carried out by *Stappaerts* [3.167, 168] and by *Kildal* and *Deutsch* [3.111].

The central dip of the high-intensity profile in Fig. 3.12 was explained as a consequence of the two-photon absorption, which pumps a significant fraction of the ground-state population density into the excited state of the two-photon resonance [3.26, 165]. For the ground state and the excited state of the two-photon resonance, the resonant contributions of the imaginary part of $\chi_T^{(3)}$ are identical, but have a different sign, see (3.16). Hence the real part of $\chi_T^{(3)}$ vanishes right on the two-photon resonance for its complete saturation. The effective total nonlinear susceptibility is therefore reduced with an increasing pump rate of the two-photon transition because of a partial cancellation of the resonant contributions according to (3.81). For the effective total nonlinear susceptibility, one has to consider also the laser-induced changes of the population densities of the other states according to the rate equations (3.76), as well as the laser-induced shift of the two-photon resonance due to the ac Stark effect. Both effects prevent a complete bleaching of the two-photon transition.

3.8.2 Self-(De)Focusing

A more careful comparison of the measurements in Fig. 3.12 with the theoretical model revealed discrepancies which could be attributed to self-defocusing [3.26]. This effect can clearly be detected from the spot size of the fundamental beam in the far field. The self-focusing or defocusing is due to field-dependent changes of the refractive index [3.169 − 171]. According to the radial intensity profile of the incoming beam, a radial gradient of the refractive index is established. The gradient acts as a focusing or defocusing lens, depending on whether the refractive index is raised or lowered on the optical axis. It is important to realize that this effect originates from the same third-order nonlinear susceptibility $\chi_{SA}^{(3)}(\omega)$ as that responsible for the destruction of the phase matching.

The self-defocusing can clearly be seen in Fig. 3.13 where the transmission through a pinhole is shown in the far field of the incident beam. The four curves in Fig. 3.13 show the transmission for the $5s^2\,{}^1S$-$5s5d\,{}^1D$ two-photon resonance in strontium, for different pinhole positions (upper and lower curves), and for different input intensities (left-hand and right-hand curves). In the upper left-hand transmission profile of Fig. 3.13, the pinhole was aligned on the optical axis. Approaching the two-photon resonance, a very pronounced defocusing occurs, which disappears right on the two-photon resonance. The intermediate maximum of the transmission profile has an origin similar to that of the central dip in Fig. 3.12, and is due to a bleaching of the two-photon transition, which removes the radial gradient of the refractive index in the vicinity of the optical axis. In the upper right-hand curve of Fig. 3.13, the input intensity was lowered by one order of magnitude. In this case the intermediate maximum disappears, indicating that the input intensity is insufficient to bleach the two-photon transition. The lower curves in Fig. 3.13 are obtained with the pinhole removed from the optical axis to the half maximum point of the incident beam. Now an expansion of the beam due to self-defocusing increases the transmission, and the transmission profiles of the lower row are effectively inverted with respect to the profiles of the upper row.

A two-photon resonant enhancement of self-defocusing was first seen in cesium by *Lehmberg* et al. [3.172,173]. *Bakhramov* et al. [3.174] reported two-photon resonant self-focusing in potassium. *Poluektov* and *Nazarkin* [3.175 − 177] have investigated a theoretical model in which they suggest the possibility of raising the conversion efficiency by exploiting an enhancement

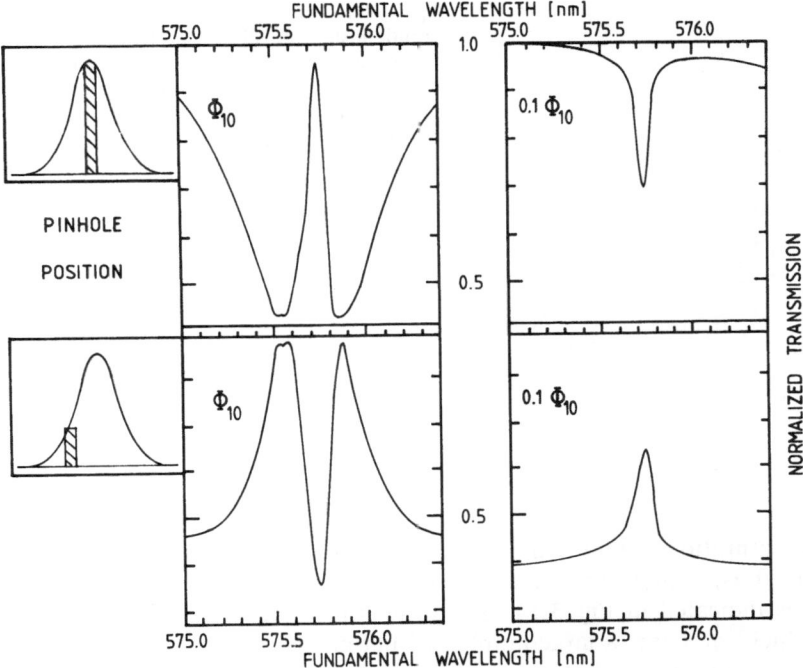

Fig. 3.13. Transmission profiles around the $5s^2\,^1S\text{-}5s5d\,^1D$ two-photon resonance for $P_{Sr} = 3.5$ Torr and $P_{Xe} = 90$ Torr as a function of pinhole position (*upper* and *lower curves*) and of input intensity (*left-* and *right-hand curves*)

of the input intensity by means of self-focusing. The model does not directly apply to the measurements of Fig. 3.13, because it has been derived for a coherent excitation with pulses which are short with respect to the phase-relaxation time. Furthermore, the results are not very conclusive because of a number of simplifying assumptions.

It should be pointed out that a theoretical analysis of the high-intensity saturation including self-defocusing is exceedingly difficult, because for a valid treatment of the four-wave mixing one then has to explicitly consider the radial derivatives of the Laplacian operator in the basic equations of nonlinear optics. We recall that (3.27) is only valid within the plane-wave approximation. For a comparison with theory it is therefore advisable to lower the column density of the nonlinear medium to a value where self-defocusing can safely be neglected, and where the theoretical model can be evaluated with a tolerable effort.

3.8.3 Redistribution of Population Densities

As indicated already in Sect. 3.7.3, two-photon absorption leads to a significant population density of the two-photon resonant state, causing a population inversion with respect to the lower-lying excited states. In the high-

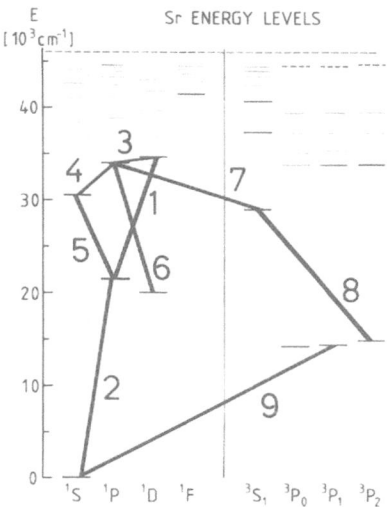

Fig. 3.14. Partial energy level diagram of strontium showing all the stimulated transitions which have been observed by pumping the $5s^2\,^1S\text{-}5s5d\,^1D$ two-photon resonance, and which together with the incident laser radiation give rise to additional parametric processes

intensity saturation regime the two-photon transition can essentially be bleached. Thus, strong stimulated emission was observed for strontium on all transitions indicated in Fig. 3.14. Since the stimulated emission is collinear with the incident wave, a large number of parametric processes can occur in which one or two laser photons combine with photons from the stimulated emission [3.26]. These parametric processes can turn out to be so efficient that their intensities are comparable to the intensity of the harmonic wave. Among the different stimulated emissions, the resonance line has also been observed. This was first reported by *Brechignac* and *Cahuzac* [3.178] who tried to explain this finding with a three-level model. This explanation was disputed in [3.26] which points out the importance of stimulated emission into metastable states by which a significant fraction of the total population density is trapped for the duration of the laser pulse. In this manner a population inversion can be established even on the resonance line. Besides the very intense parametric processes, a large number of weaker spontaneous emissions were observed which originate from cascading processes out of the ionization continuum. Similar parametric processes have also been observed in four-wave mixing experiments carried out in barium [3.179,180], mercury [3.181,182], and cadmium [3.166].

The preceding results show very clearly that for designing a monochromatic coherent vuv source, it is highly desirable to keep the electric field amplitudes of the incident waves below the high-intensity saturation to avoid the appearance of additional, perturbing waves of similar intensity.

3.8.4 Higher-Order Processes

For the electric field amplitudes of the high-intensity saturation regime, higher-order nonlinear polarizations may become important. Most of these higher-order terms represent field-dependent changes of lower-order polarizations whose physical significance has already been discussed. At the same time, a number of higher-order nonlinear polarizations can occur which no longer describe four-wave mixing processes, but which give rise to higher harmonics of the fundamental wave. This possibility was pointed out by *Harris* [3.183], and has since then been demonstrated in alkali inert-gas mixtures by the Sofia group [3.184 – 190] and in inert gases by *Reintjes* and coworkers [3.191 – 197]. The highest harmonic so far reported was the ninth harmonic [3.189]. In most experiments, picosecond pulses from a mode-locked Nd laser have been used for providing the large electric field amplitudes, except for an experiment using a mode-locked Nd laser pumping an actively mode-locked dye laser [3.190] and an experiment using nanosecond pulses from a Xe-Cl laser [3.196]. It has been found that, besides the direct generation of the nth harmonic $n\omega$ from n photons of frequency ω, cascade processes of the kind

$$n\omega = (n-2)\,\omega + \omega + \omega \tag{3.82}$$

may be very important. In the process of (3.82), a lower-order nonlinear polarization gives rise to a harmonic at the frequency $(n-2)\,\omega$, which together with two additional photons of the fundamental wave, form the final nth harmonic. The direct and the cascade processes can be comparable in intensity and can be distinguished by investigation of the phase matching. Similar to the third-order nonlinear susceptibilities, the higher-order nonlinear susceptibilities can also be resonantly enhanced. Since in practice, not all resonant denominators in (3.14) can be optimized simultaneously, *New* [3.198] has suggested performing a cascade process in a two-component system, where the first component has been selected to optimize the $(n-2)\,\omega$ wave by a proper choice of intermediate states, whereas the second component is chosen for optimizing the final cascade process according to (3.82). *Reintjes* and *She* [3.195] have shown that with suitable relative values of the nonlinear susceptibilities, and supported by favorable phase-matching conditions, the fifth harmonic can be made to exceed the third harmonic. The conversion efficiencies are so far rather low, and for practical applications the nth harmonic may have to be separated from the other perturbing harmonics.

For a quantitative description, the fundamental equations of nonlinear optics (3.27) have to be considered for all odd-order harmonics which are coupled through the direct and the cascade processes [3.199, 200]. Similar to the third harmonic analysis of *Bjorklund* [3.63], *Tomov* and *Richardson* [3.199] have investigated the phase matching for the fifth harmonic using focused beams, and considering the interference between the direct process and the cascade type process. Saturation phenomena have been observed by *Reintjes* and coworkers [3.192, 194] at high-input intensities, which are not yet fully understood. Using the Maxwell-Bloch equations, *Diels* and *Georges* [3.201] have carried out a very detailed analysis of two-photon resonant third- and fifth-harmonic generation in metal vapors with picosecond pulses. They took into account optical Stark shifts, multiphoton ionization, two-photon absorption, and the nonlinear index of refraction. These rather sophisticated model calculations have so far not been tested by experiments in any detail.

For calculation of the nth order nonlinear susceptibilities a rather large set of energy levels and transition moments may be required. According to (3.14), this generally leads to rather time-consuming calculations. In situations of this kind, *Puell* and *Vidal* [3.41] have suggested a very much faster method using a recursion relation in which, similar to the perturbation approach of the density matrix, the nth order contributions are calculated from terms of order $(n-1)$.

A different kind of nonlinear polarization is encountered if the incident wave generates a new wave inside the nonlinear medium by means of stimulated electronic Raman scattering [3.202], which in turn combines with the incident wave in a four-wave mixing process. This process is rather efficient, because the nonlinear susceptibility can be quite large due to an exact two-photon resonance. In this manner *Sorokin* and coworkers [3.203,204,1] generated tunable coherent infrared radiation in potassium, a process which was recently investigated again by *Rustagi* et al. [3.205,206]. Similar results have been reported in sodium [3.207], barium [3.179,208], and thallium [3.209].

3.9 Transient Behaviour

So far time-dependent effects in the four-wave mixing process have only been taken into account for the population densities in the rate equations (3.76), considering the two-photon resonant situation. In this case the fundamental equations of nonlinear optics (3.27) had to be integrated, taking into account the population densities which have been accumulated in a particular state of the nonlinear medium at some point in space and time. This time dependence is the most important one if the pulse duration is comparable to or longer than the radiative lifetimes of the relevant states, which are typically a few nanoseconds. In those cases where a population inversion occurs, the population densities are equilibrated by stimulated emission on a picosecond time scale, and one can frequently assume an instantaneous relaxation.

The situation is significantly different if one is dealing with ps pulses of mode-locked lasers. As in the previous discussions one then has to distinguish again between the nonresonant and the two-photon resonant situations. For the nonresonant case a perturbative approach is generally valid [3.41]. In experiments where the pulse duration was either short (7 ps) or long (300 ps) with respect to the phase relaxation time of a Rb-Xe system, *Puell* et al. [3.18] have shown the influence of a transient excitation due to adiabatic following [3.210−212], a process which is important for the highest-input intensities, where the onset of saturation occurs (Fig. 3.9).

For the two-photon resonant case on the other hand, a perturbative treatment of the nonlinear medium is no longer adequate. Using a Bloch vector model [3.213], *Elgin* and coworkers [3.214 – 220] have investigated, in a series of papers, the behavior of the nonlinear medium under the influence of ultrashort pulses. Most important for practical applications is an effect first seen by *Matsuoka* et al. [3.221] for two-photon resonant-frequency mixing in calcium. They observed that the conversion efficiency goes through a maximum if the second nonresonant pulse is delayed by several picoseconds with respect to the two-photon resonant pulse. This has also been seen in strontium [3.222 – 224]. According to *Elgin* and coworkers [3.216, 217, 219], it is helpful to consider this nonsynchronous up-conversion process in two stages. The initial pulse puts the nonlinear medium into a coherent state, which then has to be exposed to the second pulse before the coherent state has significantly decayed as a result, for example, of Doppler dephasing.

These results have recently been supported by *Benda* et al. [3.225], who performed a perturbative density matrix calculation of the nonlinear polarization in a three-level system, allowing for a time delay between the two exciting pulses. For an excitation satisfying the adiabatic following criteria, the sum frequency signal is shown to depend on the temporal overlap of the two exciting pulses, and is maximized for temporally coincident pulses. For a nonadiabatic excitation where the spectral width of the excitation has a significant overlap with one of the transition frequencies of the three-level medium, the sum frequency signal is maximized for a time delay between the two exciting pulses, as observed in the experiment.

3.10 Summary

The basic principles of four-wave mixing have been reviewed. For this purpose three regions have been distinguished which differ with respect to the electric field amplitudes and hence with respect to the nonlinear polarizations to be taken into account. In the small signal limit, the intensity of the generated wave is proportional to the product of the intensities of the incident waves, and for the case of the harmonic generation a cubic dependence for the harmonic intensity is obtained. In the region where the onset of saturation occurs, intensity-dependent changes of the refractive index start to destroy the phase-matching condition and eventually limit the conversion efficiency. Finally, in the high-intensity saturation regime, self-(de)focusing and a redistribution of the population densities have to be considered. The latter effect gives rise to strong stimulated emission and may cause numerous parametric processes.

For the design of a monochromatic light source of high efficiency, it is undesirable to work in the high-intensity saturation regime, with its additional perturbing parametric processes and its intensity-dependent beam profile. The best results are obtained in the intensity region where the onset of saturation occurs. For large input intensities, one should therefore stay away from any two-photon resonance, and two-photon resonant frequency mixing is only recommended for moderate input intensities. The nonlinear medium has to be selected according to its nonlinear susceptibilities. In order to obtain a large conversion efficiency, the nonlinear medium has to be operated with the largest possible column density, without exceeding an optical depth of the order of unity for any one of the interacting waves. This generally requires extremely careful phase matching, which can only be achieved in a highly homogeneous medium. A focusing of the incident beam is only recommended as long as one stays in the regime where the onset of saturation occurs. In this case the confocal parameter should not be smaller than the length of the nonlinear medium in order to minimize competing nonlinear processes. In Table 3.1, finally, all nonlinear systems have been summarized which so far have been investigated. The nonlinear media have been arranged according to the elements, and they are given together with the wavelength region of the sum frequency wave and the method applied.

Coherent vuv sources, based on the method of four-wave mixing in gases, have now reached a degree of perfection where they are able to extend high-resolution spectroscopy into the vuv. Work in this area is presently going on in several laboratories [3.5]. For laser spectroscopy, a linewidth of about 0.05 cm^{-1} has been reported by *Wallenstein* [3.226], where he used a dye laser oscillator amplifier system pumped by a Nd:YAG laser [3.227, 228]. Similar linewidths have been achieved by *Hutchinson* and *Thomas* [3.229] and by *Scheingraber* and *Vidal* [3.27], who used an excimer laser pumped, oscillator amplifier, dye laser system. Systems of this kind are in most cases pressure tuned in order to synchronize the various dispersive elements of the laser over the entire tuning range. Besides high resolution, accurate measurement of the wavelength is generally needed in spectroscopy. Four-wave mixing in gases offers the very attractive possibility of calibrating the coherent vuv radiation against the more accurate length and frequency standards in the visible part of the spectrum. *Scheingraber* and *Vidal* [3.27] have used the iodine spectrum as measured by *Gerstenkorn* and *Luc* [3.230] to calibrate two excimer laser pumped dye lasers which generate coherent vuv radiation with the method of two-photon resonant sum frequency mixing. The accuracy so far achieved has been, typically, 3 parts in 10^7.

For experiments in laser spectroscopy, systems which use four-wave mixing in gases generally supply sufficient intensity. The spectral brightness of four-wave mixing systems exceeds that of the synchrotron and of other known light sources by several orders of magnitude [3.231]. This is mainly a consequence of the small linewidth and the small beam divergence. For other applications where high-intensity systems are needed as, for example, in plasma diagnostics or laser chemistry, an interesting new technology has emerged which has been promoted in particular by *Rhodes* and coworkers [3.29, 31, 32]. As mentioned in the introduction, they use sum frequency mixing systems as oscillators which are amplified by a chain of excimer lasers. Systems of this kind have allowed the beam quality of the sum frequency system to be combined with the power of high-power excimer lasers. So far, however, the method is still restricted to selected wavelengths according to existing excimer lasers. In this context it is useful to refer to a recent review on extreme uv and x-ray lasers by *Sobelman* and *Vinogradov* [3.232]. Also, nonthermal recombining plasmas [3.233], or even laser-generated plasmas [3.234, 235], may eventually serve as amplifying media, thereby extending the wavelength region of high-power systems.

Table 3.1. Summary of four-wave mixing experiments in different gaseous nonlinear media which are arranged according to the elements. For every element investigated the wavelength region is given together with the method of generation. The numbers indicate the incident waves, where the numbers in italics mark those waves which are in resonance with a particular atomic transition. *1 + 1 + 2*, for example, indicates two-photon resonant sum frequency mixing. Four-wave mixing experiments in which one wave originates from stimulated electronic Raman scattering are not included

Nonlinear medium	Wavelength range [nm]	Method	References
Na	354.7	3×1	[3.17, 76, 236]
Na	330.5 − 332.1	*1 + 1 + 2*	[3.13]
Na	268 (cw)	*1 + 2 + 3*	[3.237]
Na	231	*1 + 2 + 3*	[3.60]
Na	212	5×1	[3.186, 187, 189]
Na	200	*1 + 1 + 2*	[3.117, 238]
Na	191	*1 + 1 + 1*	[3.239]
Na	151.4	7×1	[3.185, 188, 189]
Na	106	5×1	[3.188]
Na	117.7	9×1	[3.184, 189]
Rb	354.7	3×1	[3.10, 17, 18, 236]
Cs	213.4	*1 + 1 + 1*	[3.15, 132]
Be	121 − 123	*1 + 1 + 2*	[3.240]
Mg	173.5	$(2 \times 1) + 1 + 1$	[3.241]
Mg	140 − 160	*1 + 1 + 2*	[3.22, 23]
Mg	143.6 (cw)	*1 + 1 + 1*	[3.35]
Mg	115, 127	*1 + 1 + 2*	[3.242]
Mg	121.2	*1 + 1 + 2*	[3.243]
Mg	121 − 129	*1 + 1 + 2*	[3.244]
Mg II	123.6	1 + 1 + 2	[3.99]
Ca	200	*1 + 1 + 1*	[3.116]
Ca	275.8	*1 + 1 + 2*	[3.221]
Ca II	127.8	*1 + 1 + 2*	[3.98]
Sr	177.8 − 195.7	*1 + 1 + 2*	[3.14, 98, 245]
Sr	190 − 200	*1 + 1 + 2*	[3.222, 223]
Sr	165 − 166	*1 + 2 + 3*	[3.34]
Sr	192	*1 + 1 + 1*	[3.29, 126]
Sr	170	*1 + 1 + 2*	[3.224]
Sr	171.2, 169.7 (cw)	*1 + 2 + 3*	[3.33, 61]
Ba	190 − 200	*1 + 1 + 2*	[3.180]
Zn	106 − 140	*1 + 1 + 2*	[3.246]

Table 3.1 (continued)

Nonlinear medium	Wavelength range [nm]	Method	References
Cd	151.4, 117.7	$1 + 1 + 2$	[3.115, 188]
Cd	176.6	3×1	[3.115, 188]
Cd	128.7 – 135.3	$1 + 1 + 2$	[3.247]
Hg	204.0 – 249.3	$1 + 1 - 2$	[3.28]
Hg	177.3 – 210.4	$1 + 1 - 2$	[3.28]
Hg	160.9 – 187.8	$1 + 1 - 2$	[3.28]
Hg	184.9	$1 + 1 - 2$	[3.181, 182]
Hg	183.3	$1 + 1 - 2$	[3.181]
Hg	143.5, 140.1, 130.7	$1 + 1 - 2$	[3.182]
Hg	125.9, 125.0	$1 + 1 - 2$	[3.182]
Hg	124.7 – 125.5	$1 + 1 + 2$	[3.25]
Hg	125.1	$1 + 1 + 2$	[3.181]
Hg	122.8 – 123.5	$1 + 1 + 2$	[3.25]
Hg	117.4 – 122	$1 + 1 + 2$	[3.25, 248]
Hg	120.3	$1 + 1 + 2$	[3.249]
Hg	99.1 – 126.8	$1 + 1 + 2$	[3.28]
Hg	89.6	$1 + 1 + 1$	[3.250]
Hg	87.5 – 105	$1 + 1 + 2$	[3.87]
Tl	195.1	$1 + 1 + 1$	[3.130]
Eu	185.5	$1 + 1 + 2$	[3.98, 245]
Yb	194	$1 + 1 + 2$	[3.98, 245]
He, Ne, Ar, Kr, Xe	231.4	3×1	[3.6, 7]
He	53.2	5×1	[3.191 – 195]
He	38	7×1	[3.192, 194]
He	82.8	3×1	[3.91]
He, Ne	106.4	$1 + 1 + 2$	[3.194]
He, Ne	88.7	3×1	[3.194, 195]
He, Ne	76, 70.9	$4 \times 1 - 2$	[3.193, 194]
He, Ne	62.6, 59.1	$4 \times 1 + 2$	[3.193, 194]
He, Xe	49.7	5×1	[3.91]
He, Xe	35.5	7×1	[3.91]
Ne	53.2	5×1	[3.191, 193, 194]
Ne	72.05 – 73.58	3×1	[3.83]
Ne	74.3 – 74.36	3×1	[3.83]
Ne	118.2	$1 + 1 + 2$	[3.194]
Ar	120.4	$1 + 1 + 3 \times 1$	[3.190]
Ar	85.7 – 87.0	3×1	[3.82]
Ar	97.4 – 104.8	3×1	[3.82, 89]
Ar	106.7	$1 + 1 + 1$	[3.155]
Ar	102.6 – 102.8	3×1	[3.251, 252]
Ar, Kr	64	3×1	[3.79]
Ar, Kr	61.6	5×1	[3.196]
Ar	57	$1 + 1 + 1$	[3.253]
Ar, Kr	53.2	5×1	[3.194]
Kr	92.3, 94.2	$1 + 1 + 2$	[3.85]
Kr	121.6	3×1	[3.19, 21, 226, 254 – 257]
Kr	123.6	$1 + 1 + 1$	[3.155]
Kr	120.3 – 123.5	3×1	[3.19, 257]
Kr	120 – 123.5	$1 + 1 + 2$	[3.20]
Kr	110 – 116.4	$1 + 1 + 2$	[3.20, 258]

Table 3.1 (continued)

Nonlinear medium	Wavelength range [nm]	Method	References
Kr	131.2	$1+1-2$	[3.85]
Xe	155 – 220	$1+1-2$	[3.20, 259, 260]
Xe	155	$1+1-2$	[3.229]
Xe	147	$1+1+1$	[3.155]
Xe	140.3 – 146.9	3×1	[3.19]
Xe	140 – 146	$1+1+2$	[3.261]
Xe	125.4, 125.9, 126.1	$1+1-2$	[3.262]
Xe	101.5, 101.8, 102.0	$1+1+2$	[3.262]
Xe	74.8, 75, 75.2	$1+1+1$	[3.262]
Xe	113.4 – 117.0	$1+1+2$	[3.20]
Xe	117.2 – 119.2	$1+1+2$	[3.20]
Xe	126.6 – 129.4	$1+1+2$	[3.20]
Xe	163.1 – 194.6	$1+1-2$	[3.263]
Xe	118 – 147	$1+1+2$	[3.263]
Xe	118.2	3×1	[3.88, 93, 128]
Xe	106	3×1	[3.264]
Xe	82.8 – 83.3	3×1	[3.78, 91]

References

3.1 J. J. Wynne, P. P. Sorokin: Optical mixing in atomic vapors, in *Nonlinear Infrared Generation,* ed. by Y.-R. Shen, Topics Appl. Phys., Vol. 16 (Springer, Berlin, Heidelberg 1977) pp. 159 – 214
3.2 C. R. Vidal: Appl. Opt. **19**, 3897 (1980)
3.3 W. Jamroz, B. P. Stoicheff: "Generation of tunable coherent vacuum-ultraviolet radiation", in *Progress in Optics,* **20**, 326 – 380 (North Holland, Amsterdam 1983)
3.4 C. R. Vidal, P. Klopotek, H. Scheingraber: State-selective vacuum uv spectroscopy of small molecules, in *Laser Techniques in the Extreme Ultraviolet,* ed. by S. E. Harris and T. B. Lucatorto, AIP Conf. Proc. **119**, 233 – 245 (1984)
3.5 C.R. Vidal: Advances in Vacuum Ultraviolet Spectroscopy, Laser 84, San Francisco (STS Press, McLean, Va. 1985) p. 1; Adv. Atom. Molec. Phys. **23**, 1 (1988). See also further references therein.
3.6 G. H. C. New, J. F. Ward: Phys. Rev. Lett. **19**, 556 (1967)
3.7 J. F. Ward, G. H. C. New: Phys. Rev. Lett. **185**, 57 (1969)
3.8 W. G. Rado: Phys. Rev. Lett. **11**, 123 (1967)
3.9 S. E. Harris, R. B. Miles: Appl. Phys. Lett. **19**, 385 (1971)
3.10 J. F. Young, G. C. Bjorklund, A. H. Kung, R. B. Miles, S. E. Harris: Phys. Rev. Lett. **27**, 1551 (1971)
3.11 R. B. Miles, S. E. Harris: IEEE J. QE-9, 470 (1973)
3.12 P. P. Bey, J. F. Giuliani, H. Rabin: Phys. Rev. Lett. **19**, 819 (1967)
3.13 D. M. Bloom, J. T. Yardley, J. F. Young, S. E. Harris: Appl. Phys. Lett. **24**, 427 (1974)
3.14 R. T. Hodgson, P. P. Sorokin, J. J. Wynne: Phys. Rev. Lett. **32**, 343 (1974)
3.15 K. M. Leung, J. F. Ward, B. J. Orr: Phys. Rev. A**9**, 2440 (1974)
3.16 P. D. Maker, R. W. Terhune: Phys. Rev. **137**, 801 (1965)
3.17 D. M. Bloom, G. W. Bekkers, J. F. Young, S. E. Harris: Appl. Phys. Lett. **26**, 687 (1975)
3.18 H. Puell, K. Spanner, W. Falkenstein, W. Kaiser, C. R. Vidal: Phys. Rev. A**14**, 2240 (1976)

3.19 R. Hilbig, R. Wallenstein: IEEE J. QE-17, 1566 (1981)
3.20 R. Hilbig, R. Wallenstein: Appl. Opt. 21, 913 (1982)
3.21 H. Langer, H. Puell, H. Röhr: Opt. Commun. 34, 137 (1980)
3.22 S. C. Wallace, G. Zdasiuk: Appl. Phys. Lett. 28, 449 (1976)
3.23 H. Junginger, H. B. Puell, H. Scheingraber, C. R. Vidal: IEEE J. QE-16, 1132 (1980)
3.24 R. R. Freeman, R. M. Jopson, J. Bokor: "Generation of light below 100 nm in Hg vapor", in *Laser Techniques for Extreme Ultraviolet Spectroscopy*, ed. by T. J. McIlrath, R. R. Freeman, AIP Conf. Proc. 90, 422 – 430 (1982)
3.25 R. Mahon, F. S. Tomkins: IEEE J. QE-18, 913 (1982)
3.26 H. Scheingraber, C. R. Vidal: IEEE J. QE-19, 1747 (1983)
3.27 H. Scheingraber, C. R. Vidal: J. Opt. Soc. Am. B2, 343 (1985)
3.28 R. Hilbig, R. Wallenstein: IEEE J. QE-19, 1759 (1983)
3.29 H. Egger, T. Srinivasan, K. Hohla, H. Scheingraber, C. R. Vidal, H. Pummer, C. K. Rhodes: Appl. Phys. Lett. 39, 37 (1981)
3.30 H. F. Döbele, M. Röwekamp, B. Rückle: IEEE J. QE-20, 1284 (1984)
3.31 C. K. Rhodes: "Generation of extreme ultraviolet radiation with excimer lasers", in *Laser Techniques for Extreme Ultraviolet Spectroscopy*, ed. by T. J. McIlrath, R. R. Freeman, AIP Conf. Proc. 90, 112 – 116 (1982)
3.32 H. Egger, T. S. Luk, K. Boyer, D. F. Muller, H. Pummer, T. Srinivasan, C. K. Rhodes: Appl. Phys. Lett. 41, 1032 (1982)
3.33 R. R. Freeman, G. C. Bjorklund, N. P. Economou, P. F. Liao, J. E. Bjorkholm: Appl. Phys. Lett. 33, 739 (1978)
3.34 N. P. Economou, R. R. Freeman, G. C. Bjorklund: Opt. Lett. 3, 209 (1978)
3.35 A. Timmermann, R. Wallenstein: Opt. Lett. 8, 517 (1983)
3.36 N. Bloembergen: *Nonlinear Optics* (Benjamin, New York 1965)
3.37 D. C. Hanna, M. A. Yuratich, D. Cotter: *"Nonlinear Optics of Free Atoms and Molecules"*, Springer Ser. Opt. Sci., Vol. 17 (Springer, Berlin, Heidelberg 1979)
3.38 C. Cohen-Tannoudji, B. Diu, F. Laloe: *Quantum Mechanics I* (Wiley, New York 1977)
3.39 J. A. Armstrong, N. Bloembergen, J. Ducuing, P. S. Pershan: Phys. Rev. 127, 1918 (1962)
3.40 B. J. Orr, J. F. Ward: Mol. Phys. 20, 513 (1971)
3.41 H. Puell, C. R. Vidal: Phys. Rev. A14, 2225 (1976)
3.42 R. W. Terhune, P. D. Maker, C. M. Savage: Phys. Rev. Lett. 8, 404 (1962)
3.43 C. G. Bethea: Appl. Opt. 14, 2435 (1975)
3.44 R. S. Finn, J. F. Ward: Phys. Rev. Lett. 26, 285 (1971)
3.45 J. F. Ward, J. Bigio: Phys. Rev. A11, 60 (1975)
3.46 R. L. Abrams, A. Yariv, P. A. Yeh: IEEE J. QE-13, 79 (1977);
      R. L. Abrams, C. K. Asawa, T. K. Plant, A. E. Popa: IEEE J. QE-13, 82 (1977)
3.47 D. P. Shelton, A. D. Buckingham: Phys. Rev. A26, 2787 (1982)
3.48 R. W. Boyd, L.-Q. Xiang: IEEE J. QE-18, 1242 (1982)
3.49 D. J. Gauthier, J. Krasinski, R. W. Boyd: Opt. Lett. 8, 211 (1983)
3.50 R. W. Boyd, D. J. Gauthier, J. Krasinski, M. S. Malcuit: IEEE J. QE-20, 1074 (1984)
3.51 A. R. Edmonds: *Angular Momentum in Quantum Mechanics* (Princeton U. Press, Princeton 1960)
3.52 H. Scheingraber, H. Puell, C. R. Vidal: Phys. Rev. A18, 2585 (1978)
3.53 J. D. Jackson: *Classical Electrodynamics* (Wiley, New York 1962)
3.54 D. G. Steel, J. F. Lam: Opt. Lett. 4, 363 (1979)
3.55 P. P. Bey, H. Rabin: Phys. Rev. 162, 794 (1967)
3.56 A. Yariv: *Quantum Electronics* (Wiley, New York 1967)
3.57 J. A. Armstrong, J. J. Wynne: Phys. Rev. Lett. 33, 1183 (1974)
3.58 L. Armstrong, B. L. Beers: Phys. Rev. Lett. 34, 1290 (1975)
3.59 G. Alber, P. Zoller: Phys. Rev. A27, 1373 (1983)
3.60 G. C. Bjorklund, J. E. Bjorkholm, P. F. Liao, R. H. Storz: Appl. Phys. Lett. 29, 729 (1976)
3.61 G. C. Bjorklund, J. E. Bjorkholm, R. R. Freeman, P. F. Liao: Appl. Phys. Lett. 31, 330 (1977)

3.62 H. Kogelnik, T. Li: Appl. Opt. **5**, 1550 (1966)
3.63 G. C. Bjorklund: IEEE J. QE-11, 287 (1975)
3.64 Y. Mui Yiu, T. J. McIlrath, R. Mahon: Phys. Rev. A**20**, 2470 (1979)
3.65 A. Stappaerts, G. W. Bekkers, J. F. Young, S. E. Harris: IEEE J. QE-12, 330 (1976)
3.66 C. Leubner, H. Scheingraber, C. R. Vidal: Opt. Commun. **36**, 205 (1981)
3.67 B. Dick, R. M. Hochstrasser: J. Chem. Phys. **78**, 3398 (1983)
3.68 E. W. Smith, J. Cooper, W. R. Chappell, T. Dillon: J. Quant. Spectrosc. Rad. Transf. **11**, 1547 (1971)
3.69 E. W. Smith, J. Cooper, W. R. Chappell, T. Dillon: J. Quant. Spectrosc. Rad. Transf. **11**, 1567 (1971)
3.70 D. G. Steel, J. F. Lam, R. A. McFarlane: Phys. Rev. A**26**, 1146 (1982)
3.71 C. R. Vidal, J. Cooper: J. Appl. Phys. **40**, 3370 (1969)
3.72 C. R. Vidal, F. B. Haller: Rev. Scient. Instr. **42**, 1779 (1971)
3.73 H. Scheingraber, C. R. Vidal: Rev. Scient. Instr. **52**, 1010 (1981)
3.74 C. R. Vidal: J. Appl. Phys. **44**, 2225 (1973)
3.75 J. J. Wynne, P. P. Sorokin, J. R. Lankard: In *Laser Spectroscopy,* ed. by R. G. Brewer, A. Mooradian (Plenum, New York 1974) pp. 103 – 111
3.76 D. M. Bloom, J. F. Young, S. E. Harris: Appl. Phys. Lett. **27**, 390 (1975)
3.77 C. R. Vidal, M. M. Hessel: J. Appl. Phys. **43**, 2776 (1972)
3.78 H. Egger, R. T. Hawkins, J. Bokor, H. Pummer, R. Rothschild, C. K. Rhodes: Opt. Lett. **5**, 282 (1980)
3.79 H. Pummer, T. Srinivasan, H. Egger, K. Boyer, T. S. Luk, C. K. Rhodes: Opt. Lett. **7**, 93 (1982)
3.80 H. Egger, T. Srinivasan, K. Boyer, H. Pummer, C. K. Rhodes: "Generation of tunable, coherent 79 nm radiation by frequency mixing", in *Laser techniques for Extreme Ultraviolet Spectroscopy,* ed. by T. J. McIlrath, R. R. Freeman, AIP Conf. Proc. **90**, 445 – 453 (1982)
3.81 T. Srinivasan, H. Egger, H. Pummer, C. K. Rhodes: IEEE J. QE-19, 1270 (1983)
3.82 R. Hilbig, R. Wallenstein: Opt. Commun. **44**, 283 (1983)
3.83 R. Hilbig, A. Lago, R. Wallenstein: Opt. Commun. **49**, 297 (1984)
3.84 K. D. Bonin, T. J. McIlrath: Rev. Scient. Instr. **55**, 1666 (1984)
3.85 K. D. Bonin, T. J. McIlrath: J. Opt. Soc. Am. B**2**, 527 (1985)
3.86 T. B. Lucatorto, T. J. McIlrath, J. R. Roberts: Appl. Opt. **18**, 2505 (1979)
3.87 P. R. Herman, B. P. Stoicheff: Opt. Lett. **10**, 502 (1985)
3.88 H. K. Kung: Opt. Lett. **8**, 24 (1983)
3.89 E. E. Marinero, C. T. Rettner, R. N. Zare: Chem. Phys. Lett. **95**, 486 (1983)
3.90 C. T. Rettner, E. E. Marinero, R. N. Zare: J. Phys. Chem. **88**, 4459 (1984)
3.91 J. Bokor, P. H. Bucksbaum, R. R. Freeman: Opt. Lett. **8**, 217 (1983)
3.92 W. R. Gentry, C. F. Giese: Rev. Scient. Instr. **49**, 595 (1978)
3.93 A. H. Kung, J. F. Young, S. E. Harris: Appl. Phys. Lett. **22**, 301 (1973); Erratum: Appl. Phys. Lett. **28**, 239 (1976)
3.94 T. B. Lucatorto, T. J. McIlrath: Phys. Rev. Lett. **37**, 428 (1976)
3.95 T. J. McIlrath, T. B. Lucatorto: Phys. Rev. Lett. **38**, 1390 (1977)
3.96 T. B. Lucatorto, T. J. McIlrath: Appl. Opt. **19**, 3948 (1980)
3.97 J. H. Skinner: J. Phys. B**13**, 55 (1980)
3.98 P. P. Sorokin, J. A. Armstrong, R. W. Dreyfus, R. T. Hodgson, J. R. Lankard, L. H. Manganaro, J. J. Wynne: in *Laser Spectroscopy,* ed. by S. Haroche, J. C. Pebay-Peyroula, T. W. Hänsch, S. E. Harris (Springer, Berlin, Heidelberg 1975) p. 46
3.99 V. V. Lebedev, V. M. Plyasulya, B. I. Troshin, V. P. Chebotaev, Kvant. Elektron. (Moscow) **12**, 866 (1985)
3.100 F. Vallée, S. C. Wallace, J. Lukasik: Opt. Commun. **42**, 148 (1982)
3.101 F. Vallée, J. Lukasik: Opt. Commun. **43**, 287 (1982)
3.102 J. H. Glownia, R. K. Sander: Appl. Phys. Lett. **40**, 648 (1982)
3.103 L. Hellner, J. Lukasik: Opt. Commun. **51**, 347 (1984)
3.104 K. K. Innes, B. P. Stoicheff, S. C. Wallace: Appl. Phys. Lett. **29**, 715 (1976)

3.105 S. C. Wallace, K. K. Innes: J. Chem. Phys. **72**, 4805 (1980)
3.106 C. Tai, F. W. Dalby, G. L. Giles: Phys. Rev. A**20**, 233 (1979)
3.107 C. Tai: Phys. Rev. A**23**, 2462 (1981)
3.108 C. Tai, A. Tarn: Phys. Rev. A**27**, 3078 (1983)
3.109 C. Tai: Phys. Rev. A**28**, 3459 (1983)
3.110 H. Kildal, T. F. Deutsch: Opt. Commun. **18**, 146 (1976)
3.111 H. Kildal, T. F. Deutsch: IEEE J. QE-**12**, 429 (1976)
3.112 H. Kildal, S. R. J. Brueck: IEEE J. QE-**16**, 566 (1980)
3.113 C. Y. She, K. W. Billman: Appl. Phys. Lett. **27**, 76 (1975)
3.114 C. L. Pan, C. Y. She, W. M. Fairbank, K. W. Billman: IEEE J. QE-**13**, 763 (1977); Erratum: IEEE J. QE-**15**, 54 (1979)
3.115 A. H. Kung, J. F. Young, G. C. Bjorklund, S. E. Harris: Phys. Rev. Lett. **29**, 985 (1972)
3.116 A. I. Ferguson, E. G. Arthurs: Phys. Lett. A**58**, 298 (1976)
3.117 J. R. Taylor: Opt. Commun. **18**, 504 (1976)
3.118 F. S. Tomkins, D. Drapcho, R. Mahon: Opt. Lett. **6**, 27 (1981)
3.119 H. Puell, C. R. Vidal: Opt. Commun. **19**, 279 (1976)
3.120 J. J. Wynne, R. Beigang: Phys. Rev. A**23**, 2736 (1981)
3.121 B. M. Miles, W. L. Wiese: *Bibliography of Atomic Transition Probabilities,* Nat. Bur. Stand. Spec. Publ. #320 (1970)
3.122 J. R. Fuhr, B. J. Miller, G. A. Martin: *Bibliography of Atomic Transition Probabilities,* Nat. Bur. Stand. Spec. Publ. #505 (1978)
3.123 D. R. Bates, A. Damgaard: Phil. Trans. A**242**, 101 (1949)
3.124 H. B. Bebb: Phys. Rev. **149**, 25 (1966)
3.125 H. Eicher: IEEE J. QE-**11**, 121 (1975)
3.126 H. Puell, H. Scheingraber, C. R. Vidal: Phys. Rev. A**22**, 1165 (1980)
3.127 H. Puell, C. R. Vidal: IEEE J. QE-**14**, 364 (1978)
3.128 R. A. Ganeev, I. A. Kulagin, T. Usmanov, S. T. Khudalberganov: Sov. J. Quant. Electron. **12**, 1637 (1982)
3.129 H. Puell, C. R. Vidal: Opt. Commun. **27**, 165 (1978)
3.130 C. C. Wang, L. I. Davis: Phys. Rev. Lett. **35**, 650 (1975)
3.131 B. Held, G. Mainfray, C. Manus, J. Morellec, F. Sanchez: Phys. Rev. Lett. **30**, 423 (1973)
3.132 J. F. Ward, A. V. Smith: Phys. Rev. Lett. **35**, 653 (1975)
3.133 C. S. Chang, P. Stehle: Phys. Rev. Lett. **30**, 1283 (1973)
3.134 C. S. Chang: Phys. Rev. A**9**, 1769 (1974)
3.135 S. H. Autler, C. H. Townes: Phys. Rev. **100**, 703 (1955)
3.136 P. F. Liao, J. C. Bjorkholm: Phys. Rev. Lett. **34**, 1 (1975)
3.137 K. Miyazaki, H. Kashiwagi: Phys. Rev. A**18**, 635 (1978)
3.138 K. Miyazaki, T. Sato, H. Kashiwagi: Phys. Rev. Lett. **43**, 1154 (1979)
3.139 K. Miyazaki: Phys. Rev. A**23**, 1350 (1981)
3.140 K. Miyazaki, T. Sato, H. Kashiwagi: Phys. Rev. A**23**, 1358 (1981)
3.141 W. Jamroz, P. E. LaRocque, B. P. Stoicheff: Opt. Lett. **7**, 148 (1982)
3.142 D. S. Bethune, R. W. Smith, Y. R. Shen: Phys. Rev. A**17**, 277 (1978)
3.143 R. R. Freeman, J. E. Bjorkholm, R. Panock, W. E. Cooke: "Optical second harmonic generation by a single laser beam in an isotropic medium", in *Laser Spectroscopy V,* ed. by A. R. W. McKellar, T. Oka, B. P. Stoicheff, Springer Ser. Opt. Sci., Vol. 30 (Springer, Berlin, Heidelberg 1981) p. 453
3.144 H. Scheingraber, C. R. Vidal: Opt. Commun., to be submitted
3.145 M. S. Malcuit, D. J. Gauthier, R. W. Boyd: Phys. Rev. Lett. **55**, 1086 (1985)
3.146 J. S. Bakos: Phys. Rep. C**31**, 209 (1977)
3.147 S. E. Moody, M. Lambropoulos: Phys. Rev. A**15**, 1497 (1977)
3.148 P. L. Knight: Opt. Commun. **22**, 173 (1977)
3.149 A. T. Georges, P. Lambropoulos: Phys. Rev. A**18**, 587 (1978)
3.150 A. T. Georges, P. Lambropoulos, J. H. Marburger: Opt. Commun. **18**, 509 (1976)
3.151 A. T. Georges, P. Lambropoulos, J. H. Marburger: Phys. Rev. A**15**, 300 (1977)
3.152 K. Aron, P. M. Johnson: J. Chem. Phys. **67**, 5099 (1977)

3.153 R. N. Compton, J. C. Miller, A. E. Carter, P. Kruit: Chem. Phys. Lett. **71**, 87 (1980)
3.154 J. C. Miller, R. N. Compton, M. G. Payne, W. W. Garrett: Phys. Rev. Lett. **45**, 114 (1980)
3.155 J. C. Miller, R. N. Compton: Phys. Rev. A **25**, 2056 (1982)
3.156 J. H. Glownia, R. K. Sander: Phys. Rev. Lett. **49**, 21 (1982)
3.157 D. Normand, J. Morellec, J. Reif: J. Phys. B **16**, L 227 (1983)
3.158 D. J. Jackson, J. J. Wynne: Phys. Rev. Lett. **49**, 543 (1982)
3.159 D. J. Jackson, J. J. Wynne, P. H. Kes: Phys. Rev. A **28**, 781 (1983)
3.160 J. J. Wynne: Phys. Rev. Lett. **52**, 751 (1984)
3.161 M. G. Payne, W. W. Garrett: Chem. Phys. Lett. **75**, 468 (1980)
3.162 M. G. Payne, W. W. Garrett: Phys. Rev. A **26**, 356 (1982)
3.163 M. G. Payne, W. W. Garrett: Phys. Rev. A **28**, 3409 (1983)
3.164 M. Poirier: Phys. Rev. A **27**, 934 (1983)
3.165 H. Scheingraber, C. R. Vidal: Opt. Commun. **38**, 75 (1981)
3.166 J. Heinrich, K. Hollenberg, W. Behmenburg: Appl. Phys. B **33**, 225 (1984)
3.167 E. A. Stappaerts: Phys. Rev. A **11**, 1664 (1975)
3.168 E. A. Stappaerts: IEEE J. QE-**15**, 110 (1979)
3.169 S. A. Akhmanov, B. V. Khokhlov, A. P. Sukhorukov: "Self-focusing self-defocusing of laser beams", in *Laser Handbook,* ed. by F. T. Arecchi, E. O. Schulz-Dubois (North Holland, Amsterdam 1972) vol. 2, p. 1151
3.170 Y. R. Shen, Progr. Quant. Electr. **4**, 1 (1975)
3.171 J. H. Marburger: Progr. Quant. Electr. **4**, 35 (1975)
3.172 R. H. Lehmberg, J. Reintjes, R. C. Eckardt: Appl. Phys. Lett. **25**, 374 (1974)
3.173 R. H. Lehmberg, J. Reintjes, R. C. Eckardt: Phys. Rev. A **13**, 1095 (1976)
3.174 S. A. Bakhramov, U. G. Gulyamov, K. N. Drabovich, Ya. Z. Faizullaev: Sov. Phys. JETP Lett. **21**, 102 (1975)
3.175 I. A. Poluektov, A. V. Nazarkin: Sov. J. Quant. Electron. **9**, 1495 (1979)
3.176 I. A. Poluektov, A. V. Nazarkin: Sov. J. Quant. Electron. **11**, 159 (1981)
3.177 I. A. Poluektov, A. V. Nazarkin: Sov. J. Quant. Electron. **11**, 1327 (1981)
3.178 C. Brechignac, Ph. Cahuzac: J. Phys. B **14**, 221 (1981)
3.179 C. H. Skinner, H. P. Palenius: Opt. Commun. **18**, 335 (1976)
3.180 J. Heinrich, W. Behmenburg: Appl. Phys. **23**, 333 (1980)
3.181 F. S. Tomkins, R. Mahon: Opt. Lett. **6**, 179 (1981)
3.182 J. Bokor, R. R. Freeman, R. L. Panock, J. C. White: Opt. Lett. **6**, 182 (1981)
3.183 S. E. Harris: Phys. Rev. Lett. **31**, 341 (1973)
3.184 M. G. Grozeva, D. I. Metchkov, V. M. Mitev, L. J. Pavlov, K. V. Stamenov: Opt. Commun. **23**, 77 (1977)
3.185 M. G. Grozeva, D. I. Metchkov, V. M. Mitev, L. J. Pavlov, K. V. Stamenov: Phys. Lett. A **64**, 41 (1977)
3.186 D. I. Metchkov, V. M. Mitev, L. L. Pavlov, K. V. Stamenov: Opt. Commun. **21**, 391 (1977)
3.187 V. L. Doitcheva, V. M. Mitev, L. I. Pavlov, K. V. Stamenov: Opt. Quant. Electron. **10**, 131 (1978)
3.188 V. M. Mitev, L. I. Pavlov, K. V. Stamenov: J. Phys. B **11**, 819 (1978)
3.189 V. M. Mitev, L. I. Pavlov, K. V. Stamenov: Opt. Quant. Electron. **11**, 229 (1979)
3.190 S. G. Dinev, O. R. Marazov, K. V. Stamenov, I. V. Tomov: Opt. Quant. Electron. **12**, 183 (1980)
3.191 J. Reintjes, R. C. Eckardt, C. Y. She, N. E. Karangelen, R. C. Elton, R. A. Andrews: Phys. Rev. Lett. **37**, 1540 (1976)
3.192 J. Reintjes, C. Y. She, R. C. Eckardt, N. E. Karangelen, R. A. Andrews, R. C. Elton: Appl. Phys. Lett. **30**, 480 (1977)
3.193 C. Y. She, J. Reintjes: Appl. Phys. Lett. **31**, 95 (1977)
3.194 J. Reintjes, C. Y. She, R. C. Eckardt: IEEE J. QE-**14**, 581 (1978)
3.195 J. Reintjes, C. Y. She: Opt. Commun. **27**, 469 (1978)
3.196 J. Reintjes, L. L. Tankersley, R. Christensen: Opt. Commun. **39**, 334 (1981)
3.197 J. Reintjes: Appl. Opt. **19**, 3889 (1980)

3.198 G. H. C. New: Opt. Commun. **38**, 189 (1981)
3.199 I. V. Tomov, M. C. Richardson: IEEE J. QE-**12**, 521 (1976)
3.200 V. V. Rostovseva, A. P. Sukhorukov, V. G. Tunkin, S. M. Saltiel: Opt. Commun. **22**, 56 (1977)
3.201 J. C. Diels, A. T. Georges: Phys. Rev. A **19**, 1589 (1979)
3.202 N. Bloembergen: Am. J. Phys. **35**, 989 (1967)
3.203 P. P. Sorokin, J. J. Wynne, J. R. Lankard: Appl. Phys. Lett. **22**, 342 (1973)
3.204 J. J. Wynne, P. P. Sorokin: J. Phys. B **8**, L 37 (1975)
3.205 K. C. Rustagi, S. C. Mehendale, P. K. Gupta: Appl. Phys. Lett. **43**, 811 (1983)
3.206 S. C. Mehendale, P. K. Gupta, K. C. Rustagi: Appl. Phys. B **32**, 217 (1983)
3.207 W. Hartig: Appl. Phys. **15**, 427 (1978)
3.208 G. V. Venkin, G. M. Krochik, L. L. Kulyuk, D. I. Maleev, Yu. G. Khronopoulo: Sov. J. Quant. Electron. **6**, 369 (1976)
3.209 G. M. Barykinskii, V. V. Lebedev, V. M. Plyasulya: Sov. J. Quant. Electron. **12**, 312 (1982)
3.210 D. Grischkowsky: Phys. Rev. Lett. **24**, 866 (1970)
3.211 D. Grischkowsky, J. A. Armstrong: Phys. Rev. A **6**, 1566 (1972)
3.212 D. Grischkowsky, M. M. T. Loy, P. F. Liao: Phys. Rev. A **12**, 2514 (1975)
3.213 R. P. Feynman, F. L. Vernon, R. W. Hellwarth: Appl. Phys. **28**, 49 (1957)
3.214 J. N. Elgin, G. H. C. New: Opt. Commun. **16**, 242 (1976)
3.215 J. N. Elgin, G. H. C. New, K. E. Orkney: Opt. Commun. **18**, 250 (1976)
3.216 G. H. C. New: Opt. Commun. **19**, 177 (1976)
3.217 J. N. Elgin, G. H. C. New: J. Phys. B **11**, 3439 (1978)
3.218 J. N. Elgin: J. Phys. B **12**, L 261 (1979)
3.219 J. N. Elgin, G. H. C. New, P. R. C. Smith: J. Phys. B **13**, 1663 (1980)
3.220 J. N. Elgin, G. H. C. New, J. M. Catherall: J. Phys. B **13**, 3043 (1980)
3.221 M. Matsuoka, H. Nakatsuka, J. Okada: Phys. Rev. A **12**, 1062 (1975)
3.222 T. R. Royt, C. H. Lee, W. L. Faust: Opt. Commun. **18**, 108 (1976)
3.223 T. R. Royt, C. H. Lee: Appl. Phys. Lett. **30**, 332 (1977)
3.224 N. P. Economou, R. R. Freeman, J. P. Heritage, P. F. Liao: Appl. Phys. Lett. **36**, 21 (1980)
3.225 J. A. Benda, D. J. Gauthier, R. W. Boyd: Phys. Rev. A **32**, 3461 (1985)
3.226 R. Wallenstein: Opt. Commun. **33**, 119 (1980)
3.227 R. Wallenstein, T. W. Hänsch: Appl. Opt. **13**, 1625 (1974)
3.228 R. Wallenstein, H. Zacharias: Opt. Commun. **32**, 429 (1980)
3.229 H. R. Hutchinson, K. J. Thomas: IEEE J. QE-**19**, 1823 (1983)
3.230 S. Gerstenkorn, P. Luc: "Atlas du spectre d'absorption de la molécule d'iode", éditions du CNRS, 15, quai Anatole France, Paris (1978)
3.231 K. Radler, J. Berkowitz: J. Opt. Soc. Am. **68**, 1181 (1978)
3.232 I. I. Sobelman, A. V. Vinogradov: Adv. Atom. Molec. Phys. **20**, 327 (1985)
3.233 V. A. Boiko, F. V. Bunkin, V. I. Derzhiev, S. I. Yakovlenko: IEEE J. QE-**20**, 206 (1984) and further references therein
3.234 R. C. Elton, J. F. Seely, R. H. Dixon: "Population density and vuv gain measurements in laser produced plasmas" in *Laser Techniques for Extreme Ultraviolet Spectroscopy*, ed. by T. J. McIlrath, R. R. Freeman, AIP Conf. Proc. **90**, 277 (1982)
3.235 P. Jaeglé, G. Jamelot, A. Carillon, A. Klisnick, A. Sureau, H. Guennou: "Amplification of spontaneous emission in aluminum and magnesium plasmas" in *Laser Techniques in the Extreme Ultraviolet* ed. by S. E. Harris, T. B. Lucatorto, AIP Conf. Proc. **119**, 468 (1984)
3.236 Y. Ohashi, Y. Ishibashi, T. Kobayashi, H. Inaba: Jap. J. Appl. Phys. **15**, 1817 (1976)
3.237 L. T. Bolotskikh, A. L. Vysotin, Im. Tkhek-de, O. P. Podavalova, A. K. Popov: Appl. Phys. B **35**, 249 (1984)
3.238 I. N. Drabovich, V. M. Mitev, L. I. Pavlov, K. V. Stamenov: Opt. Commun. **20**, 350 (1977)
3.239 S. S. Dimov, L. I. Pavlov, K. V. Stamenov, Yu. I. Heller, A. K. Popov: Appl. Phys. B **30**, 35 (1983)

3.240 R. Mahon, T. J. McIlrath, F. S. Tomkins, D. E. Kelleher: Opt. Lett. **4**, 360 (1979)
3.241 M. H. R. Hutchinson, R. J. Manning: Opt. Commun. **55**, 55 (1985)
3.242 R. G. Caro, A. Costela, C. E. Webb: Opt. Lett. **6**, 464 (1981)
3.243 R. G. Caro, A. Costela, N. P. Smith, C. E. Webb: J. Phys. D **18**, 1291 (1985)
3.244 T. J. McKee, B. P. Stoicheff, S. C. Wallace: Opt. Lett. **3**, 207 (1978)
3.245 P. P. Sorokin, J. J. Wynne, J. A. Armstrong, R. T. Hodgson: Ann. New York Acad. Sci. **267**, 30 (1976)
3.246 W. Jamroz, P. E. La Rocque, B. P. Stoicheff: Opt. Lett. **7**, 617 (1982)
3.247 K. Miyazaki, H. Sakai, T. Sato: Opt. Lett. **9**, 457 (1984)
3.248 F. S. Tomkins, R. Mahon: Opt. Lett. **7**, 304 (1982)
3.249 K. S. Hsu, A. H. Kung, L. J. Zych, J. F. Young, S. E. Harris: IEEE J. QE-**12**, 60 (1976)
3.250 V. V. Slabko, A. K. Popov, V. F. Lukinykh: Appl. Phys. **15**, 239 (1978)
3.251 J. Reintjes: Opt. Lett. **4**, 242 (1979)
3.252 J. Reintjes: Opt. Lett. **5**, 342 (1980)
3.253 M. H. R. Hutchinson, C. C. Ling, D. J. Bradley: Opt. Commun. **18**, 203 (1976)
3.254 R. Mahon, Y. Mui Yiu: Opt. Lett. **5**, 279 (1980)
3.255 S. A. Batishche, V. S. Burakov, Yu. V. Kostenich, A. V. Mostovnikov, P. A. Naumenkov, N. V. Tarasenko, V. I. Gladushchak, S. A. Moshkalev, G. T. Razdobarin, V. V. Semenov, E. Ya. Shreider: Opt. Commun. **38**, 71 (1981)
3.256 R. Mahon, T. J. McIlrath, D. W. Koopman: Appl. Phys. Lett. **33**, 305 (1978)
3.257 D. Cotter: Opt. Commun. **31**, 397 (1979)
3.258 D. Cotter: Opt. Lett. **4**, 134 (1979)
3.259 J. Hager, S. C. Wallace: Chem. Phys. Lett. **90**, 472 (1982)
3.260 R. Hilbig, R. Wallenstein: IEEE J. QE-**19**, 194 (1983)
3.261 F. Vallée, F. DeRougemont, J. Lukasik: IEEE J. QE-**19**, 131 (1983)
3.262 Y. Mui Yiu, K. D. Bonin, T. J. McIlrath: Opt. Lett. **7**, 268 (1982)
3.263 A. H. Kung: Appl. Phys. Lett. **25**, 653 (1974)
3.264 W. Zapka, D. Cotter, U. Brackmann: Opt. Commun. **36**, 79 (1981)

4. Stimulated Raman Scattering

Jonathan C. White

With 43 Figures

This chapter describes the application of stimulated Raman scattering as a practical method for the generation of tunable, coherent radiation. A brief theoretical discussion of the stimulated Raman effect in the steady state and transient-scattering regimes is presented. Generation techniques are described in detail, with particular emphasis on molecular hydrogen and various atomic systems, as Stokes and anti-Stokes Raman scattering media. Finally, several effects that limit Raman conversion efficiencies are discussed.

4.1 Introduction

Stimulated Raman scattering was one of the first nonlinear optical processes to be observed. Since its accidental discovery in 1962, stimulated Raman scattering has been extensively studied in atomic and molecular gases, numerous liquids, and solids. The generation of tunable coherent radiation via stimulated Raman techniques is now widely employed as a straightforward method for creating intense radiation over large regions of the infrared, ultraviolet, and vacuum ultraviolet.

The stimulated Raman effect was discovered in 1962 by *Woodbury* and *Ng* [4.1] during a study of ruby laser Q-switching with a nitrobenzene Kerr cell. They detected intense infrared radiation emanating from the Kerr cell. The source of this radiation was not immediately determined. *Woodbury* and *Eckhardt* [4.2] later proposed that this radiation was due to stimulated Raman scattering in the nitrobenzene. Experimental verification of this hypothesis was provided by *Eckhardt* et al. [4.3] a short time thereafter. Subsequent to these early discoveries, stimulated Raman emission was observed in several other liquids by *Eckhardt* et al. [4.3], *Geller* et al. [4.4], *Giordmaine* and *Howe* [4.5], and by *Stoicheff* [4.6], as well as in mixtures of liquids by *Calviello* and *Heller* [4.7]. Raman emission was also observed in various solids by *Eckhardt* et al. [4.8] and in gaseous H_2, D_2, and CH_4 by *Minck* et al. [4.9]. *Hellwarth* [4.10, 11] gave an early theoretical description of the effect.

In the years since these early investigations, hundreds of experiments and theoretical treatments of the stimulated Raman effect have been published. Extensive review articles have been published by *Bloembergen* [4.12], *Eckhardt* [4.13], *Shen* and *Bloembergen* [4.14], *Wang* [4.15, 16], *Maier* [4.17], and others [4.18 – 25]. The majority of these reviews have been concerned with stimulated vibrational and rotational Raman scattering in liquids and gases as a means of generating intense coherent radiation over the visible and near infrared. Raman scattering may also occur in crystals via phonons leading to so called stimulated polariton scattering [4.26, 27]; this effect has been employed as a source of tunable far-infrared radiation [4.28]. Tunable medium to far-infrared radiation has also been generated via stimulated spin-flip Raman scattering in semiconductor crystals. This effect was first observed in InSb by *Patel* and *Shaw* [4.29]; other examples may be found in [4.30, 31]. In these devices the Raman frequency is made continuously tunable by varying the Zeeman splitting with an external magnetic field.

It is evident just from the brief introduction given thus far that a wide variety of stimulated Raman processes have been observed. In the present chapter we shall emphasize those techniques which are most widely employed in the laboratory as a means of generating tunable coherent radiation via the stimulated Raman effect. Generally, these include Stokes and anti-Stokes generation in gases (most notably H_2, D_2, CH_4), Stokes generation in liquids (e.g., N_2, O_2), and stimulated electronic Raman scattering in atomic vapors. We shall not discuss stimulated scattering via spin-flip interactions or polaritons, or the related area of stimulated Brillouin scattering [4.32].

Since we are primarily interested in a practical description of Raman generation techniques, a detailed review of the theoretical aspects of the Raman effect will not be undertaken in this chapter. Rather, an overview of the theoretical foundations will be given in Sect. 4.2. It is hoped that the simple treatment given here will permit the reader to utilize the more extensive theoretical discussions contained in the references.

4.2 Background Material

4.2.1 Basic Principles

The basic Raman process is a direct two-photon, inelastic light scattering process as shown in Figs. 4.1 a, b. Consider the most commonly encounted case Fig. 4.1 a for Stokes scattering in a medium with an initial Raman state $|i\rangle$ and a final Raman state $|f\rangle$. A photon at ω_1 incident on such a system may be absorbed while a photon at ω_2 is emitted with a corresponding material transition from the initial state $|i\rangle$, to the final state $|f\rangle$, such that $\hbar\omega_1 = \hbar\omega_2 + (E_f - E_i) = \hbar\omega_2 + \hbar\omega_v$. In the case of Fig. 4.1 a, $\omega_1 > \omega_2$; this situation in which the incident photon is red-shifted is known as Stokes scattering. The wave at frequency ω_1 is often called the pump wave ω_p, and likewise the scattered wave at ω_2 is often termed the Stokes wave ω_s.

As we shall see shortly, the strength of the Raman process is greatly enhanced by the presence of a nearby intermediate level. Generally this inter-mediate state should be of parity such that electric dipole transitions from it to

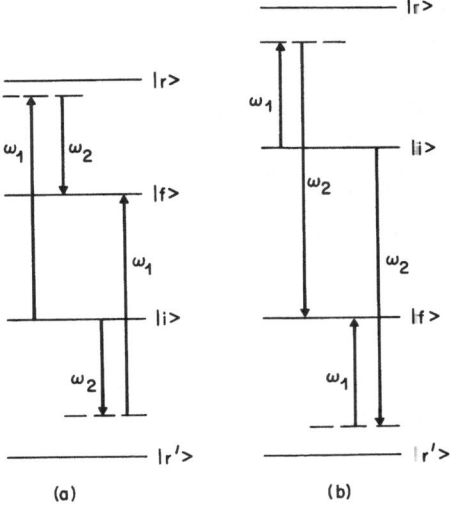

Fig. 4.1a,b. Resonance enhancement schemes for stimulated Raman scattering. (a) Stokes generation illustrating intermediate levels which lie either above or below the initial Raman state. (b) Anti-Stokes generation from the inverted, excited state $|i\rangle$ with intermediate resonances either above or below the initial state

the initial and final Raman levels are allowed and have large dipole moments. In a few cases stimulated Raman scattering has been observed where an intermediate level of the same parity as either the initial or final states is involved, in which case an electric quadrupole transition participates in the Raman process. Two types of intermediate resonances are illustrated in Fig. 4.1 a. In the case most commonly encountered, the intermediate state $|r\rangle$ lies above both the initial and final Raman levels; it is the familiar picture that one envisions in most discussions of Raman scattering. A second type of intermediate resonance exists, however, when $|r\rangle$ lies below both $|i\rangle$ and $|f\rangle$. Resonances of the second type have been considered by *Ducuing* et al. [4.33] as a method of generating far-infrared radiation on pure rotational transitions in HCl and HF. We would expect that such resonant enhancements would play an important role in systems which have been prepared in excited initial states $|i\rangle$.

When the frequency of the scattered wave is greater than the input wave (i.e., $\omega_2 > \omega_1$), the process is termed anti-Stokes scattering. This situation is illustrated in Fig. 4.1 b, and the scattered wave is usually referred to as ω_{as}. Note that the general relation $\hbar\omega_1 = \hbar\omega_2 + (E_f - E_i)$ still holds; however the initial level $|i\rangle$ must lie above the final level $|f\rangle$. As before, resonant enhancement of the scattering process involving two types of intermediate levels is possible. The usual case involves coupling the waves ω_1 and ω_2 to level $|r\rangle$ such that $\omega_2 \simeq \omega_{fr}$. Anti-Stokes scattering has been observed, however, by *Sorokin* et al. [4.34] in which resonant enhancement to a state $|r'\rangle$ was seen, where $|r'\rangle$ lies below both $|i\rangle$ and $|f\rangle$. In all cases it is necessary for the initial level to lie above the final level; and in general $|i\rangle$ must be an excited state, and some preparation of the atomic or molecular system is necessary.

In the cases illustrated in Fig. 4.1 a, b, resonant enhancements to many levels of the type $|r\rangle$ and $|r'\rangle$ contribute to the Raman scattering. Usually, however, in atomic systems ω_1 is tuned sufficiently close to a particular intermediate state so that one pathway is favored in calculating the Raman susceptibility and all the contributions of the other pathways are small enough to be ignored. Many spectroscopic applications of spontaneous Raman scattering in molecules do not make use of one dominant perturbation pathway, and a summation over possible intermediate states must be carried out [4.35].

4.2.2 Semiclassical Theory of Spontaneous Raman Scattering

The classical polarizability theory of Raman scattering was first formulated by *Placzek* [4.36]; subsequently spontaneous Raman scattering has been reviewed by several authors [4.16, 37 – 39]. We shall consider the process but briefly with the intent of reviewing those aspects relevant to our discussion of stimulated Raman scattering. If the input wave is of low intensity (generally 10^6 W/cm^2 or less) only spontaneous Raman scattering will occur. It is instructive first to consider the spontaneous Stokes process in which a vibrational excitation of the material is created or destroyed, since this is the process most often encountered (although a variety of material excitations involving rotational spin, and/or electronic transitions may occur).

Following the treatment of *Wang* [4.16], the Hamiltonian describing the interaction between a molecule and the applied electromagnetic field may be written in the dipole approximation as

$$H = H_0 - \boldsymbol{\mu} : \boldsymbol{E}_\mathrm{p} \ , \tag{4.1}$$

where H_0 is the Hamiltonian of the unperturbed molecule, and $\boldsymbol{\mu} = \alpha \boldsymbol{E}_\mathrm{p}$ is the electric dipole moment induced in the molecule by the applied electric field $\boldsymbol{E}_\mathrm{p}$. In most cases the field-interaction term $\boldsymbol{\mu} \cdot \boldsymbol{E}_\mathrm{p}$ is much smaller than H_0 and may be treated as a perturbation is the usual manner. For simplicity, the polarizability is treated as a scalar rather than a tensor quantity. Time-dependent perturbation theory may be used with the eigenstates of H_0 as the unperturbed states to calculate the molecular wave function to first order in the perturbation. The transition polarizability associated with two-photon Raman scattering process (i.e., a molecular transition from state $|i\rangle$ to state $|f\rangle$) is then

$$\alpha = \frac{1}{\hbar} \sum_r \frac{\langle i|er|r\rangle\langle r|er|f\rangle}{\omega_{ri} - \omega_\mathrm{p}} + \frac{\langle i|er|r\rangle\langle r|er|f\rangle}{\omega_{rf} + \omega_\mathrm{s}} \ . \tag{4.2}$$

Here, we have explicitly indicated the summation over all of the intermediate states $|r\rangle$ of the molecule.

For vibrational transitions the translational and rotational motion may be neglected, so that the transition polarizability may be expressed in terms of vibrational transitions only. The transition polarizability may then be written as

$$\alpha'_{vv} = \langle v | \alpha(X) | v' \rangle \quad \text{with} \tag{4.3}$$

$$\alpha(X) = \frac{1}{\hbar} \sum_n \frac{\langle o | er | n \rangle \langle n | er | o \rangle}{\omega_{no} - \omega_p} + \frac{\langle o | er | n \rangle \langle n | er | o \rangle}{\omega_{no} + \omega_s} \tag{4.4}$$

where ns are the quantum numbers associated with the electronic states. The polarizability may be expanded as a Taylor series about the equilibrium position of the vibrational coordinate Q of the molecule as [4.36]

$$\alpha(X) = \alpha_0(X) + \left(\frac{\partial \alpha}{\partial Q}\right)_0 Q + \dots \tag{4.5}$$

where $Q = \mu X^{1/2}$. The first term in (4.5) accounts for the index of refraction of the medium, while the second term allows for the coupling of the first-order Stokes (or anti-Stokes) radiation to the molecular vibration. We may now write the transition polarizability as

$$\langle v | \alpha(X) | v' \rangle = \left(\frac{\partial \alpha}{\partial Q}\right)_0 \langle v | Q | v' \rangle \tag{4.6}$$

and the calculation has been simplified so that the only dynamic molecular variable is the vibrational coordinate Q.

Under many circumstances the energy of the molecular vibration is less than kT, so that to a good approximation the majority of the molecular population lies in the ground state. The population in higher-lying levels will be reduced by the Boltzmann factor, and in general higher-order Stokes or anti-Stokes emission will be greatly reduced. It is convenient, therefore, to restrict our discussion to the first Stokes component.

Using the classical derivation for the power radiated per unit solid angle for an electric dipole oscillating at the first Stokes frequency gives

$$\frac{dP}{d\Omega} = \frac{\omega_s^4 \hbar}{8 \pi c^3 \omega_v} \left(\frac{\partial \alpha}{\partial Q}\right)_0^2 |E_p|^2 \cos^2 \Theta \tag{4.7}$$

with $\omega_s = \omega_p - \omega_v$ and Θ is the angle between incident and Stokes scattered wave vectors. The differential cross section (per molecule) for spontaneous Raman scattering is then

$$\frac{d\sigma}{d\Omega} = \frac{dP}{d\Omega} \left(\frac{4\pi}{c|E|^2}\right) = \frac{\omega_s^4 \hbar}{2 c^4 \omega_v} \left(\frac{\partial \alpha}{\partial Q}\right)_0^2 \cos^2 \Theta , \tag{4.8}$$

and the total cross section is found as

$$\alpha = \int_{4\pi} \left(\frac{d\sigma}{d\Omega}\right) d\Omega = \frac{4\pi\omega_s^4 \hbar}{3c^4\omega_v} \left(\frac{\partial\alpha}{\partial Q}\right)_0^2 .$$ (4.9)

As we shall see in Sect. 4.5, for many molecules $\sigma \simeq 10^{-30} \, cm^2$.

4.3 Stimulated Raman Scattering: Steady State Limit

As the intensity of an incident light wave is increased, an enhancement of the scattered Raman field can occur in which an initially scattered Stokes photon can further promote scattering of additional incident photons. This process in which the Stokes field grows exponentially (in the small signal limit) is known as stimulated Raman scattering (SRS). Conversion efficiencies from the incident field to the Raman field often exceed 50%, unlike the spontaneous Raman process where approximately only one photon in every 10^7 incident photons is scattered [4.16].

In the small signal limit the intensity of the Stokes field, after traversing a distance z in the Raman medium, is given by

$$I_s(L) = I_s(0) \exp(G_{ss}z) = I_s(0) \exp[g_{ss}I_p(0)z]$$ (4.10)

where G_{ss} is the Raman power gain coefficient per unit pump intensity, and I_s and I_p are the Stokes and pump intensities, respectively. As well shall shortly see, g_{ss} is proportional to the number density of molecules or atoms associated with the transition N and to the Raman susceptibility $\chi_R^{(3)}$.

Note that in (4.10) a nonzero-input intensity $I_s(0)$ at the Stokes frequency is required for gain. Although it is possible to inject radiation at ω_s and use the medium as a Raman amplifier, for most Raman devices one relies on the spontaneous Stokes photons generated at the beginning (i.e., input end) of the medium. In typical systems the gain is sufficiently high to amplify this "noise" and thereby generate $\exp(30)$ Stokes photons for each noise photon. In the small signal regime (i.e., prior to the onset of saturation effects) the intensity of the Stokes wave shows a rapid, nonlinear dependence on the input intensity. This is characteristic of the threshold behavior of all laser devices.

In the following subsections we shall consider the calculation of the Raman power gain coefficient in the so called steady state limit. The traditional coupled-wave approach will first be discussed in Sect. 4.3.1. A more recent model which makes the connection between this approach and a two-photon vector model will be outlined in Sect. 4.3.2. Higher-order Stokes generation and the parametric coupling between the Stokes and anti-Stokes orders will be addressed in Sect. 4.3.3.

4.3.1 Coupled Wave Model

We shall first consider SRS in the traditional coupled-wave or polarizability model in which the optical fields are coupled to the material vibration by the nonlinear Raman polarizability of the medium. This model has been extensively used [4.12 – 16, 20, 22 – 24, 40 – 54] and gives a good physical representation of stimulated Raman scattering.

The coupled-wave model is a semiclassical treatment in which the pump and Stokes waves are not quantized. This description is adequate for the vast majority of Raman scattering experiments and generation schemes in which the numbers of photons in the Raman modes are large. A full quantum description is necessary when the number of photons in the Raman mode is small; and although we shall not discuss quantization of the electromagnetic fields here, the technique is extensively reviewed in the literature [4.10, 11, 55 – 63].

In the coupled-wave description of SRS two sets of differential equations are used; one set describes the behavior of the two electric fields (i.e., E_p and E_s) and a second equation describes the molecular system in the framework of a damped harmonic oscillator for a coherently driven vibrational wave. These two sets of equations are linked via the nonlinear Raman polarization. For simplicity we shall only consider only first Stokes generation, with $\omega_1 > \omega_2$ as in Fig. 4.1 a.

The electric fields propagate in the medium according to the standard wave equation. Hence, with P as the induced Raman polarization, one has for the pump and Stokes waves

$$\nabla^2 E - \frac{\varepsilon}{c^2} \frac{\partial^2}{\partial t^2} E = \frac{4\pi}{c^2} \frac{\partial^2}{\partial t^2} P \ . \tag{4.11}$$

Recalling from (4.1) that the dipole moment induced in the molecule by the fields is $\mu = \alpha E$ and making use of the Fourier expansion for α of (4.5), one may readily calculate P. If N is the number density of Raman active molecules (or atoms) in the medium, we have

$$P = N \left(\frac{\partial \alpha}{\partial Q} \right)_0 Q E \ . \tag{4.12}$$

Combining (4.12, 11), the wave equation for the electric fields takes on the familiar form

$$\nabla^2 E - \frac{\varepsilon}{c^2} \frac{\partial^2}{\partial t^2} E = \frac{4\pi}{c^2} \frac{\partial^2}{\partial t^2} \left[\left(\frac{\partial \alpha}{\partial Q} \right)_0 Q E \right] \ . \tag{4.13}$$

The material vibration is described as a damped harmonic oscillator by the familiar relation

$$\frac{\partial^2}{\partial t^2} Q + 2\Gamma \frac{\partial}{\partial t} Q + \omega_v^2 Q = NF \ , \tag{4.14}$$

where ω_v is the material vibration frequency and F is driving force term. The factor Γ is the damping constant of the coherent vibrational wave and represents the phenomenological collisional dephasing rate. This term corresponds to $1/(2T_2)$, the dephasing time in the Bloch equations, and is the halfwidth at half-maximum (HWHM) for the Raman transition between the initial and final states. Stimulated Raman scattering is governed by T_2, since T_2 is usually shorter than T_1, the relaxation time of the incoherent vibrational waves. The potential energy of a molecule in the electric field is given by

$$U = -\boldsymbol{\mu} \cdot \boldsymbol{E} = -\alpha_0 \boldsymbol{E} \cdot \boldsymbol{E} - \left(\frac{\partial \alpha}{\partial Q}\right)_0 Q\boldsymbol{E} \cdot \boldsymbol{E} \tag{4.15}$$

where we have made use of (4.5). The displacement force acting on the molecules is then

$$F = -\nabla_Q U = \left(\frac{\partial \alpha}{\partial Q}\right)_0 \boldsymbol{E} \cdot \boldsymbol{E} \ . \tag{4.16}$$

Substituting (4.15) into (4.14) gives

$$\frac{\partial^2}{\partial t^2} Q + 2\Gamma \frac{\partial}{\partial t} Q + \omega_v^2 Q = N \left(\frac{\partial \alpha}{\partial Q}\right)_0 \boldsymbol{E} \cdot \boldsymbol{E} \ . \tag{4.17}$$

Equations (4.13,17) represent the starting point for the coupled-wave model.

Expressions for the electric fields and the molecular vibrational coordinate may be written using a few simplifying approximations. First, since Raman scattering normally occurs over an extended interaction length, it is normal to consider only those waves propagating along $\hat{z}$, the axis of the pump laser. This approximation is well satisfied except in the case of anisotropic or dispersive medium where phase-matching considerations can lead to Raman scattering occurring slightly off axis. We shall therefore write the fields as propagating in the forward and backward directions only. Finally, taking a single-plane wave mode pump and Stokes fields, yields:

$$E_p = \mathscr{E}_p(z, t) \exp(ik_p z - i\omega_p t) \tag{4.18a}$$

$$E_s = \mathscr{E}_s^f(z, t) \exp(ik_s z - i\omega_s t), + \mathscr{E}_s^b(z, t) \exp(-ik_s z - i\omega_s t) \tag{4.18b}$$

$$Q = Q^f(z, t) \exp(ik_v^f z - i\omega_v t) + Q^b(z, t) \exp(ik_v^b z - i\omega_v t) \ . \tag{4.19}$$

Here the $\mathscr{E}_p(z, t)$, $\mathscr{E}_s^{f,b}(z, t)$, and $Q^{f,b}(z, t)$ represent the amplitudes of the slowly varying complex fields, with the superscripts f and b standing for the forward and backward waves, respectively. In the small signal limit, the amplitude of the incident laser beam may be regarded as a constant, so $\mathscr{E}_p(z, t)$ is treated as a parameter. At high pump intensities significant depletion of the pump wave as well as conversion of the first Stokes wave into higher order Stokes and anti-Stokes components may occur. The preceding approximations must be altered accordingly [4.14, 20, 22, 64].

The growth of the Stokes field for forward scattering may be described by substituting (4.19) into (4.13, 17). We then have the frequency and phase-matching criteria

$$\omega_v = \omega_p - \omega_s \, , \tag{4.20a}$$

$$k_v^f = k_p - k_s \, , \tag{4.20b}$$

and the coupled differential equations describing the electric field and material vibration become

$$\frac{1}{v_s} \frac{\partial \mathscr{E}_s^f}{\partial t} + \frac{\partial \mathscr{E}_s^f}{\partial z} = iK_2 (Q^f)^* \, \mathscr{E}_p(z, t) \, , \tag{4.21a}$$

$$\frac{\partial (Q^f)^*}{\partial t} + \Gamma (Q^f)^* = -iK_2 \mathscr{E}_s^f \mathscr{E}_p^* (z, t) \, . \tag{4.21b}$$

In (4.21) v_s is the Stokes wave group velocity and

$$K_1 = \frac{N}{2\omega_v} \left(\frac{\partial \alpha}{\partial Q} \right)_0 \, , \tag{4.22a}$$

$$K_2 = \frac{2\pi N \omega_s^2}{c^2 k_s} \left(\frac{\partial \alpha}{\partial Q} \right)_0 \, . \tag{4.22b}$$

In a like-fashion the coupled differential equations for the growth of the backward Stokes wave may be written. The frequency and phase-matching conditions are

$$\omega_v = \omega_p - \omega_s \, , \tag{4.23}$$

$$k_v^b = k_p + k_s \tag{4.24}$$

and the coupled equations become

$$\frac{1}{v_s} \frac{\partial \mathscr{E}_s^b}{\partial t} - \frac{\partial \mathscr{E}_s^b}{\partial z} = iK_2 (Q^b)^* \, \mathscr{E}_p(z, t) \, , \tag{4.25a}$$

$$\frac{\partial (Q^b)^*}{\partial t} + \Gamma (Q^b)^* = -iK_1 \mathscr{E}_s^b \mathscr{E}_p^* (z, t) \, . \tag{4.25b}$$

Note from (4.21 b, 25 b) that the vibrational wave propagates in the forward direction for both forward and backward Stokes generation. The phonon wave vector in the forward scattering case is small, while for backward scattering the wave vector is approximately twice the wave vector of the incident pump laser.

Equations (4.21, 25) may be solved in the so-called steady state regime by some simple approximations. Generally the relaxation time for Raman scattering in liquids and solids ranges from 10^{-13} to 10^{-11} s [4.65 – 72] and in gases is of the order of 10^{-9} s [4.73]. Raman scattering with nanosecond input laser pulses may be regarded, in the absence of effects such as self-focusing, as a quasi steady state process. In that case the incident laser field may be taken as a parameter independent of distance or time (i.e., $\mathscr{E}_p(z, t) \simeq \mathscr{E}_p$) and the time derivatives of the molecular vibration and the Stokes waves may be omitted. Equations (4.21, 25) may then be solved, implying exponential growth for both the forward and backward Stokes waves as

$$\mathscr{E}_s^{f, b} = \mathscr{E}_s(0) \exp(G_{ss} z/2) \quad \text{and} \tag{4.26}$$

$$G_{ss} = \frac{2K_1 K_2 |\mathscr{E}_p|^2}{\Gamma} = \frac{2\pi \omega_s^2 N^2}{c^2 \omega_v k_s \Gamma} \left(\frac{\partial \alpha}{\partial Q}\right)_0^2 |\mathscr{E}_p|^2 . \tag{4.27}$$

Equation (4.27) is the standard formula derived for the Stokes power gain coefficient in the steady state regime.

The Stokes power gain coefficient may also be expressed in terms of the on-resonance Raman susceptibility. One then has [4.41]

$$P_s = \chi_R^{(3)} \mathscr{E}_p \mathscr{E}_p^* \mathscr{E}_s \quad \text{where} \tag{4.28a}$$

$$\chi_R^{(3)} = \frac{2n_s c K_1 K_2}{\omega_s \Gamma} \quad \text{and} \tag{4.28b}$$

$$G_{ss} = \frac{\omega_s}{n_s c} \chi_R^{(3)} |\mathscr{E}_p|^2 . \tag{4.28c}$$

Before proceeding it is useful to comment further on the conditions which must be satisfied for (4.27) to hold. As was stated before, the characteristic time scale τ_p for fluctuations in the pump laser should be long compared to the molecular relaxation time (i.e., the response time of the Raman transition, $T_2 = 1/(2\Gamma)$; this implies $\tau_p \gg T_2$). Following the analysis of *Hanna* et al. [4.35], we introduce the retarded time $\tau = t - z/v_p$, where v_p is the group velocity in the medium of the pump field. The small signal Stokes gain is then

$$I_s(\tau, z) = I_{s0} \exp[g_{ss} I_{p0}(\tau) z] \tag{4.29}$$

with $I_s = cn |\mathscr{E}_s|^2/8\pi$. One must also require that the time scale for the fluctuations of the generated Stokes field be greater than T_2. This requirement yields the inequality

$$\Gamma = \frac{1}{T_2} \gg \frac{1}{I_2(\tau, z)} \frac{\partial I_s(\tau, z)}{\partial \tau} = g_{ss} \frac{\partial I_{p0}(\tau) z}{\partial \tau} . \tag{4.30}$$

But, recall

$$\frac{\partial I_{p0}(\tau)z}{\partial \tau} = I_{p0}(\tau)z/\tau_p \tag{4.31}$$

so we have the stronger requirement that

$$\tau_p \gg g_{ss}I_{p0}(\tau)zT_2 = (G_{ss}z)T_2 . \tag{4.32}$$

Equation (4.32) may be interpreted as a statement that any pump field fluctuations are amplified by the factor $G_{ss}z$ in the Stokes field.

As an example of the use of (4.32), consider stimulated Stokes generation near threshold; that is one noise Stokes photon is amplified coherently to generate exp(30) Stokes photons. Hence $G_{ss}z = 30$ and $\tau_p \gg 30\,T_2 = 15/\Gamma$. If one assumes a value typical of gaseous Raman systems, $2\Gamma = 0.1$ cm^{-1} (4.32) implies $\tau_p \gg 1.6$ ns. This condition is usually satisfied for transform-limited pump-laser pulse widths of greater than 10 ns. Even if some laser fluctuations are present, they tend to be averaged out to a large degree.

4.3.2 Two-Photon Vector Model

In this subsection the equations of motion for stimulated Raman scattering will be derived using a more modern formalism via the so called "two-photon vector model" as given by *Takatsuji* [4.74] and *Grischkowsky* et al. [4.75]. This formalism serves as a bridge between the earlier nonlinear optics theories summarized in the proceeding section and the modern Block vector or optical resonance picture. We shall follow the derivation of *Raymer* et al. [4.76] in this subsection.

For simplicity consider again the case of Stokes scattering in the presence of an intense input laser field. The Raman medium will be an idealized three-level system, shown in Fig. 4.2, such that electric dipole transitions are

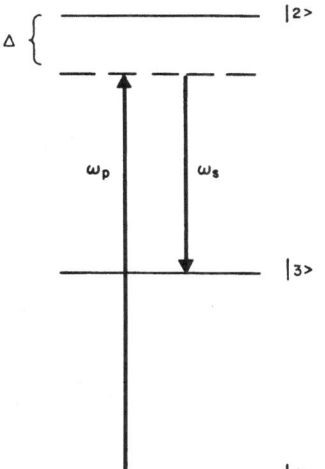

Fig. 4.2. Three-state energy level scheme for Raman scattering (Stokes-wave generation) used in the two-photon vector model

possible between states $|1\rangle$ and $|2\rangle$ and states $|2\rangle$ and $|3\rangle$. The incident pump laser is tuned near the $1-2$ transition, and the resulting Stokes field is generated near the $2-3$ transition. As before, in the small signal limit it is assumed that the incident laser field is not significantly depleted by the scattering process and, hence, may be treated as a parameter. The gain experienced by the Stokes wave may be calculated using time-dependent perturbation theory in conjunction with Maxwell's equations.

We begin by considering the interaction of the pump and Stokes fields in the three-level atom. The electric fields may again be represented in the form:

$$E_{\mathrm{p}} = \mathscr{E}_{\mathrm{p}} \exp(\mathrm{i}k_{\mathrm{p}}z - \mathrm{i}\omega_{\mathrm{p}}t) + \mathrm{c.c.} \quad , \tag{4.33a}$$

$$E_{\mathrm{s}} = \mathscr{E}_{\mathrm{s}} \exp(\pm\mathrm{i}k_{\mathrm{s}}z - \mathrm{i}\omega_{\mathrm{s}}t) + \mathrm{c.c.} \quad , \tag{4.33b}$$

where $\mathscr{E}_{\mathrm{p}}$, $\mathscr{E}_{\mathrm{s}}$ are the amplitudes of the pump and Stokes waves, respectively. The plus, minus signs refer to forward, backward propagating waves. The wave function for the atom may be written using time-dependent perturbation theory as

$$\psi = a_1 \mathrm{e}^{-\mathrm{i}\omega_1 t}\psi_1 + a_2 \mathrm{e}^{-\mathrm{i}\omega_2 t}\psi_2 + a_3 \mathrm{e}^{-\mathrm{i}\omega_3 t}\psi_3 \quad , \tag{4.34}$$

where the a_i are the slowly varying coefficients in the so called "rotating frame" [4.74 – 79]. The ψ_i are the stationary eigenstates of the unperturbed atomic Hamiltonian H_0 with energies $E_i = \hbar\omega_i$, such that

$$H_0|\psi_i\rangle = E_i|\psi_i\rangle \quad . \tag{4.35}$$

At time $t = 0$ the atom is assumed to be in the ground state, so $a_1 = 1$, $a_2 = a_3 = 0$. For $t > 0$, the atomic state evolves as

$$\frac{\partial a_1}{\partial t} = \mathrm{i}\Omega_{\mathrm{p}}a_2 \quad , \tag{4.36a}$$

$$\frac{\partial a_2}{\partial t} = \mathrm{i}\Omega_{\mathrm{p}}a_1 + \mathrm{i}\Omega_{\mathrm{s}}a_3 - \mathrm{i}\Delta a_2 \quad , \tag{4.36b}$$

$$\frac{\partial a_3}{\partial t} = \mathrm{i}\Omega_{\mathrm{s}}a_3 \quad . \tag{4.36c}$$

Here the Raman detuning is $\Delta = \omega_{21} - \omega_{\mathrm{p}}$, and the Rabi frequencies are defined in the usual manner as

$$\Omega_{\mathrm{p}} = \mu_{12}\mathscr{E}_{\mathrm{p}}/2\hbar \quad , \tag{4.37a}$$

$$\Omega_{\mathrm{s}} = \mu_{23}\mathscr{E}_{\mathrm{s}}/2\hbar \tag{4.37b}$$

where μ_{12}, μ_{23} are the dipole matrix elements connecting the $1-2$ and $2-3$ transitions, respectively. Equations (4.36) are valid for small detunings, i.e., for $\Delta \ll \omega_{21}$.

Equations (4.36) may be greatly simplified by the elimination of level $|2\rangle$. This is the basis for the "two-photon vector model" [4.74–79]. Rewriting (4.36b) we have

$$a_2(t) = \int_0^t e^{-i\Delta(t-t')} g(t') \, dt' \quad \text{with} \tag{4.38a}$$

$$g(t) = i\Omega_p a_1(t) + i\Omega_s a_3(t) \ . \tag{4.38b}$$

Integration by parts of (4.38) yields

$$a_2(t) = \frac{g(t) - e^{-i\Delta t} g(0)}{i\Delta} - \frac{\dot{g}(t) - e^{-i\Delta t} \dot{g}(0)}{(i\Delta)^2} - \cdots \tag{4.39}$$

where $a_1(0) = 1$ and $a_2(0) = 0$, $g(0) = i\Omega_p$. Now for $\Delta \gg \Omega_p$, Ω_s, one has $\dot{g}(t) \ll \Delta g(t)$ [4.76]. Hence, for large Δ

$$a_2(t) = \frac{g(t) - i\Omega_p e^{i\Delta t}}{i\Delta} \tag{4.40}$$

where the exponential term is rapidly oscillatory with respect to $g(t)$. In the framework of the rotating-wave approximation [4.76–78] this rapidly oscillating component is neglected, thereby retaining only the slowly varying amplitude, giving

$$a_2(t) = g(t)/i\Delta = (\Omega_p a_1(t) + \Omega_s a_3(t))/\Delta \ . \tag{4.41}$$

Raymer et al. [4.76] have shown that (4.41) may also be simply derived by noting that for large Δ the fields have no appreciable Fourier components at the atomic frequencies; hence, one may make the approximation $\dot{a}_2(t) \simeq 0$. This leads directly to the result of (4.41).

Substitution of (4.41) into (4.36a, b) yields two equations for the time evolution of a_1 and a_3. These expressions are similar to equations describing a one-photon excitation between a two-level system with an effective Rabi frequency $\Omega = 2\Omega_p \Omega_s/\Delta$ and an effective ac Stark shifted detuning of $\Delta' = (\Omega_p - \Omega_s)/\Delta$. These equations may be expressed in the Bloch pseudospin formalism [4.76–78] as

$$\frac{\partial U}{\partial t} = -\Delta' V - \Gamma U \ , \tag{4.42a}$$

$$\frac{\partial V}{\partial t} = \Delta' U + \Omega' W - \Gamma V \ , \tag{4.42b}$$

$$\frac{\partial W}{\partial t} = -\Omega' V \tag{4.42c}$$

where we define $U + iV = 2a_1 a_3^*$ and $W = a_3 a_3^* - a_1 a_1^*$. A phenomenological collisional dephasing rate Γ corresponding, as before, to the Raman linewidth between states $|1\rangle$ and $|3\rangle$ has been included.

The growth of the Stokes field is again calculated by solving the wave equation (4.11) with a driving polarization of the form $P = N\langle \psi | e \cdot r | \psi \rangle$. Use of (4.41) in conjunction with the slowly varying envelope approximation yields for the Stokes field

$$\frac{1}{v} \frac{\partial \mathscr{E}_s}{\partial t} \pm \frac{\partial \mathscr{E}_s}{\partial z} = -K_2 \mathscr{E}_p V \quad \text{with} \tag{4.43a}$$

$$K_2 = \frac{N \pi \omega_s v \mu_{12} \mu_{23}}{c^2 \hbar \Delta} . \tag{4.43b}$$

Here again the plus, minus signs correspond to the forward, backward propagating Stokes waves. The material excitation may be described by assuming for the inversion $W = -1$ and $\dot{W} = 0$ in (4.42). Then for exact resonance, $\Delta' = 0$, (4.42a, b, 43a, b) may be combined in conjunction with the definition $Q = U + iV$ to yield

$$\frac{1}{v} \frac{\partial \mathscr{E}_s}{\partial t} \pm \frac{\partial \mathscr{E}_s}{\partial z} = -iK_2 Q^* \mathscr{E}_p , \tag{4.44a}$$

$$\frac{\partial Q^*}{\partial t} + \Gamma Q^* = iK_1 \mathscr{E}_p^* \mathscr{E}_s , \tag{4.44b}$$

where $K_1 = \mu_{12} \mu_{23} / 2 \hbar^2 \Delta$. These two equations are identical to the coupled differential equations for the Raman process derived in the previous section. Recalling (4.28), we have another approach by which the Raman gain and susceptibility may be cast in terms of the electric dipole matrix elements and the Rabi frequencies of the three-level system.

4.3.3 Stokes/Anti-Stokes Coupling and Higher-Order Scattering

In the previous subsections, the theoretical treatment of the stimulated Raman process was simplified by assuming that only the pump and first Stokes fields were present in the medium. It is common, however, for Stokes and anti-Stokes waves (Fig. 4.1) to be generated simultaneously and to be coupled to one another and the pump field through a variety of four-wave parametric interactions. This situation complicates treatment of SRS and leads to some interesting physical phenomena. As the pump laser intensity is increased, intense higher-order Stokes, and through parametric mixing, anti-Stokes orders can be generated. Higher-order Stokes and anti-Stokes generation in H_2 is just one example of a widely employed source of tunable coherent radiation in the laboratory.

The theoretical treatment of these higher-order effects is in principle straightforward. As before one generates a series of coupled equations describing the various electric fields and their respective driving polarization terms, as by (4.11). In practice, however, the number of interacting waves can be quite large, leading to a complex physical situation whose analytical solution is intractable. Numerical solutions are available in many cases, and we shall briefly summarize some of the more important results.

Let us first consider the simplest case representing the next step in complication from the discussions of Sects. 4.3.2, 3; namely the interaction of the pump, first Stokes, and first anti-Stokes waves. The interaction of light at frequencies ω_p, $\omega_p \pm \omega_v$ must therefore be considered. For simplicity we assume that light at higher Stokes and anti-Stokes frequencies (i.e., $\omega_p \pm n\omega_v$, for $n \geqslant 2$) is absorbed so that any interactions due to these waves may be ignored. Also, although a polarization at $\omega_p + 3\omega_v$ might be created by an interaction involving $2\omega_a - \omega_s$, such a process has an unfavorable resonant denominator and yields a negligible susceptibility. The interaction is therefore one involving only the coupling of three waves: ω_p, ω_s, ω_a. Even with these approximations an analytical solution of the problem is impossible, and in general the treatment is further simplified by assuming, as before, that the pump field is not depleted and may be treated as a parameter [4.20, 22, 80].

The growth of the Stokes and anti-Stokes waves has now been reduced to the solution of the coupled-wave equations for E_s and E_a^*. Following the treatment of *Bloembergen* [4.80] and *Shen* [4.22], we assume (with no real loss in generality) that the medium is isotropic and that all the interacting waves are polarized in the same direction allowing the use of scalar Raman susceptibilities. A real, nonresonant susceptibility χ_{NR} is added to the Raman susceptibilities; the dispersion of χ_{NR} is negligible since the other electronic levels are distant compared to $\hbar\omega_a - \hbar\omega_s$.

The two coupled equations describing E_s and E_a^* may then be written as

$$\hat{e}_s \cdot \left(\nabla^2 E_s - \frac{\varepsilon_s}{c^2} \frac{\partial^2 E_s}{\partial t^2} \right) = \frac{4\pi}{c^2} \frac{\partial^2}{\partial t^2} \{ (\chi_s^{(3)} + \chi_{NR}^{(3)}) |E_p|^2 E_s$$

$$+ [(\chi_a^{(3)*} \chi_s^{(3)})^{1/2} + \chi_{NR}^{(3)}] E_p^2 E_a^* \} , \qquad (4.45a)$$

$$\hat{e}_a \cdot \left(\nabla^2 E_a^* - \frac{\varepsilon_a^*}{c^2} \frac{\partial^2}{\partial t^2} E_a^* \right) = \frac{4\pi}{c^2} \frac{\partial^2}{\partial t^2} \{ [(\chi_a^{(3)*} \chi_s^{(3)})^{1/2} + \chi_{NR}](E_p^*)^2 E_s$$

$$+ [\chi_a^{(3)*} + \chi_{NR}^{(3)}] |E_p|^2 E_a^* \} . \qquad (4.45b)$$

The relationship between $\chi_s^{(3)}$ and $\chi_a^{(3)}$ is given in [4.80], and $(\chi_s^{(3)} \chi_a^{(3)*})^{1/2} = \chi^{(3)}(\omega_s = 2\omega_p - \omega_a) = \chi^{(3)}(\omega_a = 2\omega_p - \omega_s)$. Note that in the limit of negligible dispersion $(\chi_s^{(3)} \chi_a^{(3)*})^{1/2} = \chi_s^{(3)}$.

Shen and *Bloembergen* [4.14] have solved (4.45) assuming an isotropic medium with a plane boundary at $z = 0$. Making use of the slowly varying envelope approximation for E_s and E_a, the solutions become

$$E_s = [\, \mathscr{E}_{s+} \exp(i\Delta K_+ z) + \mathscr{E}_{s-} \exp(i\Delta K_- z)]\, \exp(i\mathbf{k}_s \cdot \mathbf{r} - \alpha_s z)\ , \tag{4.46a}$$

$$E_a = [\, \mathscr{E}_{a+}^* \exp(i\Delta K_+ z) + \mathscr{E}_{a-}^* \exp(i\Delta K_- z)$$

$$\cdot \exp(-i\mathbf{k}_a \cdot \mathbf{r} - i\Delta k z - \alpha_a z)\ , \tag{4.46b}$$

where

$$k^2 = \omega^2 \varepsilon'/c^2\ ,$$

$$2k_{p(x,\,y)} = k_{s(x,\,y)} + k_{a(x,\,y)}\ ,$$

$$\Delta k = 2k_{p(z)} - k_{s(z)} - k_{a(z)}\ , \qquad k_{(i)} \equiv \mathbf{k} \cdot \hat{\mathbf{e}}_i\ ,$$

$$\Delta K_\pm = \Delta k/2 \pm [(\Delta k/2)^2 - (\Delta k)\lambda]^{1/2}\ ,$$

$$\lambda = \frac{2\pi \omega_s^2}{c^2 k_{s(z)}} (\chi_s^{(3)} + \chi_{NR}^{(3)})\,|E_p|^2\ ,$$

$$\alpha_z = \omega^2 \varepsilon''/c^2 k_z\ .$$

The general relationship between the wave vectors given by (4.46) is shown in Fig. 4.3. Taking the values of E_s and E_a at $z = 0$ as $E_s(0)$ and $E_a(0)$ we have

$$\mathscr{E}_{a\pm} / \mathscr{E}_{s\pm} = (\Delta K_\pm - \lambda)/\lambda \qquad \text{and} \tag{4.47}$$

$$\mathscr{E}_{s\pm} = [(-\Delta K_\mp + \lambda)\,\mathscr{E}_s(0) + \lambda \mathscr{E}_a^*(0)]/(\Delta K_- - \Delta K_\mp)\ .$$

The solution of (4.46, 47) leads to several interesting phenomena. First, in the limit of large phase mismatch Δk, such that $\Delta k \gg (2\pi \omega_s^2/c^2 k_{s(z)})\chi_s^{(3)}|E_p|^2$, the Stokes and anti-Stokes waves are essentially decoupled from one another. The two parts of the solution are thereby reduced to

$$\Delta K_- = \lambda\ ,$$

$$|\mathscr{E}_a^* / \mathscr{E}_s| = |\lambda/\Delta k| \ll 1\ , \tag{4.48a}$$

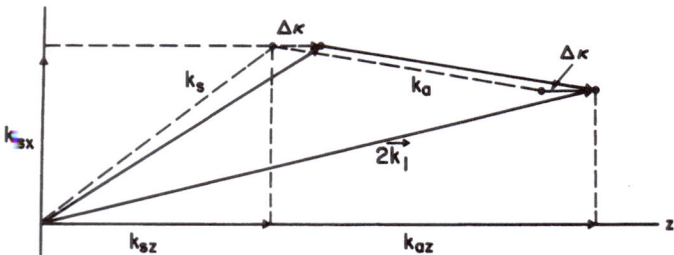

Fig. 4.3. Relationship between the Stokes, anti-Stokes, and pump laser wave vectors, as in (4.46) [4.15]

$$\Delta K_+ = -\lambda + \Delta k \; ,$$

$$|\mathscr{E}_a^* / \mathscr{E}_s| = |\Delta k / \lambda| \gg 1 \; , \qquad\qquad (4.48\,\mathrm{b})$$

where $-2\,\mathrm{Im}\{\lambda\} = G_{ss}$, the familiar steady state Raman gain coefficient. Equation (4.48 a) implies a Stokes wave with exponential gain G_{ss}, while (4.48 b) implies an anti-Stokes wave with exponential loss G_{ss}. With no coupling between the Stokes and anti-Stokes waves, the anti-Stokes wave experiences loss since it effectively couples to and amplifies the pump field.

In the case where $\Delta k = 0$ (i.e., linear phase matching), then $\Delta K_\pm = 0$ and $|\mathscr{E}_{a\pm}^* / \mathscr{E}_{s\pm}| = 1$. Although this case corresponds to exact phase matching with maximum coupling between the Stokes and anti-Stokes waves, the two fields effectively trade energy back and forth so that neither field experiences exponential gain.

Finally, for small $\Delta k \neq 0$, the positive exponential gain increases rapidly to G_{ss} and $|\mathscr{E}_a^* / \mathscr{E}_s|$ decreases to 0. These situations are summarized in Fig. 4.4. The anti-Stokes power goes through a maximum for $|\Delta k|$ near 0, as shown in Fig. 4.5. The anti-Stokes radiation generated by this parametric coupling process should appear as a double cone in k-space. In practice, however, laboratory laser sources used to generate anti-Stokes radiation have beams with a finite spread in k-space. This tends to obscure the dark band that one might expect in the anti-Stokes ring. Experimental observations [4.81] indicate that the anti-Stokes radiation is emitted with a cone angle given by the

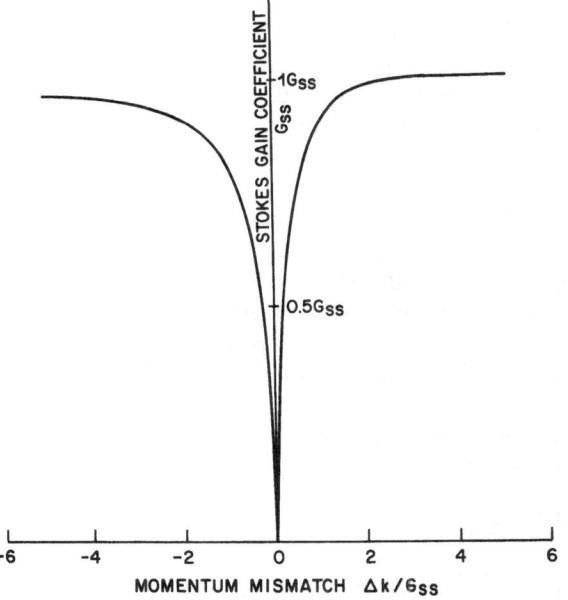

Fig. 4.4. Stokes power gain versus the normalized momentum mismatch in the $\hat{z}$ direction. The asymmetry is due to the nonresonant part of the susceptibility [4.14]

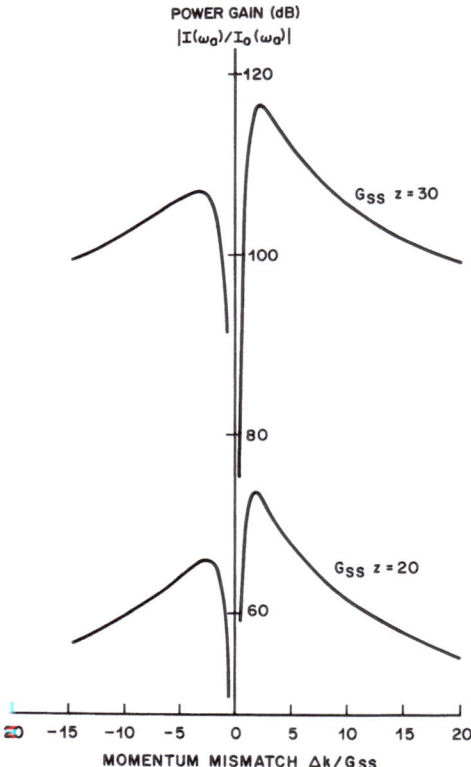

POWER GAIN (dB)
$|I(\omega_0)/I_0(\omega_0)|$

G_{ss} z = 30

G_{ss} z = 20

MOMENTUM MISMATCH $\Delta k/G_{ss}$

Fig. 4.5. Anti-Stokes intensity versus the momentum mismatch, normalized by G_{ss}. The asymmetry is again due to the nonresonant part of the susceptibility [4.15]

phase-matching condition $k_a^n = k_a^{n-1} + k_p - k_s^n$ where n is the scattering order. Additional experimental work by *Garmire* [4.82, 83] showed that two sets of cones were observable. One set occurred in the expected phase-matching directions, while a second set occurred in an anomalous direction possibly due to surface phase matching along filaments created by self-focusing of the input laser field.

As the input intensity of the input field is raised, many higher-order Stokes and anti-Stokes fields may be generated; and the previous simplified treatment for only ω_p, ω_s, and ω_a must be expanded and generalized. The higher-order waves are generated by successive, nonlinear, third-order polarizations at the appropriate frequencies. A large system of coupled differential equations for each field must be developed, and the solution is in general quite difficult.

As an example consider the driving polarization for the second Stokes wave ω_{s2} which can take the form

$$P^{(3)}(\omega_{s2}) = \chi_{s\alpha}^{(3)} E_{s1}^2 E_p^* + \chi_{s\beta}^{(3)} E_p E_{s1} E_{a1}^*$$

$$+ \chi_{s\gamma}^{(3)} |E_{s1}|^2 E_{s2} + \chi_{s\delta}^{(3)} E_{s1} E_{a1} E_{a2}^* . \tag{4.49}$$

In a like-fashion one can write polarizations responsible for the generation of the higher-order Stokes and anti-Stokes fields. The driving polarization for each field is then used in the nonlinear wave equation; the solution of the resulting coupled system is indeed formidable.

A special case in which only higher-order Stokes generation was permitted was considered by *van der Linde* et al. [4.64]. The coupled differential equations for Stokes generation were solved numerically for two limiting cases. In the first case the input pump and Stokes fields were assumed to have constant intensities over the cross section of the beam; that is, a single-mode, plane-wave interaction was assumed for all the fields. The results of this calculation are shown in Fig. 4.6. We see that as the input laser intensity is increased the first Stokes component S_1 gradually increases until an apparent threshold is reached and a rapid increase in S_1 occurs. The first Stokes power then saturates at a maximum value; note that the input pump laser is largely converted into S_1. As the pump intensity is increased further, the first Stokes field is rapidly depleted and converted into the second Stokes field S_2. This process continues to higher Stokes orders as I_p is increased.

In most laboratory situations, of course, the input laser intensity is a maximum at the beam center and decreases towards the edges. A second numerical calculation, assuming Gaussian intensity profiles for the interacting waves, is shown in Fig. 4.7. This calculation is interesting since, unlike the result of Fig. 4.6, at high-input intensities several Stokes components of approximately equal intensity are simultaneously generated.

The prediction of Fig. 4.7 was verified by *von der Linde* et al. [4.64] in a careful experiment using CS_2 as the Raman medium. Brillouin scattering in the CS_2 was effectively suppressed by choosing subnanosecond (i.e., $t_p \sim 0.5$ ns) input laser pulses, and the effects of self-focusing were controlled by using high-input intensities. The data from this experiment are plotted in Fig. 4.7, and excellent agreement between the theory and the experiment is apparent.

More recent calculations have considered the parametric interaction of the pump wave with many higher orders of the Stokes and anti-Stokes waves [4.84 – 86]. *Eimerl* et al. [4.84] have numerically solved the coupled differential equations describing the parametric Stokes, anti-Stokes interaction for Raman generation in H_2 gas, one of the most widely employed Raman medium (Sect. 4.5.1). In their analysis the multiwave equations were solved exactly in the limit that the total bandwidth was small compared to the input laser frequency. The results of the calculation are shown in Fig. 4.8, where strong parametric coupling is assumed to dominate the scattering process. For Raman scattering in H_2, this condition is easily met provided that two input waves (i.e., a strong pump and a weak Stokes or anti-Stokes line) are incident on the Raman cell [4.84]. The intensity of the nth order Stokes or anti-Stokes wave is then

$$I_n(t) = I_p(t) J_n^2(z \,|\, \beta \,|) \;, \qquad\qquad (4.50)$$

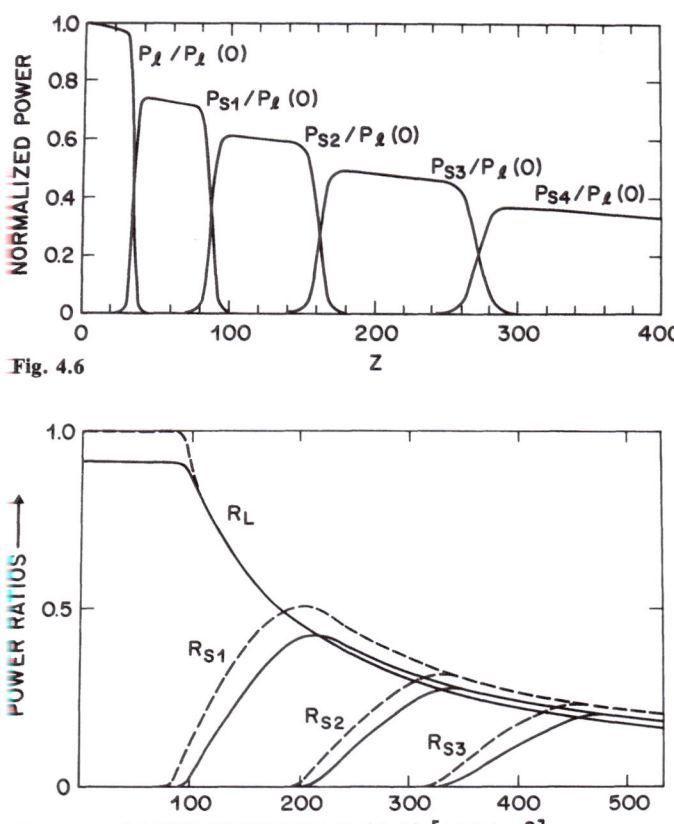

Fig. 4.6

Fig. 4.7   LASER INTENSITY $I_L(0,0)$ [MW/cm^2] ⟶

Fig. 4.6. Growth of Stokes components versus distance for a single-mode pump beam of infinite extent. The Stokes intensities are shown normalized to the incident pump intensity and are plotted as a function of the dimensionless distance parameter $z = [16\pi^3 \omega_p^2 \mathrm{Im}\{x_R\}/c^3 K_1] P_p(0) z$ [4.15]

Fig. 4.7. Calculated power ratios for the Stokes components versus incident pump intensity for a Gaussian beam. The Stokes powers are normalized to the incident pump power with a cell length of 30 cm. Linear optical absorption has been included in the calculation, with $\alpha = 3 \times 10^{-3}$ cm^{-1} (————) and $\alpha = 10^{-4}$ cm^{-1} (– – –) [4.64]

where J_n is the nth order Bessel function, and the parameter $z|\beta|$ is the parametric gain. In this limit the maximum predicted conversion efficiency to the nth order wave (either Stokes or anti-Stokes) is approximately $1/n$. In practice only the pump wave is incident on the Raman cell and pure parametric interactions with complete suppression of the so-called cascade scattering (i.e., the two-wave, nonphase-matched interaction) is not achieved. Conversion efficiencies to higher anti-Stokes lines range more typically from $(0.1)^n$ to $(0.2)^n$ for the nth order wave [4.87 – 91].

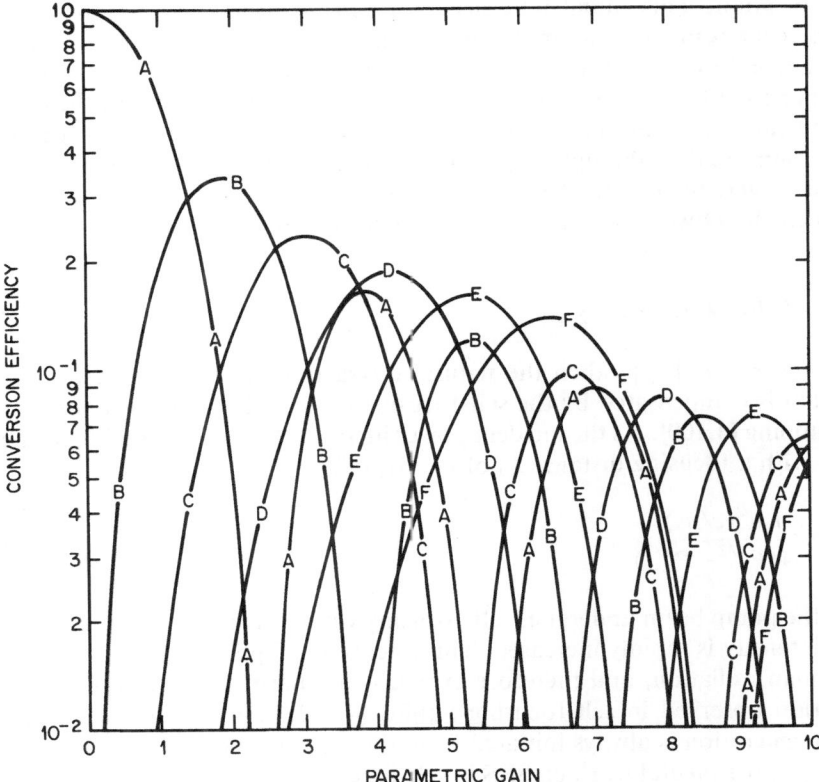

Fig. 4.8. Conversion efficiency to various Stokes and anti-Stokes orders as a function of the parametric gain, $z\,|\beta|$. The pump is labeled "A", the S_1 and AS_1 waves "B", the S_2 and AS_2 waves "C", etc. [4.84]

4.3.4 Self-Focusing Effects

Early studies of Raman scattering were for the most part conducted using liquids with large Kerr constants, such as nitrobenzene and CS_2. Several anomalous effects were soon evident in these early studies. First the experimentally observed pump intensity threshold for stimulated Raman scattering was often a factor of 10 or more below the theoretically predicted value [4.92 – 94]. In addition an asymmetry in the forward-backward Stokes waves [4.6], anomalous anti-Stokes rings [4.6, 82, 83, 95, 96], and spectral broadening of the output [4.6] was observed. These effects were subsequently understood as resulting from self-focusing of the pump laser beam [4.97 – 99].

Self-focusing is due to the interaction of an input laser beam of finite diameter with a medium of positive field induced refractive index. For most purposes the incident field may be modeled as having a radial Gaussian intensity distribution such that $I(r) = I_0 \exp(-r^2/2a^2)$, a being the HWHM intensity point. This radial intensity variation describes a variation in the refractive index in the medium that the beam experiences; the intense center of the beam induces a larger refractive index and propagates slower than the outer portion of the beam. This variation in refractive index refracts the laser beam towards the propagation axis. The laser beam creates its own lens within the medium; and as the light begins to concentrate, the transverse gradients become larger, further accelerating the effect.

When the powder density of the incident laser reaches a critical level P_c given by

$$P_c = 5.763 \ \lambda^2 c/16 \pi^3 n_2 \ , \tag{4.51}$$

where $n = n_0 + n_2 | \mathcal{E}_p |^2$, then the natural diffraction effect which defocuses the beam is compensated by the self-focusing effect. This effect is known as self-trapping [4.100]. As the incident power increases beyond P_c, self-focusing occurs with a focusing distance L_f of [4.101, 102]

$$L_f = \frac{n_0 a^2 (c/n_2)^{1/2}}{4(P^{1/2} - P_c^{1/2})} \ . \tag{4.52}$$

When the pump beam undergoes self-focusing within the Raman medium, the pump intensity is rapidly increased. This intense focal point readily generates the Raman radiation, and therefore explains the sharp threshold for Stokes generation observed in self-focusing liquids [4.102, 103]. In these cases the Raman generation is always initiated from the self-focused region.

Early experimental work on SRS in self-focusing liquids also uncovered an anomalous asymmetry in the power generated in the forward and backward Stokes waves. As we saw in Sect. 4.3.1, simple theoretical predictions imply that the gain coefficient for the forward and backward Stokes waves should be equal. This was found not to be the case, however, as is illustrated in Fig. 4.9, where the generated power in the forward and backward directions in a toluene cell was measured [4.104]. This effect is likewise explained in terms of self-focusing of the pump beam. If, for example, the pump beam experiences strong self-focusing near one end of the Raman medium, then the forward and backward generated waves will have different effective interaction lengths and will experience different overall Raman gain. The problem of forward-backward scattering is further complicated since the Stokes wave generated in the forward direction travels along in the medium with the pump field that is being depleted with distance. In contrast, the backward Stokes wave continues to encounter a "fresh" undepleted pump beam. This effect can lead to pulse compression in backward scattering [4.48, 105].

Spectral broadening of the pump laser and the generated Raman output was also observed in several early experiments [4.106, 107]. The strong self-modulation of the incident laser caused by self-focusing leads to the observed

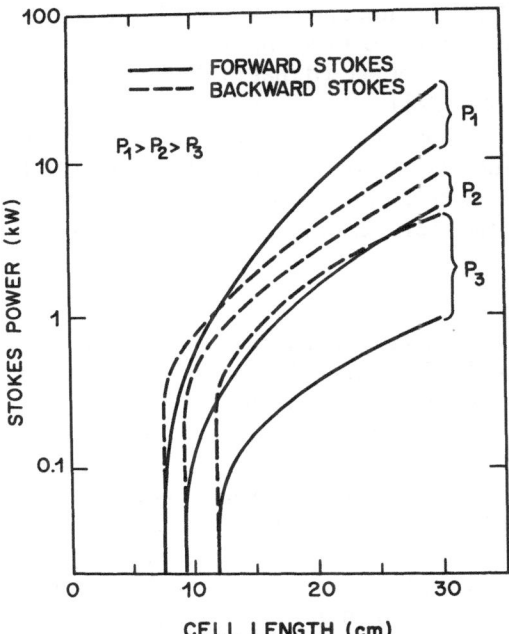

spectral broadening. Effects related to self-focusing and the intensity dependent refractive index are addressed in greater detail by *Bloembergen* [4.12].

4.4 Stimulated Raman Scattering: Transient Limit

When the pulse width of the incident laser is shorter than the material dephasing time, i.e., $\tau_p \ll T_2$, the steady state solutions for the Raman gain coefficient derived in Sect. 4.3 are no longer valid. In this case transient effects become important, and the scattered wave must be treated as a function of both space and time. As will become evident shortly, the transient Raman effect is characterized by a reduction in the gain seen by the Stokes wave, a time delay between the maximum intensity points of the pump and scattered waves, and a reduction in the pulse width of the scattered wave.

The theory of transient SRS has been addressed by many authors [4.16, 17, 20, 46 – 48, 108 – 110]. For ultrashort pulses, the material relaxation time of the medium is no longer important, and a variety of interesting coherent effects such as self-induced transparency and pulse breakup can occur. A discussion of these later effects is beyond the scope of this chapter, however various aspects of ultrashort scattering are addressed by several authors [4.74, 111 – 115].

In the following subsections we shall examine a fundamental difference for transient SRS for the forward and backward scattered waves. In the absence of strong dispersive effects, the pump and Stokes waves travel in the medium with approximately equal group velocities, and as such the intensity of the forward Stokes component can never exceed that of the incident pump intensity. The backward generated Stokes component, however, always encounters the full, undepleted pump pulse. This may lead to substantial pulse sharpening, and the intensity of the leading edge of the Stokes pulse can be amplified to many times that of the incident pump pulse. These considerations and some experimental studies of the transient limit are discussed next.

4.4.1 Forward Scattering

The theory of forward transient Raman scattering has been discussed in great detail by many groups, in particular by *Carmen* et al. [4.109] and *Akhmanov* et al. [4.108]. The forward scattered case can be treated analytically for a medium of negligible dispersion, and is important as being the dominant scattering process even in the limit of ultrashort input pulses. Following the formalism of *Carmen* et al. [4.109], we shall treat the incident laser field as a prescribed, time-dependent parameter, and the population difference of the initial and final Raman states is assumed to be constant. These assumptions imply that effects due to pump field depletion or saturation of the Raman medium will be ignored. Finally, in order to simplify the analysis, only the coupling between the pump and first Stokes fields will be considered.

The starting point for the analysis is as before, the coupled equations describing the evolution of the electric fields and the material excitation. The electric fields are again described via the nonlinear wave equation (4.11), and the material vibration is given by (4.14). For single-mode, plane-wave fields, à la (4.18, 19), the coupled equations take on the familiar form of (4.21), derived in a former section.

If the Raman medium has negligible dispersion, we may assume to good approximation that the pump pulse and resulting Stokes field propagate with the same group velocity. This implies

$$\mathscr{E}_p(z, t) = \mathscr{E}_p(t - z/v_S) = \mathscr{E}_p(t') \ . \tag{4.53}$$

Now introduce the spatial coordinate transformation

$$z' = z - v_{ph}t \ , \tag{4.54}$$

where v_{ph} is the group velocity of the optical phonon. The term v_{ph} is small since $v_{ph}/c \sim 10^{-10}$, and may be ignored yielding $z' \simeq z$. Equation (4.21) may be written in terms of the variables t' and z' yielding

$$\frac{\partial \mathscr{E}_s^f}{\partial z'} = iK_2(Q^f)^* \, \mathscr{E}_p(t') \tag{4.55a}$$

$$\frac{\partial (Q^f)^*}{\partial t'} + \Gamma(Q^f)^* = -iK_1 \mathscr{E}_s^f \mathscr{E}_p^*(t') \ . \tag{4.55b}$$

One may eliminate either $(Q^f)^*$ or $\mathscr{E}_s$ from (4.55), such that both $(Q^f)^*$ and $\mathscr{E}_s^f \mathscr{E}_p^{-1}(t')$ obey the same second-order, hyperbolic partial differential equation, namely [4.109]

$$\frac{\partial^2 F}{\partial t' \partial z'} + \Gamma \frac{\partial F}{\partial z'} = K_1 K_2 |\mathscr{E}_p(t')|^2 F = 0 \qquad (4.56)$$

where F stands for either $(Q^f)^*$ or $\mathscr{E}_s^f \mathscr{E}_p^{-1}(t')$. Making yet another variable change, $F = U\exp(-\Gamma t')$, (4.56) becomes

$$\frac{\partial U}{\partial t' \partial z'} - K_1 K_2 |\mathscr{E}_p(t')|^2 U = 0 \ . \qquad (4.57)$$

We may reduce (4.57) to an equation with constant coefficients by the transformation to a new variable τ, as

$$\tau(t') = \int_{-\infty}^{t'} |\mathscr{E}_p(t'')|^2 dt'' \quad \text{and} \qquad (4.58a)$$

$$\frac{d\tau}{dt'} = |\mathscr{E}_p(t')|^2 \ . \qquad (4.58b)$$

The variable τ is a measure of the integrated pump pulse energy up to the time t'. This last transformation casts (4.57) in a standard hyperbolic form, given by

$$\frac{\partial^2 U}{\partial \tau \partial z'} - K_1 K_2 U = 0 \ . \qquad (4.59)$$

Equation (4.59) may be solved for arbitrary initial conditions using Riemann's method [4.16]. In the present analysis, however, we are interested in the special case in which there is no material excitation at the beginning of the laser pulse i.e., $\partial \mathscr{E}_s^f / \partial z' = Q^{f*}(z') = 0$ for $t \to -\infty$. Furthermore if the Stokes wave at the entrance to the Raman medium $z' = z = 0$ is described by $\mathscr{E}_s^f(0, t')$, then the solution to (4.55 – 59) may be written as

$$\mathscr{E}_s^f(z, t') = \mathscr{E}_s^f(0, t') + (K_1 K_2 z)^{1/2} \mathscr{E}_p(t') \int_{-\infty}^{t'} \exp[-\Gamma(t' - t'')]$$

$$\times [\mathscr{E}_p^*(t'') \mathscr{E}_s^f(0, t'') [\tau(t') - \tau(t'')]^{1/2}$$

$$\times I_1(2\{K_1 K_2 z' [\tau(t') - \tau(t'')]\}^{1/2})] dt'' \ , \qquad (4.60a)$$

$$(Q^f)^* = iK_1 \int_{-\infty}^{t'} \exp[-\Gamma(t' - t'')] [\mathscr{E}_p^*(t'') \mathscr{E}_s^f(0, t'')$$

$$\times I_0(2\{K_1 K_2 z' [\tau(t') - \tau(t'')]\}^{1/2})] dt'' \ . \qquad (4.60b)$$

Here τ is defined by (4.58) and $I_n(x)$ is the nth order Bessel function of imaginary argument. The above solutions are valid for arbitrary shapes of the incident laser and Stokes pulses. A numerical solution of (4.60) is necessary for most cases; however, the maximum Stokes gain from (4.60a) is governed by the asymptotic value of the Bessel function I_n. For $x \to \infty$, $I_n(x) \to (2\pi x)^{-1/2} e^x$, so that in the limit of large transient power gain $\exp(G_T)$, the transient gain coefficient is

$$G_T = \ln |(\mathscr{E}_s^f)_{max} / \mathscr{E}_s^f(0)|^2 \simeq 4 \left[K_1 K_2 z \int_{-\infty}^{t'} |\mathscr{E}_p(t'')|^2 dt'' \right]^{1/2}. \qquad (4.61)$$

This result implies that the transient gain is proportional to the square root of both the length of the Raman medium and the integrated laser energy. This is in contrast to the steady-state solution in which the Raman gain was proportional to only the instantaneous pump intensity.

An example of the evolution of the Stokes pulse and material excitation is shown schematically in Fig. 4.10 for a rectangular pump pulse of duration t_p. In the limit of large gain, the evolution of the Stokes field is given by [4.109]

$$\mathscr{E}_s^f(z, t') \simeq \mathscr{E}_s^f(0, t') \exp(4 K_1 K_2 z |\mathscr{E}_p|^2 t')^{1/2} \qquad (4.62)$$

for $0 < t < t_p$. The Stokes field grows rapidly from zero as the leading edge of the laser pulse passes through the medium. Immediately following the trailing edge of the laser pulse, the Stokes pulse drops rapidly to zero; the maximum Stokes amplitude occurs for $t = t_p$. Notice that the peak of the Stokes pulse always occurs after the peak of the pump pulse; the Stokes pulse width at half-maximum is given by $(2 \ln 2) t_p / G_T$. The material excitation follows a similar dependence at the leading edge of the laser pulse, but decays as $\exp(-\Gamma t)$ after the laser pulse passes through the medium. This "flywheel" effect of the molecular excitation is important when considering the amplification of a broadband input pulse.

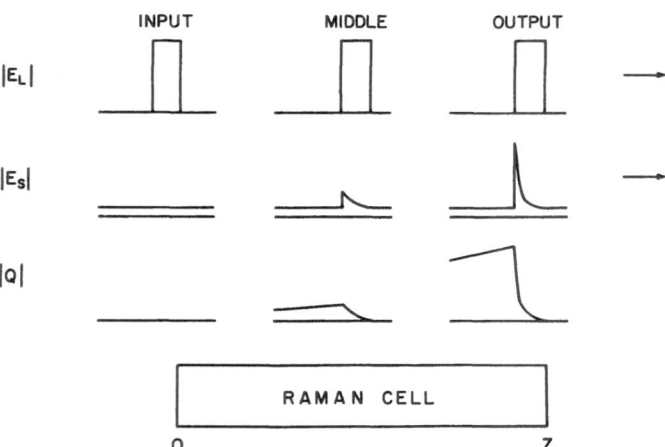

Fig. 4.10. Schematic diagram illustrating the spatial evolution of the pump laser, Stokes, and phonon wave amplitudes in parts of the Raman cell for transient forward scattering [4.109]

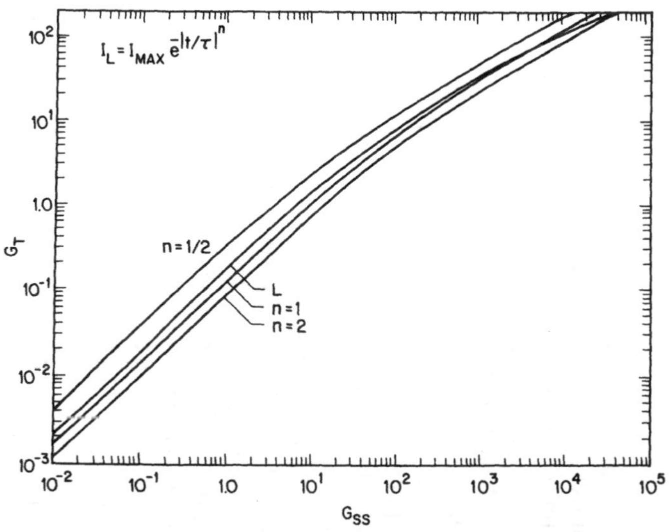

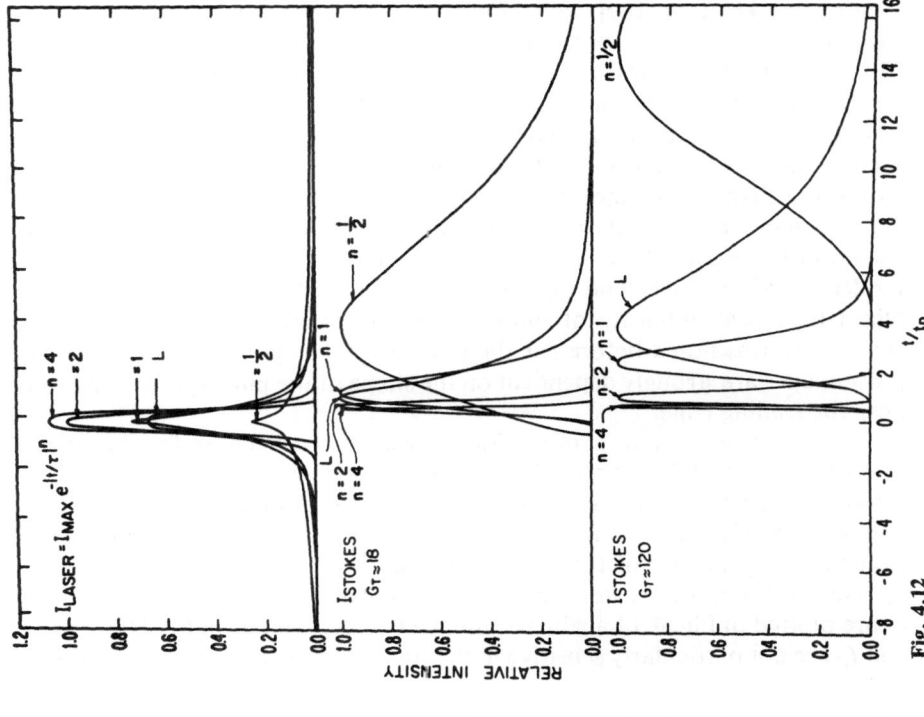

Fig. 4.11. Transient Raman gain coefficient G_T versus the steady state gain G_{ss}, which would be calculated for a constant pump intensity equal to the maximum pulse intensity. The different shaped pulses are normalized to have constant total energy and equal pulse width at half-maximum amplitude [4.109]

Fig. 4.12. Stokes pulse shapes for different laser pulse shapes (at constant laser energy) shown at top. Stokes pulse shapes for moderate gain are shown in the middle, and for high gain at the bottom [4.109]

Carmen et al. [4.109] have solved (4.60) numerically for a variety of incident laser pulse shapes. The transient Raman gain coefficient for a incident pulse of the form $I_p(t') = I_{max} \exp(-|t'/\tau|^n)$ is shown in Fig. 4.11. Here the pump pulse width was chosen as $t_p = 0.1/\Gamma$, and the laser pulses were normalized for equal integrated energy. A Lorentzian pulse shape was also included in the calculation. Note that the transient gain G_T is largely insensitive to the shape of the incident laser pulse.

The shapes of the Stokes pulses resulting from this type of pump pulse are shown in Fig. 4.12, where again $t_p = 0.1/\Gamma$. The Stokes pulses are shown for moderate gain $G_T = 18$ and high gain $G_T = 120$; the pulses have been normalized to a peak intensity of unity to aid the comparison of their relative shapes. As was noted before for the case of a rectangular input, the Stokes pulse shapes are strongly dependent on the shape of the pump pulse, especially near the trailing edge.

The transient Raman gain coefficient was also calculated for the special case of an incident Gaussian pulse (i.e., $n = 2$) with constant total energy but different pulse widths t_p. These results are shown in Fig. 4.13. For $t_p \leqslant \Gamma^{-1}$, one sees that G_T is less than G_{SS}. At high gain, Fig. 4.13 shows the transition to the $z^{1/2}$ dependence of the gain coefficient.

The variations of the resulting Stokes pulse width t_s and relative time delay t_D are plotted in Fig. 4.14 again assuming a Gaussian input pulse. Note that t_s and t_D are not particularly sensitive to the transient gain coefficient for $G_T < 1$.

4.4.2 Backward Scattering

Backward transient stimulated Raman scattering has very different behavior compared to the forward scattering case and leads to some interesting physical phenomena. For incident laser pulses of the order $t_p \sim 10^{-10} - 10^{-9}$ s such that the pulse duration is comparable to the length of the Raman medium (in a typical laboratory case), part of the backward Stokes wave can be amplified to intensities much larger than that of the pump field. This effect was first observed by *Maier* et al. [4.105], and subsequently the theory of this effect was developed by the same group [4.48]. As the incident pulse length is further shortened to ultrashort durations the backward Stokes wave ceases altogether, since the interaction length becomes too small.

We shall briefly examine the problem of backward transient scattering following the formalism of *Maier* et al. [4.48]. The starting point of this discussion is again the coupled-wave equations for the backward generated wave (4.25). Since the backward propagating Stokes wave always sees a high pump intensity, it is reasonable to expect that the pump beam may be largely converted into Stokes light. It is necessary, therefore to add a third equation describing the depletion of the pump field. The coupled system is then

$$\frac{1}{v_s} \frac{\partial \mathscr{E}_s^b}{\partial t} - \frac{\partial \mathscr{E}_s^b}{\partial z} = iK_2 (Q^b)^* \mathscr{E}_p(z, t) \; , \qquad (4.63a)$$

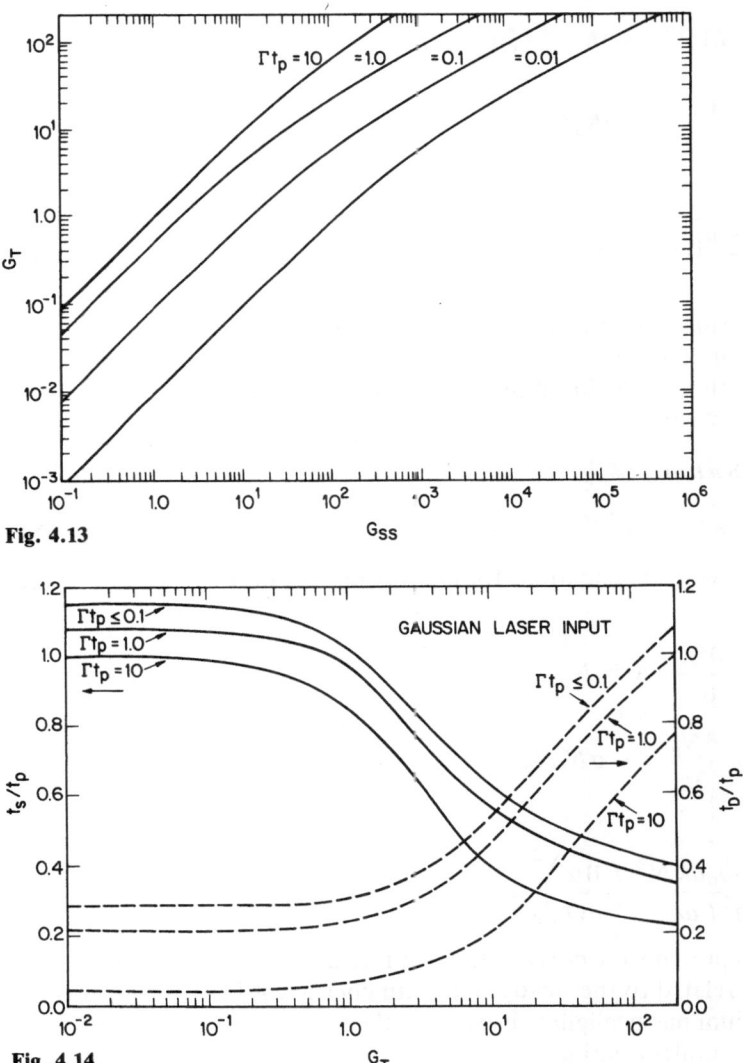

Fig. 4.13

Fig. 4.14

Fig. 4.13. Transient gain coefficient G_T for Gaussian input pulses at constant total energy but with different pulses widths [4.109]

Fig. 4.14. Variation of the Stokes pulse width t_s and delay t_D versus transient gain coefficient, plotted for Gaussian input pulses of width t_p in terms of the optical phonon dephasing time $1/\Gamma$ [4.109]

$$\frac{\partial (Q^b)^*}{\partial t} + \Gamma (Q^b)^* = iK_1 \mathscr{E}_s^b \mathscr{E}_p^* (z, t) \; , \tag{4.63 b}$$

$$\frac{1}{v_p} \frac{\partial \mathscr{E}_p}{\partial t} + \frac{\partial \mathscr{E}_p}{\partial z} = -iK_3 Q^b \mathscr{E}_s^b \tag{4.63 c}$$

where

$$K_3 = \frac{2 \pi N \omega_p^2}{c^2 k_p} \left(\frac{\partial \alpha}{\partial Q} \right)_0 . \tag{4.64}$$

If the molecular vibrations are heavily damped, such that $\partial Q / \partial t \ll \Gamma Q$, the vibrational amplitude can be eliminated from (4.63). We now express the relations for the fields in terms of the photon flux density (i.e., flux $\mathrm{cm}^{-2} \mathrm{s}^{-1}$), N, where

$$N_p = (cn/8 \pi \hbar \omega_p) \, | \mathscr{E}_p |^2 \; , \tag{4.65 a}$$

$$N_s = (cn/8 \pi \hbar \omega_s) \, | \mathscr{E}_s^b |^2 \; . \tag{4.65 b}$$

This yields two coupled equations for the flux density in the pump and Stokes pulses as

$$\frac{1}{v_s} \frac{\partial N_s}{\partial t} - \frac{\partial N_s}{\partial z} = \sigma N_s N_p \; , \tag{4.66 a}$$

$$\frac{1}{v_p} \frac{\partial N_p}{\partial t} + \frac{\partial N_p}{\partial z} = -\sigma N_s N_p \tag{4.66 b}$$

where

$$\sigma = \frac{8 \pi^2 \hbar \omega_p \omega_S N^2}{c^2 n^3 \Gamma \omega_v} \left(\frac{\partial \alpha}{\partial Q} \right)_0^2 . \tag{4.67}$$

The factor σ represents the cross section for interaction of a pump and Stokes photon and is related to the steady state gain coefficient by $\sigma = G_{SS}/N_p$.

If the medium has negligible dispersion, then $v_s \simeq v_p = c/n$. We introduce the coordinate transformation

$$\varrho = t - nz/c \; , \tag{4.68 a}$$

$$\varepsilon = t + nz/c \tag{4.68 b}$$

thereby casting (4.66) into the form

$$\frac{\partial N_s}{\partial \varrho} = \left(\frac{\sigma c}{2n} \right) N_S N_p \; , \tag{4.69 a}$$

$$\frac{\partial N_p}{\partial \varepsilon} = -\left(\frac{\sigma c}{2n} \right) N_S N_p \; . \tag{4.69 b}$$

Because N_p may be eliminated, and by rearranging and interchanging the order of integration, one obtains

$$\frac{\partial}{\partial \varrho}\left[\frac{\partial \ln N_\mathrm{S}}{\partial \varepsilon}+\left(\frac{\sigma c}{2n}\right)N_\mathrm{S}\right]=0 \ . \tag{4.70}$$

Equation (4.70) can be integrated in two steps, first giving

$$\frac{1}{N_\mathrm{S}}\frac{\partial N_\mathrm{S}}{\partial \varepsilon}+\left(\frac{\sigma c}{2n}\right)N_\mathrm{S}=f(\varepsilon) \tag{4.71}$$

and then

$$N_\mathrm{S}(\varrho,\varepsilon)=\frac{\partial h/\partial \varepsilon}{(\sigma/2n)h(\varepsilon)+g(\varrho)} \ . \tag{4.72a}$$

Here

$$\partial h/\partial \varepsilon=\exp\left[\int_0^\varepsilon f(y)\,dy\right]$$

where $f(\varepsilon)$ and $g(\varrho)$ are arbitrary functions dependent on the initial conditions. The solution for N_p is obtained upon substitution of (4.72a) into (4.69b) to give

$$N_\mathrm{p}(\varrho,\varepsilon)=-\left(\frac{2n}{\sigma c}\right)\frac{\partial g(\varrho)/\partial \varrho}{(\sigma c/2n)h(\varepsilon)+g(\varrho)} \ . \tag{4.72b}$$

Equations (4.72) are the general solutions for transient backward scattering as given by *Maier* et al. [4.48] for input pulses of arbitrary shape.

The functions h and g may be determined by setting the boundary conditions of the system. The appropriate boundary conditions for Raman scattering are: (1) $N_\mathrm{S}=0$, $N_\mathrm{L}=N_\mathrm{L}(\varrho,0)$ for $\varepsilon\leqslant0$, and (2) $N_\mathrm{L}=0$, $N_\mathrm{S}=N_\mathrm{S}(0,\varepsilon)$ for $\varepsilon>0$. This choice fixes the pump laser photon density encountered by the leading edge of the Stokes flux, and likewise the Stokes photon density seen by the leading edge of the pump flux.

Substituting these conditions into (4.72a, b) leads to the results

$$h(\varepsilon)=\frac{2n}{\sigma c}\left[A\exp\left(\frac{\sigma c}{2n}D_\mathrm{S}(\varepsilon)\right)-g(0)\right] , \tag{4.73a}$$

$$g(\varrho)=g(0)+A\left[\exp\left(-\frac{\sigma c}{2n}D_\mathrm{L}(\varrho)\right)-1\right] \tag{4.73b}$$

where

$$D_\mathrm{S}(\varepsilon)=\int_0^\varepsilon N_\mathrm{S}(0,y)\,dy \ ,$$

$$D_\mathrm{L}(\varrho)=\int_0^\varrho N_\mathrm{L}(y,0)\,dy \ .$$

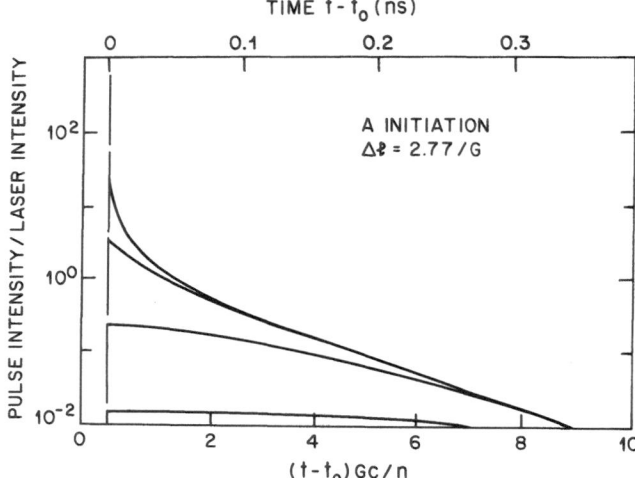

TIME $t - t_0$ (ns)

A INITIATION
$\Delta \ell = 2.77/G$

PULSE INTENSITY / LASER INTENSITY

$(t - t_0)$ Gc/n

Fig. 4.15. Calculated normalized Raman pulse intensity as a function of time t for a step-function excitation, where t_0 is the leading edge of the pulse. The pulse development is shown at length intervals of $\Delta l = 2.77/G$ [4.48]

Equations (4.73 a, b) may then be used in (4.72 a, b) yielding the final solution for the Stokes and pump photon flux densities, as

$$N_S(\varrho, \varepsilon) = \frac{N_S(0, \varepsilon) \exp[(\sigma c/2 n) D_S(\varepsilon)]}{\exp[(\sigma c/2 n) D_S(\varepsilon)] + \exp[-(\sigma c/2 n) D_L(\varrho) - 1]} \qquad (4.74\,a)$$

$$N_L(\varrho, \varepsilon) = \frac{N_L(\varrho, 0) \exp[-(\sigma c/2 n) D_L(\varrho)]}{\exp[(\sigma c/2 n) D_S(\varepsilon)] + \exp[-(\sigma c/2 n) D_L(\varrho) - 1]} \qquad . \qquad (4.74\,b)$$

Numerical solutions for (4.74) have been carried out by *Maier* et al. [4.48] for a variety of incident pump pulse shapes. A complete description of their results is beyond the space limitations of this discussion, however the following example is illustrative of the behavior encountered in transient backward SRS. The computer calculations for a step-function input Stokes pulse, such that $N_S(z = 0, t) = N_{S0}$ for $t > 0$ and $N_S(z = 0, t) = 0$ for $t < 0$ are shown in Fig. 4.15. Here the pump pulse was taken as $N_L(0, \varepsilon) = N_{L0}$ for $\varepsilon > 0$ and $N_L(0, \varepsilon) = 0$ for $\varepsilon \leqslant 0$. In this figure, t_0 is the leading edge of the pulse, and the pulse development is shown at intervals of length $\Delta l = 2.77/G$, where G is the Raman gain coefficient. Notice that in portions of the Stokes beam the intensity exceeds the pump intensity by a factor of 100. Naturally, however, the total energy in the Stokes pulse cannot exceed the total energy in the pump pulse.

4.5 Generation Techniques

Due largely to the ease with which stimulated Raman scattering occurs and a large selection of Raman media, stimulated Raman scattering is a widely employed technique for generating coherent radiation. An additional asset of the SRS method is that the output light is tunable by tuning the input laser source. This fact permits large regions of the spectrum which might otherwise be difficult to access to be covered with relative ease by SRS. (In most cases tuning is achieved by tuning the input field; a notable exception is the Raman spin-flip laser [4.29 – 31] in which the level splitting, and not the input field, is tuned.)

In most situations the apparatus necessary for Raman generation is simple, as illustrated in Fig. 4.16. The most complex component of such a system is usually the input laser itself and not the Raman cell. In order to achieve the high-input intensity necessary for Raman threshold, a focusing lens is normally used before the Raman medium. A collecting lens is sometimes used to recollimate the generated light, which can then be separated from the pump field with a prism or other dispersive element. Gaseous or liquid Raman media can be contained in fairly simple cells with appropriate input and output windows to accommodate the frequencies or static pressures involved.

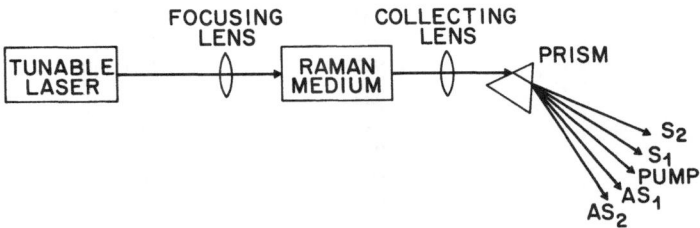

Fig. 4.16. Schematic diagram for basic apparatus used for Raman generation experiments

Apart from the simplicity of the SRS apparatus, the wide variety of Raman active materials makes the effect particularly useful. A sample of known Raman media is given in Table 4.1 for various liquids, solids, and gases. Detailed data for some of the more widely employed materials is given in Table 4.2, with measurements of the Raman linewidth, differential scattering cross section, and stimulated gain coefficient g_{SS}. Although data for some solid materials is given, we shall confine the present discussion to gaseous and liquid media.

Stimulated electronic Raman generation in a number of molecular and atomic vapors has not been included in these tables, since the Raman shift and gain coefficient depends on the particular choice of input laser wavelength and the material state. These considerations will be addressed in some of the following sections.

4.5.1 Molecular Systems

a) Gaseous H_2

Of the dozens of Raman materials studied and listed in Tables 4.1, 2, without question the most widely studied and utilized for practical Raman generation

Table 4.1. Some known Raman materials

Material	Frequency shift [cm^{-1}]	Reference
Liquids		
Bromoform	222	[4.116]
Tetrachloroethylene	447	[4.117]
CCl$_4$	460	[4.118]
Ethyl iodide	497	[4.119]
Hexafluorobenzene	515	[4.118]
Bromoform	539	[4.116]
Trichloroethylene	640	[4.116]
CS$_s$	655.6	[4.65, 116, 120]
Chloroform	667	[4.116]
o-Xylene	730	[4.5]
α-Dimethylphenethylamine	836	[4.121]
Dioxane	836	[4.116]
Morpholine	841	[4.118]
Thiophenol	916	[4.118]
Nitromethane	927	[4.118]
Deuterated benzene	944	[4.3, 14]
1,3-Dibromobenzene	990	[4.117]
Benzene	992	[4.3, 14, 16]
Pyridine	992	[4.3]
Aniline	997	[4.14]
Styrene	999	[4.14]
m-Toluidine	999	[4.118]
Bromobenzene	1 000	[4.14]
Chlorobenzene	1 001	[4.118]
Benzaldehyde	1 001	[4.117]
Benzonitrile	1 002	[4.14]
Ethylbenzene	1 002	[4.5, 14]
γ-Butylbenzene	1 002	[4.117]
Toluene	1 004	[4.3, 14]
Fluorobenzene	1 012	[4.14]
γ-Picoline	1 016	[4.118]
m-Cresol	1 029	[4.118]
m-Dichlorobenzene	1 030	[4.118]
1-Fluoro-2-chlorobenzene	1 034	[4.117]
Iodobenzene	1 070	[4.118]
Benzaldehyde	1 086	[4.118]
Benzoylchloride	1 086	[4.118]
Anisole	1 097	[4.118]
Pyrrole	1 178	[4.118]
Furan	1 180	[4.118]
Styrene	1 315	[4.14, 122]
Nitrobenzene	1 344	[4.3, 14, 16]
1-Bromonaphthalene	1 363	[4.3, 14]
1-Chloronaphthalene	1 368	[4.14]
Naphthalene	1 380	[4.14]

Table 4.1 (continued)

Material	Frequency shift [cm^{-1}]	Reference
Liquids		
2-Ethylnaphthalene	1 382	[4.117]
m-Nitrotoluene	1 389	[4.118]
Quinoline	1 427	[4.118]
Bromocyclohexane	1 438	[4.119]
Furan	1 522	[4.118]
Methylsalicylate	1 612	[4.118]
Cinnamaldehyde	1 624	[4.14]
Styrene	1 631	[4.14, 122]
3-Methylbutadiene	1 638	[4.14, 123]
1,3-Pentadiene	1 655	[4.14, 123]
Isoprene	1 792	[4.121]
1-Hexyne	2 116	[4.117, 118]
o-Dichlorobenzene	2 202	[4.118]
Benzonitrile	2 229	[4.14]
Acetonitrile	2 250	[4.119]
1,2-Dimethylaniline	2 292	[4.118]
Methylcyclohexane	2 817	[4.118]
Methanol	2 831	[4.116]
cis,trans-1,3-Dimethylcyclohexane	2 844	[4.117]
Tetrahydrofuran	2 849	[4.14]
Cyclohexane	2 852	[4.3, 14, 122]
cis-1,2-Dimethylcyclohexane	2 854	[4.117]
α-Dimethylphenethylamine	2 856	[4.121]
Dioxane	2 856	[4.116]
Cyclohexanone	2 863	[4.5, 14]
Cyclohexane	2 863	[4.3, 116]
cis,trans-1,3-Dimethylcyclohexane	2 870	[4.117]
cis-1,4-Dimethylcyclohexane	2 873	[4.117]
Cyclohexane	2 884	[4.3, 116]
Dichloromethane	2 902	[4.118]
Morpholine	2 902	[4.118]
2-Octene	2 908	[4.117]
Limonene	2 910	[4.121]
2,3-Dimethyl-1,5-hexadiene	2 910	[4.5, 14]
o-Xylene	2 913	[4.5, 14]
1-Hexyne	2 915	[4.5, 117]
Methane	2 916	[4.14]
Mesitylene	2 920	[4.121]
cis-2-Heptene	2 920	[4.117]
2-Bromopropane	2 920	[4.121]
Acetone	2 921	[4.5, 116]
Ethanol	2 921	[4.116]
cis-1,2-Dimethylcyclohexane	2 921	[4.117]
Carvone	2 922	[4.121]
Dimethylformamide	2 930	[4.116]

Table 4.1 (continued)

Material	Frequency shift [cm^{-1}]	Reference
Liquids		
2-Chloro-1-methylbutane	2931	[4.117]
2-Octene	2931	[4.117]
cis,trans-1,3-Dimethylcyclohexane	2931	[4.117]
Piperidine	2933	[4.14]
m-Xylene	2933	[4.5]
1,2-Diethyltartrate	2933	[4.5]
o-Xylene	2933	[4.5]
1,2-Diethylbenzene	2934	[4.117]
1-Bromopropane	2935	[4.117]
2-Chloro-2-methylbutene	2935	[4.117]
Tetrahydrofuran	2939	[4.14]
Piperidine	2940	[4.5]
Cyclohexanone	2945	[4.5, 14]
2-Nitropropane	2948	[4.117]
1,2-Diethylcarbonate	2955	[4.118]
1,2-Dichloroethane	2956	[4.118]
trans-Dichloroethylene	2956	[4.116]
1-Bromopropane	2962	[4.117]
2-Chloro-2-methylbutane	2962	[4.117]
α-Dimethylphenethylamine	2967	[4.121]
Dioxane	2967	[4.116]
α-Picoline	2982	[4.118]
o-Dichlorobenzene	2982	[4.118]
p-Chlorotoluene	2982	[4.118]
Bromocyclopentane	2982	[4.118]
Cyclohexanol	2982	[4.118]
Cyclopentane	2982	[4.118]
Cyclopentanol	2982	[4.118]
1,1,2,2-Tetrachloroethane	2984	[4.14]
p-Xylene	2988	[4.5]
o-Xylene	2992	[4.5]
Dibutylphthalate	2992	[4.118]
1,1,1-Trichloroethane	3018	[4.116]
Ethylene chlorhydrin	3022	[4.118]
Isophorone	3022	[4.118]
Nitrosodimethylamine	3022	[4.118]
Propylene glycol	3022	[4.118]
Cyclohexane	3038	[4.118]
Styrene	3056	[4.14, 122]
Benzene	3064	[4.3, 14]
tert-Butylbenzene	3064	[4.117]
1-Fluoro-2-chlorobenzene	3084	[4.117]
Turpentine	3090	[4.118]

Table 4.1 (continued)

Material	Frequency shift [cm^{-1}]	Reference
Liquids		
Acetic acid	3 162	[4.118]
Acetonylacetone	3 162	[4.118]
Methylmethacrylate	3 162	[4.118]
γ-Picoline	3 182	[4.118]
Aniline	3 300	[4.14]
H$_2$O	3 651	[4.118]
Gases		
Oxygen	1 552	[4.14]
Nitrogen	2 326.5	[4.14]
Methane	2 916	[4.10, 14]
Deuterium	2 991	[4.10, 14]
Hydrogen	4 155	[4.10, 14]
Solids		
Quartz	128	[4.124]
Lithium niobate	152	[4.18]
Lithium tantalate	201	[4.125]
α-Sulfur	216	[4.14]
Lithium niobate	256	[4.26, 28]
Quartz	466	[4.124]
α-Sulfur	470	[4.14]
Lithium niobate	637	[4.26, 28]
Barium sodium niobate	650	[4.125]
Calcium tungstate	911	[4.14]
Stilbene	997	[4.121]
Polystyrene	1 001	[4.116]
Calcite	1 084	[4.116]
Diamond	1 332	[4.14]
Naphthalene	1 380	[4.14]
Stilbene	1 591	[4.121]
Triglycine sulfate	2 422	[4.118]
Triglycine sulfate	2 702	[4.118]
Triglycine sulfate	3 022	[4.118]
Polystyrene	3 054	[4.116]

in the laboratory is high-pressure H$_2$ gas. There are several reasons why SRS in H$_2$ gas has achieved this prominence. First, H$_2$ is an inexpensive medium, readily available in a highly pure form. Being a gas at room temperature implies Raman cells of simple design and construction as well as easy control of the gas pressure, and hence number density of the Raman medium. Further-

Table 4.2. Quantitative data on selected Raman materials

Material (reference)	Frequency shift $[cm^{-1}]$	Line width $\Gamma\,[cm^{-1}]$	Cross section $d\sigma/d\Omega \times 10^8$ $[cm^{-1}\,ster^{-1}]$	Gain $g_{ss} \times 10^3$ $[cm/MW]$
CS_2 [4.120]	655.6	0.50	7.55	24
Benzene [4.120]	992	2.15	3.06	2.8
Bromobenzene [4.71]	1 000	1.9	1.5	1.5
Chlorobenzene [4.71]	1 002	1.6	1.5	1.9
Toluene [4.65, 92]	1 003	1.94	1.1	1.2
Nitrobenzene [4.71]	1 345	6.6	6.4	2.1
Liquid O_2 [4.120]	1 552	0.177	0.48	14.5
Liquid N_2 [4.120]	2 326.5	0.067	0.29	17
Gaseous H_2 [4.73]	4 155	0.2	–	1.5

more, the $4\,155\,cm^{-1}$ Raman shift for the Q_{01} (1) branch in H_2 is one of the largest available in a molecular system. The reasonably large gain cross section (Table 4.2) along with its low dispersion make H_2 an attractive Raman medium over a wide range of input wavelengths. Finally, H_2 gas has negligible absorption over large parts of the uv and vuv spectral regions, allowing for generation of tunable radiation via four-wave anti-Stokes generation to wavelengths in the deep vuv.

Stimulated Raman scattering in H_2 gas was first observed by *Minck* et al. [4.9] using a Q-switched ruby laser as a pump source. Since that time SRS in H_2 has been studied extensively, both in an effort to understand the fundamental physics of SRS and as a convenient laboratory source of coherent radiation. Much of the early work on SRS in H_2 was concerned with measurements of the Raman gain cross section and the Raman linewidth Γ. Gain measurements were conducted by *Lallemand* et al. [4.126] to investigate the Raman linewidth in the transition region in which Γ changes from being dominated by Doppler broadening to be determined by collisional broadening. In their experiments the linewidth for the Q_{01} (1) line was measured using back-scattered Stokes radiation, since the Doppler broadening for this case is some six times larger than for forward scattering. The experiment was conducted by taking the back-scattered Stokes radiation from an H_2 generator cell and then amplifying this radiation in a separate H_2 amplifier cell. The gain may then be calculated from the ratio of the amplified to incident Stokes amplitudes. Since in H_2 the Raman frequency is a function of pressure [4.126], the gain in the amplifier cell could be measured as a function of frequency by changing the pressure of the generator cell. The gain curve has the same frequency dependence as the spontaneous Raman transition, and as such the gain profile provides a direct measure of Γ. The results of their experiments are shown in Fig. 4.17. At low pressure the Raman linewidth is determined by Doppler broadening. As the pressure is increased the width of the backward scattered radiation shows a distinct line narrowing at about 10 amagat. This motional narrowing effect was first predicted by *Dicke*

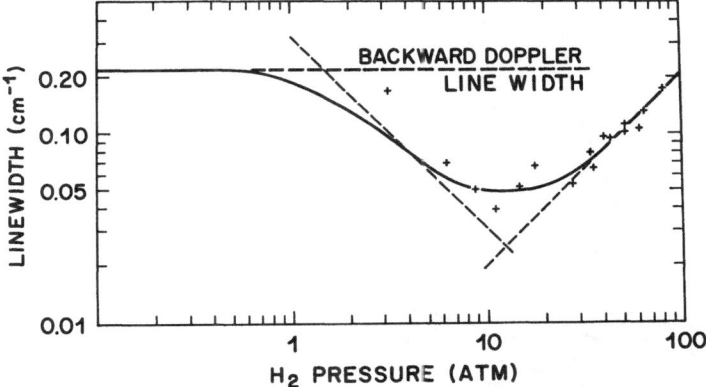

Fig. 4.17. Measured linewidth of the Q_{01} (1) vibrational Raman transition in H_2 as function of the gas pressure [4.126]

[4.127]. At still higher pressures, the Raman linewidth is determined by collisional dephasing, and Γ increases with pressure. The solid curve in Fig. 4.17 is a theoretical calculation by *Galatry* [4.128] accounting for these effects, and the experimental points are in good agreement with the calculation. Several authors have addressed other aspects of the collisional narrowing effect [4.129, 130].

Raman scattering in H_2 has been employed as a frequency conversion technique from the vuv to ir spectral regions. Figure 4.18 is illustrative of the versatility and wide tuning range possible for SRS with a tunable dye laser as an input pumping source. As one example, the output energies available in the various Stokes and anti-Stokes orders are shown [4.131]. The input source for these measurements was a dye laser operating at 560 nm with an energy of 85 mJ in an approximately 10 ns pulse. The excitation source for the dye laser was the second harmonic of a Q-switched Nd:YAG laser.

Tunable near and mid-ir generation using SRS in H_2 has been investigated by many authors. *Schmidt* and *Appt* [4.132] reported generation of tunable Raman scattering with a ruby laser pumped dye laser first in 1972. In later experiments [4.133] they extended the technique for generation in the near infrared. In that case a 580 nm pump source was utilized and S_1, S_2, S_2 were generated at wavelengths 0.76, 1.12 and 2.10 μm, respectively.

Shortly thereafter *Frey* and *Pradere* [4.134] generated tunable infrared radiation by scattering the output of a ruby laser pumped dye laser in H_2 and CH_4. Continuously tunable light was generated from 14 000 to 2 300 cm^{-1} (i.e., 0.71 to 4.3 μm). Output powers of 20 – 60 MW between 14 000 and 9 700 cm^{-1}, 5 – 20 MW between 9 700 and 5 500 cm^{-1}, and 0.5 – 2 MW between 5 500 and 2 300 cm^{-1} were generated. The linewidth of the Stokes radiation was 0.07 cm^{-1} with a beam divergence of 0.2 mrad. Conversion efficiencies ranging from 10% – 20% for the S_1 line were obtained by using a

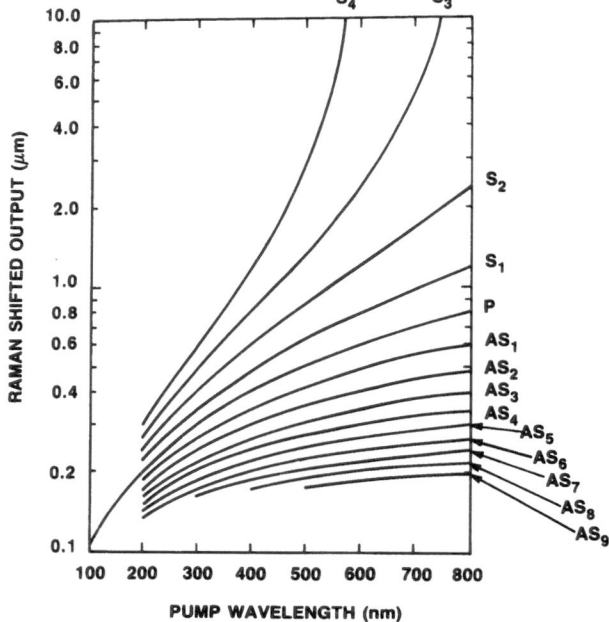

WAVELENGTH (nm)	ENERGY (mJ)	PRESSURE (psi)
195 (AS$_8$)	0.0031	125
213 (AS$_7$)	0.0091	125
234 (AS$_6$)	0.024	110
259 (AS$_5$)	0.054	115
290 (AS$_4$)	0.10	145
330 (AS$_3$)	0.26	160
382 (AS$_2$)	0.78	190
454 (AS$_1$)	2.1	200
730 (S$_1$)	17	90
1048 (S$_2$)	6.2	300
1855 (S$_3$)	0.60	275

Fig. 4.18. Stokes and anti-Stokes output wavelengths versus pump laser wavelength for Raman generation in high-pressure H_2 gas. The wavelengths and energies achieved for the various Raman components using an 85 mJ, 560 nm pump source are summarized in the table [4.131]

second H_2 Raman amplifier cell after the first H_2 cell, and amplification factors of 10^3 were achieved. This amplification technique is noteworthy as early evidence that high conversion efficiencies were readily obtainable without the need for external cavities or Raman oscillators.

Continuously tunable radiation from 3.5 to 13 μm was generated by *Brosnan* et al. [4.135] a few years later. In their work the 1.06 μm output from a Q-switched Nd:YAG laser was used to drive a LiNbO$_3$ optical parametric oscillator. With 100 mJ incident energy at 1.06 μm, the OPO generated a

10 mJ output tunable from 1.4 to 2.12 μm. This infrared OPO output was then utilized as the pump wave to a 1 m long H_2 Raman cell operating at a pressure of 20 atm. Output powers ranging near 10 kW in the ir were reported.

In an effort to increase the effective interaction length while maintaining a high pump intensity, several authors have investigated the use of a waveguide Raman cell [4.136 – 138]. *Rabinowitz* et al. [4.136] utilized a 51 cm long glass waveguide with 500 μm bore. This hollow waveguide was placed in an external H_2 pressure cell, and the input light was coupled into the waveguide through windows in this external cell. With a pump intensity of 1.35 GW/cm^2 at 1.064 μm a quantum efficiency of 0.6 was measured for the S_1 line at 1.9 μm. At the S_2 line (9.2 μm) a quantum efficiency of 0.014 was reported.

A waveguide Raman cell design was also employed by *Hartig* and *Schmidt* [4.137], and a schematic of their experimental apparatus is shown in Fig. 4.19. A 532 nm source (i.e., Nd:YAG second harmonic) was used to pump a tunable dye laser oscillator-amplifier system. The output of the dye laser was 10 ns in duration and tunable over 550 – 720 nm with a pulse energy between 20 and 35 mJ. This light was focused into the H_2 cell consisting of a 600 μm bore diameter, 75 cm long glass waveguide. Tunable ir radiation spanning 0.7 to 7 μm was generated by using the first three Stokes outputs. Typical output energies were 3 – 4 mJ at S_1, 2 mJ at S_2, and 0.4 mJ at S_3 for an input pulse energy of 30 mJ.

Finally, *May* and *Sibbett* [4.138] have studied the regime of transient stimulated Raman scattering using H_2 and CH_4 gases and a waveguide Raman cell. A passively mode-locked dye laser-amplifier system was utilized to generate 250 fs pulses with a peak power of ~2 GW tunable over the range 612 – 625 nm. A capillary with bore diameter 0.2 mm and total length 1 m was operated up to pressures of 20 atm. Output energies of 16 μJ and 0.2 μJ were

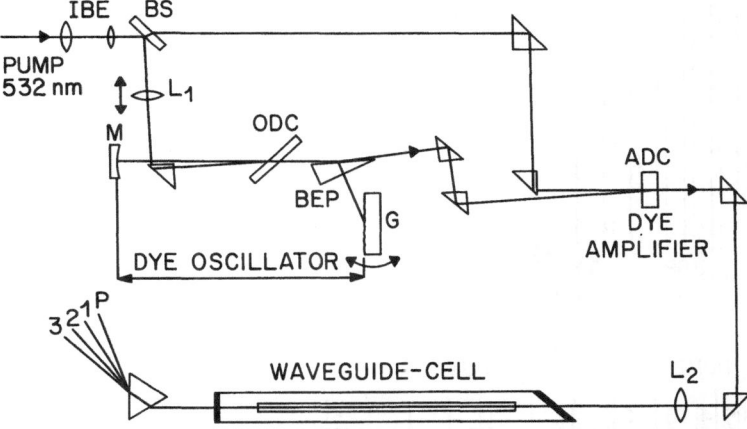

Fig. 4.19. Experimental apparatus for a dye laser pumped, waveguide Raman laser. The 532 nm pump laser excites a dye laser oscillator-amplifier, whose output is focused with lens L_2 into the waveguide Raman cell [4.137]

observed for S_1 (0.83 μm) and S_2 (1.26 μm), respectively. Pulse shortening of the Stokes waves by a factor of between 2.5 and 2.9 was seen, consistent with predictions for transient Raman scattering (Sect. 4.4.1), implying peak powers in excess of 270 MW for S_1 and 160 MW for S_2.

Stimulated Raman scattering in H_2 has also been utilized as a tunable source over large regions of the vuv, uv, and visible. *Wilke* and *Schmidt* [4.87] studied Raman generation using a Nd : YAG pumped rhodamine 590 dye laser system as the pump source. Either the dye laser fundamental or its second harmonic (created using an ADP frequency doubling crystal) were focused into a 20 cm long cell, run typically at pressures from $10-20$ atm. The dye laser pulse energy was 42 mJ in an 8.5 ns duration pulse at 560 nm; the energy conversion efficiency to the second harmonic was ~50% at 280 nm. In this configuration eight anti-Stokes and three Stokes orders were generated with the undoubled dye laser and were tunable over the $549-578$ nm tuning range of the dye laser. The shortest wavelength seen with the dye laser pump was AS_8 at 194 nm. With the doubled uv radiation five anti-Stokes and five Stokes orders were generated. The shortest wavelength seen in this case was 175 nm for AS_5. The output energies and tuning ranges for the various Raman orders are shown in Fig. 4.20.

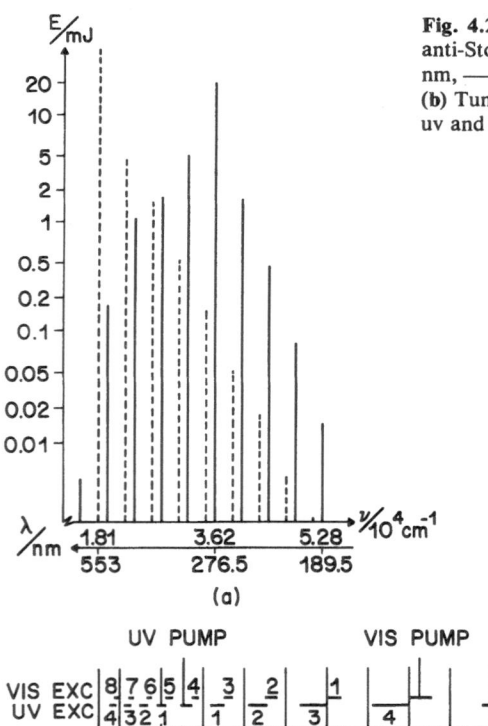

Fig. 4.20. (a) Pulse energies of various Stokes and anti-Stokes orders and pump energies for uv (276.5 nm, ———) and visible (553 nm, $- - -$) excitation. (b) Tuning ranges for the various Raman orders for uv and visible excitation. After [4.87]

In an extension of their earlier work, *Wilke* and *Schmidt* [4.139] employed a Nd: YAG excited dye laser system in which three different laser dyes were employed: rhodamine 590, rhodamine 610, and rhodamine 640. Again the dye laser second harmonic was also utilized to pump the Raman process. The pump energies ranged between 27 and 42 mJ per pulse at the dye fundamental wavelengths. Figure 4.21 shows the spectral coverage achieved using the three dye lasers and the second harmonic outputs. Note that the spectral region from 185 – 880 nm is covered without any gaps; the output powers ranged from approximately 1 kW to 1 MW per pulse with conversion factors between successive orders of 0.3. The pressure dependence of the output energies of the various anti-Stokes orders is shown in Fig. 4.22. These curves illustrate that the output for a particular AS order can be maximized by adjusting the H_2 pressure. Note also that with increasing AS order, the optimum gas pressure shifts downward due to the dispersion of H_2. The refractive index dispersion limits to a large degree the generation of still higher AS orders (see [4.139] for a complete discussion).

Related studies were also conducted by *Döbele* et al. [4.91] and *Schomburg* et al. [4.90]. A fluoresceine 27 dye laser tunable from 540 – 560 nm with peak pulse energies of 100 mJ was used as the Raman pump source. A standard Raman cell was employed and operated at pressures from 2 – 3 atm in order to optimize AS conversion. A total of 13 anti-Stokes orders were observed with 138 nm being the shortest wavelength seen. Energies in the long-wavelength vuv near 190 nm were 50 μJ per pulse corresponding to a conversion factor of 5×10^{-4}.

High-power, ultraviolet excimer laser systems have also been used by many authors as pump sources for H_2 Raman scattering. *Loree* et al. [4.140] initially studied Raman conversion with ArF (193 nm) and KrF (248 nm) pump sources. These lasers were operated with an unstable cavity resonator in an effort to improve the beam divergence and mode quality. Focal intensities from 1 – 5 GW/cm^2 were produced with this configuration. Hydrogen gas pressures from between 10 and 80 atm were used, and Raman orders from AS_2 at 206 nm to S_6 at 650 nm were observed with the KrF pump. Energy conversion efficiencies of 50% were measured at high pressures (~ 80 atm). Shortly thereafter *Loree* et al. [4.89] investigated Raman generation in a variety of simple molecular media, including H_2, with four pump wavelengths: ArF (193 nm), KrCl (223 nm), KrF (248 nm), and XeCl (308 nm). A variety of Stokes and anti-Stokes orders were generated ranging in wavelength over the 175 – 425 nm region. The observed conversion factor for each of the successive Stokes order was between 0.1 and 0.2.

Other studies of Raman shifting the ArF excimer laser were performed by *Hargrove* and *Paisner* [4.88], and *Döbele* and *Rückle* [4.141]. In [4.88] an unstable cavity resonator ArF oscillator-amplifier system was used to pump a 1 m H_2 cell operating at pressures up to 40 atm. With an input energy of 60 mJ per pulse, the intensity at the focus of the Raman cell exceeded 10^{11} W/cm^2. Energy conversion efficiencies of 30% were readily obtained at both S_1

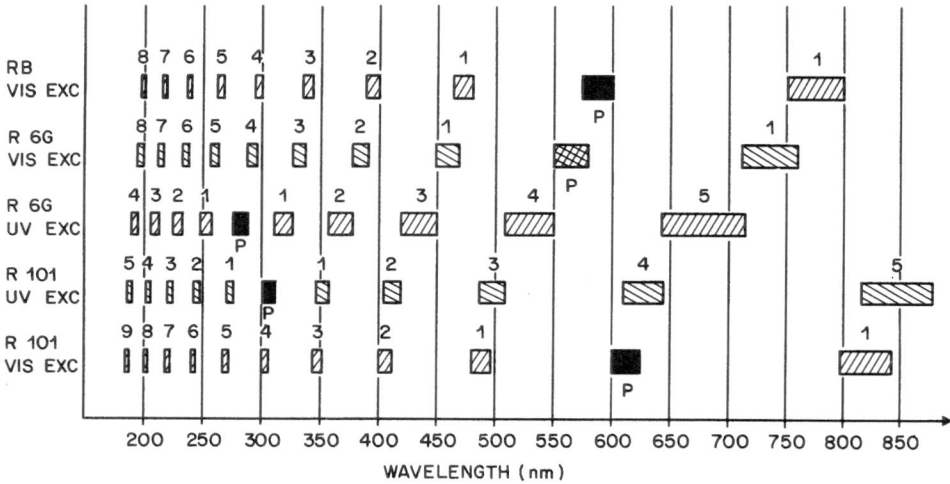

RB VIS EXC
R 6G VIS EXC
R 6G UV EXC
R 101 UV EXC
R 101 VIS EXC

WAVELENGTH (nm)

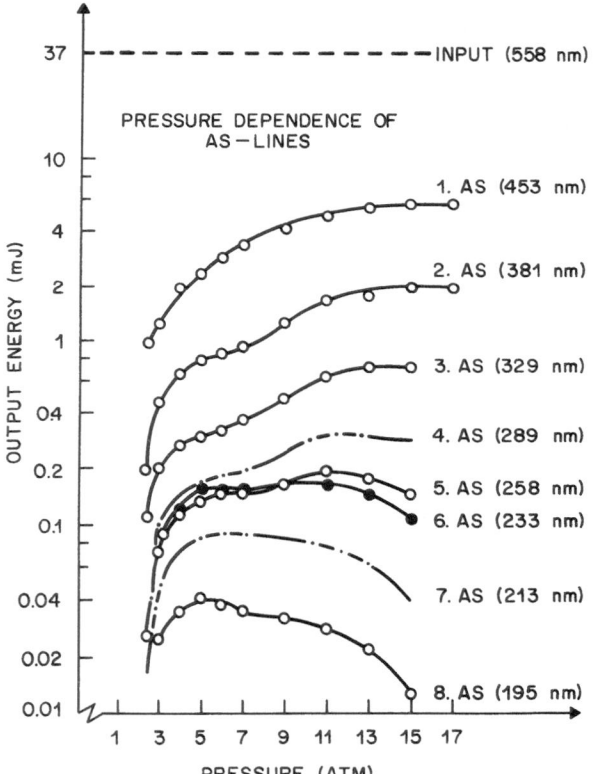

Fig. 4.21. Tuning ranges of Raman orders (*cross-hatched rectangles*) for different uv and visible excitation sources (*solid rectangles*) [4.139]

PRESSURE DEPENDENCE OF AS–LINES

INPUT (558 nm)

1. AS (453 nm)
2. AS (381 nm)
3. AS (329 nm)
4. AS (289 nm)
5. AS (258 nm)
6. AS (233 nm)
7. AS (213 nm)
8. AS (195 nm)

OUTPUT ENERGY (mJ)

PRESSURE (ATM)

Fig. 4.22. Pressure dependence of the anti-Stokes orders with visible excitation (pump: 558 nm) as a function of the H_2 gas pressure in the Raman cell [4.139]

(210 nm) and S_2 (230 nm); 5% conversion to AS_1 (179 nm) and weaker output at AS_2 (166 nm) was also seen.

In the experiments of *Döbele* and *Rückle* [4.141] a low divergence (<0.5 mrad) ArF system capable of 60 mJ per pulse was scattered in H_2. Four anti-Stokes and seven Stokes components were identified using a 1.5 m, 5 atm Raman cell. Output powers generated in the vuv were: (a) over 100 kW at AS_1 (179 nm), (b) approximately 20 kW at AS_2 (166 nm), and (c) 3 kW at AS_3 (156 nm).

Raman scattering of excimer laser radiation H_2 has also received attention as a possible high-efficiency, high-power blue-green source (i.e., in 400 – 500 nm region) for underwater submarine communication systems. *Trainor* et al. [4.142] studied production of S_1 (414 nm) and S_2 (499 nm) using a XeF pump laser operating at 353 nm. The excimer laser was a 1 m long, e-beam excited system, producing 0.6 J in a 0.4 µs duration pulse. This radiation was focused into a 40.6 cm long H_2 cell operated at a maximum pressure of 40 atm. An overall conversion into the first two Stokes orders of 35% was measured with a ratio of energy (S_1)/energy $(S_2) \approx 5$.

Komine et al. [4.143] also studied XeF laser scattering as a high-power, blue-green source. In their experiments an H_2 oscillator-amplifier system was employed in an attempt to optimize S_2 production at 499 nm. The output of a 7.5 J, 0.7 µs pulse XeF laser was divided into two beams in order to pump the Raman oscillator-amplifier system. Approximately 1 J of pump radiation was used for the 2 m oscillator cell which was operated at a pressure between 7.8 and 9.2 atm. The Stokes orders thus produced were sent into a 5.8 m amplifier cell along with the remaining XeF pump beam. At an H_2 pressure of 7.8 atm, 1.2 J per pulse was produced at S_1 (414 nm); and 1.7 J per pulse was produced at S_2 (499 nm). This corresponds to 34% energy and 43% peak power conversion into the desired S_2 wavelength.

Finally, the use of a XeCl (308 nm) pump source for blue-green laser development was considered by *Komine* and *Stappaerts* [4.144]. A Raman oscillator-amplifier arrangement was again employed. A 1 m long H_2 oscillator cell at 6 atm pressure was used; the amplifier cell was 2 m long and operated at 6 atm as well. Simultaneous generation of S_1 (353 nm), S_2 (414 nm), and S_3 (499 nm) was observed at energy conversion efficiencies of 18%, 22%, and 14%, respectively. Pump depletion of 90% was noted.

Anti-Stokes Raman scattering of tunable dye laser radiation has also been employed as an injection source for excimer laser amplifiers in an effort to improve the linewidth, beam mode quality and divergence, and tunability of the excimer system. This technique, although applicable at almost any excimer laser wavelength, has received the most attention in connection with the ArF (193 nm) laser, due primarily to the difficulties inherent in producing reasonable power at this wavelength by other methods. *White* et al. [4.145] performed the first experiments in which the output of a frequency-doubled rhodamine 590 dye laser was Raman shifted in H_2 at a pressure of 6 atm. The tunable AS_4 line at 193 nm was used to injection lock an ArF laser amplifier with excellent results. Subsequently, *Schomburg* et al. [4.146] employed anti-Stokes generation with a fluoresceine 27 dye laser to generate 50 μJ per pulse at AS_8 (193 nm) with a linewidth of 0.8 cm^{-1}.

The previous discussions of SRS in H_2 gas have all addressed the process in which the Q_{01} (1) branch transition (Raman shift: 4155 cm^{-1}) is used. Other Raman transitions are possible in H_2, and we shall briefly discuss some of the more interesting variations. *Audibert* and *Lukasik* [4.147] studied SRS involving the Q_{12} (1) vibrational transition. In this case scattering originates from the $v = 1$ to the $v = 2$ vibrational states with a Raman shift of 3920 cm^{-1}. The population in the $v = 1$ level was prepared by Raman scattering on the Q_{01} (1) branch with the output of a Q-switched ruby laser at 694.3 nm. Once the population in the $v = 1$ level was sufficiently large, stimulated Raman scattering of the ruby laser radiation was observed on the Q_{12} (1) transition leading to S_1 (953.9 nm) and AS_1 (545.8 nm). At an incident pump laser intensity of 1 MW, up to 3 kW of power was generated at 953.9 nm.

The use of rotational transitions in H_2 for medium wavelength infrared generation has also received considerable attention. *Byer* [4.148] proposed the use of stimulated rotational Raman scattering of a tunable CO_2 laser in gaseous para-H_2 as a means of efficiently generating a high power, tunable 16 μm source for laser driven separation of uranium isotopes [4.149]. *Sorokin* et al. [4.150] utilized a four-wave parametric Raman process with a narrow linewidth ruby laser and a broadband CO_2 laser. These two pump lasers were overlapped spatially and temporally in the 2 m Raman cell. The lasers were passed through a $\lambda/4$ plate and converted to a circular polarization prior to the cell in order to drive the $J = 0$ to $J = 2$ rotational transition at 354.3 cm^{-1}. A schematic of the four-wave parametric Raman process is shown in Fig. 4.23. The narrowband ruby laser v_L generates its rotational Raman shifted Stokes photon v_s. These two fields then mix with the CO_2 laser photon v_i, thereby generating the desired 16 μm output at v_o. In this way the linewidth of the 16 μm output is controlled by v_L and not the comparatively broadband output of the high-power TEA CO_2 laser. With the CO_2 laser tuned to the $R(30)(00°1 - 100°)$ transition at 982.1 cm^{-1}, approximately 40 μJ per pulse was generated at 627.8 cm^{-1}. The incident laser energies were 2.7 J and 0.3 J per pulse for the ruby and CO_2 lasers, respectively.

This four-wave mixing technique was also studied by *Trutna* and *Byer* [4.151,152] using a Q-switched Nd:YAG laser at 1.06 μm in place of the ruby laser described above. The Raman cell in these experiments consisted of a multipass configuration, yielding 25 passes through the para-H_2 gas for an effective interaction length of 377 m. A peak energy of 50 mJ per pulse at 16.95 μm was generated by this technique [4.152].

Finally, *Rabinowitz* et al. [4.153,154] demonstrated direct rotational SRS in para-H_2 using a sophisticated CO_2 laser oscillator followed by two CO_2 amplifier sections as a pump source. The pump laser was tuned to the $R(22)(00°1 - 100°)$ band at 977.2 cm^{-1}; the incident energy was 2.9 J per

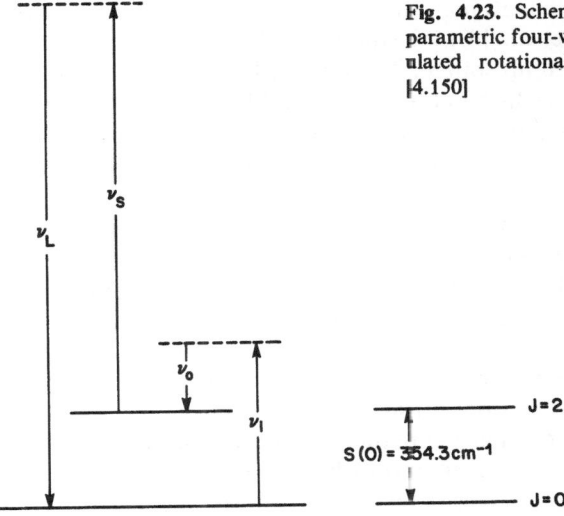

pulse. A multipass cell design was employed with 21 passes yielding an effective interaction length of about 78 m. The energy generated at the first Stokes field was 1.6 J/pulse; the first anti-Stokes wave was also seen with an output between $20-50$ mJ.

Deuterium gas, not too surprisingly, has also been employed as a Raman medium with a modest shift of $2\,991$ cm^{-1}. D_2 can be utilized in basically the same manner as H_2 [4.10]. In one example, Raman shifting of ArF (193 nm), KrF (248 nm), XeCl (308 nm), and KrCl (223 nm) excimer lasers in D_2 was studied by *Loree* et al. [4.89, 140], where at least 3 Stokes orders and 1 anti-Stokes order were observed in each case. Rotational Raman scattering in D_2 with circularly polarized light was studied by *Minch* et al. [4.155].

Loree et al. [4.89] also investigated the use of mixed-media Raman scattering, i.e., a Raman cell filled with two or more Raman active materials. In their work a mixture of H_2 and D_2 gas was used in conjunction with 193 and 248 nm pump lasers. Many combination lines were observed corresponding to a wide variety of possible scattering and four-wave mixing processes. The converted photons were shifted into so many lines, however, that good conversion into a particular desired frequency was difficult.

b) Other Raman Systems

In this subsection we shall briefly discuss a few of the more commonly employed molecular Raman media. These include CH_4, cryogenic materials, liquid N_2 (LN$_2$) and liquid O_2 (LO$_2$), and a very brief look at far infrared generation.

Being a gas at room temperature, CH_4 offers many of the advantages found with H_2. Raman-cell design is straightforward, and a variety of cell configurations can be used. The Raman shift is $2\,916\ \mathrm{cm}^{-1}$, and is very nearly that of D_2 [4.10]. *Duardo* et al. [4.156] studied Stokes production using a Q-switched ruby laser as a pump source. The CH_4 cell was 12.5 cm long and operated at gas pressures exceeding 8 atm. At an input laser of 35 mW several Stokes orders were observed. Infrared generation over the range 11 000 to $4\,500\ \mathrm{cm}^{-1}$ was studied by *Frey* and *Pradere* [4.134, 157] using a dye-laser pumped, CH_4 Raman system.

More recently CH_4 has been studied as a medium for short pulsed ir generation by *May* and *Sibbett* [4.138]. Methane is an attractive candidate for very short pulse excitation since, although its Raman shift is less than H_2, its phonon lifetime of 3.5 ps [4.158] is more than 50 times smaller, implying enhanced transient Raman gain. This was in fact observed using a 250 fs pulse duration dye system operating with 0.4 mJ/pulse at 617 nm. As with the H_2 case discussed in the previous section, a 1 m long waveguide gas cell was used to contain CH_4 at 20 atm. A capillary bore diameter of 0.2 nm was used. The S_1 (752 nm), S_2 (964 nm), and S_3 (1.34 µm) orders were observed with pulse energies of 8 µJ, 9 µJ, and 0.5 µJ, respectively. Pulse shortening by a factor of 1.7 was observed implying peak powers in excess of 80 mW for S_1, 90 mW for S_2, and 5 mW for S_3. Five orders of anti-Stokes radiation were also observed over the spectral range of 325 – 522 nm.

The use of CH_4 as a shifting medium with excimer laser pump sources has been investigated by *Loree* et al. [4.49, 140], and KrF (248 nm) and XeCl (308 nm) were used as pump sources. The methane cell was 1 m long and operated at a pressure of 10 atm or greater. With the KrF laser source a total of five Stokes and two anti-Stokes orders were observed. The XeCl pump resulted in three Stokes and one anti-Stokes order. The output energy as a function of XeCl pump laser energy for the various Raman lines is shown in Fig. 4.24. A maximum energy of 10 mJ at S_1 (338 nm) was seen at a pump energy of about 60 mJ.

Liquid nitrogen and liquid oxygen have been extensively studied as Raman media for frequency conversion from the uv to the mid-ir. These media are attractive due to a combination of factors, including the high molecular density of a liquid, a single Raman-active mode, the narrow Raman linewidth, and high Raman gain. In addition LN_2 is almost completely transparent throughout the uv to the far ir [4.159]. There are only three absorption bands in LN_2: (a) $v_1 = 3\,300\ \mathrm{cm}^{-1}$ due to H_2O absorbed in LN_2, (b) $v_2 = 2\,330\ \mathrm{cm}^{-1}$ coincident with the vibrational frequency of N_2, and (c) a band at $v_3 = 1\,550\ \mathrm{cm}^{-1}$ corresponding to O_2, a common impurity in LN_2. Of course careful attention should permit the removal of the impurity absorption bands. Liquid O_2 shows a discrete spectrum of absorption bands corresponding to 1.26 µm and 1.06 µm at the long wavelength end. At shorter wavelengths the absorption bands are repeated every $1\,550\ \mathrm{cm}^{-1}$ [4.160].

Both of these cryogenic liquids have been studied extensively as Raman media. The Raman linewidth [4.65, 161], the Raman gain [4.120], the non-linear coefficients for self-focusing [4.162], and the parameters describing

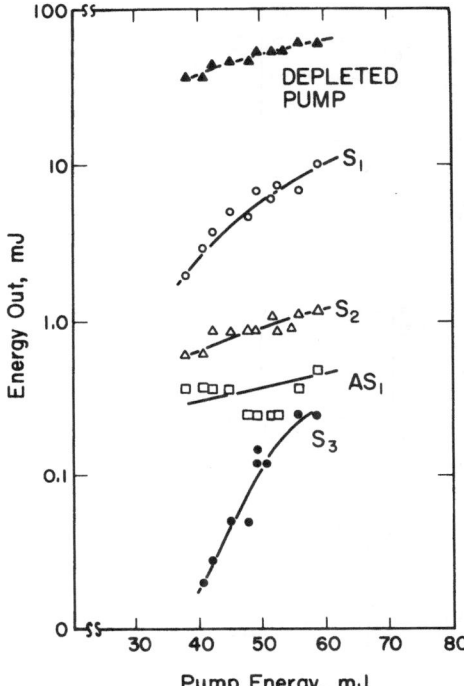

Fig. 4.24. Observed output energy for the first three Stokes and first anti-Stokes orders versus pump energy for SRS of a XeCl laser in methane gas [4.89]

laser-induced breakdown [4.163] have been measured. The critical power for self-focusing has been measured as 7 mW and 5 mW for LN_2 and LO_2, respectively [4.164]; these values are considerably greater than in other Raman active liquids. Both of these media have been employed primarily as Raman media for mid-infrared generation and for frequency-shifting excimer lasers. In the infrared, these cryogenic liquids have been used with very good results, and a good review of the topic may be found in *Grasiuk* and *Zubarev* [4.165].

In one of the first detailed studies of mid-ir generation in LN_2, *Zubov* et al. [4.166] investigated Stokes and anti-Stokes generation using a Q-switched ruby laser pump source. A single-pass LN_2 Raman cell 8 cm long was used. The ruby laser radiation, with energy up to 0.4 J/pulse, was focused with a 15 cm lens into the Raman cell. Four Stokes and three anti-Stokes components were observed. The growth of the various Raman orders as a function of the energy of the pump laser is shown in Figs. 4.25a, b for the Stokes and anti-Stokes waves, respectively. Note that the energy conversion efficiency into the first two Stokes waves is very good, being 17% for S_1 and 32% for S_2. An energy conversion factor of 2.5% was seen for AS_1 as well.

Large Nd: glass laser systems have also been used as pump sources for mid-ir generation with LN_2 [4.24, 167]. *Bocharov* et al. [4.167] scattered a 240 J Nd: glass laser pulse, resulting in efficient generation of S_1 at 1.40 μm

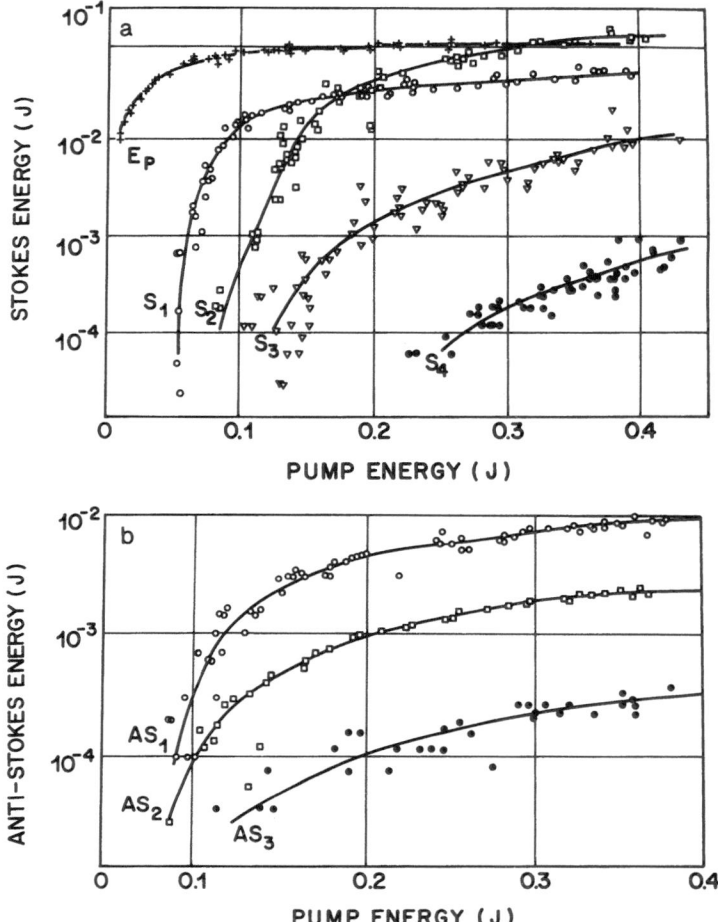

Fig. 4.25. (a) Energy generated on various Stokes lines versus incident ruby laser energy for Raman generation in liquid nitrogen; E_p is the energy remaining in the pump beam at the end of the Raman cell. (b) Energy generated for the anti-Stokes components in liquid nitrogen versus incident ruby laser energy. After [4.166]

and S_2 at 2.08 μm. An energy of 36 J/pulse was observed in S_1, while 6 J/pulse was seen in the S_2 beam.

A tunable, rhodamine 590 dye laser system was utilized by *Brueck* and *Kildal* [4.168] to pump a single-pass, 10 cm long LN_2 cell. The LN_2 was condensed into the Raman cell from high-purity gaseous N_2 and was maintained at 77 K with a separate LN_2 cooling bath. This technique resulted in a quiescent liquid of good optical quality and low-impurity content. Tunable Raman radiation at S_1 (644 nm), S_2 (758 nm), S_3 (920 nm), and AS_1 (495 nm) was monitored. Using a long focal length lens for the pump radiation (i.e., a collimated beam geometry) helped to reduce parametric coupling to higher Stokes orders and resulted in a quantum efficiency of 92% for S_1 production.

Infrared generation in the 4 μm region has been addressed by several groups [4.159, 169, 170]. *Grasiuk* [4.169] described a two-step system in which a Nd: glass laser was used to pump a gaseous H_2 Raman laser yielding 1.89 μm radiation, which in turn pumped a nonresonant LN_2 Raman cell. Pulse energies as high as 100 mJ were obtained for S_1 (3.39 μm) from the LN_2 cell with a Nd: glass laser pump energy of some 35 J, for an overall efficiency of about 0.3%. In experiments by *Efimovsky* et al. [4.159] comparable overall conversion efficiency was obtained by pumping the LN_2 Raman cell directly with the output of a Nd: glass laser and selecting the S_3 line at 4.07 μm. Finally, *Marquardt* et al. [4.170] generated radiation near 4 μm far more efficiently by using a Ho: YLF laser operating at 2.05 μm [4.171] to pump a single-pass LN_2 cell. Photon conversion efficiency of 95% was seen to S_1 at 3.92 μm using a 12 cm long path-length Raman cell. This corresponds to ~22% energy conversion to the first Stokes wave.

A LN_2 Raman converter tunable in the range of 15 – 18 μm has been described by *Frey* et al. [4.172]. A pump source tunable near 3.4 μm was used having an energy of 40 mJ in a 2 ns duration pulse. This radiation was focused into a 30 cm long cell which was configured for three passes through the gain medium. The first Stokes output was tunable from 15 – 18 μm with an energy of 1.7 mJ in the 2 ns pulse for a quantum efficiency for conversion of 20%.

The use of LN_2 as a medium for frequency-shifting excimer lasers has also been investigated. *Baranov* et al. [4.173] used XeF (351 nm) and XeCl (308 nm) excimer sources to pump a 50 cm long single-pass cell. The excimer laser produced uv radiation with a pulse duration of 25 ns and a beam divergence of 10 mrad; this light was focused into the LN_2 cell with a 40 cm focal length lens. At a XeF input energy of 220 mJ, approximately 26 mJ at S_1 (382 nm) and 6 mJ at S_2 (420 nm) were observed. The XeCl pump laser (incident energy 170 mJ) produced 13 mJ at S_1 (332 nm) and 3.5 mJ at S_2 (359 nm). Raman shifting the ArF (193 nm) and KrF (248 nm) lasers in LN_2 was addressed by *Loree* et al. [4.89]. In both cases three Stokes lines were seen.

The effects of thermal blooming in LN_2 have been examined as a mechanism limiting the usable repetition rate for the Raman cell. Severe beam distortions due to thermal effects were noted by *Wild* and *Maier* [4.174] for repetition rates exceeding 1 Hz in a loosely-focused, collimated beam geometry Raman experiment. The beam distortions were due to thermal lensing within the LN_2 caused by the vibrational relaxation of the upper Raman level to the ground molecular state. This vibrational energy is converted into heat yielding a change in the refractive index within the interaction region.

Brueck and *Kildal* [4.168] showed that good beam quality could be maintained even at high repetition rates (e.g., 10 Hz) by using a tightly-focused beam geometry. In their work thermal blooming was avoided since the laser spot size was much less than the thermal diffusion length l_D for the 100 ms interval τ between laser pulses. This diffusion length is given by $l_D = (8\lambda\tau/\varrho_0 c_p)^{1/2}$, where $\lambda = 0.14$ W/mK is the thermal conductivity, $\varrho_0 = 0.8$ g/cm^3 is the density, and $c_p = 2$ J/g K is the heat capacity at constant pressure. In the tightly-focused case any thermal lens generated after each input pulse has diffused away prior to the arrival of the next laser pulse. Obviously, thermal blooming effects are not a problem for single-shot experiments (with a ns duration pulse), since the time for development of the thermal lens is much greater than the incident pulse length.

In the final part of this section, we shall take a brief glimpse at the generation of coherent far-infrared (fir) radiation via SRS. The application of resonantly enhanced Raman scattering for fir generation was first proposed by *Ducuing* et al. [4.175]. Shortly thereafter *Frey* et al. [4.176] successfully demonstrated tunable, resonantly enhanced fir generation via SRS over the spectral range from 60 to 160 µm for the Q(J) transitions in gaseous HCl. The excitation source for the Raman process operated between $3.3 - 3.4$ µm and was created by Raman scattering a tunable dye laser in H$_2$ to create the second Stokes wave. This pump light was focused into a 50 cm long waveguide Raman cell with a 3 mm internal diameter. The Stokes output from the HCl cell was tunable approximately 2.3 cm^{-1} around the J = 2 to 7 transitions, and a photon conversion efficiency of 12% was measured giving peak powers around 80 kW.

In an improvement of their earlier work, *De Martino* et al. [4.177] studied fir generation in a gaseous HF waveguide Raman cell. The pump source at 2.5 µm was created through two Raman shifts in H$_2$ of a ruby pumped dye laser. The 6 mm diameter by 50 cm long teflon waveguide HF cell was operated at a pressure around 60 Torr. Raman tunability of 5 cm^{-1} around several Q(J) transitions (J = 0 – 5) was seen corresponding to step-tunable radiation over the range of $50 - 256$ µm. Output powers up to 300 kW were measured for a photon conversion efficiency of 18%.

A significant improvement in these techniques was made by *Mathieu* and *Izatt* [4.178]. They employed a high-pressure CO$_2$ laser to pump a ^{12}CH$_3$F Raman cell. Tunable fir radiation resulted from Raman scattering on the R branch transitions of the ^{12}CH$_3$F ν_3 band. With an input energy of 150 mJ/pulse from the CO$_2$ laser, ~ 0.2 mJ was produced and was step tunable over 85% of the $220 - 400$ µm spectral region. The authors further suggest that the use of ^{13}CH$_3$F should extend the tunable spectrum to $\sim 1\,200$ µm [4.179].

c) Dimer Lasers

Bound-bound electronic transitions in simple, diatomic molecules have been extensively studied as coherent radiation sources in the visible and near infrared. Literally hundreds of laser transitions have been studied in such molecules as Li$_2$, Na$_2$, K$_2$, Bi$_2$, S$_2$, Te$_2$, I$_2$, and Br$_2$, with both pulsed and cw

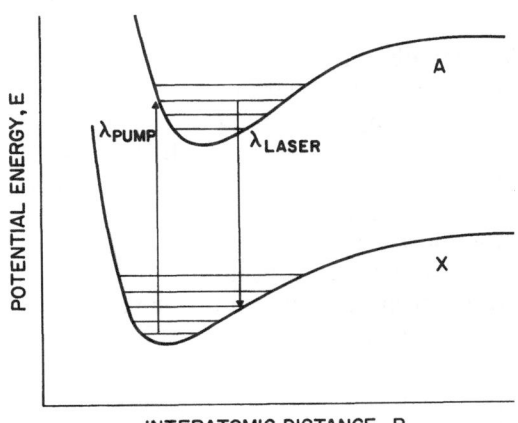

POTENTIAL ENERGY, E

λ_{PUMP} λ_{LASER}

A

X

INTERATOMIC DISTANCE , R

Fig. 4.26. Schematic level diagram for molecular dimer lasers, showing the Stokes shifted laser created by optical excitation of the ground X state to vibrational levels of the excited A state

operation in region of $400-1\,350$ nm having been achieved. An excellent review of the topic was given by *Wellegehausen* [4.180].

A schematic representation of these dimer laser systems is illustrated in the energy level diagram of Fig. 4.26. We shall consider optical excitation from low-lying vibrational levels of the ground X electronic state to the vibrational states of, for instance, the first excited electronic state A. Once excited to the A state vibrational manifold, a variety of processes may occur. Direct, resonant laser transitions from A→X may occur along pathways with large Franck-Condon factors. Additionally, the excited A states may vibrationally relax prior to radiating to the X state levels. Although these examples, and indeed many of the laser transitions observed in the literature, utilize resonant excitation and emission, the process of Fig. 4.26 is very suggestive of a three-level Raman laser system. In fact many of the properties of dimer laser systems are due to the occurrence of coherent, Raman scattering processes [4.180]. For this reason we shall undertake a short review of dimer lasers as potential near-resonant Raman media.

Laser down-conversion in optically pumped I_2 was first observed using pulsed excitation by *Byer* et al. [4.181]. Laser radiation on more than 150 lines spanning the spectral region from 544 to $1\,335$ nm was observed for excitation by the 530.6 nm second harmonic wavelength of a Q-switched Nd:YAG laser. The Stokes shifted output occurred on selected vibrational-rotational levels of the $B\,^3\Pi_{0u}^+ \rightarrow X\,^1\Sigma_g^+$ electronic transition. Output powers as high as 1 W in a single line (618 nm) were obtained at a 0.5% conversion efficiency.

In I_2, cw excitation and oscillation was achieved a few years later by two groups. *Koffend* and *Field* [4.182] obtained laser emission from 570 nm in the $(v' = 43,\ v'' = 9)$ band to $1\,027$ nm in the $(v' = 43,\ v'' = 56)$ band by exciting the B←X electronic transition with a cw, Ar^+ laser at 514.5 nm. Some 30 laser transitions were seen, and a maximum conversion efficiency of 0.14% at an output power of 0.3 mW was achieved. *Wellegehausen* et al. [4.183] also studied pumping the B←X states with a cw, Ar^+ laser. Continuous oscillation was seen for various transitions from 583 to 1 338 nm; for pump powers of 3 W, output powers up to 250 mW were obtained.

Pulsed excitation of Br_2 with the 532 nm second harmonic wavelength of a Q-switched Nd:YAG laser was examined by *Wodarczyk* and *Schlossberg* [4.184]. Absorption of the pump radiation occurs along the $B\,^3\Pi_{0u}^+ \rightarrow X\,^1\Sigma_g^+$ transition from $v'' = 0$ to $v' \geqslant 25$. Output lasers tunable from 550 to 750 nm were observed with a conversion efficiency of about 10^{-3}.

Molecular sodium is perhaps the most extensively studied of the dimer laser systems, due in part to the ease with which Na heat pipes may be constructed and to the variety of excitation techniques which may be employed. The first experiments with optically excited Na_2 were performed by *Henesian* et al. [4.185] and *Itoh* et al. [4.186] in 1976. The frequency-doubled output of various lines from a Q-switched, Nd:YAG laser was used by *Henesian* et al. [4.185] to pump the A←X and B←X transitions. Laser lines from 785 to 808 nm were seen from the excited A states, while emission over the range of 529 – 549 nm was seen from the B band. In the experiments of *Itoh* et al. [4.186] a pulsed dye laser near 600 nm was used to pump the A←X band resulting in laser emissions from 790 – 810 nm.

A short time thereafter [4.187], cw operation of the Na_2 laser was achieved using the blue lines of an Ar^+ laser as the excitation source to excite the $B\,^1\Pi_u \leftarrow X\,^1\Sigma_g^+$ transition. Output powers up to 3 mW were observed for pump powers of 0.5 W. A total of 23 laser lines between 525 and 560 nm were seen at a Na_2 density of about $10^{16}\,cm^{-3}$. Ring laser cavities were later employed [4.188, 189] in an effort to lower the laser threshold and increase the tuning range. In one experiment [4.188] a Na_2 supersonic beam was used to increase the fraction of dimer molecules in the gas from the normal equilibrium value of 4% to greater than 10%. Thresholds as low as 1 mW of pump power at 476.5 nm and output powers up to 0.6 mW were achieved. Finally, excitation along with C←X transition was studied by *Wellegehausen* et al. [4.190] using the 350.7 nm line from a Kr^+ laser as the excitation source. Laser oscillation around 2.5 μm on the $C\,^1\Pi_u \rightarrow (3)\,^1\Sigma_g^+$ was seen, as well as a variety of cascade transitions to lower molecular states.

Raman tuning the Na_2 system has been investigated by *Man-Pichot* and *Brillet* [4.191] and *Luhs* and *Wellegehausen* [4.192]. In both cases a ring oscillator design was used with a Na heat pipe oven. Raman tuning of approximately 75 GHz was seen using a cw rhodamine 590 dye laser pump beam to excite the A←X transition [4.192]. This tuning range was limited by the available pump power.

Other alkali dimer systems have also been studied. The Li_2 laser has been pumped on the B←X transition with the 496.5 nm line of cw, Ar^+ laser. A total of 22 resonant and 4 Raman lasers were observed. The K_2 dimer (as well as Na_2 and I_2) has been pumped with a 25 mW, 632.8 nm He-Ne laser, yielding powers of 0.1 mW from 685 – 700 nm [4.193].

A variety of other dimer systems have been studied. Laser action on the $B\,^3\Sigma_u^- \rightarrow X\,^3\Sigma_g^-$ transition of S_2 was achieved by optically pumping with a frequency-doubled dye laser and a nitrogen laser [4.194]. The resulting lasers were line tunable from 365 to 570 nm with an energy conversion efficiency of 2%. More than 30 laser doublets in the range of 387 – 709 nm were observed for the Se_2 system following excitation of the $BO_u^+ \leftarrow XO_g^+$ transition with an Ar^+ laser at 351.1 nm [4.195]. Continuous wave output powers up to 3 mW per laser line were reported. Excitation of the BO_u^+ state of Te_2 was also carried out in two separate experiments [4.196, 197]. When pumped with the 406.7 and 413.1 nm lines of a Kr^+ laser, greater than 50 laser transitions were identified from 460 – 780 nm [4.196]. An Ar^+ laser pump at 476.5 nm was also employed for Te_2 [4.197], yielding over 80 laser transitions over the range of 550 – 660 nm with an overall power conversion efficiency of 5%. Finally, the Bi_2 system has been examined using both cw [4.197] and pulsed [4.198] excitation of the $A \leftarrow X$ transition. With cw excitation at 514.5 nm from an Ar^+ laser, output powers up to 0.35 W were seen in the range of 593 – 748 nm for an input power of 3.5 W [4.197]. An efficiency of 0.2% was noted following pulsed excitation of Bi_2 with a dye laser operating between 540 and 580 nm [4.198]; a very dense dimer emission spectrum between 660 and 710 nm was observed.

4.5.2 Atomic Systems

In order to overcome the comparatively low again often encountered in molecular systems, SRS near electronic resonances in atomic systems is widely used. The use of atomic systems has several advantages. First, the Raman shifts tend to be much larger (generally from $10\,000 - 30\,000$ cm^{-1}), and are therefore more suitable for ir generation with tunable dye laser pump sources. Also, the large electronic matrix elements of atomic systems imply large Raman gain cross sections and large single-pass Raman gain at modest vapor densities. The advantage of higher gain in electronic Raman scattering is achieved, however, at the expense of decreased tuning range, since it is generally necessary to tune the pump laser near a local resonance in order to minimize the Raman detuning Δ.

In this subsection we shall review generation techniques based on electronic resonances in atomic systems; this process is often termed stimulated electronic Raman scattering (SERS). We shall first consider Stokes generation in which the incident pump beam is frequency down-converted via the Raman process. This approach has been widely studied for creating high-power, tunable ir sources; and more recently has been applied to shifting powerful uv lasers (e.g., the excimer systems, ArF, KrF, etc.,) to new wavelengths in the uv and visible spectra. The most recent advances in SERS have occurred in the area of stimulated anti-Stokes generation, in which an electronically inverted atomic system is used to up-convert the frequency of the incident pump laser. Stimulated anti-Stokes generation is particularly attractive as a means of generating tunable, higher-power vuv lasers by Raman scattering existing uv sources.

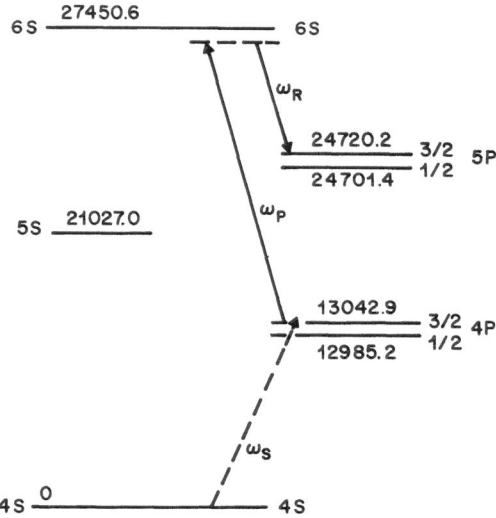

6S   27450.6   6S

ω_R

24720.2   3/2 5P
24701.4   1/2

5S   21027.0 ω_P

13042.9   3/2 4P
12985.2   1/2

ω_S

4S   0   4S

Fig. 4.27. Potassium energy level diagram for SERS. The $4p$ initial Raman level population is created by optical pumping ($- - -$), and Raman scattering on the $4p-5p$ transition driven by a pump source at frequency ω_p [4.200]

Stimulated electronic Raman scattering was first observed by *Rokni* and *Yatsiv* [4.199,200] and a short time later by *Sorokin* et al. [4.201]. The experiments of Rokni and Yatsiv illustrate several of the important characteristics of SERS. In their experiments, Raman down-conversion in K vapor was studied. The energy level scheme for the process is shown in Fig. 4.27. Using a Q-switched ruby laser as the pump source, they studied Raman scattering between the K($4p\ ^2P^0_{3/2}$) and K($5p\ ^2P^0_{3/2}$) levels. This scheme is attractive since the ruby laser wavelength ($\hbar\omega_L = 14\,398$ cm^{-1}) is detuned a mere 10 cm^{-1} from the signle photon $4p\ ^2P^0_{3/2}-6s\ ^2S_{1/2}$ transition, implying a large resonant enhancement of the Raman gain. The Raman shift in this case is the energy gap between the $5p\ ^2P^0_{3/2}-4p\ ^2P^0_{3/2}$ states (i.e., $11\,677$ cm^{-1}) yielding a Stokes wavelength of about 3.7 μm with the ruby laser pump source. In order to utilize this system, however, the initial $4p$ state had to be populated with a sufficiently high density to ensure reasonable Raman gain. This was accomplished by optically pumping the $4s\ ^2S_{1/2}-4p\ ^2P^0_{3/2}$ transition with radiation created by Stokes shifting a ruby laser in a liquid nitrobenzene Raman cell. The S_1 emission from the liquid cell was then within 10 cm^{-1} of the $4s\ ^2S_{1/2}-4p\ ^2P^0_{3/2}$ transition. Potassium vapor pressures of several Torr were used, and self-broadening of the resonance transition was large enough to ensure absorption of this radiation in the wings of the transition. In this manner the Stokes output from the liquid Raman cell populated the initial $4p\ ^2P^0_{3/2}$ Raman level, while the remaining light at the ruby laser fundamental wavelength pumped the intended electronic Raman process in the excited K atoms. By varying the temperature of the ruby laser rod, the authors were able to tune the pump wavelength over 3 cm^{-1}. As expected, the Stokes field at 3.7 μm tuned by the same amount.

It is useful at this point to make order of magnitude comparisons between gains observed in stimulated vibrational-rotational Raman scattering in molecules and that seen with SERS in atoms. From Table 4.2, we see that the Raman gain for H_2 (the most commonly used molecular system) is $g_{ss} = 1.5$ cm/GW at gas pressures above about 20 atm. Using the relations for K_1 and K_2 derived in Sect. 4.3.2, we may readily calculate g_{ss} for the K system studied by Rokni and Yatsiv. Data on the pertinent dipole matrix elements is usually readily available. More often, however, considerable line broadening occurs in SERS, and accurate knowledge of the Raman linewidth Γ is difficult. If we assume $\Gamma = 0.4$ cm^{-1} and that a $4p\,^2P_{3/2}^0$ population density of 10^{15} excited states/cm^3 was achieved by optical pumping (i.e., about 10% of the ground state K population), then the gain factor for SERS in K is of the order of $g_{ss} \sim 10^2$ cm/GW. Pump power densities in excess of 100 MW/cm^2 are readily obtainable with Q-switched ruby lasers, so that small signal gains of $\exp(100)$ are possible in a single-pass, 10 cm long K Raman cell.

One should note that the large gains for SERS occur at number densities many orders of magnitude lower than in molecular systems. This dramatically illustrates the advantage of near-resonant pumping to achieve small Raman detunings Δ. In addition, the dipole matrix elements in atomic systems are concentrated along the electronic transitions and not distributed among the many vibrational-rotational states of a molecular system. This enhancement is achieved at the expense of tunability, however. In molecular systems the pump frequency is generally detuned far from any intermediate resonances, so that the Raman gain coefficient is a slowly varying function of pump-laser wavelength. The gain in SERS is inversely proportional to the detuning squared, implying a more limited tuning range. Even so, we shall see shortly that tuning ranges greater than 1 000 cm^{-1} are routinely seen by using tunable pump lasers (e.g., dye and excimer lasers) to exploit local atomic resonances.

a) Stokes Generation

Laser down-conversion via SERS has been widely employed as a source of tunable, coherent radiation. The spectral range from $1 - 20$ μm is effectively covered by Raman scattering tunable dye laser systems tuned near the principle series resonance lines of the alkali metals. Photon conversion efficiencies in the mid ir as high as 50% are seen. Efficient frequency conversion of high-power, uv excimer lasers from the uv to the visible spectral region has also been extensively studied.

As with the molecular systems, the apparatus for SERS is straightforward. A wide variety of atomic species have vapor pressures of a few Torr or more at reasonable temperatures so that construction of a gas phase Raman cell is not overly complex. In most cases heat pipe ovens are used to contain the metal vapor. Such ovens are convenient since they may be operated for long periods without condensation of the metal vapor on the cold cell windows. In this case, however, the very uniform vapor density achieved with the heat pipe oven that is so necessary for efficient four-wave mixing applications (Chap. 3) is not vital in SERS, since there is no phase-matching condition to be satisfied.

Widely tunable ir generation via SERS was first obtained by *Sorokin* et al. [4.201] using a K vapor Raman medium. The pertinent energy levels for this system are shown in Fig. 4.28. A pulsed, nitrogen laser pumped dye laser system was used as the Raman pump source. This laser was tuned near the $4s\,^2S_{1/2} - 5p\,^2P^0_{1/2,3/2}$ transitions at ~404 nm. Stokes emission was observed to the $5p\,^2P^0_{1/2}$ state, corresponding to a Raman shift of 21 027 cm^{-1}. The dye laser beam had a maximum power of 1 kW and was focused with a 50 cm focal length lens into a 30 cm long metal-vapor cell. The K cell was operated at vapor pressures ranging from 2 – 15 Torr. Tunable ir Stokes radiation was generated between 2.62 and 2.78 μm or over a range of ~210 cm^{-1}. The Stokes radiation was subsequently used as one of the driving waves for a four-wave Raman mixing process [4.201], thereby increasing the tuning range covered to include 2.1 – 5.4 μm. In a subsequent experiment [4.202], a higher-power (i.e., 80 kW, multimode) dye laser was used as the Raman pump, extending the tuning range to 1 087 cm^{-1} from 2.40 to 3.25 μm.

This basic scheme for mid-ir generation in K vapor was refined in the experiments of *Cotter* et al. [4.203]. In their studies the dye laser pump was improved yielding a power of 20 – 30 kW in a nearly diffraction limited beam. The resulting Stokes radiation was tunable over about 1 000 cm^{-1}, covering the range from 2.56 – 3.5 μm, and was limited on the short-wavelength end by the dye-laser tuning range. The output power exceeded 100 mW over the entire range, and was a high as 1 kW over the central 250 cm^{-1} range. Typical S_1 energy of 4 μJ was observed at an incident pump energy of 120 μJ. The divergence of the Raman output was measured to be 25 mrad.

Cotter et al. [4.203] further explored the usefulness of this ir source for spectroscopic applications, by studying the absorption spectrum of CO_2 gas in the 2.7 μm region. The spectrum observed in this manner is shown in Fig. 4.29; these data imply a Raman linewidth and stability of about 0.4 cm^{-1}. This observation was somewhat surprising, since the dye laser linewidth was of the order of 0.1 – 0.2 cm^{-1} and the Doppler width of the $4s - 5s$ transition was calculated to be only ~0.06 cm^{-1} [4.35]. This was the first verification that line broadening was occurring. We shall discuss the Raman linewidth and line-broadening effects in Sect. 4.6.3.

Other aspects of SERS in K vapor have been studied by several authors. Saturation effects influencing the 2.7 μm Stokes laser were studied by *Tyler* et al. [4.204]. Buffer gas effects were addressed by *Glushko* et al. [4.205]. Finally, Raman scattering to higher-lying electronic states (e.g., $4s - 7s$ or $8s$) was observed by *Wynne* and *Sorokin* [4.206].

The analogous SERS process was studied in Rb vapor by *Wynne* and *Sorokin* [4.202]. In this case the input dye laser beam was tuned near the $6p$ resonance lines at 420.2 and 421.5 nm, inducing Raman scattering from the ground $5s$ to the $6s$ level. The Raman shift was therefore 20 134 cm^{-1} so that the Stokes wavelength fell near 2.7 μm (3 700 cm^{-1}). A tuning range of about

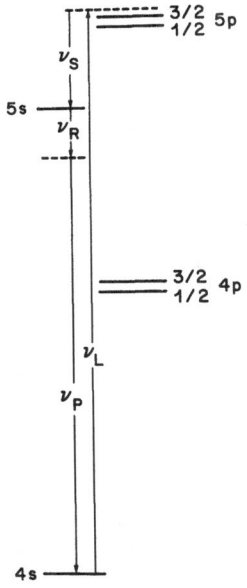

Fig. 4.28. Potassium energy level scheme for tunable ir Stokes generation via parametric four-wave interaction. The pump and Stokes waves are denoted by ν_L and ν_S, respectively. A second pump laser at a different frequency ν_p may be added to generate the difference frequency, ν_R [4.201]

300 cm^{-1} was observed using a 60 cm long Raman cell at 7 Torr pressure and an input dye laser power of ~40 kW. A maximum power conversion efficiency of 2.8% was seen at S_1 (2.704 μm) at an input power of 70 kW. Spectroscopic measurements of CO_2 vapor was again used as an indirect measure of the Stokes linewidth. With an input laser linewidth of ~0.4 cm^{-1}, an effective Stokes linewidth of 0.4 cm^{-1} was measured.

Raman scattering using the $6s-7s$ transition in Cs has been studied by *Cotter* et al. [4.207], *Cotter* and *Hanna* [4.208], and *Kung* and *Itzkan* [4.209]. A 20 kW peak power, nitrogen pumped dye laser was tuned near the Cs ($6s\,^2S_{1/2}-7p\,^2P^0_{1/2,3/2}$) transitions. Stokes emission down to the $7s\,^2S_{1/2}$ level was observed corresponding to a Raman shift of 18 535 cm^{-1}. Tunable ir

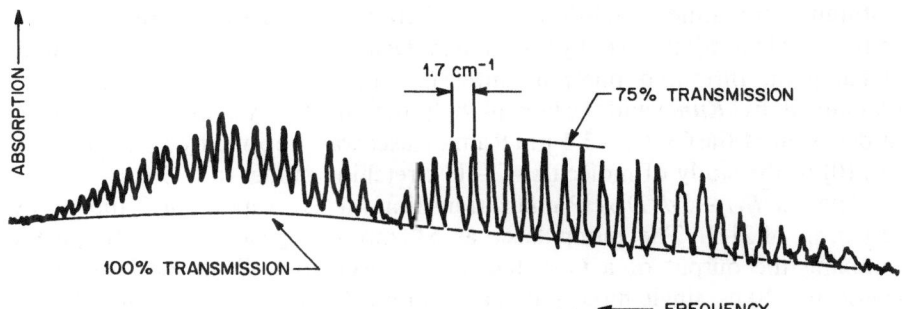

Fig. 4.29. Absorption spectrum of CO_2 gas (50 Torr, 10 cm path length) near 2.7 μm obtained using SERS in potassium vapor as the source of the ir light [4.203]

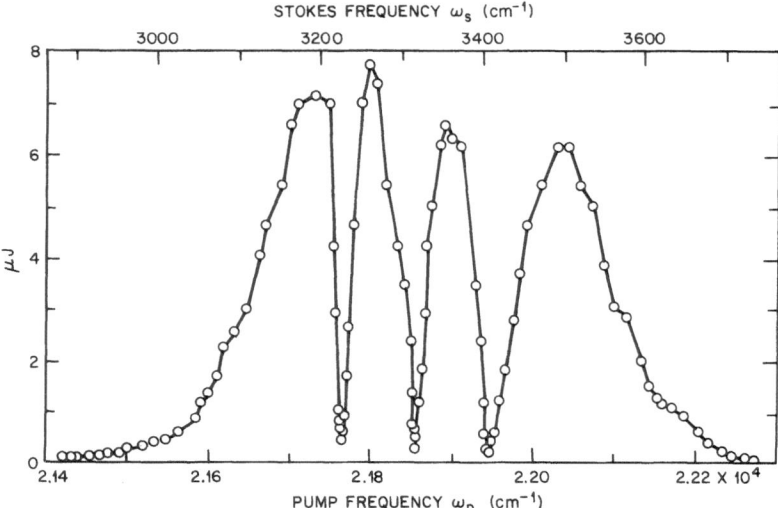

Fig. 4.30. Infrared Stokes output energy observed as a function of dye laser tuning for SERS in cesium vapor. The minima correspond to resonances of the pump wavelength with single-photon transitions in the atom [4.208]

radiation was generated between 2.67 and 3.47 μm (i.e., a 860 cm^{-1} tuning range), with peak powers up to 1.5 kW and photon conversion efficiencies as high as 50%. The Stokes output energy and tuning range were characterized as a function of dye laser focusing and Cs vapor pressure. Optimum output power and tunability were obtained at a Cs pressure of 10 Torr and with the pump beam focused confocally in the 25 cm long metal vapor region. The ir Stokes energy as a function of pump wavelength is shown in Fig. 4.30 for the above conditions. Three distinct minima are apparent; the outer two correspond to tuning of the incident dye laser into resonance with the $7p\,^2P^0_{1/2,3/2}$ doublet. *Cotter* and *Hanna* [4.208] suggest that the minima are due to competition between SERS and other absorption processes such as single-photon absorption and resonantly enhanced multiphoton effects. The optimum operating conditions were dictated primarily by the need to minimize absorption losses by Cs$_2$ dimers. Detailed measurements of the small signal gain, threshold pumping intensity, and conversion efficiency were conducted by *Kung* and *Itzkan* [4.209] for the Cs system. Spectroscopic application of the Cs $2.7 - 3.5$ μm Raman laser was reported by *Peterson* et al. [4.210] to the study of molecular C − H stretching modes in HCN.

The $6s\,^2S_{1/2} - 5d\,^2D_{5/2}$ Raman transition in Cs vapor was studied by *Hodgson* [4.211]. The pump laser at 532 nm was generated by frequency doubling the output of a Q-switched Nd:YAG laser. With 2 mJ of pump energy in a 10 ns, single-mode pulse, a quantum efficiency of 7% to the 2.38 μm Stokes wave was measured. In agreement with the theoretical calculations for the Raman gain (Sect. 4.3.1), the stimulated Raman threshold was seen to be independent of the pump laser linewidth (over the range of $0.1 - 1$ cm^{-1}). With a 0.1 cm^{-1} linewidth green beam, the Stokes linewidth was ~0.4 cm^{-1}. As the incident laser linewidth was increased to 1 cm^{-1}, the Raman linewidth matched that of the green pump beam.

Barium vapor was studied as a medium for SERS in the 2.92 μm region by *Carlsten* and *Dunn* [4.212]. The Raman transition was from the $6s^2\,^1S_0$ ground state to the metastable $6s\,5d\,^3D_{1,2}$ triplet states. The pump laser beam was tuned near the 791.1 nm intercombination line to the $6s\,6p\,^3P_1$ level. The generated Stokes beam energy as a function of dye-laser tuning for an incident energy of 250 mJ/pulse is shown in Fig. 4.31. For these data an 80 cm long Ba vapor cell was used. At a Ba pressure of 2 Torr as much as 30 mJ of Stokes light was produced for a photon conversion efficiency of 40%. The tuning profiles evident in Fig. 4.31 are typical for Stokes (and as we shall see anti-Stokes) Raman generation. At low vapor densities the tuning behavior is more or less symmetric about the intermediate state frequency and falls off with increasing detuning due to the Δ^{-2} factor in the Raman gain. At higher vapor pressures a distinct dip in the tuning profile occurs when the incident laser is tuned into resonance with the intermediate state. Again, this drop in Stokes output is due to competition between SERS and single-photon resonant absorption of the dye laser light. Carlsten and Dunn also studied depletion and saturation effects with this Ba system, and their results are discussed in Sect. 4.5.1.

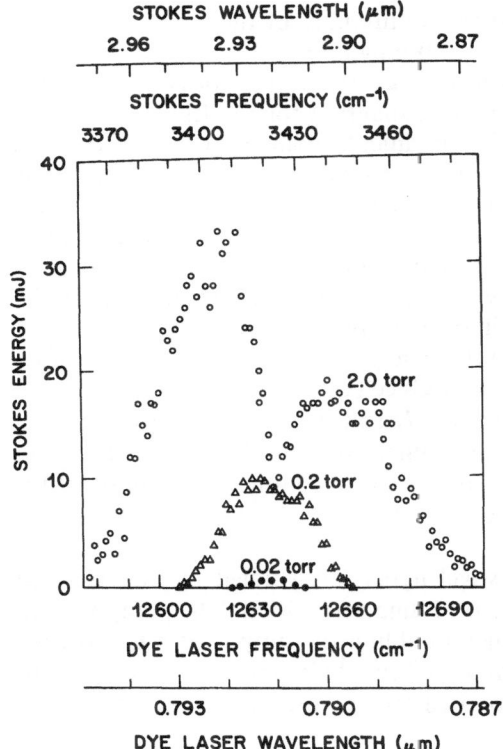

Fig. 4.31. Tuning profile of the Stokes signal generated near 2.9 μm via SERS in barium vapor at various vapor pressures [4.212]

Generation of deeper-infrared radiation is possible by utilizing Raman transitions with higher-lying Rydberg states. As the principle quantum number n is increased, the frequency spacing between the ns and np states decreases as $1/n^3$. *Lau* et al. [4.213] considered the use of transitions between Rydberg levels for fir frequency conversion processes. Two general techniques have been exploited in an effort to utilize Rydberg transitions for the generation of ir radiation. Raman transitions from the ground atomic state to higher-lying Raman levels may be pumped using laser sources deeper in the uv. Until recently, convenient, high-power, tunable uv pump sources were difficult to construct. In addition, the Rydberg levels of the lighter alkali atoms occur at higher energies than for the heavier alkali atoms. For these reasons, most experiments for infrared SERS have been conducted using the heavier alkali atoms.

An alternate approach to mid-ir generation is to induce Raman scattering to Rydberg levels using an excited state, rather than the ground atomic state, as the initial Raman level. The pump-laser wavelength needed to drive the Raman process is then a visible or near-uv source. A disadvantage of this technique is that the initial Raman level (now an excited state) must be populated somehow. This is often accomplished by means of a separate laser beam or a discharge. In the alkali metals, Raman transitions originating from the first excited p states (i.e., the resonance levels) have received considerable investigation. These states are readily populated, and at atomic vapor pressures of a few Torr or more, the effective lifetimes of these states are often several microseconds due to radiation trapping effects.

Tunable ir generation using Cs atoms was studied by *Cotter* et al. [4.207] and *Kärkkäinen* [4.214]. The energy level diagram for the Raman processes studied is shown in Fig. 4.32. In their experiments tunable dye lasers were used as pump sources for SERS from the initial, ground Cs level to the higher lying $7s$, $8s$, and $9s$ states. Resonant enhancement for the Raman process was assured by tuning the dye lasers near the $7p$, $8p$, and $9p$ transitions, respectively. The 200 mJ second harmonic of a pulsed ruby laser was used to pump a dye laser oscillator-amplifier. This dye system could be tuned in the regions near 450, 385, and 360 nm in order to pump the $6s-7s$, $6s-8s$, and $6s-9s$ Raman transitions, respectively. The pump beam was focused with a 50 cm focal length lens into a 35 cm long Cs cell operated at pressures ranging from 3 – 30 Torr. SERS terminating in the $7s$ state was observed over a Stokes timing range of 2.5 – 4.75 μm at a maximum power of 25 kW. Scattering from the $8s$ state resulted in S_1 generation from 5.67 – 8.65 μm at a power of 7 kW, while ir radiation with powers up to 2 kW from 11.7 – 15 μm was generated using the $9s$ final level.

Infrared generation at longer wavelengths was studied by *Sorokin* and *Lankard* [4.215] using the Cs $(6s-10s)$ Raman transition. The pump laser at 347.0 nm was generated by frequency doubling the output of a Q-switched ruby laser. Intense Raman emission at 500 cm^{-1} (20 μm) was observed; this process was expected to have a particularly favorable Raman gain cross

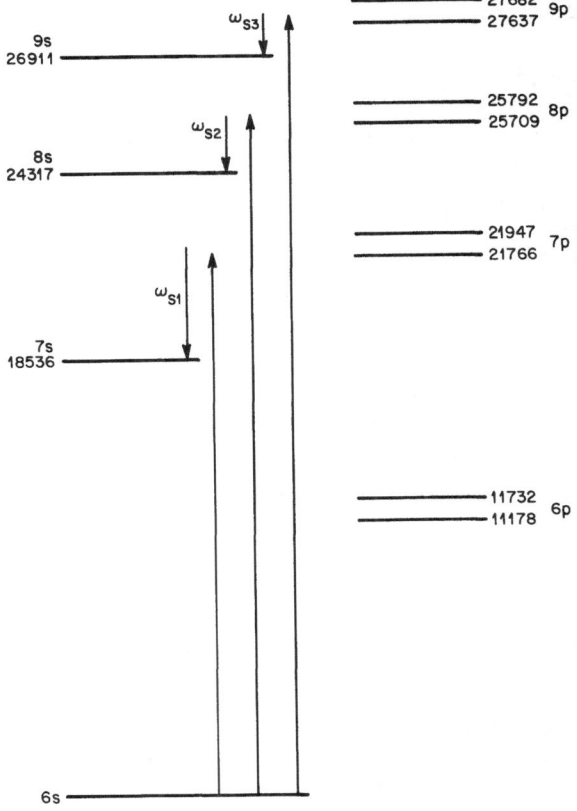

Fig. 4.32. Energy level diagram for SERS using Rydberg states in atomic cesium [4.207]

section due to the proximity $(46\,\mathrm{cm}^{-1})$ of the pump beam frequency to the $10p\,{}^2P^0_{3/2}$ intermediate state. A variety of other parametric processes initiated by the SERS to the $10s$ level were also observed.

Mid-ir generation using excited atomic states as the initial Raman level has been studied by several researchers. As mentioned earlier, the first experimental observation of SERS conducted by *Rokni* and *Yatsiv* [4.199, 200] involved such a process in K vapor (Fig. 4.26). The $\mathrm{K}(4p\,{}^2P^0_{1/2,3/2})$ levels were populated by direct optical pumping from the ground state. Raman transitions from the excited $4p$ level to higher lying Rydberg states were then possible with visible lasers.

As a means of generating Stokes radiation in the mid-ir, SERS from the $4p$ levels of K has been extensively studied. *Kung* and *Itzkan* [4.216] created 8.5 and 16 μm radiation using the $4p-6p$ and $4p-7p$ Raman transitions respectively (Fig. 4.33). The $4p$ population was prepared by tuning a pump laser near the $4s-5p$ transition; several cascade processes including SERS from $5p-5s$ served to funnel population into the $4p$ resonance level. This method was chosen, rather than direct excitation at 770 nm, due to the

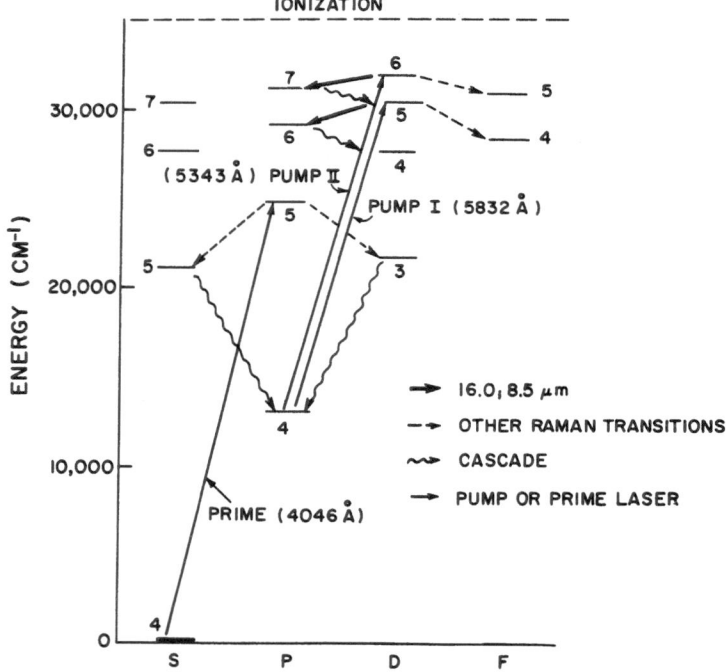

Fig. 4.33. Potassium energy level diagram for SERS from the excited $4p$ states. A priming laser at 404.6 nm is used to optically pump the $5p$ level; cascade transitions downward from this state populate the $4p$ initial Raman level. Two pump wavelengths at 534.3 and 583.2 nm drive the $4p-7p$ and $4p-6p$ Raman transitions, respectively [4.216]

availability of a convenient pump laser near 400 nm. The effective lifetime for the $4p$ level may be calculated using the radiation-trapping formalism of *Holstein* [4.217, 218], and in the case of K vapor at pressures of several Torr, lifetimes on the order of a few tens of microseconds are predicted. Kung and Itzkan estimated a $4p$ lifetime of approximately 3.3 μs when the K cell was operated at 3 Torr with 1.5 Torr of He buffer gas. As the He buffer gas pressure was increased to 31 Torr, the $4p$ lifetime decreased to ~1 μs. These lifetimes are somewhat shorter than expected based on radiative trapping alone, but are consistent with the observation that effective trapping times are often reduced due to collisional quenching of the excited states with other species [4.219]. With a pump laser tuned near the $4p-5d$ transition at 583.2 nm, a Stokes wave at 8.5 μm to the $6p$ state was generated. A pump source at 534.3 nm (near the $4p-6d$ transition) was also used to generate 16 μm radiation to the $7p$ level. Typically 10 mW of 16 μm radiation was observed with a 1 kW pump laser pulse; the Stokes frequency was tunable over ~4 cm^{-1}.

This $4p-7p$ Raman transition in K was also used by *Grischkowsky* et al. [4.220, 4.221] for 16 μm generation experiments. In their work the $4p$ population necessary for the Raman scattering was prepared using an electrical discharge, rather than by optical pumping. The K vapor was contained in a heat pipe oven constructed of a ceramic outer tube and a split interior wick which acted as the electrodes for a cw glow discharge. The cell was operated at a K pressure of 1 Torr (i.e., ~370 °C) and a discharge current of 40 mA. The ground state K number density was 1.6×10^{16} cm^{-3}, and the excited $4p$ number density was estimated to be 6×10^{11} cm^{-3} over the 100 cm long distance region. The $4p-7p$ Raman transition was driven with a nitrogen pumped dye laser tuned near the $6d$ intermediate levels resulting in Stokes generation at 16 μm. An output energy of 1 μJ/pulse was measured at 16 μm for a 0.5 mJ dye laser pump pulse.

Grischkowsky et al. [4.220, 221] noted that the 1 μJ output energy observed for the 16 μm Stokes beam implied that as many as 10 times the number of Stokes photons were generated as the number of $4p$ atoms present in the beam path at the beginning of the pump pulse. They concluded that rapid cascading (i.e., amplified spontaneous emission) of the atoms from the final $7p$ Raman level back down to the $4p$ levels effectively recycled the population to the $4p$ states many times during the pumping pulse. This recycling phenomenon to some degree compensates for the added complexity of using an excited state as the initial Raman level, however the lower excited state number density implies that saturation effects will become important at low pump intensities (Sect. 4.6.1).

Bethune et al. [4.222] have shown that SERS may be utilized to generate broad-band ir radiation for single-shot spectroscopy experiments. When operated in a dispersionless medium, the Raman conversion efficiency and threshold pumping intensity are independent of the band width of the pump laser. This effect was utilized to generate 5 ns duration Stokes beams with a spectral width of ~400 cm^{-1} by Raman scattering a very broad band, ASE dye laser in various alkali vapors. The resulting broad-band ir radiation was transmitted through an absorbing species and then up-converted into the visible by means of another nonlinear mixing cell. The ir spectra thus obtained could be recorded on photographic film with 5 ns temporal resolution. The authors named this technique TRISP, for time-resolved infrared spectral photography. In a typical K Raman cell, a 200 kW broad-band pump laser pulse was converted yielding 2 kW in the first Stokes. Using a variety of alkali atom resonances and dye laser pump wavelengths, the spectral region from $2-11$ μm can be covered continuously by this technique [4.222, 227]. A large number of molecular species have been studied using TRISP [4.222 – 227].

As an efficient means for shifting the output from uv laser sources to longer wavelengths, SERS has also been extensively studied. Excimer lasers have received considerable attention as pump sources, due primarily to their high efficiency, high peak powers, and ease of operation. In addition, excimer lasers provide coverage over large portions of the uv spectral region, and can be run as narrow bandwidth, low divergence, tunable coherent sources. Raman shifted excimer lasers are becoming increasingly important for a wide variety of applications, ranging from underwater communications to photo-chemistry.

The first experiment with Raman shifting an excimer laser via SERS was performed by *Djeu* and *Burnham* [4.228]. Barium vapor was used to convert the 351 nm output of a XeF laser to the first Stokes at 585 nm. The $6s^2\,^1S_0 - 5d\,^1D_2$ Raman transition was used for a 11 395 cm^{-1} down-shift of the incident photon energy. The pump source was within ~ 70 cm^{-1} of the intermediate $6p\,^1P^0$ resonance level of the Ba atom, thereby insuring a large resonant gain enhancement. A photon conversion efficiency of 80% was achieved by using a frequency-narrowed, high mode quality XeF pump laser.

Raman shifting the XeCl laser output (308 nm) in Ba has been studied by *Cotter* and *Zapka* [4.229]. By utilizing the $6s^2\,^1S_0 - 5d\,^1D_2$ Raman transition, the incident 308 nm laser was shifted to blue-green Stokes radiation at 475 nm. In this case the input laser frequency lies within ~ 125 cm^{-1} of the $7p\,^1P_1^0$ intermediate level. In spite of the fact that the excimer laser beam was of poor spatial quality, with an angular beam divergence more than 25 times greater than the diffraction limit, a photon conversion efficiency of 20% was measured; and an output energy of 5 mJ/pulse was seen at 475 nm. *Burnham* and *Djeu* [4.230] have also studied XeCl excitation of Ba vapor, and they report a photon conversion efficiency as large as 31% to the S_1 beam at a Ba vapor pressure of 10 Torr.

Raman shifting of the XeCl laser (308 nm) in Pb vapor has been studied by several groups. *Burnham* and *Djeu* [4.230] reported efficient generation of the first Stokes wave at 459 nm in a 15 cm long Pb vapor cell. In this case the $6p^2\,^3P_0 - 6p^2\,^3P_2$ Raman transition yields a Raman shift of 10 650 cm^{-1}. At a Pb vapor pressure of 30 Torr, an energy conversion efficiency of 40% (corresponding to a photon conversion efficiency of 60%) was measured. Under these conditions the 50 mJ XeCl pump beam was converted to 20 mJ of light at 459 nm. This Pb system has also been examined by *Brosnan* et al. [4.231] using a 2 J/pulse, nearly diffraction limited XeCl pump source. An output energy of nearly 1 J/pulse was observed at 459 nm for an energy conversion efficiency of 50%.

The XeCl (308 nm) laser has also been exploited as an efficient pump source with both bismuth and thallium vapor Raman cells [4.230]. In Bi the $6p^3\,^4S_{3/2}^0 - 6p^3\,^2D_{3/2}^0$ Raman transition was used; this provided a Raman shift of 11 419 cm^{-1} yielding an output Stokes beam at 475 nm. The optimum Bi pressure for Raman conversion was 10 Torr, where approximately 10% of the incident pump energy was converted. At higher pressures, absorption at 308 nm due to Bi$_2$ dimer molecules severely limited the Raman process. In Tl the 7 793 cm^{-1} splitting of the ground $^2P^0$ states was used as the Raman transition. Stimulated electronic Raman scattering in Tl on the S_1 line at 405 nm was observed at vapor pressures greater than 1 Torr (for a 15 cm path length Raman cell). As the Tl vapor pressure was increased, the S_2 component at 593 nm appeared; and at vapor pressures exceeding a few Torr, the 593 nm output dominated the Raman emission. By adjusting the Tl pressure, up to

10% of the 308 nm energy could be converted into the S_1 or S_2 beams or into a combination of both.

Tin vapor has been studied by *Djeu* [4.232] as an efficient medium for Raman shifting the KrF laser output at 248 nm to 285 nm in the uv. A diffraction-limited KrF laser producing 10 μJ in an 8 ns pulse was focused to an intensity of 100 MW/cm^2 in a 5 cm long Sn cell. At a metal-vapor pressure of 3 Torr, a peak power conversion efficiency of 40% to S_1 (285 nm) was observed.

Finally, *Manners* [4.233] investigated the use of Ba vapor as a means of shifting a XeCl (308 nm) laser to the ir. Stokes radiation at 2.36 μm was produced using the 28 230 cm^{-1} shift of the $6s^2\,^1S_0 - 7s\,^1S_0$ Raman transition. Photon conversion efficiencies up to 1% were observed.

b) Anti-Stokes Generation

Stimulated anti-Stokes generation from inverted atomic (or molecular) species has recently attracted considerable attention. Anti-Stokes stimulated electronic Raman scattering (ASERS) is of particular interest as a potential means for generating high-power, tunable, vuv lasers from existing uv sources. The typical anti-Stokes system is shown schematically in Fig. 4.34a, where $|i\rangle$ and $|f\rangle$ represent the initial and final Raman levels, respectively, $|r\rangle$ is the intermediate state, ω_1 is the pump laser frequency, and ω_2 is the anti-Stokes frequency. It is first necessary to prepare the Raman medium in an excited state, such that a population inversion exists between the $|i\rangle$ and $|f\rangle$ Raman states. Therefore, ASERS differs significantly from anti-Stokes mixing processes (cf., high-pressure H_2 gas), in which a parametric four-wave interaction is responsible for the anti-Stokes radiation. In the latter case anti-Stokes emission occurs in cones due to phase-matching requirements; whereas no phase-matching condition is required for ASERS. Several of these lasers have recently been realized in atomic systems, where the process is often termed an anti-Stokes Raman laser (ASRL).

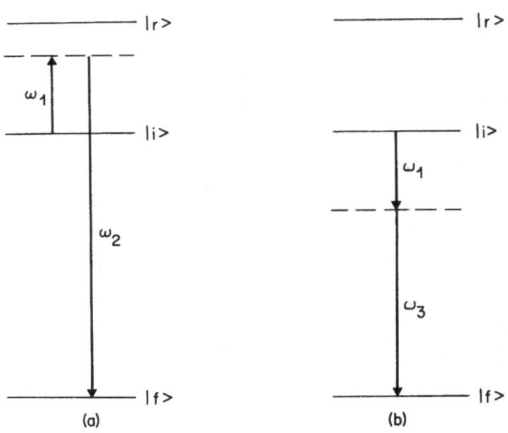

Fig. 4.34. Schematic energy level diagram for two-photon amplification schemes for (a) anti-Stokes laser generation and (b) two-photon emission

Anti-Stokes lasers are of interest for several reasons. First, being Raman processes, the anti-Stokes output light in inherently tunable (to the extent that the pump laser tunes). Also the necessary population inversion occurs on the two-photon $|i\rangle \rightarrow |f\rangle$ transition. If no single-photon allowed states occur between the initial and final Raman levels, as shown in Fig. 4.34a, then the Raman inversion is stored in the metastable $|i\rangle$ state. This is a great advantage for many applications, since in principle, the population inversion may be stored slowly over an extended period of time and then rapidly extracted via ASERS with a fast pump pulse. The pump photon is thereby up-converted to a shorter wavelength. Note, however, that the maximum anti-Stokes energy gain over the pump laser energy is limited to ω_2 / ω_1.

The anti-Stokes laser is in several ways related to the so-called two-photon laser (TPL), shown in Fig. 4.34b. In a TPL device it is also necessary to prepare the system such that a population inversion exists along the $|i\rangle \rightarrow |f\rangle$ two-photon transition. The stored energy is extracted from the system with an intense trigger pulse ω_1 which induces stimulated two-photon emission at the complementary frequency ω_3. Here $\hbar\omega_1 + \hbar\omega_3$ equals the energy splitting of the $|i\rangle \rightarrow |f\rangle$ transition. Unlike the ASERS process, the two-photon laser is not ultimately limited by the pump laser energy; and once initiated will proceed until the metastable population inversion has been exhausted. Note that in a TPL both the input field ω_1 and the generated field ω_3 see gain; this is in contrast to the anti-Stokes process in which ω_2 is amplified and the pump field ω_1 is depleted. *Carmen* [4.234] has proposed that such devices would be useful as high-energy lasers for laser initiated thermonuclear fusion efforts.

Although two-photon lasers were proposed several years ago [4.235, 236], the efforts to construct such a laser have been unsuccessful [4.234, 237, 238]. One difficulty lies in the conflicting requirements for the device. On the one hand, the TPL storage level should be metastable with no single-photon allowed intermediate states $|r\rangle$ between the initial and final states. However, in order to ensure a large two-photon gain cross section, one would like an intermediate state to lie between $|i\rangle$ and $|f\rangle$, thereby resonantly enhancing the susceptibility for the process. The anti-Stokes laser, however, can effectively couple to an intermediate level which lies above the $|i\rangle$ state, and as such, is a much easier process to initiate. In spite of these difficulties some success has been realized on the TPL problem using the technique of adiabatic rapid passage to obtain two-photon inversions [4.239, 240]. Additional analysis of the TPL problem has been given by several researchers [4.241 – 243].

Carmen [4.234] has performed a detailed analysis of the interplay and competition between the TPL and ASERS in a two-photon system. This analysis shows that under certain special conditions it is possible for the TPL to dominate the anti-Stokes process. However, even when the TPL predominates in the early stages, parametric coupling between the fields will eventually initiate the anti-Stokes process. The ASERS will then dominate in most situations, unless the population inversion is depleted very early in the laser pulse. For these reasons, anti-Stokes lasers are far easier to create, and in recent years several successful devices have been operated.

The first observation of ASERS was made by *Sorokin* et al. [4.244] in potassium vapor. The output of a Q-switched ruby laser was Stokes shifted using a liquid nitrobenzene Raman cell to within 11 cm^{-1} of the $K(4s\,^2S_{1/2} - 4p\,^2P^0_{3/2})$ resonance line frequency. Anti-Stokes emission with a 58 cm^{-1} energy up-shift was observed. This Raman shift corresponds to the spin-orbit splitting of the $4p\,^2P^0$ doublet states; where the $4p\,^2P^0_{3/2}$ level acts as the initial Raman level, and the $4p\,^2P^0_{1/2}$ level is the final Raman state. Resonant enhancement of the Raman gain was provided by the proximity of the input laser frequency to the strongly allowed resonance transition (downward) to the $4s$ ground state. This type of resonance enhancement process is illustrated in Fig. 4.1 b. The necessary population inversion between the $4p\,^2P^0_{3/2}$ and $4p\,^2P^0_{1/2}$ Raman states was achieved by optical pumping into the wings of the $4s\,^2S_{1/2} - 4p\,^2P^0_{3/2}$ transition with the shifted ruby laser light.

Vinogradov and *Yukov* [4.245] first proposed the use of the $7\,603 \text{ cm}^{-1}$ spin-orbit splitting between the iodine $5p\,^2P^0_{1/2} - 5p\,^2P^0_{3/2}$ ground state doublet. The population inversion would be created between the metastable $^2P^0_{1/2}$ and the ground $^2P^0_{3/2}$ states. Studies of atomic I vapor as an anti-Stokes gain medium were conducted by *Carmen* and *Lowdermilk* [4.246, 247]. The $5p\,^2P^0_{1/2}$ metastable inversion was created by flash photolysis of CF_3I vapor, yielding a peak inversion density from $(2-5) \times 10^{16}$ atoms/cm^3. Gain from ASERS scattering was measured by pumping the population inversion with the 1.06 µm fundamental wavelength of a Nd:YAG laser and probing with a broadband dye laser at 588.4 nm (the anti-Stokes wavelength). An exponential gain of $\exp(7)$ over the 50 cm path length was measured in good agreement with theory. In their experiments, however, Carmen and Lowdermilk were unable to initiate anti-Stokes oscillation, due primarily to the unfavorably large (i.e., $40\,000 \text{ cm}^{-1}$) Raman detuning with the intermediate Raman level.

The first successful operation of an anti-Stokes laser was performed by *White* and *Henderson* [4.248] using a metastable population inversion in thallium vapor. The Raman transition occurred between the $Tl(6p\,^2P^0_{3/2} - 6p\,^2P^0_{1/2})$ spin-orbit split doublet states, and the Raman up-shift in this case was $7\,793 \text{ cm}^{-1}$. The metastable $Tl(6p\,^2P^0_{3/2})$ inversion was created by selective photodissociation of a suitable Tl donor molecule [4.249]. In the first experiment a vapor of TlCl was selectively photodissociated using the 193 nm output of an ArF excimer laser to yield a high-density inversion of $6p\,^2P^0_{3/2}$ atoms. At a TlCl density of 7×10^{16} molecules/cm^3, an inversion density of 4×10^{16} atoms/cm^3 was created over the 25 cm active zone of the TlCl cell. Raman scattering from the $^2P^0_{3/2}$ level was driven using the second (532 nm) and third (355 nm) harmonic frequencies of a Q-switched Nd:YAG laser, as shown in Fig. 4.35a and b, respectively. At a 532 nm pump energy of 25 mJ, a pulse energy of 1.8 mJ in ~7 ns was seen at the 376 nm AS_1 wavelength. Also, an anti-Stokes energy of 2.5 mJ in a 5 ns duration pulse was measured at 278 nm using a 100 mJ, 355 nm pump pulse. Additional studies were

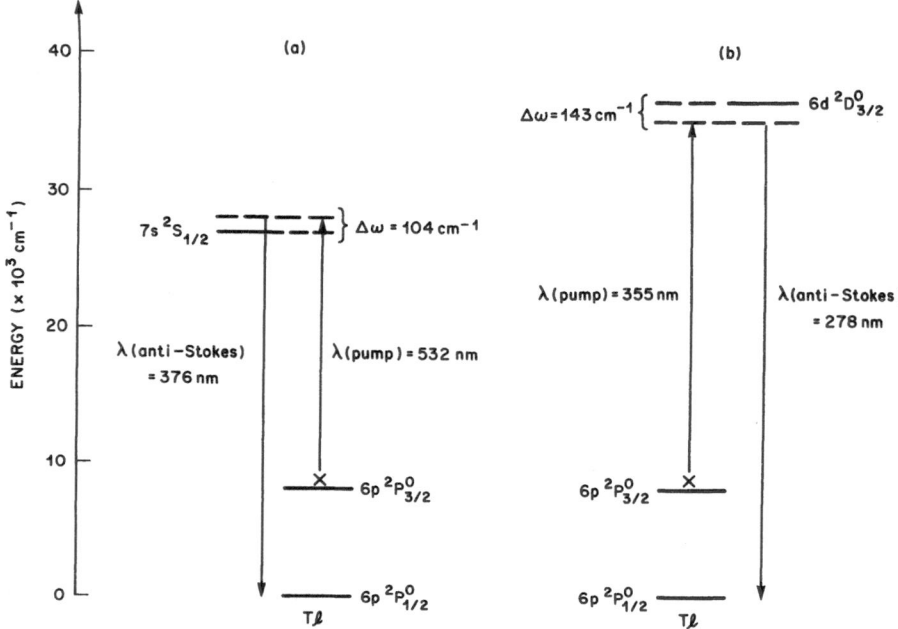

Fig. 4.35. Caption see opposite page

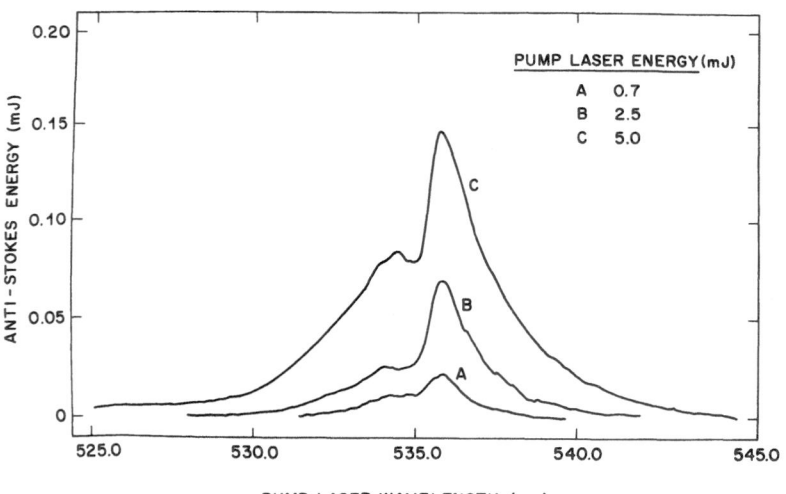

Fig. 4.36. Caption see opposite page

performed [4.250] indicating that Tl $(6p\,^2P^0_{3/2})$ population inversions $>2\times10^{15}$ atoms/cm^3 could be created using a 248 nm KrF laser to selectively dissociate either TlCl or TlI molecules.

Subsequent to those early studies, *White* and *Henderson* utilized a tunable dye oscillator-amplifier system to pump the Tl anti-Stokes laser [4.251, 252]. A tuning range in excess of $1\,100\,\mathrm{cm}^{-1}$ was observed about the $7s\,^2S_{1/2}$ intermediate Raman level for a single-pass Raman cell; this tuning range was limited primarily by the available energy from the dye laser (10 mJ in these experiments). The anti-Stokes output energy as a function of pump laser wavelength is shown in Fig. 4.36 for various pump laser intensities. The $6p\,^2P^0_{3/2}$ inversion was $\sim1.8\times10^{15}$ atoms/cm^3 for these measurements. At low-input intensities just above pump threshold for the Raman process, the tuning profile was symmetric about the $7s\,^2S_{1/2}$ intermediate resonance (occurring at 535 nm). As the pump laser intensity was increased, a noticeable dip occurred in the anti-Stokes tuning spectrum as the pump laser was tuned through the $7s$ intermediate state. This dip is due to competition of the ASERS process with single-photon absorption and pump depletion arising from resonant excitation of the $7s$ level. *Ludewigt* et al. [4.253] have suggested that the asymmetry in the anti-Stokes tuning profile is due to self-focusing/defocusing effects when the pump laser is tuned near the intermediate resonance. On the red-wavelength side of the intermediate state, defocusing of the pump laser results in slightly higher output energy as the volume of Tl$(6p\,^2P^0_{3/2})$ atoms pumped is increased in the saturated regime. Similar tuning profiles were obtained in [4.253] using the $7s\,^2S_{1/2}$ and $6d\,^2D^0_{3/2}$ levels as intermediate states for anti-Stokes generation near 377 and 277 nm, respectively. Energy conversion efficiencies to AS$_1$ (377 nm) of up to 80% have been achieved in the Tl system [4.254]. It should come as no surprise that the anti-Stokes tuning profiles of Fig. 4.36 are very similar in appearance to typical Stokes tuning profiles, as illustrated in Fig. 4.31. The effects of collisions on the Tl system were considered by *Cunningham* et al. [4.255].

Inverted, metastable indium vapor was employed by *White* and *Henderson* [4.256] as an anti-Stokes medium. The In system is structurally very similar to the previously studied Tl system in that the Raman inversion is prepared between the spin-orbit split $5p\,^2P^0_{3/2}$ metastable and $5p\,^2P^0_{1/2}$ ground states. The Raman up-shift in In is a more modest $2\,212\,\mathrm{cm}^{-1}$, however. The $^2P^0_{3/2}$ inversion is again prepared by selective photodissociative pumping of an In compound; and for the experiment reported in [4.256], InI vapor was selectively dissociated with a 248 nm KrF laser to yield an excited state

---

Fig. 4.35 a, b. Pertinent energy level for anti-Stokes Raman lasing from inverted, metastable thallium atoms. (a) Raman emission at 376 nm using a 532 nm pump source. (b) Raman emission at 278 nm using a 355 nm pump source [4.248]

Fig. 4.36. Anti-Stokes laser tuning profile versus pump laser wavelength for Raman scattering in inverted, metastable thallium vapor. Curves *A*, *B*, and *C* were taken at pump energies of 0.7, 2.5, and 5.0 mJ, respectively [4.252]

inversion density of about 4×10^{14} atoms/cm^3. A tunable, 451 nm dye laser was used as a pump source to couple the Raman levels through the $6s\,^2S^0_{1/2}$ intermediate level. Approximately 15 μJ/pulse was generated at AS$_1$ (410 nm) with a tuning range of 35 cm^{-1}. The low-output energy and limited tuning range were due to the somewhat low population inversion achieved. The effective storage time of the In $(5p\,^2P^0_{3/2})$ state was measured by delaying the Raman pump pulse relative to the photodissociation pulse. At an InI density of $\sim 10^{16}$ molecules/cm^3, the effective storage time was about 120 ns and was independent of the buffer gas pressure used in the cell over the range of $10 - 100$ Torr. This effective lifetime is probably due to quenching collisions with the parent InI species.

Ludewigt et al. [4.253] have studied population inversions in tin vapor as a medium for ASERS. The relevant energy level diagram and pump, anti-Stokes wavelengths are shown for the Sn system in Fig. 4.37. Population inversions between the $^1S_0$ and $^1D_2$ states and the $^3P_{0,1,2}$ fine structure components of the ground state were prepared by selective photodissociation of SnBr$_2$ vapor with KrF laser radiation at 248 nm. The various pump laser wavelengths employed, and the generated anti-Stokes lasers are tabulated in Table 4.3. The output energies at the anti-Stokes lines were generally <100 μJ/pulse, and the emissions were limited to narrow tuning ranges

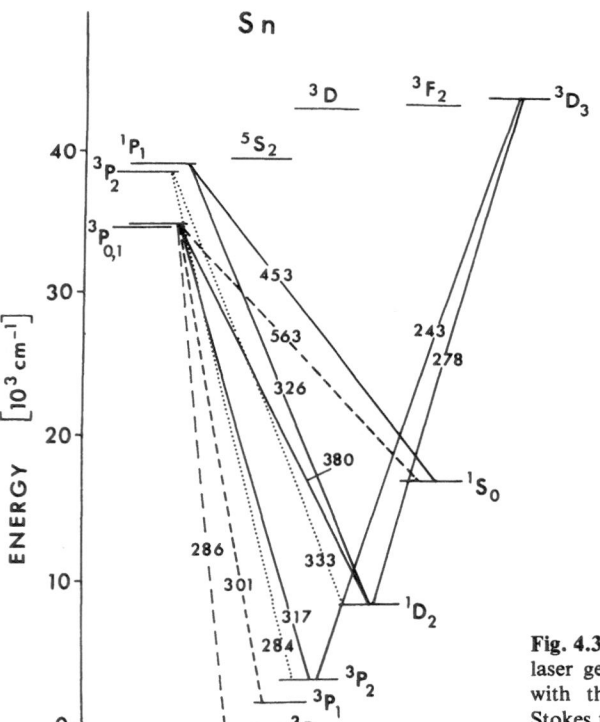

Fig. 4.37. Level scheme for anti-Stokes laser generation in inverted tin vapor with the observed pump and anti-Stokes transitions [4.253]

Table 4.3. Data on Sn anti-Stokes Raman lasers, after [4.253]

Pump		Anti-Stokes	Transition	Inter-mediate
Wave-length [nm]	Energy [mJ]	wave-length[a] [nm]		state
563	10	301	$5p\,^1S_0 - 5p\,^3P_0$	$6s\,^3P_1$
		286	$5p\,^1S_0 - 5p\,^3P_1$	$6s\,^3P_1$
453	10	326	$5p\,^1S_0 - 5p\,^1D_2$	$6s\,^1P_1$
380	5	317	$5p\,^1D_2 - 5p\,^3P_2$	$6s\,^3P_1$
333	4	284	$5p\,^1D_2 - 5p\,^3P_2$	$6s\,^3P_2$
278[b]	1.5	243	$5p\,^1D_2 - 5p\,^3P_2$	$5d\,^3D_3$

Experimental conditions: Photodissociation of $SnBr_2$ by KrF (249 nm) laser. Photodissociation energy typically 60 mJ. Heat pipe: vapor zone: 10 cm, temperature: 520 K.
[a] At present all lines oscillate with low output energy (<100 μJ) just around resonance (tuning range <0.5 nm).
[b] Frequency-doubled dye.

(<0.5 nm) near the intermediate resonances. The authors suggested that these effects were due to a relatively low inversion density created by the two-step photodissociation process and by incomplete recombination and quenching of the excited states by other photo-fragments. Still, the Sn system is attractive for further research, since a Raman shift up to 17 163 cm^{-1} was observed with the $^1S_0 - {}^3P_0$ transition. This is currently the largest energy shift that has been observed in an operational anti-Stokes laser.

Two-step photodissociation of PbI_2 has been used by the same group [4.254] as a means of preparing Pb inversions along the $6p^2\,^1S_0 - 6p^2\,^1D_2$ and $6p^2\,^3P_2 - 6p^2\,^3P_0$ Raman transitions. Laser radiation at 248 nm from a KrF excimer source was used to dissociate the PbI_2; a PbI_2 cell temperature of 450°C was used. A tunable 500.5 nm dye laser was used to pump the $^1S_0$ inversion, resulting in anti-Stokes radiation at 357.3 nm. The Pb ($7s\,^1P_1^0$) level served as the intermediate level in this case. Further investigations of anti-Stokes emission on the $^3P_2 - {}^3P_0$ ground-state inversion are underway at present.

Anti-Stokes laser experiments with metastable inversions in halogen atoms were conducted by *White* and *Henderson* [4.257, 258]. Inverted iodine vapor was used to construct the first vuv anti-Stokes laser [4.257]. The ground-state I ($5p^5\,^2P_{1/2}^0 - 5p^5\,^2P_{3/2}^0$) splitting was used as the Raman transition yielding a 7 603 cm^{-1} up-shift. The I ($5p^5\,^2P_{1/2}^0$) metastable inversion was created by selective photodissociation of NaI with the 248 nm output of a KrF laser. NaI was chosen as the I-donor molecule, since the ground state Na atom liberated in the dissociation step would not strongly absorb the pump laser or anti-Stokes wavelengths. A tunable 206 nm source was used to drive the anti-Stokes process near the $6s\,^2P_{3/2}$ intermediate state. Anti-Stokes laser emission at 178 nm was seen, and an energy conversion efficiency of ~50% was achieved. By delaying the arrival of the pump laser pulse with respect to the dissociation pulse, an effective I storage time up to 0.5 μs was observed. The 178 nm laser could be tuned $5-6$ cm^{-1} about the $6s\,{}^2P_{3/2}$ intermediate state. Although the output power was limited to 7 kW, the authors point out that the I system might be improved by using one of the many organic I-donor molecules that have been so successfully employed in the 1.3 μm I laser development [4.259 – 262]. Furthermore, the use of a Raman shifted KrF laser (i.e., AS$_2$ of 248 nm in high pressure H$_2$ gas) should permit 178 nm output energies of a few mJ or more.

The deepest, vuv anti-Stokes laser created to date was reported by *White* and *Henderson* [4.258] in a medium of inverted bromine atoms. As in the previous I example, the population inversion was created in the upper doublet level of the spin-orbit split ground state. This Br $(4p^5\,{}^2P^0_{1/2}-4p^5\,{}^2P^0_{3/2})$ Raman transition at 3 685 cm^{-1} was used to up-shift the 157 nm output of an F$_2$ discharge laser to 149 nm. Selective dissociation of NaBr with 250 nm photons was used to prepare the Br $(4p^5\,{}^2P^0_{1/2})$ inversion. Raman scattering of the 157 nm input laser was particularly favorable in the Br case, since the laser frequency was a mere 8 cm^{-1} from the $5s\,{}^2P_{3/2}$ intermediate level. Up to 0.1 mJ of 149 nm light was created in a 5 ns duration pulse, corresponding to an energy conversion efficiency of ~20%. Since the F$_2$ pump laser is not tunable, the 149 nm laser emission could not be tuned.

Two proposals for future anti-Stokes laser development aimed at generating high-power uv and vuv sources were given by *White* [4.263, 264]. The possibility of utilizing the metastable inversions available in Tl, In, I, and Br atoms (as discussed in previous paragraphs) to up-convert various excimer laser wavelengths was considered in [4.263]. A summary of possible anti-Stokes systems, the pump laser wavelengths, the generated anti-Stokes wavelengths, and the Raman gain cross sections is given in Table 4.4. Many new lasers in the 200 – 325 nm region should be possible with this technique. Note, that the 149 nm Br laser and the 178 nm I laser have already been demonstrated [4.257, 258].

Metastable inversions using the ${}^1S_0$ states of the group VI elements (oxygen, sulfur, and selenium) have been suggested as host media for a new class of high-power, vuv anti-Stokes lasers [4.264]. The group VI elements have been extensively studied as potential high-energy lasers using the low-gain ${}^1S_0-{}^1D_2$ ("auroral") and ${}^1S_0-{}^3P_{0,1}$ ("transauroral") transitions; a good review of these studies and the underlying physics is given by *Murray* and *Rhodes* [4.265]. The ${}^1S_0$ inversions are again created by photodissociation of selected molecular species. Early work in this area showed that efficient, pulsed electron-beam machines could be used to excite rare-gas dimer fluorescence of an appropriate wavelength to drive the dissociation processes with quantum yields >0.9 in each case. Metastable inversion densities greater than 10^{16} atoms/cm^3 could be created and stored for 1 μs or more. These species could be utilized as anti-Stokes laser media in two ways. In one case

Table 4.4. Possible anti-Stokes up-converters for excimer lasers using Tl, In, I, or Br population inversions, after [4.263]

Medium	Initial state	Final state (Energy)	Pump laser (Wavelength)	Anti-Stokes wavelength	Intermediate state (Energy)	Raman detuning, $\Delta\nu$ [cm^{-1}]	Gain cross section [cm^4/W]
Br	$4p^5\,^2P^o_{1/2}$	$4p^5\,^2P^o_{3/2}$ (3685 cm^{-1})	F$_2$ (157 nm)	149 nm	$5s\,^2P_{3/2}$ (65743 cm^{-1})	8	8.6×10^{-23}
I	$5p^5\,^2P^o_{1/2}$	$5p^5\,^2P^o_{3/2}$ (7603 cm^{-1})	Xe$_2$ (172 nm)	152 nm	$6s\,^2D_{3/2}$ (65743 cm^{-1})	600	2.0×10^{-26}
I	$5p^5\,^2P^o_{1/2}$	$5p^5\,^2P^o_{3/2}$ (7603 cm^{-1})	ArF (193 nm)	169 nm	$6s\,^2P_{1/2}$ (63187 cm^{-1})	3878	2.0×10^{-27}
Tl	$6p\,^2P^o_{3/2}$	$6p\,^2P^o_{1/2}$ (7793 cm^{-1})	XeBr (282 nm)	231 nm	$9s\,^2P_{1/2}$ (43166 cm^{-1})	113	2.2×10^{-26}
In	$5p\,^2P^o_{3/2}$	$5p\,^2P^o_{1/2}$ (2212 cm^{-1})	XeBr (282 nm)	265 nm	$7s\,^2S_{1/2}$ (36301 cm^{-1})	1397	1.6×10^{-25}
Tl	$6p\,^2P^o_{3/2}$	$6p\,^2P^o_{1/2}$ (7793 cm^{-1})	XeCl (308 nm)	248 nm	$7d\,^2D_{1/2}$ (42011 cm^{-1})	1750	3.0×10^{-27}
In	$5p\,^2P^o_{3/2}$	$5p\,^2P^o_{1/2}$ (2212 cm^{-1})	XeCl (308 nm)	288 nm	$5d\,^2D_{3/2}$ (32892 cm^{-1})	1788	1.0×10^{-24}
Tl	$6p\,^2P^o_{3/2}$	$6p\,^2P^o_{1/2}$ (7793 cm^{-1})	XeF (353 nm)	277 nm	$6d\,^2D_{3/2}$ (36117 cm^{-1})	11	1.3×10^{-21}
In	$5p\,^2P^o_{3/2}$	$5p\,^2P^o_{1/2}$ (2212 cm^{-1})	XeF (351 nm)	326 nm	$5d\,^2D_{3/2}$ (32892 cm^{-1})	2198	6.6×10^{-25}
I	$5p^5\,^2P^o_{1/2}$	$5p^5\,^2P^o_{3/2}$ (7603 cm^{-1})	AS$_2$ (KrF) (206 nm)	178 nm	$6s\,^2P_{3/2}$ (56093 cm^{-1})	77	7.4×10^{-24}

the $^1S_0 - {}^1D_2$ Raman transition could be utilized yielding up-shifts of ~2 eV. The pump laser could also be tuned near intercombination transitions so that a 3 – 4 eV photon energy gain from the $^1S_0 - {}^3P_{0,1,2}$ Raman transition would be realized. Many new anti-Stokes lasers in the 100 – 200 nm region should be possible.

To date, all of the successful anti-Stokes lasers have utilized metastable population inversions which have been prepared via selective photodissociation of donor molecules with uv lasers. Future research on anti-Stokes devices will no doubt center on the characterization of new media and other more efficient or convenient means for preparing excited state inversions. We have already mentioned the use of a high-energy, pulsed e-beam to excite rare-gas dimer fluorescence as a means of pumping selective molecular dissociation to create the excited state inversions [4.264]. *Rinke* [4.266] has studied the use of incoherent, flash-lamp excitation to drive the TlI dissociation process; but to date, only modest metastable inversions have been achieved. The possibilities for creating metastable population inversions for the ASERS process via selective chemical reactions are also very exciting. Chemically driven, 1.3 μm I lasers have already been demonstrated [4.265 – 270]. These methods may be directly applicable to the I anti-Stokes laser, although special attention will have to be given to the potential problem of species with large vuv absorptions within the chemical reaction vessel. Chemically excited molecular CO has also been proposed as a potential vuv anti-Stokes media by *White* et al. [4.271]. Electrical discharges might also yield suitable population inversions, and *Komine* and *Byer* [4.272] proposed one such system in Hg vapor. Ionic inversions have recently been demonstrated based on selective inner-shell ionization [4.273], and such techniques look promising for anti-Stokes laser applications.

4.5.3 Backward Raman Scattering and Pulse Compression

The possibility that the energy in a pump laser pulse could be compressed and shifted by means of transient backward Raman scattering was examined theoretically in Sect. 4.4.2. Recently, there has been renewed interest in backward Raman scattering as a means of compressing in time high-power optical pulses. Much of the current research has been motivated by the emergence of the rare-gas halide lasers and their potential application to laser-driven thermonuclear fusion experiments. Unfortunately, while several of the rare-gas halide lasers show efficiencies of several percent, their pulse lengths are generally tens of nanoseconds and are too long to be suitable for present laser fusion plans. Backward Raman pulse compression may be an efficient means of reshaping rare-gas halide laser pulses to sub-nanosecond durations.

The basic design of a backward Raman laser system was first considered by *Glass* [4.274], and these basic principles are illustrated in Fig. 4.38. A pump laser beam of pulse length τ is sent into the Raman cell of length $\tau/2$ and propagates from the right to the left. A short pulse Stokes beam is injected into the cell (at the left end) just as the leading edge of the pump pulse begins to exit the Raman cell. As the Stokes pulse propagates from left to right in the Raman cell, the Stokes beam is amplified. Note, that in this process the Stokes beam continually encounters a "fresh", undepleted pump pulse. After a time $\tau/2$ the Stokes pulse emerges from the Raman cell in the backward direction with respect to the pump pulse. Since the seed Stokes pulse can be much shorter than the pump pulse, the energy available from the pump laser can be effectively compressed.

A very useful feature inherent in backward Raman scattering is that a high-quality Stokes beam (i.e., a well collimated beam with low divergence and smoothly varying intensity profile) can be amplified by a pump beam with high divergence and large intensity variations. This effect may be understood by considering an isolated point on the Stokes pulse as it propagates through the Raman amplifier. Since the Stokes beam experiences gain proportional to the local intensity of the pump beam, a nonuniform pump intensity profile will cause different gains and intensity ripples on the Stokes beam. These intensity fluctuations will be averaged out so long as different points on the Stokes pulse see the same integrated gain over the length of the Raman cell. In this sense, the backward Raman amplifier may be thought of as an

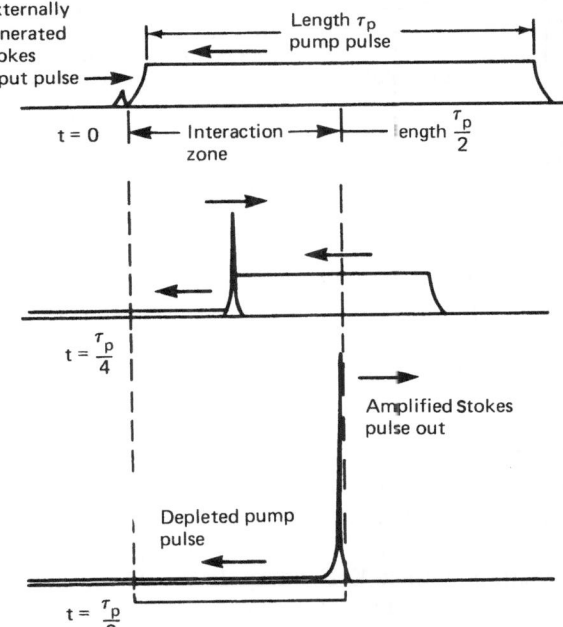

Externally
generated
Stokes
input pulse →

Length τ_p
pump pulse

t = 0 |← Interaction →|← length $\dfrac{\tau_p}{2}$
zone

$t = \dfrac{\tau_p}{4}$

Amplified Stokes
pulse out

Depleted pump
pulse

$t = \dfrac{\tau_p}{2}$

Fig. 4.38. Schematic representation of pulse compression by backward Raman scattering. A short duration Stokes pulse is amplified by a longer counter-propagating pump pulse [4.276]

energy storage laser in which the energy is stored in the input pump pulse rather than being stored in a material excitation. Calculations of gain and saturation effects can be addressed in a manner similar to that used to describe conventional laser amplifiers. The intensity averaging and beam clean-up effects rely, of course, on the conditions that the pump and Stokes pulses overlap everywhere in the amplifier, and that the intensity fluctuations in the pump beam must be sufficiently random that coherence effects due to any regular intensity variations are unimportant.

One of the first experimental studies of backward Raman amplifiers was published by *Maier* et al. [4.105]. In their experiment a superfluorescent backward Raman Stokes beam was observed in a ruby laser pumped, CS_2 Raman cell. The backward Raman pulses generated were gain sharpened to a duration of 30 ps and were approximately 10 times more intense than the incident laser pulse. *Maier* et al. [4.48] subsequently developed the basic theory of the backward Raman effect, and it is their treatment which has been used extensively in the discussion of the effect in Sect. 4.4.2.

Experiments with backward-wave amplification in high-pressure H_2 gas were conducted by *Culver* et al. [4.275]. A Q-switched, 10 ns duration ruby laser pulse was focused near the end of an H_2 cell operated at a pressure of ~7 atm. The back-scattered Stokes pulse was 0.3 ns in duration, and significant pump depletion was observed.

Various aspects of Raman pulse compression of excimer laser sources have been addressed in an excellent review article by *Murray* et al. [4.276]. These

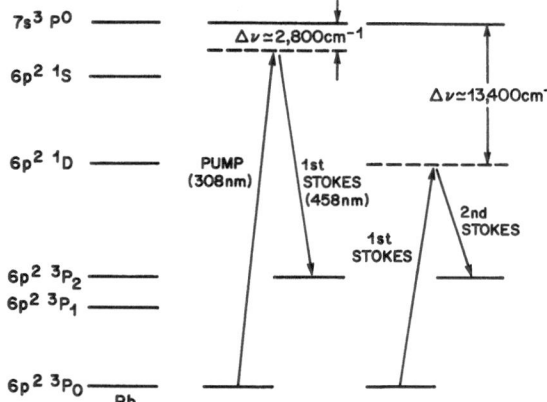

Fig. 4.39. Lowest-lying energy levels for atomic lead and their relation to the pump and first Stokes wavelengths. Note that Raman generation of S_2 from the first Stokes wave is unlikely, due to the relatively large Raman detuning [4.279]

authors were concerned with the application of efficient, uv excimer sources (and the 248 nm KrF laser, in particular) as driver lasers for inertial confinement, thermonuclear fusion experiments. A comparison of various Raman media was given, and high-pressure CH_4 gas was identified as an attractive medium in conjunction with a KrF pump laser. Their analysis showed that in general the most serious limitation to the performance of a Raman compressor is the generation of a backward second Stokes wave driven by the backward first Stokes component. The production of this second Stokes component is a *forward* scattering process and can cause significant depletion of the desired S_1 field.

Backward Raman gain measurements and characterization of saturation effects were conducted for CH_4 by *Murray* and co-workers [4.277, 278]. An input KrF laser was converted to S_1 (268 nm) light in a 2 m long cell operated at a CH_4 pressure of 27 atm. Pump depletion of 30% was obtained with an intensity gain of 2 in the 1 ns duration Stokes pulse.

Backward Raman compression of a XeCl (308 nm) laser in lead vapor was investigated by *Djeu* [4.279]. Lead vapor was chosen in an effort to avoid the parasitic second Stokes generation from the S_1 wave. A schematic energy level diagram for Raman generation in Pb is shown in Fig. 4.39. Note that although the Raman gain for S_1 generation is resonantly enhanced by the nearby $7s\,^3P^0$ levels, the gain cross section for S_2 generation is significantly reduced by the much larger Raman detuning. Generation of the second Stokes can be effectively repressed using such schemes. A 1 m long Pb cell operated at vapor pressure of 20 Torr was used to scatter a 20 ns duration, 20 mJ excimer laser pulse. A compression factor in excess of 6.7 was seen for the first Stokes pulse with a photon conversion efficiency of 42%.

4.5.4 Stimulated Hyper-Raman Scattering

Stimulated hyper-Raman scattering (SHRS) is the term used to describe the next higher-order Raman process involving three, rather than two, photons.

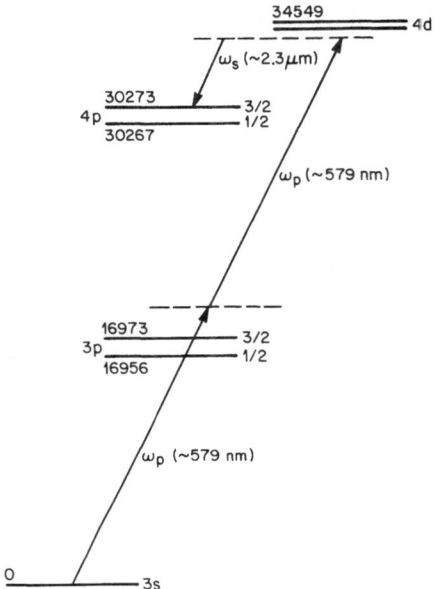

Fig. 4.40. Schematic energy level diagram for stimulated hyper-Raman scattering in atomic sodium using the $3s-4p_{3/2}$ Raman transition [4.292]

In its most widely studied form, SHRS is initiated by the absorption of two pump photons resulting in the emission of a single Raman output photon. The energy level diagram for SHRS in atomic Na is shown in Fig. 4.40; this type of level structure is typical of that used for hyper-Raman experiments. The atomic or molecular system makes a Raman transition between initial and final states of opposite parity. The hyper-Raman effect is described by a fifth-order nonlinear susceptibility, and the gain calculation is a straightforward extension of our earlier theoretical work. Note that the gain coefficient is inversely proportional to the square of two resonance detuning terms, corresponding physically to the two intermediate resonant states in the hyper-Raman process. Although all present experimental work on SHRS has involved Stokes generation, one could in principle generate stimulated anti-Stokes radiation from excited states with this technique.

The hyper-Raman effect was first observed as a weak spontaneous emission process by *Terhune* et al. [4.280]. Subsequently, SHRS was observed for the first time by *Yatsiv* et al. [4.281] in a somewhat complex arrangement using an excited state as the initial Raman level. A considerable amount of theoretical work has been published on various aspects of SHRS [4.282–290], and the reader is referred to the references for further details.

Hyper-Raman generation has been studied in a number of atomic systems. *Vrehen* and *Hikspoors* [4.291] studied SHRS in Cs vapor pumped by two photons at 1.06 μm from a Q-switched Nd:YAG laser. In their experiments the Cs atoms were excited from the $6s\,^2S_{1/2}$ ground state to the final $6p\,^2P^0_{3/2}$ Raman level thereby emitting a 1.416 μm Stokes photon. A relatively high pump laser intensity ($2.5\,\text{GW/cm}^2$) was required to reach threshold due primarily to the somewhat large $254\,\text{cm}^{-1}$ detuning of the two-photon excitation from the final intermediate Raman level. Vrehen and Hikspoors noted that these high pumping intensities produced noticeable optical Start shifting of the initial and final states and broadening of the Raman output. At an input pump laser power of 60 MW, a maximum Stokes power of 10 kW was generated.

Cotter et al. [4.292] used the Na system shown in Fig. 4.40 to generate tunable radiation near 2.3 μm. A tunable rhodamine 590 dye laser with a peak power of 500 kW was used as the pump source. By tuning the dye laser near the $3s-4d$ two-photon resonance (~ 579 nm), a maximum Stokes energy of 5 kW was obtained. The observed maximum output energy of 37 μJ was limited primarily by depletion of the Na ground state, nevertheless this represents a photon conversion efficiency of 2%. The Stokes output could be tuned approximately $160\,\text{cm}^{-1}$, from $4240-4400\,\text{cm}^{-1}$ in the ir.

Tunable 16 μm radiation was generated by SHRS in Sr vapor by *Reif* and *Walther* [4.293]. A 20 kW rhodamine 590 dye laser was tuned to ~ 576 nm to produce hyper-Raman scattering along the $5s^2\,{}^1S_0 - 6p\,{}^1P_1^0$ transition. The scattering was resonantly enhanced by tuning near the $5d\,{}^1D_2$ intermediate level. An ir output power of 20 mW was obtained, and this value was limited by the relatively low pump laser power available. The low incident power likewise limited the tuning range of the 16 μm Stokes beam to about $4\,\text{cm}^{-1}$.

Finally, cw hyper-Raman generation was demonstrated for the first time by *Krökel* et al. [4.294] using the same energy level scheme in Na vapor as in the work of *Cotter* et al. [4.292] (Fig. 4.40). In order to increase the pump intensity seen by the Sr atoms, the metal-vapor Raman cell was placed in a Fabry-Perot resonator for the dye radiation. A maximum timing range of 30 GHz was seen for the 2.3 μm Stokes beam.

4.6 Further Considerations and Limiting Processes

4.6.1 Saturation

Aside from any of the competing processes or loss mechanisms, stimulated Raman scattering processes (both Stokes and anti-Stokes) are, in general, limited by two inherent saturation effects. These saturation effects are related to the finite number of available pump photons (pump depletion) and the limited number density present in the initial Raman level (atomic or molecular depletion) [4.35]. Pump depletion occurs when number of Raman photons generated approaches the number of pump laser photons, and in this regime the intensity of the Raman output is linearly proportional to the pump-laser intensity. The second form of saturation is related to the depletion of the Raman medium itself and occurs when the relaxation time of the final Raman level is long compared to the pulse width of the incident laser. In that case the

Raman output is limited by the number of initial Raman states that are available to scatter the pump photons.

Both saturation mechanisms have been studied for Stokes and anti-Stokes generation schemes by several investigators. The experiments of *Carlsten* and *Dunn* [4.212] on stimulated Stokes generation in Ba vapor are representative of the results observed in many situations. As we described in Sect. 4.5.2a, a 791.1 nm dye laser was Raman scattered using the Ba $(6s^2\,^1S_0 - 5d\,^3D_{1,2})$ transition thereby creating tunable Stokes light at 2.9 μm. The observed 2.9 μm Stokes output energy is plotted as a function of the input dye laser energy for two different Ba vapor pressures in Fig. 4.41. We shall first examine the data corresponding to a Ba vapor pressure of 0.2 Torr. Carlsten and Dunn observed that as the dye laser energy was increased from a low value, the Stokes output rose rapidly to a maximum photon conversion efficiency of 40%. At higher dye laser energies (say, between 10 – 100 mJ), the Stokes output increased linearly with the pump-laser intensity. This regime corresponds to the "pump-depletion" mode in which conversion to the Stokes wave is limited not by the number of Ba atoms in the initial state, but by the absolute number of pump photons available for the Raman process. At still higher input energies (at constant Ba vapor pressure) the Stokes energy was seen to saturate at a constant value. This observation is consistent with the "atom-depletion" regime and corresponds to the situation in which the number of initial state Ba atoms in the beam path is approximately equal to the number of generated Stokes photons. In this case every available Ba atom has participated in the Raman process, and the Stokes output saturates at a constant value. The above explanation assumes that the relaxation time of the final Raman level is long compared to the duration of the input laser pulse; an assumption that is certainly valid for the Ba $(5d\,^3D_{1,2})$ metastable levels.

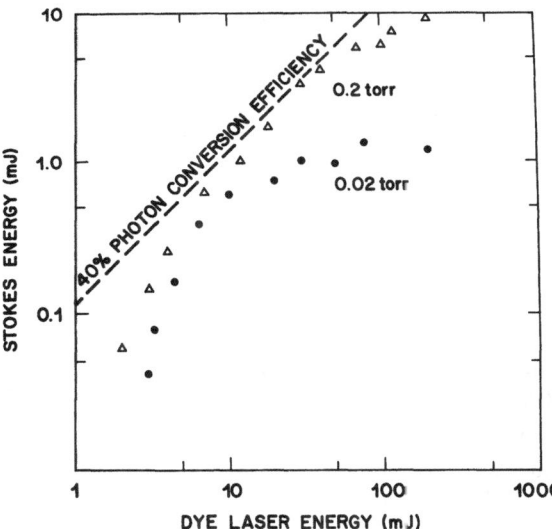

Fig. 4.41. Observed Stokes output energy versus incident dye laser energy for SERS in barium vapor, taken at two different metal-vapor pressures. The *dashed line* indicates a 40% photon conversion efficiency [4.212]

The data of Fig. 4.41 taken at the lower vapor pressure of 0.02 Torr exhibit a similar behavior. Threshold for SERS was seen to occur at a slightly higher dye-laser energy, since the number density of atoms was lower. In the range of 10 mJ of incident laser light, the Stokes energy was again seen to be linearly proportional to the dye laser energy. The onset of "atom depletion" occurred, as expected, at a lower dye laser energy, since fewer pump photons were required to exhaust the population in the initial Raman level.

These effects may be illustrated in a slightly different way by plotting the Stokes output energy as a function of the Ba vapor pressure, as in Fig. 4.42. At constant dye-laser energy, the Stokes output was seen to rise rapidly and then increase linearly with the Ba vapor pressure. This corresponds to the "atom-depletion" mode where the output is limited by the availability of atoms to Raman scatter the dye laser. At a pressure of ~0.2 Torr, however, the number of atoms is comparable to the number of pump photons and "pump depletion" occurs. Further increases in Ba vapor pressure have no effect on the Stokes output (ignoring any effects due dimer absorption, etc.), since all of the available pump photons have been converted.

Similar saturation behavior has been demonstrated in a variety of Raman systems. Atom-depletion effects have been observed by *Cotter* and *Hanna* [4.295] in Raman generation experiments using Cs vapor. Both pump- and atom-depletion effects have been demonstrated for the anti-Stokes Raman laser by *White* and *Henderson* [4.252] in a Tl Raman system. Similar results for the ASRL in the Tl were also obtained by *Ludewigt* et al. [4.253].

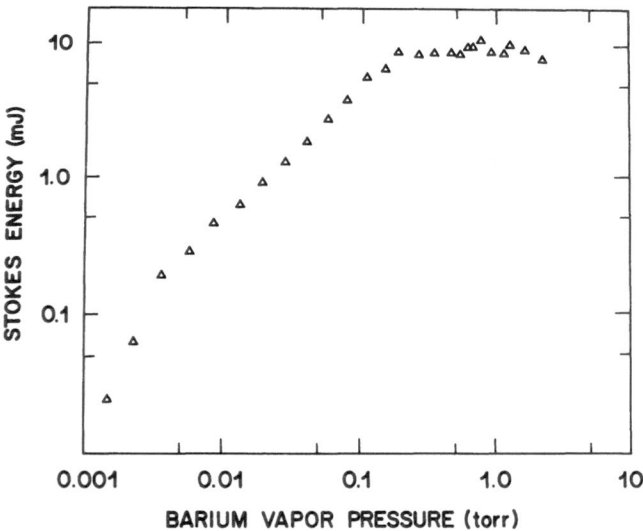

Fig. 4.42. Observed Stokes output energy versus vapor pressure for SERS in atomic barium vapor. Note the linear rise up to about 0.2 Torr and saturation at pressures greater than 0.2 Torr [4.212]

4.6.2 Broad-Band Pumping and Dispersion

In Sect. 4.3.1 we briefly discussed the conditions under which the steady-state Raman gain calculation was valid. In most Raman generation schemes there is only one effective intermediate state, in which case (4.27) predicts that the steady-state gain coefficient is inversely proportional to the square of the Raman detuning. Under steady-state conditions the pump intensity necessary to reach threshold for SRS should increase as Δ^2; and at a fixed input intensity and number density of the medium, this dependence will define the available tuning range of the Raman laser. As for the case of SERS in atoms, the gain coefficient can be enhanced resonantly by tuning the pump laser near a strongly-allowed intermediate resonance. Near these intermediate states, however, the index of refraction of the medium is changing rapidly, and the influence of group-velocity dispersion must be considered. These considerations are manifested in some interesting and important physical behavior.

Theoretical analysis of the behavior of SRS using a multimode, stochastically varying pump field has been given by *Akhmanov* and co-workers [4.108, 296]. These authors have shown that the Raman-gain coefficient for a multimode pump field is simply $G_{ss}\bar{I}_p(t)$, $\bar{I}_p(t)$ being the pump intensity averaged over the fluctuations. This implies that the gain coefficient is independent of the pump-laser bandwidth; this effect has been used with great success to generate broad-band Stokes light for spectroscopy (see discussion of TRISP in Sect. 4.5.2a).

As the characteristic time scale for changes in the pump intensity becomes shorter, however, group-velocity dispersion effects become important. One requires that the most rapid fluctuations of the pump pulse and the resulting Raman scattered pulse remain in synchronism as the waves travel down the medium. This defines an effective coherence length for Raman scattering [4.41]. *Akhmanov* and co-workers [4.108, 296] have shown that the effects of group-velocity dispersion may be neglected under the condition

$$\left| \left(\frac{d\omega}{dk} \right)_p^{-1} - \left(\frac{d\omega}{dk} \right)_R^{-1} \right| < \frac{G_{ss}\bar{I}_p(t)}{2\Delta\omega_p} , \qquad (4.75)$$

where $(d\omega/dk)_{p,R}$ are the pump laser and Raman group velocities and $\Delta\omega_p$ is the pump-laser bandwidth. Equation (4.75) implies that there is a range of pump-laser frequencies around the intermediate resonance for which the Raman threshold is independent of detuning. The extent of the region of constant threshold intensity is linearly proportional to the pump-laser bandwidth, and for many systems extents over tens of cm^{-1} to either side of the resonance. *Trutna* et al. [4.41] have shown that the preceding analysis is equivalent to the statement that dispersion effects must be considered when the bandwidth of the pump source exceeds the effective phase − matching bandwidth of the Raman medium.

Several experimental investigations have verified these predictions. The importance of group-velocity dispersion effects has been studied by *Mikhailov*

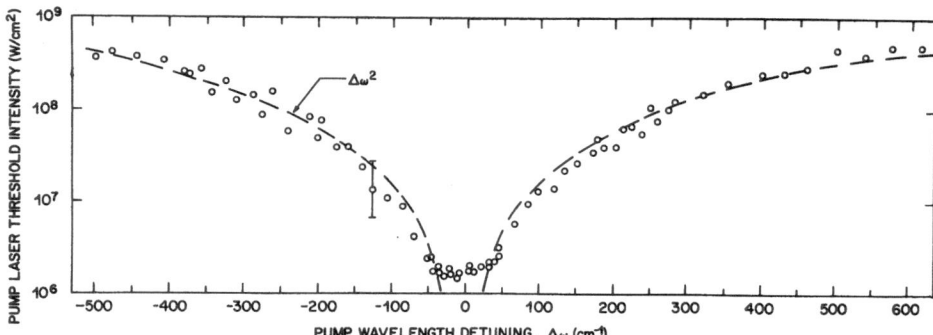

Fig. 4.43. Pump-laser threshold intensity for anti-Stokes lasing in inverted thallium vapor as a function of the detuning of pump laser from the intermediate Raman state. Positive detuning corresponds to tuning the pump laser to the high-energy (i.e., blue-wavelength) side of the intermediate state [4.251]

et al. [4.297] and *Korolev* et al. [4.298] for SERS in Rb vapor. Similar effects for anti-Stokes laser generation were demonstrated by *White* and *Henderson* [4.251] in a medium of inverted, metastable Tl vapor. The data from this experiment are shown in Fig. 4.43 where the pump laser threshold intensity is plotted as a function of the detuning of the pump source from the intermediate Raman state, at constant inversion density. At large detunings the pump threshold intensity varies as Δ^2 symmetrically about the intermediate resonance. Near the intermediate resonance (i.e., $|\Delta| \lesssim 50\,\mathrm{cm}^{-1}$), dispersion effects become important, and a constant threshold intensity is observed. These findings are in excellent agreement with the simple theoretical treatment given above [4.251].

4.6.3 Broadening Processes

Accurate prediction of the pump laser threshold requirements and the expected tuning range of Raman devices is often complicated by uncertainty in the value of Γ to be used in the calculation of the gain coefficient. Experimental results discussed in previous sections have shown that considerable line broadening has been observed in many Raman generation schemes. Since the Raman threshold (and likewise the anticipated tuning range) varies as $1/\Gamma$, any large error in the value of Γ used for the calculations will have a significant effect on the calculated device performance.

Generally speaking, the use of Doppler-broadened values for the Raman linewidth Γ have lead to significant discrepancies between theoretical and experimental values for Raman threshold and tuning [4.214]. In SERS generation, for example, the Stokes linewidth typically varies from 0.3 to 1.0 cm^{-1} using 0.1 cm^{-1} line width pump sources. Under these conditions, the calculated Raman linewidth assuming a predominant Doppler-broadening mechanism ranged between $0.03 - 0.1 \, \text{cm}^{-1}$. Although the theory of Raman generation with a multimode pump source [4.296] shows that the observed Stokes bandwidth can broaden up to the band width of the pump laser, this broadening alone is not sufficient to account for the majority of the experimental observations.

Wyatt [4.299] has performed some detailed measurements of line broadening observed in SERS using the Cs $(6s-9s)$ Raman transition, with the aid of a $0.15 \, \text{cm}^{-1}$ resolution monochromator. These measurements are indicative of the complexity of the problem. In one experiment of $0.15 \, \text{cm}^{-1}$ dye laser with a power of $4 \, \text{kW}$ was used to pump the $6s-9s$ Raman transition; the generated Stokes light had a line width of $0.3 - 0.4 \, \text{cm}^{-1}$ for Cs pressures between $5-30$ Torr. As the intensity of the pump laser was increased to $250 \, \text{kW}$, the observed Stokes linewidth grew to $0.7 - 0.9 \, \text{cm}^{-1}$. Similar behavior was observed with the Cs $(6s-8s)$ Raman transition. The measured Stokes linewidth at a pump intensity $2 \, \text{MW/cm}^2$ was $\sim 0.4 \, \text{cm}^{-1}$; while at a pump intensity of $10 \, \text{GW/cm}^2$ a Stokes linewidth of $1.65 \, \text{cm}^{-1}$ was seen.

Subsequent studies of line broadening were conducted on the Cs $(6s-7s)$ Raman transition by *Cotter* and *Hanna* [4.208]. The Raman pump source was a nitrogen pumped dye laser system with pulse energies between $14-17 \, \text{kW}$ and a linewidth of $\sim 0.1 \, \text{cm}^{-1}$. The resulting Stokes line width was measured using a monochromator with a $0.1 \, \text{cm}^{-1}$ resolution, and the line width varied from $0.25 - 0.55 \, \text{cm}^{-1}$. An increase of a factor of two in Stokes line width was observed when the intensity of the pump source was increased by a factor of 100. Tuning the pump laser closer to the $7p$ intermediate levels also increased the observed line width, as did increasing the Cs vapor pressure from $1 - 30$ Torr.

A number of line-broadening mechanisms are of possible consequence in Raman generation. Unfortunately, meaningful calculations of the expected broadening are often difficult, due in part to the dynamic interactions possible between the broadening mechanisms and the Raman scattering process itself. Several possibilities, including the ac Stark shifting of the Raman levels, Start broadening due to ions and electrons created via multiphoton ionization, and the effects of resonance broadening and the dynamics of the populations in the Raman levels have been considered by *Wyatt* [4.299]. As one example, an ac Stark shift of $0.1 \, \text{cm}^{-1}$ is expected for Raman generation on the K $(4s-5s)$ transition using a pump laser of intensity $50 \, \text{MW/cm}^2$ detuned $10 \, \text{cm}^{-1}$ from the $5p$ intermediate resonance. Raman generation using higher-lying final states of alkali vapors can be further complicated by strong amplified spontaneous emission (ASE) from the final Raman level (generally an ns state) to lower $(n-1)p$ levels. This ASE can markedly reduce the effective lifetime of the final Raman level, leading to broadening of the Raman transition. The growth of the population in the lower $(n-1)p$ state due to the ASE can further influence Raman broadening. Radiative trapping of this population may lead to resonance broadening (i.e., dipole-dipole collisional broadening) of the Raman transition. It should be clear, even from this limited discussion, that an understanding of broadening processes in SRS is a rich area for future research.

4.6.4 Diffraction Effects

The analysis in Sect. 4.3 for the Raman gain coefficient was carried out for plane-wave interactions. In most experimental situations, however, the pump radiation and the generated Raman output are beams of finite extent with a nonuniform radial intensity. A more complete analysis of Raman gain, including the effects of focusing the pump beam into the Raman medium, leads to the observation that the growth of the shifted output can be adversely affected by diffraction effects. This is particularly true for Stokes generation in which the scattered wavelength is much longer than the incident wavelength. In that case, diffraction losses can lead to scattering of the Stokes wave outside the active volume of the pump radiation, resulting in a significant reduction in the Stokes gain.

An estimate of the relative importance of diffraction effects in Stokes generation can be made by considering the ratio $K = \omega_s/\omega_p$ for the process, where ω_s is the Stokes frequency and ω_p is the pump frequency. Generally speaking, diffraction effects are unimportant for K near unity; that is, for Raman generation in media with shifts that are small compared to the energy of the pump-laser photon. In many ir generation schemes, however, visible or uv lasers are scattered using high-lying transitions and large Raman shifts; K often ranges from $10^{-3} - 10^{-1}$ in these cases. Diffraction losses of the Stokes wave may be substantial, and the Raman gain and tunability of the device will be reduced.

A quantitative analysis of Raman gain and the effect of diffraction has been given by *Hanna* et al. [4.35] and by *Cotter* et al. [4.300] for a pump beam with a Gaussian radial profile, and we shall follow their formalism. Assume that the pump beam travels in the positive $\hat{z}$ direction and this focused into the Raman medium at $z = f$ with a spot size of w and confocal parameter b. The Raman medium is taken to extend from $z = 0$ to $z = L$. The spatially-dependent Raman gain is then

$$G_{ss}(\delta, r) = G_{ss}(0,0) \frac{\exp[-2r^2/w^2(1+\delta^2)]}{1+\delta^2} \,, \qquad (4.76)$$

where $\delta = 2(z-f)/b$ and $G_{ss}(0,0)$ represents the peak, on-axis Raman gain coefficient at the focal point. Expression (4.76) for the Raman gain may then be substituted in the wave equation for the Stokes component in which the transverse derivatives are retained. Solution of the resulting equation for the Stokes field can be simplified by approximating the exponential term in (4.76) as $[1 - 2r^2/w^2(1+\delta^2)]$. The Stokes wave then experiences a parabolic radially-dependent gain, and an exact solution of the wave equation is possible.

Following this procedure *Cotter* et al. [4.300] find that the Stokes power at the end of the medium, $P_s(L)$ is

$$P_s(L) = P_s(0) \exp[(\mathscr{P}_p - 2\,\mathscr{P}_p^{1/2}) \tan^{-1}(L/b)/K] \tag{4.77}$$

where $\mathscr{P}_p = G_{ss}(0,0)\,bK$ is a dimensionless factor proportional to the incident pump-laser power.

The physical interpretation of the various terms in (4.77) is straightforward. The $\tan^{-1}(L/b)$ term expresses the dependence of the Stokes gain on the focusing of the pump radiation. The role of diffraction losses in the Stokes gain is contained in the $2\,\mathscr{P}_p^{1/2}$ term. If diffraction losses are large such that $\mathscr{P}_p \leqslant 4$, the Stokes wave will not experience any net gain from the Raman process.

The role of diffraction losses in Stokes generation imposes certain limitations on the usefulness of optical resonators for many far-ir generation schemes. When K is much less than unity, placing mirrors around the Raman medium to resonate the pump and/or Stokes waves produces disappointing results, since the power gain from (4.77) decreases more rapidly with $\mathscr{P}_p$ than does the $2\,\mathscr{P}_p^{1/2}$ term governing the diffraction loss. Although optical resonators have been used successfully for K near unity [4.133], far better results for small K have been achieved using waveguide Raman cells to contain the Stokes radiation [4.136 – 138], as described in Sect. 4.5.1 a.

References

4.1   E. J. Woodbury, W. K. Ng: Proc. IRE **50**, 2347 (1962)
4.2   See E. J. Woodbury, G. M. Eckhardt: US Patent No. 3,371,265 (February 27, 1968)
4.3   G. Eckhardt, R. W. Hellwarth, F. J. McClung, S. E. Schwartz, D. Weiner, E. J. Woodbury: Phys. Rev. Lett. **9**, 455 (1962)
4.4   M. Geller, D. P. Bortfeld, W. R. Sooy: Appl. Phys. Lett. **3**, 36 (1963)
4.5   J. A. Giordmaine, J. A. Howe: Phys. Rev. Lett. **11**, 207 (1963)
4.6   B. P. Stoicheff: Phys. Lett. **7**, 186 (1963)
4.7   J. A. Calviello, Z. H. Heller: Appl. Phys. Lett. **5**, 112 (1964)
4.8   G. Eckhardt, D. P. Bortfeld, M. Geller: Appl. Phys. Lett. **3**, 137 (1963)
4.9   R. W. Minck, R. W. Terhune, W. G. Rado: Appl. Phys. Lett. **3**, 181 (1963)
4.10 R. W. Hellwarth: Phys. Rev. **130**, 1850 (1963)
4.11 R. W. Hellwarth: Appl. Opt. **2**, 847 (1963)
4.12 N. Bloembergen: Am. J. Phys. **35**, 989 (1967)
4.13 G. Eckhardt: IEEE J. QE-2, 1 (1966)
4.14 Y. R. Shen, N. Bloembergen: Phys. Rev. **137**, A1787 (1965)
4.15 C.-S. Wang: Phys. Rev. **182**, 482 (1969)
4.16 C.-S. Wang: in *Quantum Electronics,* Vol. 1, Part A, ed. by H. Rabin, C. L. Tang (Academic, New York 1975) pp. 447 – 472
4.17 M. Maier: Appl. Phys. **11**, 209 (1976)
4.18 N. Bloembergen, G. Bret, P. Lallimand, A. Pine, P. Simova: IEEE J. QE-3, 197 (1967)
4.19 A. Anderson (ed.): *The Raman Effect* (Marcel Dekker, New York 1971)
4.20 W. Kaiser, M. Maier: In *Laser Handbook,* Vol. 2, ed. by F. T. Arecchi, E. O. Schulz-Dubois (North Holland, Amsterdam 1972) pp. 1077 – 1150

4.21 M. Schubert, B. Wilhelmi: Sov. J. Quant. Electron. **4**, 575 (1974)
4.22 Y.-R. Shen: In *Light Scattering in Solids,* ed. by M. Cardona, Topics Appl. Phys., Vol. 8 (Springer, New York 1975) pp. 275 – 328
4.23 Y.-R. Shen: Rev. Mod. Phys. **48**, 1 (1976)
4.24 A. Z. Grasiuk, I. G. Zubarev: In *Tunable Lasers and Applications,* ed. by A. Mooradian, T. Jaeger, P. Stokseth, Springer Ser. Opt. Sci., Vol. 3 (Springer, Berlin, Heidelberg 1976) pp. 88 – 95
4.25 N. G. Basov (ed.): *Stimulated Raman Scattering* (Consultants Bureau, New York 1982)
4.26 S. K. Kurtz, J. A. Giordmaine: Phys. Rev. Lett. **22**, 792 (1969)
4.27 M. A. Piedstrup, R. N. Fleming, R. H. Pantell: Appl. Phys. Lett. **26**, 418 (1975)
4.28 J. Gelbwachs, R. H. Pantell, H. E. Puthoff, J. M. Yarborough: Appl. Phys. Lett. **14**, 258 (1969)
4.29 C. K. N. Patel, E. D. Shaw: Phys. Rev. Lett. **24**, 451 (1970)
4.30 M. J. Colles, C. R. Pidgeon: Rep. Prog. Phys. **38**, 329 (1975)
4.31 S. D. Smith, R. B. Dennis, R. G. Harrison: Prog. Quantum. Electron. **5**, 205 (1977)
4.32 N. Goldblatt: Appl. Opt. **8**, 1559 (1969)
4.33 J. Ducuing, R. Frey, F. Pradère: In *Tunable Lasers and Applications,* ed. by A. Mooradian, T. Jaeger, P. Stokseth, Springer Ser. Opt. Sci., Vol. 3 (Springer, Berlin, Heidelberg 1976) pp. 81 – 87
4.34 P. P. Sorokin, N. S. Shiren, J. R. Lankard, E. C. Hammond, T. G. Kazyaka: Appl. Phys. Lett. **10**, 44 (1967)
4.35 D. C. Hanna, M. A. Yuratich, D. Cotter: *Nonlinear Optics of Free Atoms and Molecules,* Springer Ser. Opt. Sci., Vol. 17 (Springer, Berlin, Heidelberg 1979) pp. 187 – 250
4.36 G. Placzek: In *Marx Handbuch der Radiologie,* Vol. 6, Part II, ed. by E. Marx (Akademische Verlagsgesellschaft, Leipzig 1934) pp. 205 – 374
4.37 S. Bhagavantam: *Scattering of Light and Raman Effect* (Chemical Publ., New York 1942)
4.38 M. Born, K. Huang: *Dynamic Theory of Crystal Lattice* (Oxford U. Press, London 1956)
4.39 B. P. Stoicheff: In *Methods of Experimental Physics: Molecular Physics,* Vol. 3, ed. by D. Williams (Academic, New York 1962) pp. 111 – 264
4.40 R. L. Carmen, F. Shimizu, C. S. Wang, N. Bloembergen: Phys. Rev. A **2**, 60 (1970)
4.41 W. R. Trutna, Jr., Y. K. Park, R. L. Byer: IEEE J QE-**15**, 648 (1979)
4.42 J. Eggleston, R. L. Byer: IEEE J. QE-**16**, 850 (1980)
4.43 J. Goldhar, J. R. Murray: IEEE J. QE-**18**, 399 (1982)
4.44 A. T. Georges: Opt. Commun. **41**, 61 (1982)
4.45 G. C. Valley: IEEE J. QE-**18**, 1370 (1982)
4.46 N. M. Kroll: J. Appl. Phys. **36**, 34 (1965)
4.47 C. L. Tang: J. Appl. Phys. **37**, 2945 (1966)
4.48 M. Maier, W. Kaiser, J. A. Giordmaine: Phys. Rev. **177**, 580 (1969)
4.49 V. T. Platonenko, R. V. Khokhlov: Sov. Phys. JETP **19**, 378 (1964)
4.50 V. T. Platonenko, R. V. Khokhlov: Sov. Phys. JETP **19**, 1435 (1964)
4.51 V. T. Platonenko, R. V. Khokhlov: Opt. Spectrosc. **18**, 211 (1965)
4.52 J. A. Giordmaine, W. Kaiser: Phys. Rev. **144**, 676 (1966)
4.53 C. L. Tang, T. F. Deutsch: Phys. Rev. **138**, A1 (1965)
4.54 H. Hans, P. L. Kelley, H. Zeiger: Phys. Rev. **138**, A690 (1965)
4.55 N. Bloembergen, Y. R. Shen: Phys. Rev. Lett. **13**, 720 (1964)
4.56 N. Bloembergen, Y. R. Shen: Phys. Rev. **133**, A37 (1964)
4.57 C. L. Tang: Phys. Rev. **134**, A1166 (1964)
4.58 V. M. Fain, E. G. Yashchin: Sov. Phys. JETP **19**, 474 (1964)
4.59 G. Rivoire: Compt. Rend. **258**, 4001 (1964)
4.60 K. Grob: Phys. Lett. **10**, 52 (1964)
4.61 K. Grob: Z. Physik **201**, 59 (1965)
4.62 J. S. Margolis, G. Birnbaum: J. Appl. Phys. **36**, 726 (1965)
4.63 M. G. Raymer, J. Mostowski: Phys. Rev. A **24**, 1980 (1981)
4.64 D. von der Linde, M. Maier, W. Kaiser: Phys. Rev. **178**, 11 (1969)
4.65 W. R. L. Clements, B. P. Stoicheff: Appl. Phys. Lett. **12**, 246 (1968)

4.66 Y. Kato, H. Takuma: J. Chem. Phys. 54, 5398 (1971)
4.67 G. G. Bret, M. M. Demariez: Appl. Phys. Lett. 8, 151 (1966)
4.68 D. von der Linde, A. Laubereau, W. Kaiser: Phys. Rev. Lett. 26, 954 (1971)
4.69 A. Laubereau, D. von der Linde, W. Kaiser: Phys. Rev. Lett. 27, 802 (1971)
4.70 A. Laubereau, D. von der Linde, W. Kaiser: Phys. Rev. Lett. 28, 1162 (1972)
4.71 J. G. Skinner, W. G. Nilsen: J. Opt. Soc. Am. 58, 1131 (1968)
4.72 W. D. Johnston, Jr., I. P. Kaminow: Phys. Rev. 168, 1045 (1968)
4.73 E. E. Hagenlocker, R. W. Minck, W. G. Rado: Phys. Rev. 154, 226 (1967)
4.74 M. Takatsuji: Phys. Rev. A11, 619 (1975)
4.75 D. Grischkowsky, M. M. T. Loy, P. F. Liao: Phys. Rev. A12, 2514 (1975)
4.76 M. G. Raymer, J. Mostowski, J. L. Carlsten: Phys. Rev. A19, 2304 (1979)
4.77 L. Allen, J. H. Eberly: Optical Resonance and Two-level Atoms (Wiley, New York 1975)
4.78 J. D. Macomber: The Dynamics of Spectroscopic Transitions (Wiley, New York 1976)
4.79 E. Courtens: In Laser Handbook, ed. by F. T. Arecchi, E. O. Schulz-Dubois (North-Holland, Amsterdam 1972)
4.80 N. Bloembergen: Nonlinear Optics (Benjamin/Cummings Reading, MA 1965)
4.81 R. Y. Chiao, B. P. Stoicheff: Phys. Rev. Lett. 12, 290 (1964)
4.82 E. Garmire: Phys. Lett. 17, 251 (1965)
4.83 E. Garmire: In Physics of Quantum Electronics, ed. by P. L. Kelley, B. Lax, P. E. Tannenwald (McGraw-Hill, New York 1966)
4.84 D. Eimerl, R. S. Hargrove, J. A. Paisner: Phys. Rev. Lett. 46, 651 (1981)
4.85 R. B. Andreev, V. A. Gorbunov, S. S. Gulidov, S. B. Papernyi, V. A. Serebryakov: Sov. J. Quant. Electron. 12, 35 (1982)
4.86 V. I. Odintsov: Opt. Spectrosc. 54, 299 (1983)
4.87 V. Wilke, W. Schmidt: Appl. Opt. 16, 151 (1978)
4.88 R. S. Hargrove, J. A. Paisner: In Digest of Topical Meeting on Excimer Lasers (Opt. Soc. Am., Washington, D.C. 1979) paper ThA6
4.89 T. R. Loree, R. C. Sye, D. L. Barker, P. B. Scott: IEEE J. QE-15, 337 (1979)
4.90 H. Schomburg, H. F. Döbele, B. Rückle: Appl. Phys. B30, 131 (1983)
4.91 H. F. Döbele, M. Röwekamp, B. Rückle: In Laser Techniques in the Extreme Ultraviolet, ed. by S. E. Harris, T. McIlrath (Am. Inst. Phys., Washington, D.C. 1984)
4.92 F. J. McClung, W. G. Wagner, D. Weiner: Phys. Rev. Lett. 15, 96 (1965)
4.93 G. Bret: Compt. Rend. Acad. Sci. 259, 2991 (1964)
4.94 G. Bret: Compt. Rend. Acad. Sci. 260, 6323 (1965)
4.95 R. W. Terhune: Solid State Design 4, 38 (1964)
4.96 H. J. Zeiger, P. E. Tannenwald, S. Kern, R. Burendeen: Phys. Rev. Lett. 11, 419 (1963)
4.97 P. Lallemand, N. Bloembergen: Phys. Rev. Lett. 15, 1010 (1965)
4.98 G. Hauchecorne, G. Mayer: Compt. Rend. Acad. Sci. 261, 4014 (1965)
4.99 Y. R. Shen, Y. J. Shakam: Phys. Rev. 15, 1008 (1965)
4.100 R. Y. Chiao, E. Garmire, C. H. Townes: Phys. Rev. Lett. 13, 479 (1964)
4.101 P. L. Kelley: Phys. Rev. Lett. 15, 1005 (1965)
4.102 C. C. Wang: Phys. Rev. Lett. 16, 344 (1966)
4.103 M. M. T. Loy, Y. R. Shen: Appl. Phys. Lett. 19, 285 (1971)
4.104 Y. R. Shen, Y. A. Shaham: Phys. Rev. 163, 224 (1967)
4.105 M. Maier, W. Kaiser, J. A. Giordmaine: Phys. Rev. Lett. 17, 1275 (1966)
4.106 Y. R. Shen, M. M. T. Loy: Phys. Rev. A3, 2099 (1971)
4.107 G. K. L. Wong, Y. R. Shen: Appl. Phys. Lett. 21, 163 (1972)
4.108 S. A. Akhmanov, K. N. Drabovich, A. P. Sukhorukov, A. S. Chirkin: Sov. Phys. JETP 32, 266 (1971)
4.109 R. L. Carmen, F. Shimizu, C. S. Wang, N. Bloembergen: Phys. Rev. A2, 60 (1970)
4.110 N. M. Kroll, P. L. Kelley: Phys. Rev. A4, 763 (1971)
4.111 N. Tan-No, T. Shirahata, K. Yokoto, H. Inaba: Phys. Rev. A14, 2225 (1976)
4.112 E. M. Belenov, I. A. Poluektov: Sov. Phys. JETP 29, 754 (1969)
4.113 I. A. Poluektov, Yu. M. Popov, V. S. Roitberg: JETP Lett. 20, 243 (1974)
4.114 J. N. Elgin, T. B. O'Hare: J. Phys. B12, 159 (1979)

4.115 V. A. Gorbunov: Sov. J. Quant. Electron. **12**, 98 (1982)
4.116 S. Kern, B. Feldman: MIT Lincoln Lab. Solid State Res. Rept. **3**, 18 (1964)
4.117 J. J. Barrett, M. C. Tobin: J. Opt. Soc. Am. **56**, 129 (1966)
4.118 M. D. Martin, E. L. Thomas: IEEE J. QE-2, 196 (1966)
4.119 M. A. El-Sayed, F. M. Johnson, J. Duardo: J. Chimie Physique **1**, 227 (1967)
4.120 J. B. Grun, A. K. McQuillan, B. P. Stoicheff: Phys. Rev. **180**, 61 (1969)
4.121 D. L. Weinberg: MIT Lincoln Lab. Solid State Res. Rept. **2**, 31 (1965)
4.122 D. P. Bortfeld, M. Geller, G. Eckhardt: J. Chem. Phys. **40**, 1770 (1964)
4.123 V. A. Subov, M. M. Sushchinskii, I. K. Shuvalton: J. Exp. Theor. Phys. (USSR) **47**, 784 (1964)
4.124 P. E. Tannenwald, J. B. Thaxter: Science **134**, 1319 (1966)
4.125 W. D. Johnston, Jr., I. P. Kaminow, J. G. Bergman, Jr.: Appl. Phys. Lett. **13**, 190 (1968)
4.126 P. Lallemand, P. Simova, G. Brett: Phys. Rev. Lett. **17**, 1239 (1966)
4.127 R. H. Dicke: Phys. Rev. **89**, 472 (1953)
4.128 L. Galatry: Phys. Rev. **122**, 1218 (1961)
4.129 F. DeMartini, F. Simoni, E. Santamato: Opt. Commun. **9**, 176 (1973)
4.130 D. Robert, J. Bonamy, F. Marsault-Herail, G. Levi, J. P. Marsault: Chem. Phys. Lett. **74**, 467 (1980)
4.131 G. Bjorklund, IBM Research Center, San Jose, CA (private communication)
4.132 W. Schmidt, W. Z. Appt: Z. Naturforsch. **27A**, 1373 (1972)
4.133 W. Schmidt, W. Z. Appt: IEEE J. QE-10, 792 (1974)
4.134 R. Frey, F. Pradere: Opt. Commun. **12**, 98 (1974)
4.135 S. J. Brosnan, R. N. Fleming, R. L. Herbst, R. L. Byer: Appl. Phys. Lett. **30**, 330 (1977)
4.136 P. Rabinowitz, A. Kaldor, R. Brickman, W. Schmidt: Appl. Opt. **15**, 2005 (1976)
4.137 W. Hartig, W. Schmidt: Appl. Phys. **18**, 235 (1979)
4.138 P. G. May, W. Sibbett: Appl. Phys. Lett. **43**, 624 (1983)
4.139 V. Wilke, W. Schmidt: Appl. Phys. **18**, 177 (1979)
4.140 T. R. Loree, R. C. Sze, D. L. Barker: Appl. Phys. Lett. **31**, 37 (1977)
4.141 H. F. Döbele, B. Rückle: Appl. Opt. **23**, 1040 (1984)
4.142 D. W. Trainor, H. A. Hyman, I. Itzkan, R. M. Heinrichs: Appl. Phys. Lett. **37**, 440 (1980)
4.143 H. Komine, E. A. Stappaerts, S. J. Brosnan, J. B. West: Appl. Phys. Lett. **40**, 551 (1982)
4.144 H. Komine, E. A. Stappaerts: Opt. Lett. **7**, 157 (1982)
4.145 J. C. White, J. Bokus, R. R. Freeman, D. Henderson: Opt. Lett. **6**, 293 (1981)
4.146 H. Schomburg, H. F. Döbele, B. Rückle: Appl. Phys. B**28**, 201 (1982)
4.147 M. Audibert, J. Lukasik: Opt. Commun. **21**, 137 (1977)
4.148 R. L. Byer: IEEE J. QE-12, 732 (1976)
4.149 R. J. Jensen, J. G. Marinuzzi, G. P. Robinson, S. D. Lockwood: Laser Focus **12**, 51 (1976)
4.150 P. P. Sorokin, M. M. T. Loy, J. R. Lankard: IEEE J. QE-13, 871 (1977).
4.151 R. L. Byer, W. R. Trutna: Opt. Lett. **3**, 144 (1978)
4.152 W. R. Trutna, R. L. Byer: Appl. Opt. **19**, 301 (1980)
4.153 P. Rabinowitz, A. Stein, R. Brickman, A. Kaldor: Opt. Lett. **3**, 147 (1978)
4.154 P. Rabinowitz, A. Stein, R. Brickman, A. Kaldor: Appl. Phys. Lett. **35**, 739 (1979)
4.155 R. W. Minck, E. E. Hagenlocker, W. G. Rado: Phys. Rev. Lett. **17**, 229 (1966)
4.156 J. A. Duardo, L. J. Nugent, F. M. Johnson: J. Chem. Phys. **46**, 3585 (1967)
4.157 R. Frey, F. Pradere: Infrared Phys. **16**, 117 (1976)
4.158 G. V. Mikhailov: Opt. Spectrosc. **12**, 361 (1962)
4.159 S. V. Efimovsky, I. G. Zubarev, A. V. Kotov: Sov. J. Quant. Electron. **7**, 1155 (1977)
4.160 V. I. Dianov-Klokov: Optika i Specktroskopiya **4**, 448 (1958)
4.161 M. J. Clouter, H. Kiefte: J. Chem. Phys. **66**, 1736 (1977)
4.162 H. Kildal, S. R. J. Brueck: J. Chem. Phys. **73**, 4951 (1980)
4.163 S. R. J. Brueck, H. Kildal: J. Appl. Phys. **52**, 1004 (1981)
4.164 P. D. McWane, D. A. Sealer: Appl. Phys. Lett. **8**, 278 (1966)
4.165 A. Z. Grasiuk, I. G. Zubarev: Appl. Phys. **17**, 211 (1978)
4.166 V. A. Zubov, A. V. Kraiskii, K. A. Prokhorov, M. M. Sushchinskii, I. K. Shuvalov: Sov. Phys. JETP **28**, 231 (1969)

4.167 V. V. Bocharov, A. Z. Grasiuk, I. G. Zubarev, A. V. Kotov, V. G. Smirov: Sov. J. Quant. Electron. **4**, 1216 (1975)
4.168 S. R. J. Brueck, H. Kildal: IEEE J. QE-**18**, 310 (1982)
4.169 A. Z. Grasiuk: Appl. Phys. **21**, 173 (1980)
4.170 C. L. Marquardt, M. E. Storm, I. Schneider, L. Esterowitz: J. Appl. Phys. **54**, 5645 (1983)
4.171 D. Gettemy, N. P. Barnes, E. Griggs: Rev. Sci. Instrum. **51**, 1194 (1980)
4.172 R. Frey, F. Pradere, J. Lukasik: Opt. Commun. **22**, 355 (1977)
4.173 V. O. Baranov, V. M. Borisov, A. Z. Grasyuk, S. V. Efimovskii, S. G. Mamonov, V. G. Smirnov, Yu. Yu. Stepanov: Sov. Tech. Phys. Lett. **6**, 127 (1980)
4.174 E. Wild, M. Maier: J. Appl. Phys. **51**, 3078 (1980)
4.175 J. Ducuing, R. Frey, F. Pradère: In *Tunable Lasers and Applications,* ed. by A. Mooradian, T. Jaeger, P. Stokseth, Springer Ser. Opt. Sci., Vol. 3 (Springer, Berlin, Heidelberg 1976)
4.176 R. Frey, F. Pradère, J. Ducuing: Opt. Commun. **23**, 65 (1977)
4.177 A. DeMartino, R. Frey, F. Pradère: Opt. Commun. **27**, 262 (1978)
4.178 P. Mathieu, J. R. Izatt: Opt. Lett. **6**, 369 (1981)
4.179 D. G. Biron, B. G. Danly, B. Lax, R. J. Temkin: J. Opt. Soc. Am. **70**, 674 (1980)
4.180 B. Wellegehausen: IEEE J. QE-**15**, 1108 (1979)
4.181 R. L. Byer, R. L. Herbst, H. Kildal, M. D. Levenson: Appl. Phys. Lett. **20**, 463 (1972)
4.182 J. B. Koffend, R. W. Field: J. Appl. Phys. **48**, 4468 (1977)
4.183 B. Wellegehausen, K. H. Stephan, D. Friede, H. Welling: Opt. Commun. **23**, 157 (1977)
4.184 F. J. Wodarczyk, H. R. Schlossberg: J. Chem. Phys. **67**, 4476 (1977)
4.185 M. A. Henesian, R. L. Herbst, R. L. Byer: J. Appl. Phys. **47**, 1515 (1976)
4.186 H. Itoh, H. Uchiki, M. Matsuoka: Opt. Commun. **18**, 271 (1976)
4.187 B. Wellegehausen, S. Shahdin, D. Friede, H. Welling: Appl. Phys. **13**, 97 (1977)
4.188 P. J. Jones, U. Gaubatz, U. Hefter, K. Bergmann, B. Wellegehausen: Appl. Phys. Lett. **42**, 222 (1983)
4.189 B. Wellegehausen, H. H. Heitmann: Appl. Phys. Lett. **34**, 44 (1979)
4.190 B. Wellegehausen, W. Luhs, A. Topouzkhanian, J. d'Incan: Appl. Phys. Lett. **43**, 912 (1983)
4.191 C. N. Man-Pichot, A. Brillet: IEEE J. QE-**16**, 1103 (1980)
4.192 W. Luhs, B. Wellegehausen: Opt. Commun. **46**, 121 (1983)
4.193 W. Luhs, M. Hube, U. Schottelius, B. Wellegehausen: Opt. Commun. **48**, 265 (1983)
4.194 S. R. Leone, K. G. Kosnik: Appl. Phys. Lett. **30**, 346 (1977)
4.195 B. Wellegehausen, A. Topouzkhanian, C. Effantin, J. d'Incan: Opt. Commun. **41**, 437 (1982)
4.196 A. Topouzkhanian, B. Wellegehausen, C. Effantin, J. d'Incan: Laser Chem. **1**, 195 (1983)
4.197 B. Wellegehausen, D. Friede, G. Steger: Opt. Commun. **26**, 391 (1978)
4.198 W. P. West, H. P. Broida: Chem. Phys. Lett. **56**, 283 (1983)
4.199 M. Rokni, S. Yatsiv: Phys. Lett. **24 A**, 277 (1967)
4.200 M. Rokni, S. Yatsiv: IEEE J. QE-**3**, 329 (1967)
4.201 P. P. Sorokin, J. J. Wynne, J. R. Lankard: Appl. Phys. Lett. **22**, 342 (1973)
4.202 J. J. Wynne, P. P. Sorokin: In *Nonlinear Infrared Generation,* ed. by Y. R. Shen, Topics Appl. Phys., Vol. 16 (Springer, Berlin, Heidelberg 1977) pp. 170 – 174
4.203 D. Cotter, D. C. Hanna, P. A. Kärkäinen, R. Wyatt: Opt. Commun. **15**, 143 (1975)
4.204 I. L. Tyler, R. W. Alexander, R. J. Bell: Appl. Phys. Lett. **27**, 346 (1975)
4.205 B. A. Glushko, M. E. Movsesyan, T. O. Ovakimyan: Opt. Spectrosc. **52**, 458 (1982)
4.206 J. J. Wynne, P. P. Sorokin: J. Phys. B**8**, L37 (1975)
4.207 D. Cotter, D. C. Hanna, R. Wyatt: Opt. Commun. **16**, 256 (1976)
4.208 D. Cotter, D. C. Hanna: Opt. Quant. Electron. **9**, 509 (1977)
4.209 R. T. V. Kung, I. Itzkan: IEEE J. QE-**13**, 73 (1977)
4.210 A. B. Peterson, I. M. W. Smith, D. C. Hanna: J. Opt. Soc. Am. **68**, 655 (1978)
4.211 R. T. Hodgson: Appl. Phys. Lett. **34**, 58 (1979)
4.212 J. L. Carlsten, P. C. Dunn: Opt. Commun. **14**, 8 (1975)
4.213 A. M. F. Lau, W. K. Bischel, C. K. Rhodes, R. M. Hill: Appl. Phys. Lett. **29**, 245 (1976)

4.214 P. A. Kärkkäinen: Ph. D. Thesis, University of Southampton (1975)
4.215 P. P. Sorokin, J. R. Lankard: IEEE J. QE-9, 227 (1973)
4.216 R. T. V. Kung, I. Itzkan: Appl. Phys. Lett. 29, 780 (1976)
4.217 T. Holstein: Phys. Rev. 72, 1212 (1947)
4.218 T. Holstein: Phys. Rev. 83, 1159 (1951)
4.219 L. Krause: Adv. Chem. Phys. 28, 1159 (1975)
4.220 D. R. Grischkowsky, J. R. Lankard, P. P. Sorokin: IEEE J. QE-13, 392 (1977)
4.221 D. Grischkowsky, P. P. Sorokin, J. R. Lankard: Opt. Commun. 18, 205 (1976)
4.222 D. S. Bethune, J. R. Lankard, P. P. Sorokin: Opt. Lett. 4, 103 (1979)
4.223 D. S. Bethune, J. R. Lankard, M. M. T. Loy, P. P. Sorokin: IBM J. Res. Dev. 23, 556 (1979)
4.224 Ph. Avouris, D. S. Bethune, J. R. Lankard, J. A. Ors, P. P. Sorokin: J. Chem. Phys. 74, 2304 (1981)
4.225 D. S. Bethune, J. R. Lankard, P. P. Sorokin, A. J. Schnell-Sorokin, R. M. Plecenik, Ph. Avouris: J. Chem. Phys. 75, 2231 (1981)
4.226 Ph. Avouris, D. S. Bethune, J. R. Lankard, P. P. Sorokin, A. J. Schnell-Sorokin: J. Photochem. 17, 227 (1981)
4.227 D. S. Bethune, A. J. Schnell-Sorokin, J. R. Lankard, M. M. T. Loy, P. P. Sorokin: Adv. Laser Spectrosc. 2, 1 (1983)
4.228 N. Djeu, R. Burnham: Appl. Phys. Lett. 30, 473 (1977)
4.229 D. Cotter, W. Zapka: Opt. Commun. 26, 251 (1978)
4.230 R. Burnham, N. Djeu: Opt. Lett. 3, 215 (1978)
4.231 S. J. Brosnan, H. Komine, E. A. Stappaerts, M. J. Plummer, J. B. West: Opt. Lett. 7, 154 (1982)
4.232 N. Djeu: In *High-Power Lasers and Applications,* ed. by K. L. Kompa, H. Walther, Springer Ser. Opt. Sci., Vol. 9 (Springer, Berlin, Heidelberg 1979) pp. 176−177
4.233 J. Manners: Opt. Commun. 44, 366 (1983)
4.234 R. L. Carmen: Phys. Rev. A 12, 1048 (1975)
4.235 P. P. Sorokin, N. Braslan: IBM J. Res. Dev. 8, 177 (1964)
4.236 A. M. Prokhorov: Science 149, 828 (1965)
4.237 E. B. Gordon, Yu. L. Moskvin: Sov. Phys. JETP 43, 901 (1976)
4.238 D. S. Bethune, J. R. Lankard, P. P. Sorokin: J. Chem. Phys. 69, 2076 (1978)
4.239 D. Grischkowsky, M. M. T. Loy: Phys. Rev. A 12, 1117 (1975)
4.240 M. M. T. Loy: Phys. Rev. Lett. 41, 473 (1978)
4.241 H. P. Yuen: Appl. Phys. Lett. 26, 505 (1975)
4.242 L. M. Narducci, W. W. Edison, P. Furcinitti, D. C. Eteson: Phys. Rev. A 16, 1665 (1977)
4.243 S. Yatsiv, M. Rokni, S. Barak: Phys. Rev. Lett. 20, 1282 (1968)
4.244 P. P. Sorokin, N. S. Shiren, J. R. Lankard, E. C. Hammond, T. G. Kazyaka: Appl. Phys. Lett. 10, 44 (1967)
4.245 A. V. Vinogradov, E. A. Yukov: JETP Lett. 16, 447 (1972)
4.246 R. L. Carmen, W. H. Lowdermilk: Phys. Rev. Lett. 33, 190 (1974)
4.247 R. L. Carmen, W. H. Lowdermilk: IEEE J. QE-10, 706 (1974)
4.248 J. C. White, D. Henderson: Phys. Rev. A 25, 1226 (1982); 25, 3430 (E) (1982)
4.249 N. J. A. Van Veen, M. S. DeVries, T. Batter, A. E. DeVries: Chem. Phys. 55, 371 (1981)
4.250 J. C. White, D. Henderson: IEEE J. QE-18, 941 (1982)
4.251 J. C. White, D. Henderson: Opt. Lett. 7, 517 (1982)
4.252 J. C. White, D. Henderson: Opt. Lett. 8, 15 (1983)
4.253 K. Ludewigt, K. Birkmann, B. Wellegehausen: Appl. Phys. B 33, 133 (1984)
4.254 Wellegehausen, CLEO (1984)
4.255 D. C. Cunningham, D. Denvir, I. Duncan, T. Morrow: Optica Acta 31, 249 (1984)
4.256 J. C. White, D. Henderson: IEEE J. QE-20, 462 (1984)
4.257 J. C. White, D. Henderson: Opt. Lett. 7, 204 (1982)
4.258 J. C. White, D. Henderson: Opt. Lett. 8, 520 (1983)
4.259 A. M. Pravilov: Sov. J. Quant. Electron. 11, 847 (1981)
4.260 K. J. Witte, E. Fill, G. Brederlow, H. Baumbacker, R. Volk: IEEE J. QE-17, 1809 (1981)

4.261 J. B. Koffend, S. R. Leone: Chem. Phys. Lett. **81**, 136 (1981)
4.262 E. E. Fill, W. Skrlac, K.-J. Witte: Opt. Commun. **37**, 123 (1981)
4.263 J. C. White: IEEE J. QE-**20**, 185 (1984)
4.264 J. C. White: Opt. Lett. **9**, 38 (1984)
4.265 J. R. Murray, C. K. Rhodes: J. Appl. Phys. **47**, 5041 (1976)
4.266 G. Rinke: Appl. Phys. B **32**, 83 (1983)
4.267 W. E. McDermott, N. R. Pchelkin, D. J. Benard, P. R. Bousek: Appl. Phys. Lett. **32**, 469 (1978)
4.268 R. J. Richardson, C. E. Wiswall, P. A. G. Carr, F. E. Hovis, H. V. Lilenfeld: J. Appl. Phys. **52**, 4962 (1981)
4.269 G. N. Hays, G. A. Fisk: IEEE J. QE-**17**, 1823 (1981)
4.270 J. Bachar, S. Rosenwaks: Appl. Phys. Lett. **41**, 16 (1982)
4.271 J. C. White, D. Henderson, T. A. Miller, M. Heaven: In *Laser Spectroscopy VI,* ed. by H. P. Weber, W. Lüthy, Springer Ser. Opt. Sci., Vol. **40** (Springer, Berlin Heidelberg 1983) pp. 407–411
4.272 H. Komine, R. L. Byer: Appl. Phys. Lett. **27**, 300 (1975)
4.273 W. T. Silfvast, J. J. Macklin, O. R. Wood II: Opt. Lett. **8**, 551 (1983)
4.274 A. J. Glass: IEEE J. QE-**3**, 516 (1967)
4.275 W. H. Culver, J. T. A. Vanderslice, V. W. T. Townsend: Appl. Phys. Lett. **12**, 189 (1968)
4.276 J. R. Murray, J. Goldhar, D. Eimerl, A. Szöke: IEEE J. QE-**15**, 342 (1979)
4.277 J. R. Murray, J. Goldhar, A. Szöke: Appl. Phys. Lett. **32**, 551 (1978)
4.278 J. R. Murray, J. Goldhar, D. Eimerl, A. Szöke: Appl. Phys. Lett. **33**, 399 (1978)
4.279 N. Djeu: Appl. Phys. Lett. **35**, 663 (1979)
4.280 R. W. Terhune, P. D. Maker, C. M. Savage: Phys. Rev. Lett. **14**, 681 (1965)
4.281 S. Yatsiv, M. Rokni, S. Barak: IEEE J. QE-**4**, 900 (1968)
4.282 D. A. Long, L. Stanton: Proc. Roy. Soc. Lond. A **318**, 441 (1970)
4.283 Yu. A. Il'inski, V. D. Tarankukin: Sov. J. Quant. Electron. **4**, 997 (1976)
4.284 G. S. Argawal: Opt. Commun. **31**, 325 (1979)
4.285 A. Guzman de Garcia, P. Meystre, M. Sargent III: Opt. Commun. **43**, 364 (1982)
4.286 S. J. Petuchovsky, J. D. Oberstar: Phys. Rev. A **20**, 529 (1979)
4.287 D. J. Kim, P. D. Coleman: IEEE J. QE-**16**, 300 (1980)
4.288 G. D. Willenberg, J. Heppner, F. B. Foote: IEEE J. QE-**18**, 2060 (1982)
4.289 J. A. Hermann, B. V. Thompson: J. Phys. B: **14**, 2961 (1981)
4.290 M. Ahmed el Deberky, B. V. Thompson: J. Phys. B: **15**, L779 (1982)
4.291 Q. H. F. Vrehen, H. M. J. Hikspoors: Opt. Commun. **21**, 127 (1977)
4.292 D. Cotter, D. C. Hanna, W. H. F. Tuttlebee, M. A. Yuratich: Opt. Commun. **22**, 190 (1977)
4.293 J. Reif, H. Walther: Appl. Phys. **15**, 361 (1978)
4.294 D. Krökel, D. Frölich, W. Klische, A. Feitisch, H. Welling: Opt. Commun. **48**, 57 (1983)
4.295 D. Cotter, D. C. Hanna: IEEE J. QE-**14**, 184 (1978)
4.296 S. A. Akhmanov, Yu. E. D'Yakov, L. I. Pavlov: Sov. Phys. JETP **39**, 249 (1974)
4.297 V. A. Mikhailov, V. I. Odintsov, L. F. Rogacheva: JETP Lett. **25**, 138 (1977)
4.298 F. A. Korolev, V. A. Mikhailov, V. I. Odintsov: Opt. Spectrosc. **44**, 535 (1978)
4.299 R. Wyatt: Ph. D. Thesis, University of Southampton (1976)
4.300 D. Cotter, D. C. Hanna, R. Wyatt: Appl. Phys. **8**, 333 (1975)

5. Urea Optical Parametric Oscillator for the Visible and Near Infrared

Kevin Cheng, Mark J. Rosker, and Chung L. Tang

With 11 Figures

A singly resonant OPO based on urea crystals is described. When pumped with 2 MW, 7 ns pulses at 354.7 nm (the third harmonic of Nd: YAG), the OPO produces coherent light tunable from 498 nm to 1.23 μm. Output-pulse energies of several mJ, potentially scalable to ~1 J, are allowed by conversion efficiencies of up to 35%. Second-harmonic generation in a separate urea crystal has extended the tuning range to 249 nm, and pumping with 266 nm (the fourth harmonic of Nd: YAG) should extend tuning of the OPO fundamental to ~300 nm.

5.1 Background Material

The optical parametric oscillator (OPO), first introduced in 1965 [5.1], has long held promise as a solid-state source of coherent light tunable over wide spectral regions [5.2, 3]. The OPO possesses a number of potential advantages over the pulsed dye laser. Most notably, in order to span a large wavelength region, pulsed dye lasers require expensive pump circulation systems and numerous expensive dyes. Furthermore, dye lasers only work well in a limited spectral range mainly in the visible, are of low gain, or are short-lived. In comparison, the OPO is compact, inexpensive, easy to operate, and can deliver tunable radiation (with good efficiency, mode quality, and linewidth) over a tuning range many times that of a single dye.

Unfortunately, a variety of material problems have limited the use of OPOs in the past. In order to be useful, a prospective OPO material must display numerous demanding material properties, including (1) large nonlinearity to provide for suitable parametric gain, (2) transparency in the vicinity of both the pump wavelength and the anticipated tuning range, (3) adequate birefringence to allow phase matching in the desired spectral region, (4) growth and polishing methods which yield a long length (typically, ≥10 nm) crystal of high optical quality, (5) suitable mechanical and thermal properties, and (6) high optical damage threshold in order for the crystal to tolerate the high-intensity fields generally necessary to achieve oscillation. It is also preferable for the material to be nonhygroscopic. The optical damage problem has been particularly limiting, as, for example, in the case of one of the previously most useful crystals, namely, $LiNbO_3$.

Recent advances in nonlinear materials, however, have renewed interest in parametric oscillators as tunable sources. In the $1.5 - 18 \, \mu m$ range, for instance, the $AgGaS_2$ crystal has demonstrated great promise [5.4]. For the generation of coherent radiation at shorter wavelengths, urea is well suited for use in an OPO [5.5 – 7]. Urea is an organic crystal [5.8] within the space group class $\bar{4}2\,m$, the same group as the ADP isomorphs. It is optically clear from 200 nm to $1.4 \, \mu m$, consistent with OPO tunability in the visible and the near infrared. Further, its birefringence $[(n_e - n_o)/n_o \approx 6.6 \times 10^{-2}]$ is considerably greater than that of ADP ($\approx 3.1 \times 10^{-2}$), and this leads to OPO tunability at shorter wavelengths than possible with many other nonlinear crystals. The nonlinear coefficient of urea, which has been measured to be 2.5 times that of ADP ($d_{14} = 1.4 \times 10^{-12} \, m/V$), is large enough to provide adequate parametric gain.

Mechanically, urea is a relatively soft crystal, but it can be optically polished to less than a wavelength using methods similar to that commonly used for ADP. The thermal characteristics of urea are very good. The temperature dependence of the phase-matching angle of urea is much smaller than that for KDP. The measured single-shot damage threshold of urea is on the order of $1 \, GW/cm^2$ in the visible and ir. The multiple-shot (for example, more than 10^5 shots of 10 ns pulses) damage threshold is considerably lower, however. At 355 nm, it is approximately $180 \, MW/cm^2$, which is still much better than $LiNbO_3$. The most difficult aspect of using urea is that the growth of this crystal remains a difficult task, but high-quality crystals of length greater than 20 mm in the $(1\bar{1}0)$ direction have been grown in our laboratory. It has been reported most recently that crystals up to 50 mm long in the same direction can now be produced commercially, using the liquid-phase solution growth method.

In this chapter, we review the results obtained in earlier work relating to the urea OPO, as well as discuss recent work aimed at determining the high-power limitations of such oscillators. It will be shown that with suitable pump sources, it is reasonable to expect the urea optical parametric oscillator to generate continuously tunable coherent radiation throughout the visible and near ir, with output pulse energies up the Joule level.

5.2 Optical Parametric Oscillator

5.2.1 Experimental Setup and Parametric Oscillator Design Considerations

In the urea OPOs described below, we made use of the fact that the urea crystal is positive uniaxial, in contrast to $LiNbO_3$, ADP, and many other nonlinear crystals. This characteristic is of great importance for the design of an angle-tuned OPO. By utilizing type II(o→o + e) phase matching and resonating the ordinary wave, we are then able to reduce greatly the degree of Poynting vector walk-off of the signal from the pump that is caused by double refraction. Furthermore, for a positive uniaxial crystal, the effective d-coefficient is maximum at 90°, which allows for noncritical phase matching.

The first OPO constructed [5.5] contained a small (12.7 mm length) crystal which was collinearly pumped in the near-field of the third harmonic of an Nd:YAG (Quanta-Ray DCR-1) laser, where the third harmonic was in the form of 2 MW, 7 ns pulses at 354.7 nm, with a 10 Hz repetition rate. The tuning range was, however, very limited, extending only from 500 to 509 nm. The OPO was a singly-resonant oscillator (SRO), i.e., only one of the two generated frequencies was resonated within the oscillator. This feedback was accomplished using flat, dichroic mirrors.

Very high pump intensity was required in order to achieve oscillation threshold, mainly because of the fact that only ~12 round trips through the oscillator were made during the 7 ns pump pulse. The high intensity posed serious difficulties, due to optical damage of the dichroic mirror surfaces of the oscillator. The 7 ns pump pulse duration is substantially shorter than that used in many previous OPOs. Nevertheless, it is typical of modern commercial Nd:YAG systems, and is thus more widely available, if somewhat less suitable, than sources with longer pulse widths. In an OPO, gain in the active medium is present only when the pump field is on. This is different from, for example, a dye laser, where there is gain storage because of the finite lifetime of the upper laser state. Thus, the threshold and efficiency of the OPO are very sensitive to the pump pulse duration.

A second urea OPO was built [5.6] which utilized a larger (23 mm) crystal pumped by the same harmonic of the same laser as in the first experiment. This OPO was pumped in the far-field of the laser, a choice which leads to two main advantages. First, the Airy profile of the far-field of a Nd:YAG laser is known to reduce the oscillation threshold as compared to the typical doughnut shape in the near-field of some commercial Nd:YAG lasers. More importantly, "hot-spots" in the near-field, which can damage the dielectric coatings of the dichroic mirrors, are dispersed in the far-field, so that higher pump energies can be accommodated without optical damage.

The experimental setup of this second OPO is shown in Fig. 5.1. To produce the desired far-field spatial profile, the pump laser's 6 mm beam diameter was reduced ~3× by the collimating telescope and allowed to propagate 4 m. A pinhole was used to stop all but the lowest-order Airy maximum, which was measured to have a diameter of 1 mm and a peak intensity of 50 MW/cm^2 at the crystal. Notice that this intensity is several times lower than the multiple-shot damage threshold (180 MW/cm^2) mentioned above. A portion of the pump beam was split off and measured with a reference diode (Si p-i-n).

The parametric oscillator was again of a simple, singly resonant design. The mirrors M_1 and M_2 were both flat and dichroic. The crystals was located in the OPO within a cell filled with n-hexane, which acted as an index-matching liquid (IML). The reasons for surrounding the crystal with an IML were to eliminate crystal degradation from atmospheric moisture, to reduce

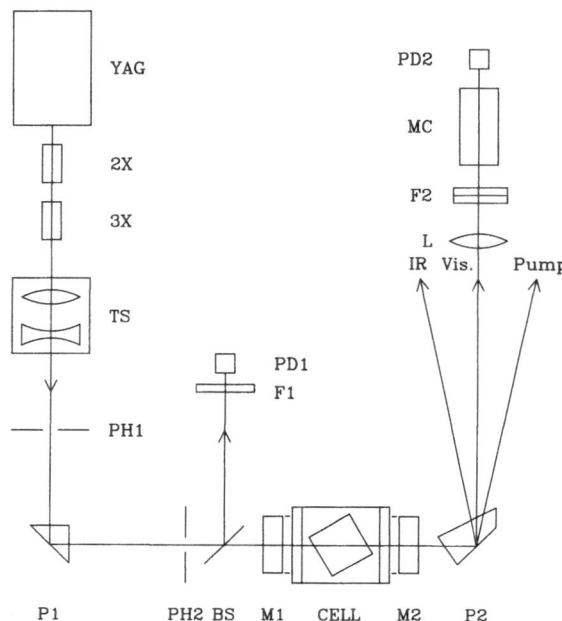

Fig. 5.1. Experimental schematic for OPO tuning-curve measurements. (Key: TS, $3\times$ collimating telescope; PH1 and PH2: pinholes; P1: uv quartz $90°$ tuning prism; BS: beam splitter; M1 and M2: OPO dichroic mirrors; CELL: urea crystal mounted inside a cell with uv quartz windows fulled with hexane; P2: uv quartz Brewster's angle separating prism; L: collimating lens; F1 and F2: ND filters; MC: monochromator; and PD1 and PD2: Si p-i-n photodiodes)

the intracavity Fresnel losses, and to facilitate alignment of the OPO as the crystal was angle tuned. Hexane was selected mainly for its uv transparency, although certain properties of hexane, including an absorption feature at 1.1 μm and a relatively large difference in the index of refraction from that of urea, suggest the need to identify a more superior index-matching liquid. Decalin can also be used. In fact, decalin provides a better index matching than hexane, but there is a strong self-defocusing effect in decalin which makes it unusuable at high intensities. The angle tuning was accomplished by rotating the crystal within the cell about the [100] axis with a precision rotation stage. The optical beams propagated in and out of the cell through uncoated uv quartz windows.

The optical length of the OPO, 100 mm, was the shortest possible in order to maximize the number of passes through the crystal during the finite pump pulse length. This length was consistent with location of the mirrors outside the cell and with a cell large enough to allow for crystal rotation for tuning. Because of the short pump pulse duration, the OPO losses were dominated by the effective rise-time losses rather than the mirror transmission losses. Thus, a single set of dichroic mirrors with a reflectivity product as low as 10% was successfully used over the entire tuning range. Should high-reflectivity mirrors (for example, $R > 95\%$) have been needed for oscillation, it would have been necessary to use several sets of dichroic mirrors for M_1 and M_2, to cover the enormous wavelength range of the resonated wave. The reflectivity product at the nonresonated wavelength was less than 1%; thus, the OPO was acting as a singly resonant oscillator. At no time did we observe temporal and spectral instabilities that typify doubly resonant oscillators. The reflectivity of the output coupler M_2 was typically less than 50% at the resonant wavelength; thus, the intensity of the resonant wave outside the OPO cavity was actually quite large. Finally, the output of the OPO was sent uncollimated through a uv quartz prism, in order to separate the pump wave from the resonated and nonresonated waves.

5.2.2 Tuning and Linewidth Characteristics

The measured OPO tuning curve, which relates the wavelengths of the two generated waves as a function of internal crystal angle, is shown in Fig. 5.2; the corresponding calculated and measured efficiency curves are shown in Fig. 5.3. (Figure 5.2 also shows the result of second harmonic generation in a separate urea crystal.) Note the very broad tuning range. The resonated wave was tuned continuously from 498 nm to 640 nm, corresponding to 1.23 μm to 790 nm for the nonresonated wave. The gap from 640 nm to 790 nm was not reached because of mechanical obstruction at large crystal angles, due to the dimensions of the particular crystal used, but the limitation is not an intrinsic one. Thus, with a 355 nm pump, a urea OPO is potentially tunable over the *entire* range from 498 nm to 1.23 μm.

The linewidth of the resonated wave was measured at several points in the tuning curve. It was approximately 1.2 Å, roughly independent of the crystal angle for angles near 90°. The spatial profile of the resonated wave was very

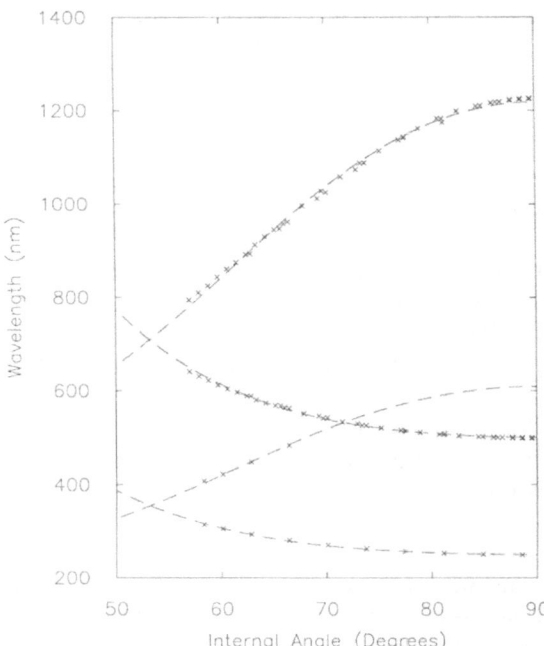

Fig. 5.2. OPO and SHG tuning curves. The uppermost branch is the OPO resonated wave. The lowest two branches represent the output of the OPO/SHG combination. All angles refer to the internal propagation angle in the urea crystal. The lines are calculated from the Sellmeier equations (5.1,2) assuming collinear phase matching. All curves refer to a 3547 Å OPO pump wavelength

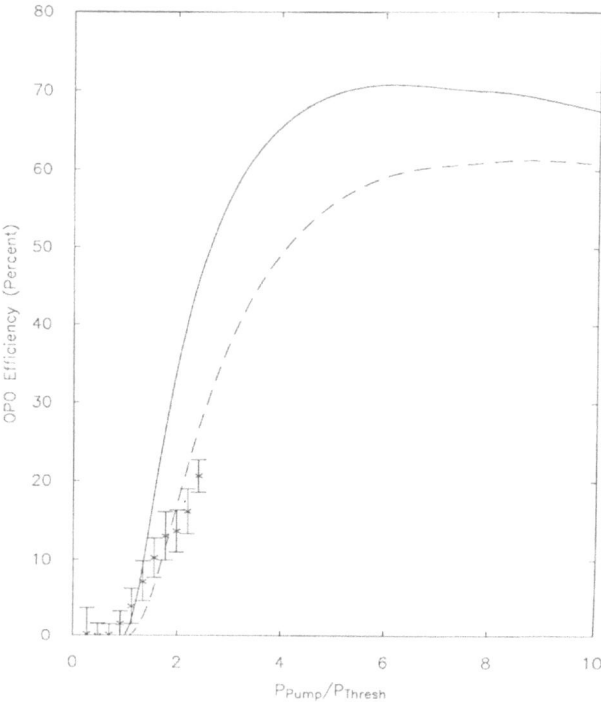

Fig. 5.3. Calculated and measured OPO efficiency for a singly resonant oscillator (SRO). The solid line represents calculated efficiency for a spatially Gaussian cw oscillator [5.9], and the dotted line is for the pulsed case. The weighting factor used is that same as in [5.10]. The error bars shown for the experimental data represent root-mean-square deviations

nearly a TEM_{00} Gaussian wave. The maximum total efficiency (sum of both generated waves relative to the pump power coupled into the cavity) of the OPO was 23%.

These experiments demonstrate that wide tunability in the visible and the near ir can be achieved using the urea OPO. Nevertheless, one major obstacle in further developing such an oscillator is that the growth of large, optically clear urea crystals is still very difficult. The urea crystals used in the experiments reported in this chapter were all grown from a methanol solution, and the growth procedure is long and involved. The 23 mm crystal used required a rather long growth time. It is, therefore, important to demonstrate that efficient parametric oscillation can be achieved in as short a crystal as possible.

The far-field experiment was repeated with a urea crystal measuring $13 \times 5 \times 15$ mm^3, with the 13 mm dimension along the direction of propagation at 90°, and the 15 mm direction along the crystal z-axis. A crystal of this size can be grown relatively easily in about two months. For this experiment, the dielectric mirrors acted as the IML cell windows, in order to minimize the resonator length. The optical path length of the cavity was thus reduced to 60 mm, allowing ~17 passes during the 7 ns pump pulse duration. The shorter cavity length partially compensated for the shorter crystal length used in this experiment. The pump beam diameter was about the same as before (~1 mm). The measured tuning curve for this oscillator with a 13 mm crystal was virtually identical to that shown in Fig. 5.2 for the 23 mm crystal.

5.2.3 Threshold and Efficiency Considerations

Figure 5.4 shows the calculated OPO threshold pump energy as a function of optical cavity length for various crystal lengths, including the 13 and 23 mm lengths actually used. The measured thresholds shown there agree well with the calculated values. These results also show the importance of reducing the optical cavity length. The calculated results show that for a cavity length of 100 mm, a reduction of the crystal length from 23 to 13 mm would have meant an increase of the oscillator threshold from almost 5 to 15 mJ/pulse. The measured increase was only from 5 to 8.5 mJ/pulse because of the shorter cavity length (60 mm) used with the shorter crystal.

Since the initial experiments on the urea OPO have all been conducted at relatively low energy/pulse and low average power level, the question arises of

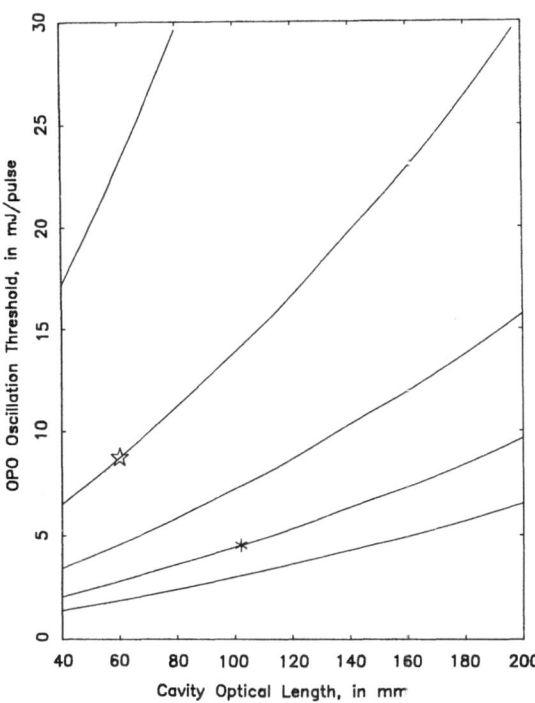

Fig. 5.4. Oscillation threshold of a urea OPO as a function of cavity length for various crystal lengths. The pump is assumed to be Gaussian in shape temporally, with a width of 7 ns. From top to bottom, the *solid lines* represent calculations assuming 8, 13, 18, 23, and 28 mm crystal lengths, respectively. (Star: OPO with 13 mm crystal; asterisk: OPO with 23 mm crystal)

its potential for high-power applications. A series of experiments were carried out that allowed us to extrapolate the high-power limitations.

As previously shown in Fig. 5.3, the measured total efficiency for the 23 mm urea OPO agrees well with the calculated results in the range of pump power studied. The calculated results show that the total efficiency can be well over 50% with a spatially Gaussian pump beam. With a pump beam of optimum spatial and temporal shape, this efficiency can in principle be close to 100% at sufficiently high pumping levels.

A second experiment was conducted to measure the efficiency of the 23 mm urea OPO, this time up to the maximum allowable pump power. Once again, the OPO was pumped in the far-field of the Nd : YAG pump laser, and the pump-field diameter at the OPO was approximately 1 mm. Nevertheless, by using a fixed crystal orientation (and thereby sacrificing tunability), and again using the mirrors as cell windows, it was possible to reduce the cavity length to 31 mm. The measured efficiency versus pump power is shown in Fig. 5.5. Note the flattening of the efficiency curve shown in Fig. 5.5 above 180 MW/cm^2, which we believe is because of the degradation of the crystal, and which yields a maximum efficiency of about 35%. At this point, we do not know whether this long-term damage is associated with the solvents trapped in the crystal during growth, or whether better quality crystals can withstand higher fields at 355 nm. The damage threshold of the mirrors used was at approximately 220 MW/cm^2.

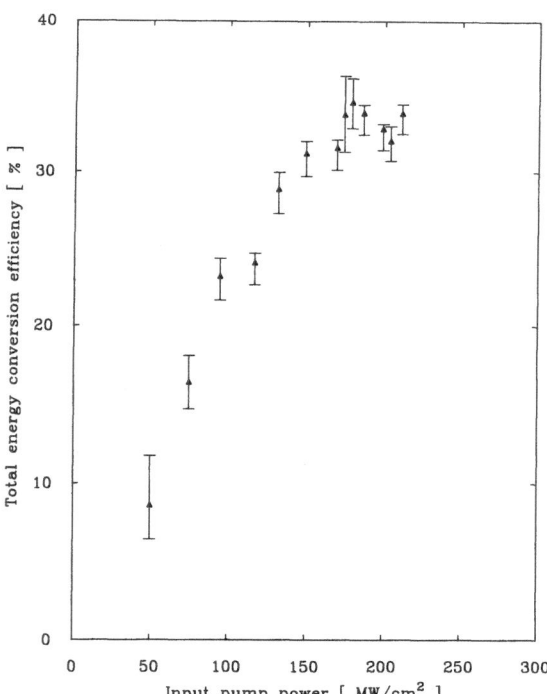

Fig. 5.5. Total energy output of urea OPO using the far-field of Quanta-Ray DCR-2A Nd : YAG laser (i.e. roughly Gaussian spatial profile). Pump diameter (0.8 ± 0.1) mm, pump duration $(8 \pm 1$ ns)

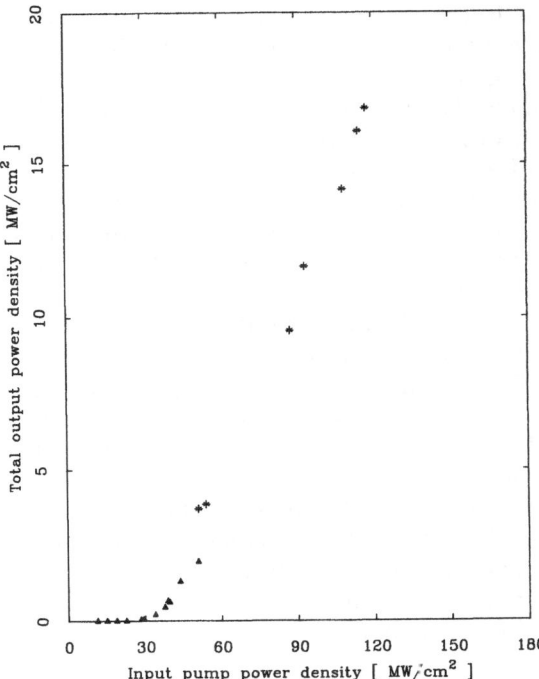

Fig. 5.6. Total power output characteristic of urea OPO obtained from combining the results obtained in the near-field of the Nd:YAG laser with and without reducing telescope. See text for explanation

Since the efficiency of the OPO was basically pump-field limited, to achieve high energy/pulse or high average power, one can simply scale up the beam size. Based on the results shown in Fig. 5.5 for a 1 mm diameter pump beam, it should be possible to scale the OPO into the Joule/pulse range, assuming a suitable high-energy source is available, and as long as the pump intensity is kept below the damage level. Preliminary experimental results, indicating possible high-energy operation of the device, are shown in Fig. 5.6. The OPO was the same as for Fig. 5.5, but it was operated in the near-field of the Nd:YAG pump laser. The triangles correspond to the case where no beam reducing optics were used, and the asterisks correspond to the case where the pump beam diameter was reduced by 2.8× by a telescope. The intensity was limited to about 120 MW/cm^2 by the potential for mirror damage associated with hot spots in the near-field. The two sets of data mesh well, indicating that scaling up the intensity for the same beam diameter should work well. Thus, with a large diameter high-energy pump beam in the far-field and suitable optics, very high-energy operation at an efficiency of up to 35% should be possible.

5.3 Spontaneous Parametric Fluorescence

Spontaneous parametric fluorescence from the urea was also examined. Because the phase-matching condition is the same for the spontaneous and stimulated parametric processes, the tuning characteristics of the spontaneous parametric fluorescence yield the tuning curves of the corresponding optical parametric oscillator. This is important because it implies that one can experimentally determine the tuning characteristics of an oscillator before it is actually constructed. Thus, study of parametric fluorescence is often the first step in assessing the feasibility of a particular parametric oscillator. Furthermore, because parametric fluorescence is linear with the incident power, while second harmonic generation is sensitive to the spatial profile of the fundamental beam, it is more accurate to measure the second-order optical susceptibility of the crystal using the parametric process.

The experimental setup used to study the parametric fluorescence is identical to that shown in Fig. 5.1, except that the dichroic mirrors M_1 and M_2 were removed. The parametric fluorescence pattern was easily visible to the unaided eye. A color photograph of spontaneous parametric fluorescence in urea with a 355 nm pump incident on the crystal at $\theta = 90°$ (where θ is the angle between the pump beam direction and the crystal z-axis) is shown in Fig. 5.7. At each of several values of θ, the wavelength of the fluorescence was measured as a function of emission angle by translating a small monochromator perpendicularly to the direction of pump propagation. The resulting fluorescence tuning curves are compared in Fig. 5.8 with the curves calculated from the Sellmeier equations given in [5.5, 6]:

$$n_o^2(\lambda) = 2.1548 + \frac{0.0131}{(\lambda^2 - 0.0318)} \, , \tag{5.1}$$

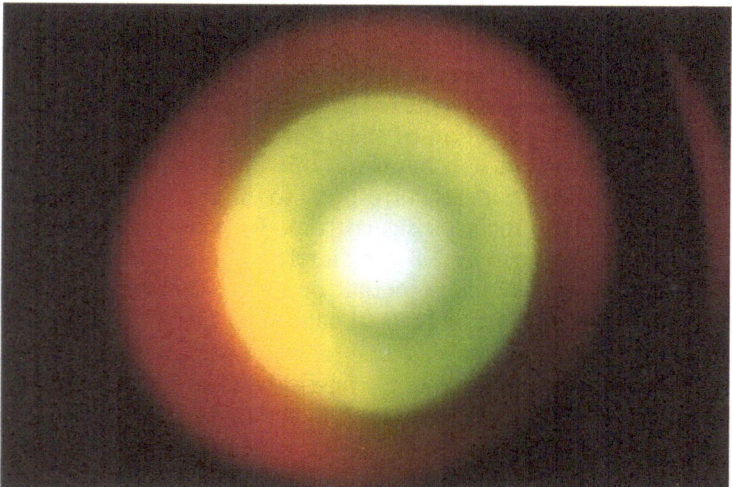

Fig. 5.7. Color photograph of spontaneous parametric fluorescence in urea with 355 nm pump incident on the crystal at $\theta = 90°$. The fluorescence extends into the near infrared, but the photographic emulsion is sensitive only to visible light

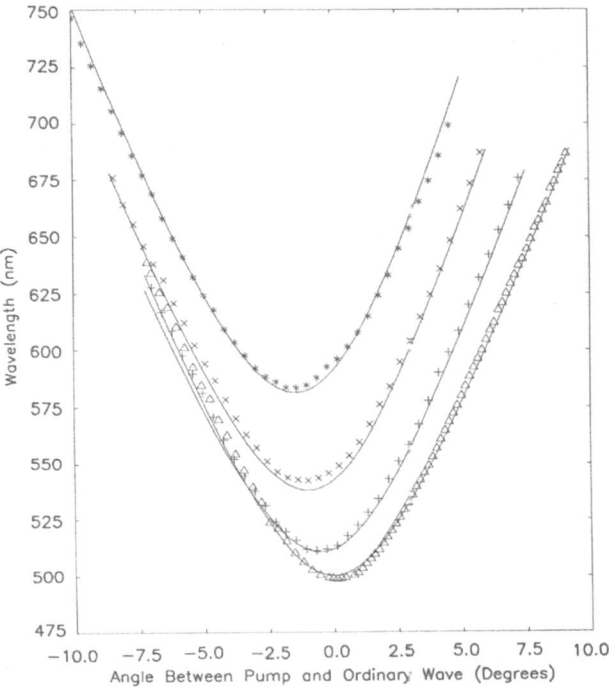

Fig. 5.8. Tuning curves for spontaneous parametric fluorescence in urea at several crystal orientations. Key: $\triangle$, $\theta = 89.72°$; $+$, $\theta = 78.28°$; $\times$, $69.28°$; $*$, $\theta = 62.28°$. The *solid lines* are calculations made using the Sellmeier equations (5.1,2)

$$n_e^2(\lambda) = 2.5527 + \frac{0.01784}{(\lambda^2 - 0.0294)} + 0.0288 \frac{(\lambda - 1.50)}{[(\lambda - 1.50)^2 + 0.03371]} , \quad (5.2)$$

where λ is in μm.

A Type I (o→e+e) phase-matched urea OPO pumped at the fourth harmonic (266 nm) of the Nd:YAG laser output is potentially tunable almost continuously from 300 nm to over 1.4 μm. To explore this possibility, the parametric fluorescence wavelength was measured as a function of the emission angle for various values of θ (θ defined above), with a 266 nm pump. A small but significant difference was observed between the collinear data and the predicted fluorescence curves calculated from the Sellmeier equations (5.1,2). On the basis of the 266 nm data, we have deduced new Sellmeier equations better suited for this spectral region:

$$n_o^2(\lambda) = 2.1823 + \frac{0.0125}{(\lambda^2 - 0.0300)} , \quad (5.3)$$

$$n_e^2(\lambda) = 2.51527 + \frac{0.0240}{(\lambda^2 - 0.0300)} + 0.0202 \frac{(\lambda - 1.520)}{[(\lambda - 1.52)^2 + 0.08771]} , \quad (5.4)$$

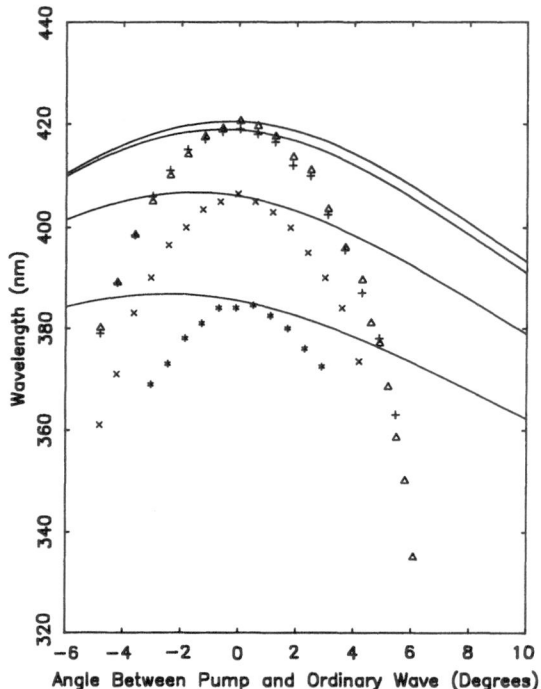

Fig. 5.9. Tuning curves for spontaneous parametric fluorescence in urea at several crystal orientations. Key: △ for $\theta = 86.7°$; + for $\theta = 84.2°$; × for $\theta = 75.2°$; and * for $\theta = 66.3°$. The solid lines are calculations made using the Sellmeier equations (5.3,4) for each appropriate value of θ. The pump wavelength is 266 nm

where λ is in μm. These revised Sellmeier equations do not significantly affect the tuning curve obtained with the 355 nm pump, but at higher frequencies the predictions of (5.3,4) fit the collinear data better than those of (5.1,2).

The measured fluorescence tuning curves for a 266 nm pump are shown in Fig. 5.9. Even with the modified Sellmeier equations (5.3,4), the data were not found to be in good agreement with calculation for *noncollinear* fluorescence. The shape of the fluorescence tuning curve can be shown to be related to the slope of the crystal dispersion curve. But a Sellmeier equation with one or two absorption terms only approximates the real physical dispersion function, so it is not surprising that the local slope of the dispersion curve predicted by the Sellmeier equation can be in error. This argument is especially important as the uv resonances of the crystal are approached. For instance, the calculated fluorescence tuning curves based on (5.1,2) agree better with experiment at 355 nm than at 266 nm. This indicates that a more complex form of the Sellmeier equations (i.e., with more parameters) is needed in order to account for the measured noncollinear parametric fluorescence tuning curves.

If only the collinear point of the fluorescence data is considered, however, then the magnitude (and not the local slope) of the dispersion curve is the parameter of significance. When the angle between the pump and ordinary wave is fixed at zero and the corresponding parametric fluorescence is plotted

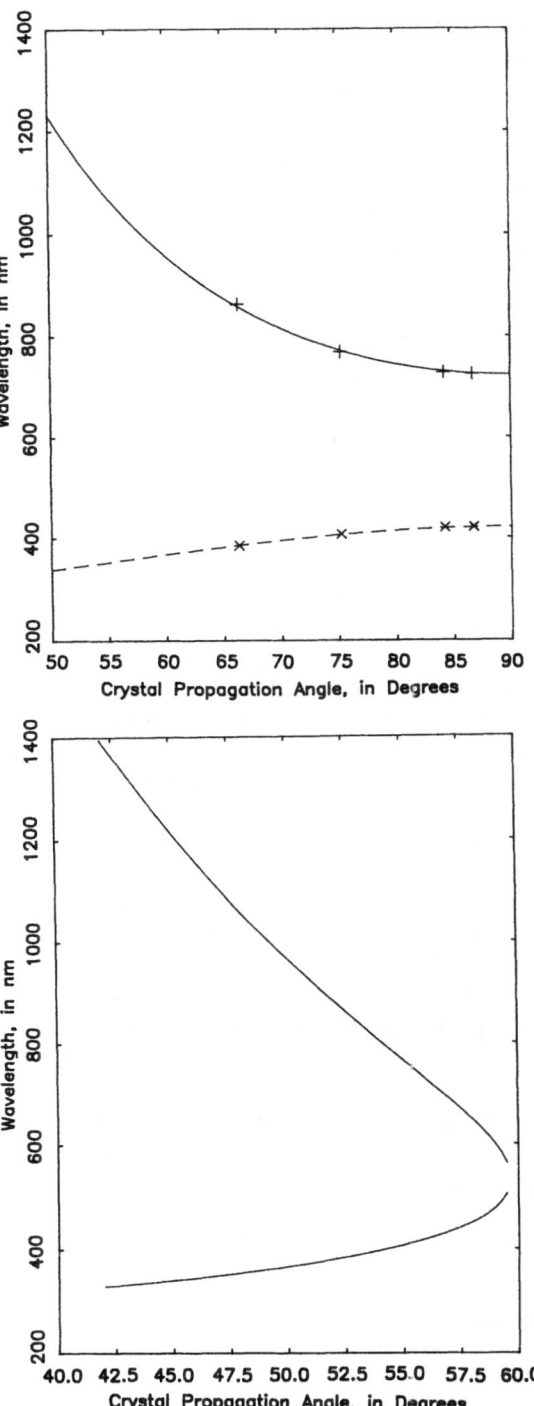

Fig. 5.10. Calculated OPO tuning curve for a pump wavelength of 266 nm for Type II phase matching (o→o+e). The calculations were made using the Sellmeier equations (5.3,4). The wave with extraordinary polarization is shown as a *solid line*, while the one with ordinary polarization is shown as a *dotted line*

Fig. 5.11. Calculated OPO tuning curve for a pump wavelength of 266 nm for Type I phase matching (o→e+e). The calculations were made using the Sellmeier equations (5.3,4)

as a function of crystal rotation angle (θ), the result should be identical to an OPO tuning curve for a collinear interaction. In Figs. 5.10, 11, the calculated OPO tuning curves using (5.3, 4) are shown for both Type-I and Type-II phase matching with a 266 nm pump. The data points indicated for the Type-II case are derived from the fluorescence data, and are in good agreement with calculation. For Type-II phase matching, the additional spectral range accessible with a urea OPO beyond the 355 nm pumped oscillator is fairly limited, but for Type-I phase matching, the entire region from 330 nm to 1.4 μm could be conveniently produced.

There are several considerations beyond simple tunability for operation with a 266 nm pump. For instance, the lower optical damage thresholds of the crystal, the index-matching liquid, and the dichroic optics would probably compel the use of longer pulse length pump sources (>10 ns), in order to reduce the oscillation threshold safely below the onset of optical damage.

5.4 Conclusion

We have demonstrated that the urea optical parametric oscillator is a convenient device for the generation of tunable, coherent radiation in the visible and the near infrared. The device is compact ($3 - 8$ cm long, depending on the tuning range and hence the size of the crystal required), and remarkably easy to align and to frequency tune. For urea, in contrast to previous nonlinear materials used in parametric devices, the optical damage threshold does not pose particularly demanding practical constraints. Parametric conversion efficiencies as high as 35% have been obtained with 7 ns pump pulses. With longer pulses, well over 60% efficiency is possible. Further, the beam characteristics of the OPO output, such as its linewidth, mode quality, and temporal behavior, are suitable for many applications.

While the growth of high-quality urea crystals remains troublesome, we have shown that a useful OPO, capable of producing light tunable from 498 nm $-$ 1.23 μm, can be constructed using a crystal as short as 13 mm, which can be grown relatively easily. We have observed parametric fluorescence pumped at 266 nm, suggesting the possibility of obtaining an even broader tuning curve, from 300 nm $-$ 1.4 μm, by pumping the urea OPO with the fourth harmonic of Nd : YAG.

We believe that the practical advantages of the optical parametric oscillator in general will eventually lead to its widespread use for a variety of applications. The urea optical parametric oscillator is only one specific example.

References

5.1 J. A. Giordmaine, R. C. Miller: Phys. Rev. Lett. **14**, 973 – 976 (1965)
5.2 R. L. Byer, R. L. Herbst: "Parametric oscillation and mixing", in *Nonlinear Infrared Generation,* ed. by Y. R. Shen, Topics Appl. Phys., Vol. 16 (Springer, Berlin, Heidelberg 1977) pp. 81 – 137

5.3 R. L. Byer: "Optical parametric oscillators", in *Treatise in Quantum Electronics,* ed. by H. Rabin, C. L. Tang (Academic, New York 1973) pp. 587–702

5.4 Y. X. Fan, R. C. Eckardt, R. L. Byer, R. K. Route, R. S. Feigelson: Appl. Phys. Lett. **45**, 313–315 (1984)

5.5 W. R. Donaldson, C. L. Tang: Appl. Phys. Lett. **44**, 25–27 (1984)

5.6 M. J. Rosker, C. L. Tang: J. Opt. Soc. B **2**, 691–696 (1985)

5.7 M. J. Rosker, K. Cheng, C. L. Tang: IEEE J. QE-**21**, 1600 (1985)

5.8 J. M. Halbout, S. Blit, W. Donaldson, C. L. Tang: IEEE J. QE-**15**, 1176–1180 (1979)

5.9 J. E. Bjorkholm: IEEE J. QE-**7**, 109–118 (1971)

5.10 S. J. Brosnan, R. L. Byer: IEEE J. QE-**15**, 415–431 (1979)

6. Color Center Lasers

Linn F. Mollenauer

With 46 Figures

Color centers — electrons trapped by defects in insulating crystals — can be used to make optically pumped lasers, broadly tunable over various bands in the ultraviolet, the visible, and most importantly, in the $0.8-4$ μm range of the infrared. Almost all are capable of cw and cw-model-locked operation. In this chapter, the background color-center physics is surveyed, and laser construction and performance are detailed. Included is the soliton laser, a color-center laser whose mode-locked pulses, controlled by the length of optical fiber in its feedback loop, can be as short as a few tens of femtoseconds.

6.1 Background Material

Color centers are electron (or hole) trapping defects in insulating crystals. Since the immediately surrounding crystal usually provides most of the potential experienced by the trapped particle or particles, the electronic states of the center are tightly coupled to the crystal phonons. Thus, just as detailed in Sect. 1.3, optical transitions in color centers usually occur as wide ($\Delta \nu \sim 1000$ cm^{-1}), vibrationally broadened bands, and represent a nearly perfect case of *homogeneous* broadening. Additionally, the lowest energy, or fundamental, transition (in both absorption and Stokes-shifted emission) usually has a large ($f > 0.1$) oscillator strength, leading to correspondingly large ($\sigma \gtrsim 10^{-16}$ cm^2) absorption and gain cross sections. Finally, self-absorption (Sect. 1.4) rarely occurs with the fundamental transition of color centers. Thus, in terms of the most fundamental requirements, many color centers are nearly ideal for the creation of efficient, optically pumped, broadly tunable lasers.

Color centers have been intensively studied and are best known [6.1] in the alkali halides, where they are most easily created. There the prototypical center is the F center (F from the German *Farbe*, meaning color), a single electron trapped in an anion (halide ion) vacancy. Easily created by radiation damage or by chemical means, the F center can be combined with itself or with other defects to form a large variety of centers, many of them laser active [6.2, 3]. (As will be shown later, however, the ordinary F center is not itself laser active.)

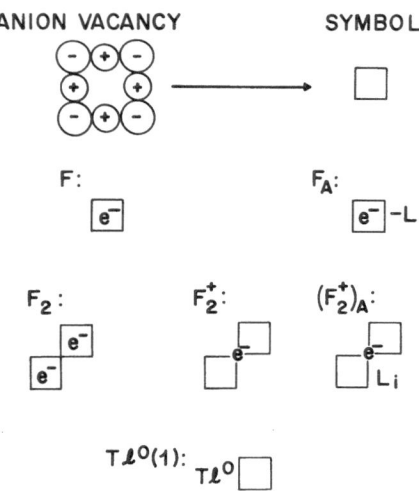

ANION VACANCY   SYMBOL

F: F$_A$:

F$_2$:   F$_2^+$:   (F$_2^+$)$_A$:

Tℓ^0(1): Tℓ^0

Fig. 6.1. The F center and associated laser-active centers

Several laser-active centers or their progenitors are shown in Fig. 6.1, along with the F center. As shown there, the F$_A$ center is an F center for which one of the six immediately surrounding alkali-metal ions is foreign, as for example, Li$^+$ in a potassium halide. As will be detailed later, the subgroup known as F$_A$(II) contains important and stable laser-active centers. Two nearest neighbor F centers along a [110] axis form the laser-active F$_2$ center, which is, however, too easily ionized to be of practical interest. Thus, its singly ionized counterpart, the F$_2^+$ center, (itself experiencing more subtle problems of instability) is the more practical form. Stability of the pure F$_2^+$ center can be further improved through association with a foreign metal ion (F$_2^+$)$_A$, or other defects (F$_2^+$)*, (F$_2^+$)**, and F$_2^+$:O^{2-}. (For simplicity, these stabilized F$_2^+$ centers are not shown in Fig. 6.1.) The Tl^0(1) center is yet another stable and important laser-active center. Although derived from the association of an F center with a foreign Tl$^+$ ion, the Tl^0(1) center cannot be regarded as an F$_A$ center. The reason is that the electron resides on the Tl$^+$ ion (making it a neutral atom), and not in the vacancy, thus radically altering its physics from that of a true F$_A$ center. Finally, note that the F$^+$ center in the alkaline-earth oxides, the analog of the alkali halide F center, has been shown to be laser active. The physics of each of the center types mentioned above will be related in detail in the following section.

The best performance to date, as well as figures relating the more usually attained performance, are listed in Table 6.1 for examples of each of the known laser-active center types. Most examples were selected because of their practical import, although a few are listed primarily for exceptional performance of one kind or another. First, note that (except for the F$^+$ center in CaO) the listed tuning ranges are all in the 0.8 – 4 μm region of the near infrared. This fact reflects the most significant contribution of color center lasers, viz., that they cover the near infrared. This region, where dye lasers fail, is one of

Table 6.1. Performance of color center lasers

Host Center	CaO F^+	LiF F_2^+	NaF $(F_2^+)^*$	KF F_2^+	NaCl $F_2^+ : O^{2-}$	KCl:Tl $Tl^0(1)$
Pump [μm/W]	0.351 0.11	0.647 4	0.87 1	1.064 5	1.064 9	1.064 6
Tuning range [μm]	0.36 – 0.40	0.82 – 1.05	0.99 – 1.22	1.22 – 1.50	1.40 – 1.75	1.4 – 1.6
Max. power Output, cw	20 mW	1.8 W	400 mW	2.7 W	1 W	1.1 W
Ref.	[6.4]	[6.5]	[6.6]	[6.5]	[6.7]	[6.8]
Host Center	KCl:Na $(F_2^+)_A$	KCl:Li $(F_2^+)_A$	KCl:Na $F_B(II)$	KCl:Li $F_A(II)$	RbCl:Li $F_A(II)$	KI:Li $(F_2^+)_A$
Pump [μm/W]	1.34 0.1	1.34 0.15	0.595 1.5	0.647 2.9	0.647/0.676 2.2	1.73 6 mj (pulsed)
Tuning range [μm]	1.62 – 1.91	2.0 – 2.5	2.22 – 2.74	2.3 – 3.1	2.5 – 3.65	2.38 – 3.99
Max. power Output, cw	12 mW	10 mW	60 mW	280 mW	92 mW	0.3 mj (pulsed)
Ref.	[6.9]	[6.10]	[5.11]	[6.11]	[6.11]	[6.12]

great scientific and technological importance. Second, note that the listed figures are almost all for cw operation. (In general, pulsed operation is more easily attained and does not represent nearly as severe a test of center stability.) Finally, note that light-to-light conversion efficiencies are often high, and in a few instances are as great as ~ 50%. Closely associated with such high efficiencies are (cw) output powers of ~1 W or more.

Tunable color-center lasers were conceived as a solid state analog to the dye laser [6.13,14], and it is thus perhaps not surprising that the two laser types have many features in common. (To be sure, pulsed laser action had been obtained with color centers a decade earlier [6.15]. Nevertheless, the laser was not tuned, and the efficiency was very low. Thus, [6.15] remained something of a curiosity until the first demonstration of cw, tunable laser action described in [6.13].) Figure 6.2 shows a cw color center laser, typical of the lasers used to obtain the results of Table 6.1. Note that it uses the folded, astigmatically compensated cavity, with beam waist tightly focused into a thin gain medium, just like the cavity commonly used for cw dye lasers (Fig. 1.3). Once again, this configuration is chosen to maximize pump intensity in the color center crystal, and to allow the crystal slab to be oriented at Brewster's angle to the beam. (Although it may be merely a convenient option in other lasers, the elimination of anti-reflection coatings is a virtual necessity where color center crystals are concerned. Such coatings would almost certainly be destroyed by the various processes used to produce color centers.)

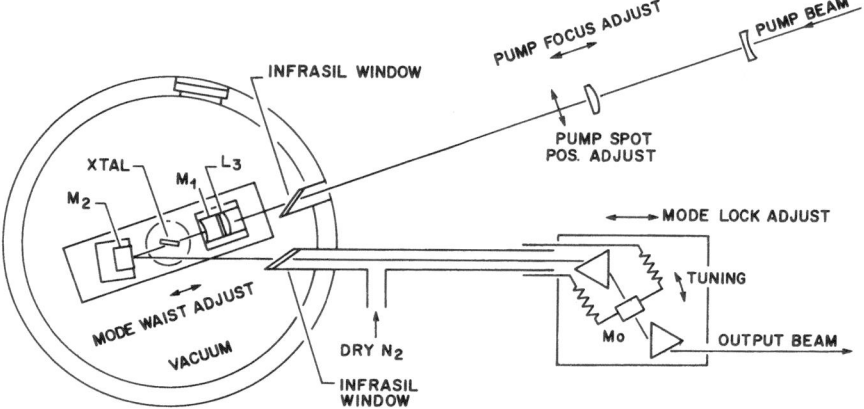

Fig. 6.2. Typical color center laser. Mirror radii: M_1, 25 mm; M_2, 50 mm; M_0, ∞

To prevent destruction of the centers during laser action, and also to maintain them during off periods, the crystal is almost always thermally anchored to a cold finger. The cold finger (for simplicity, not shown in Fig. 6.2), is usually maintained at ~ 77 K by a small storage Dewar of liquid nitrogen. The vacuum chamber shown in Fig. 6.2 is thus to provide thermal isolation for the cold finger, and to prevent condensation of vapors onto the crystal surfaces. The need for cryogenics is often cited as the major disadvantage of color center lasers, but thus far is not to be avoided. Nevertheless, small and inexpensive Dewars with several days hold time are now easily produced by using "space age" technology, and thus the cryogenic aspect of color center lasers is not as formidable as it may at first seem.

In color center lasers the gain medium is static, as opposed to the flowing and turbulent jet stream of a dye laser. This fact makes it much easier to obtain extremely narrow linewidths in color center lasers than in dye lasers. For example, a cw color center laser with ~ 1 kHz linewidth and long-term frequency definition to better than one part in 10^{10} was recently used at the US National Bureau of Standards as one link in a chain of lasers employed in the direct measurement of optical frequencies [6.16]. More generally, the capacity for precise frequency definition has made certain color center lasers an ideal source for atomic [6.17 – 25] and molecular [6.26 – 32] spectroscopy, especially for the study of bands (in the $\sim 2.5 – 3.5$ μm wavelength region) associated with fundamental O – H and C – H bond stretch frequencies.

Of equal importance, the large gain cross sections of color centers (combined with their large homogeneous bandwidths) make it easy to create mode-locked lasers, especially through the convenient technique of synchronous pumping [6.5, 8] (Sect. 1.8). Mode-locked color center lasers have been of particular importance for the study of picosecond pulse propagation and solitons in optical fibers [6.33 – 38]. (Thus far they have been the only source of sufficient power, tunability, and frequency definition to be truly useful in such pulse studies.) The most important wavelength regions for such studies are those near 1.3 and 1.5 μm, where the fibers combine low loss with the necessary dispersive properties for solitons. As dye lasers fail completely at those wavelengths, the need of a source for fiber pulse studies was a principal driving force behind the development of color center lasers.

Finally, there is an interesting example here of science enhancing the art that enabled its initial exploration. That example lies in the soliton laser [6.39–41], a mode-locked color center laser in which pulse shaping and solitons in a fiber are used to gain precise control of the laser pulses. The soliton laser will be discussed in considerable detail in the last section of this chapter.

6.2 The Basic Physics and Materials Science of Laser-Active Color Centers

This section relates the basic physics and methods for creation of the various known laser-active color centers. Included are certain other center types, which, although not themselves laser active, nevertheless have significant roles to play.

6.2.1 The F Center

As noted in the introduction, the F center in the alkali halides forms a fundamental building block for the more complex, laser-active centers. Therefore this section and the two that immediately follow provide important background for the later sections dealing with the laser-active centers themselves.

In the F center, the vacancy provides a simple square well potential for the electron; Fig. 6.3 shows the corresponding energy level diagram. The transition from the s-like ground state to the p-like first excited state is known as the "F" band. With oscillator strength near unity, the F band dominates the absorption spectrum; transitions to higher excited states ("K" and "L" bands) are more than 10 times weaker. The energies and wavelengths corresponding to the F band peak are plotted in Fig. 6.4 as a function of lattice constant for the various alkali halides; also shown there are the corresponding emission band energies. Note the wide range (3:1) of absorption (or emission) energies, reflecting a corresponding variation of potential well depths and widths. Many other color centers exhibit similar, extensive tuning of the bands with change of host. This simple fact is, of course, of great benefit to the laser art.

The F center exhibits an exceptionally large degree of relaxation following optical excitation, and in this sense it deviates considerably from its status as the "prototypical" color center. The relaxation takes place as follows: in response to the lower electron density of the excited state, the surrounding (positive) ions move outward by mutual repulsion, thereby raising the bottom

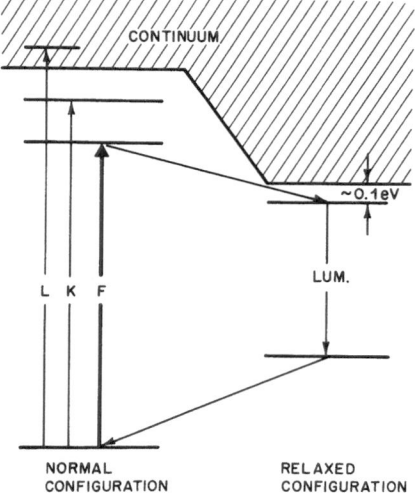

Fig. 6.3. Energy levels and optical transitions of the F center

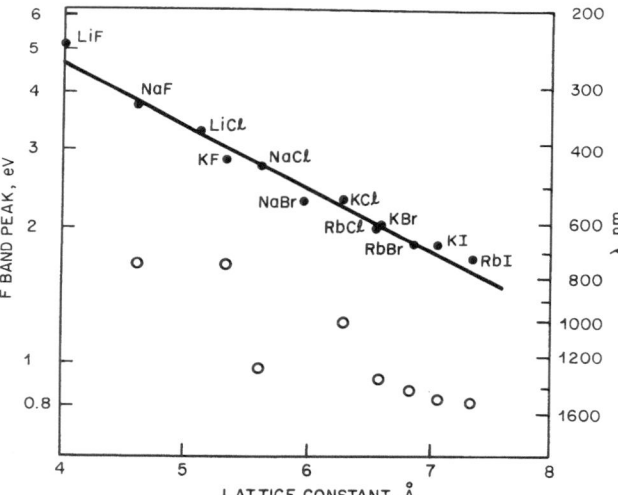

Fig. 6.4. Energies and wavelengths of F center absorption (● ● ●) and emission bands (○ ○ ○) as a function of alkali-halide lattice parameter

of the potential well. In response, the excited state wavefunction becomes even more diffuse, causing further reaction of the ions. The process thus feeds on itself until halted by restoring forces supplied by the rest of the crystal [6.42].

The wavefunction of the resultant relaxed-excited state (RES) is spatially diffuse (mean radius of several lattice constants) and lies close to the edge of the conduction band. Thus, the RES of the F center is similar to a shallow donor state in a semiconductor. This behavior is in strong contrast to that of the $1s$ and $2p$ states of the normal configuration (and to the terminal state of the emission), where in all three cases the electron is largely confined to the vacancy.

The diffuse nature of the RES has two important consequences. First, the optically excited F centers are thermally ionized when the crystal is at or near room temperature. The resultant empty (and mobile) anion vacancies can then be used in the formation of the more complex and laser-active centers. Second, the emission cross section is small ($\sigma < 10^{-17}$ cm^2), primarily because of poor overlap between the RES wavefunction and that of the terminal state. Thus, when absorption from the RES itself and other losses are taken into account, F centers tend to show a net optical loss rather than gain [6.2, 3].

6.2.2 The F$^-$ Center

As its name implies, the F$^-$ center is an F center containing an additional trapped electron. (This two-electron center is sometimes referred to by its older name F'.) The F$^-$ center is also of some importance to the laser art, although it too is not laser active. In most alkali-halide hosts, the F$^-$ center has no bound excited state, and its sole absorption band represents a strong ($f \sim 1$) transition into the continuum. The band begins abruptly at a low-energy limit (in KCl at about 1 eV) and then declines only slowly in strength over the next several electron volts [6.43].

The F$^-$ center is of significance in two ways. First, it is often an important, albeit often temporary, repository for trapped electrons. In this capacity it plays an important role in the formation (to be detailed shortly) of the more complex ("F-aggregate") centers. Second, its absorption band is sometimes in a position to interfere with the optical gain from other centers. Fortunately, the F$^-$ center is easily bleached through optical excitation.

6.2.3 Creation of F Centers Through Additive Coloration

An alkali-halide crystal containing F centers is chemically equivalent to a perfect crystal plus a stoichiometric excess of the alkali metal. Additive coloration is based on this fact, and involves bringing the crystal into equilibrium with a bath of the alkali vapor. Thus, in this process, the F centers are formed at the crystal surfaces and fill up the body of the crystal through diffusion. The equilibrium density N_0 of F centers is determined by a simple solubility equation,

$$N_0 = \alpha N' , \qquad (6.1)$$

where N' is the metal vapor density, and the solubility constant α is only weakly temperature dependent [6.44]. For KCl colored in K vapor at 600 °C, $\alpha = 2.3$.

The approach to equilibrium, however, is determined by the diffusion rate. At the beginning of coloration, the F centers are concentrated at the crystal surfaces. For a thin slab, the behavior at later times ($t > \tau_D$) is given approximately by

$$N(x, t) = N_0 \left[1 - \frac{4}{\pi} \sin \frac{\pi x}{l} \exp \left(\frac{-t}{\tau_D} \right) \right] , \tag{6.2}$$

where x is the distance in from one surface of the slab, l is its thickness, D is the diffusion constant, and where τ_D is the characteristic diffusion time given by the expression

$$\tau_D = \frac{l^2}{\pi^2 D} . \tag{6.3}$$

The diffusion constant for F centers involves a large activation energy. Its empirically determined behavior can be fit, over the range of interest, to the following:

$$D(T) = D_0 \exp(-T_0/T) . \tag{6.4}$$

For example, in KCl, $D_0 = 1.22 \times 10^2 \, \text{cm}^2/\text{s}$ and $T_0 = 14\,400\,\text{K}$.

Additive coloration is usually carried out at temperatures well below the melting point, but high enough to make the coloration time conveniently short. For example, in KCl (mp 768 °C), $D = 8 \times 10^{-6} \, \text{cm}^2/\text{s}$ at 600 °C. Hence for a slab of 2 mm thickness (typical of laser crystals), $\tau_D = 8.4$ min, and a coloration time of ~30 min would then result in an essentially uniform coloration. The 600 °C temperature is also significantly greater than that (in KCl, $T \leqslant 400$ °C) at which colloid formation is favored [6.45].

An apparatus [6.46] for additive coloration, allowing for precise control of the F center density and for convenient loading and unloading of the crystals, is shown in Fig. 6.5. It is based on the principle of the heat pipe. Liquid metal is confined entirely to the wick (made of fine stainless steel mesh). The dividing line between the region of pure metal vapor and the buffer gas (N_2 or Ar) occurs at that level on the wick where the temperature is at the dew point. Since the gases remain well separated, the metal vapor pressure is always equal to that of the buffer gas.

The operating cycle is as follows: with the ball valve closed, the crystal container is loaded into the air-lock space and the air there replaced with N_2 (or Ar) of the correct pressure. The valve is then opened and the crystal lowered into the coloration zone. A set of baffles in the crystal container prevents condensation of vapor onto the crystal while the container temperature rises through the dew point. For removal of the crystal, this procedure is simply reversed. While it is still in the air-lock space, the sample is cooled rapidly to room temperature by a stream of dry N_2 gas. Such handling has made it possible to color an optically polished sample with little or no loss of surface quality. This has considerably improved the preparation of laser-quality crystals.

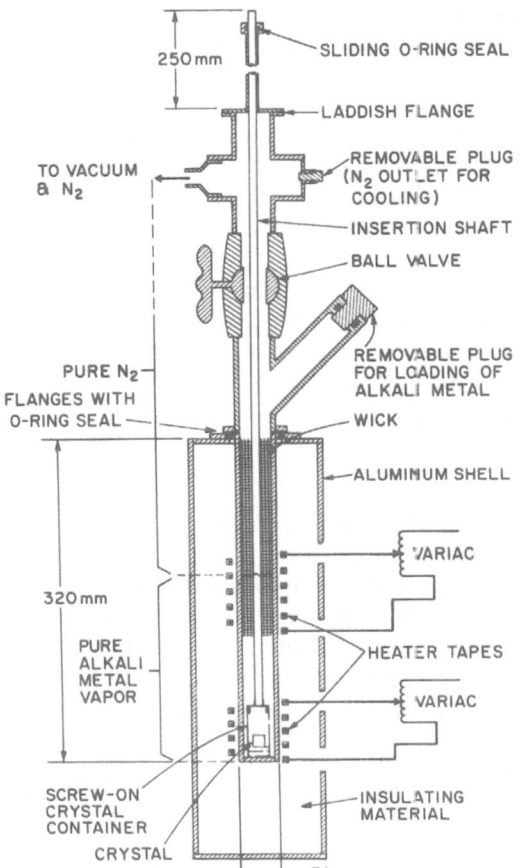

250 mm

TO VACUUM
& N₂

PURE N₂

FLANGES WITH
O-RING SEAL

320 mm

PURE
ALKALI
METAL
VAPOR

SCREW-ON
CRYSTAL
CONTAINER

CRYSTAL

SLIDING O-RING SEAL

LADDISH FLANGE

REMOVABLE PLUG
(N₂ OUTLET FOR
COOLING)

INSERTION SHAFT

BALL VALVE

REMOVABLE PLUG
FOR LOADING OF
ALKALI METAL

WICK

ALUMINUM SHELL

VARIAC

HEATER TAPES

VARIAC

INSULATING
MATERIAL

31 mm

Fig. 6.5. Apparatus for additive coloration of laser-quality crystals (see text)

6.2.4 Center Creation by Radiation Damage

Radiation damage is easily produced in the alkali halides, with F centers a principal end product. The damage is thought to take place as follows: first, no matter what the source of damage, whether it be ultraviolet light, x-rays, γ-rays, or high-energy electrons, its primary effect is to remove electrons from anions, thereby producing electron-hole pairs. Recombination of the electron-hole pairs (which takes place always at anion sites) then leads to the formation of so-called self-trapped excitons [6.47].

In the F center formation that can result from decay of such excitons, a neutral halide atom is ejected into an interstitial site, while the electron remains trapped at the newly created anion vacancy. The ultimate fate of the excess halide is not known with certainty, but it is thought [6.48] to become stabilized through interaction with dislocations and the formation of X_2 molecules (X stands for halide).

Radiation damage of laser crystals is usually carried out at temperatures below $-50\,°C$ to prevent, or at least to control, formation of the more complex centers. At the same time, it is often necessary to produce high levels of damage, as for example, when the goal is to produce large densities of F_2^+ centers. Usually, the only source intense enough to create the necessary damage in a reasonably short time (<1 h) is an electron beam. Typically, energies are $\sim 1-2$ MeV, high enough to allow penetration through several millimeters of crystal, and current densities are about $2-5\,\mu m/cm^2$. It has become common practice [6.49] to perform the radiation as follows: $\sim 1-2$ mm thick crystal slabs, shaped and polished for the laser, are sealed in single layer of household Al foil ($\sim 12\,\mu m$ thick), and are exposed to the electron beam on one or both of their large faces while cooled to $T \sim -100\,°C$ by a stream of dry N_2 gas. The required exposures depend on the particular host and the desired end product, but typically range from a few tens to many hundreds of $\mu A\,min/cm^2$.

6.2.5 Aggregation of F Centers

As already mentioned, F centers are readily combined with each other or with other defects through a simple process of aggregation [6.50]. Optical excitation of the F centers at or near room temperature can begin the process. (The excitation is often carried out with an unfiltered tungsten microscope lamp.) Thermal ionization of the optically excited F centers then results in the formation of pairs of F^- centers and empty vacancies. At high enough temperature ($T > -50\,°C$), the empty vacancies wander through the lattice until they meet either an F center or a foreign metal ion. Recapture of an electron from optically ionized F^- centers by the vacancy then leads to formation of F_2 centers in the first instance, or to formation of F_A (or F_B) centers in the second.

If the foreign-metal ion concentration is several orders of magnitude greater than that of the F centers, a nearly complete conversion can be carried out, with F_A centers as the nearly exclusive end product. Thus, the creation of F_2 centers is most efficient in crystals of low impurity concentration. Nevertheless, F_2 centers are always accompanied by substantial amounts of higher aggregates, such as F_3 and F_4 centers.

Aggregation also occurs during radiation damage at or near room temperature, or on warm-up of crystals following radiation damage at lower temperatures. Then it is not necessary to excite the crystals optically, as the radiation damage itself produces many empty anion vacancies. (The corresponding electrons are trapped by various defects or impurities.) Thus, considerable densities of aggregate centers, such as F_2 and F_2^+, are automatically produced in radiation damaged crystals.

6.2.6 The F_A Centers

According to [6.51, 52], F_A centers can be classified according to the way they relax following optical excitation. In this respect, those known as "type-I" behave almost indistinguishably from the ordinary F center. Hence, with their similarly diffuse RES and low-emission cross sections, the type-I F_A centers are equally unsuitable for laser action. Nevertheless, in the other (and rarer) class are the F_A centers of "type-II". These relax to a radically different configuration following optical excitation. As indicated earlier (Table 6.1), F_A (II) centers make very practical lasers.

The normal and relaxed configurations of the F_A (II) center are shown in Fig. 6.6, along with the associated energy levels. Note that the relaxed configuration is a symmetrical double well. The wavefunctions for the excited and ground states of the relaxed center are made up of antisymmetric and symmetric combinations, respectively, of a single-well s-state. The resultant oscillator strength for an electric dipole transition is nearly the allowed maximum $(f \sim 0.2 - 0.35)$. In combination with the gain bandwidth of $\sim 600 \, \text{cm}^{-1}$, that oscillator strength makes the gain cross section large $(\sigma \sim 3 \times 10^{-16} \, \text{cm}^2)$ at band center. Figure 6.7 shows laser output versus wavelength for the most useful type-II centers [6.11].

The luminescence quantum efficiency η of type-II centers is strongly temperature dependent. For example, for F_A (II) centers in KCl: Li, η is about 50% for $T = 1.6 \, \text{K}$ and decreases almost linearly with increasing temperature, until it becomes less than $\sim 20\%$ at 200 K [6.3]. Nevertheless, laser action has been obtained with F_A (II) centers in KCl: Li for T as large as 200 K.

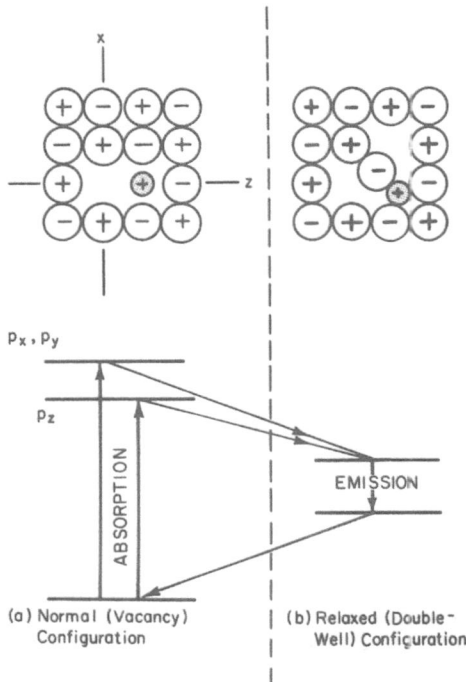

(a) Normal (Vacancy) Configuration

(b) Relaxed (Double-Well) Configuration

Fig. 6.6. Normal (single-vacancy) and relaxed (double-well) configurations of the F_A(II) center, and associated energy levels

TUNING CURVE

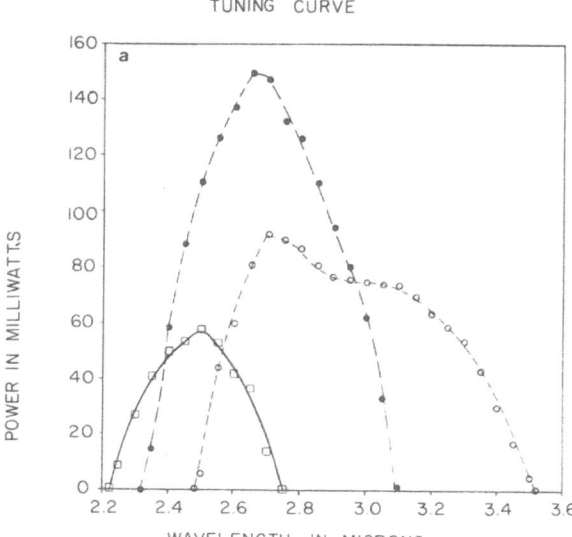

Fig. 6.7. (a) cw output power as a function of wavelength for $F_A(II)$ and $F_B(II)$ center lasers, grating tuned, and with 9−12% output coupling [6.11]. ● ● ●: KCl: Li $F_A(II)$, pump 2 W at 610 nm; ○ ○ ○:   RbCl: Li $F_A(II)$, pump 1.3 W at 660 nm; □□□: KCl: Na $F_B(II)$, pump 1.6 W at 595 nm

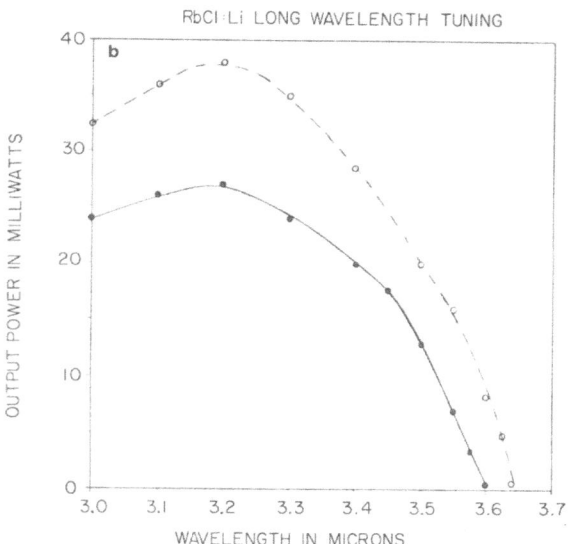

Fig. 6.7. (b) cw output power in extended long-wavelength tuning of RbCl: Li $F_A(II)$ center laser pumped with 1.3 W total at 647 and 676 nm, and using a grating with 3% output coupling [6.11]. *Upper curve:* simultaneous operation on several longitudinal modes; *lower curve:* single-frequency operation

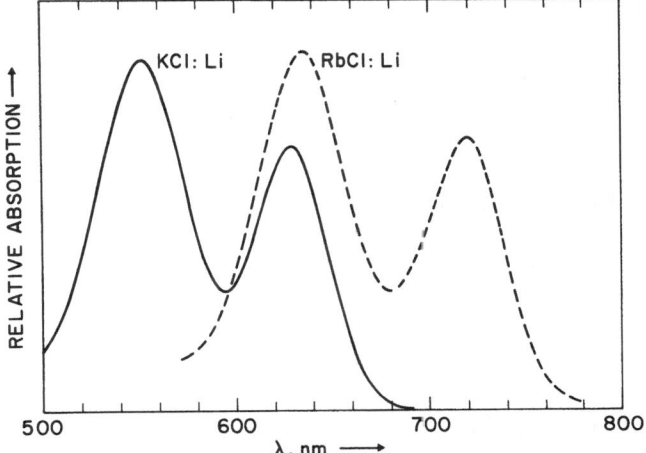

Fig. 6.8. Fundamental absorption (laser pump) bands of F_A(II) centers in the hosts KCl: Li and RbCl: Li (Courtesy of CRC Press)

The huge Stokes shift of F_A(II) centers (nearly 5:1 ratio of pump to emission photon energies) limits laser output power. In the first place, pump input power is limited to ~1 or 2 W because 80% or more of that power must be dissipated as heat in the crystal. Furthermore, even in a laser of 100% quantum efficiency, power output can be no more than ~20% of the input power. Thus, output powers of cw F_A(II) center lasers tend to be no more than a few hundred mW or less (Fig. 6.7).

In the normal configuration of F_A centers, the presence of the foreign ion causes p_z orbitals to be of lower energy than p_x and p_y orbitals. Thus, transitions to the p_z orbitals result in a longer-wavelength band that is often well resolved from the main absorption band (Figs. 6.6,8). In laser applications, this extra band considerably increases the probability of wavelength overlap with a convenient (laser) pump source.

6.2.7 Reorientation of F_A (II) Centers

Note from Fig. 6.6 that with probability 1/2, the halide ion separating the two wells will move into the original F_A center vacancy on completion of the optical pumping cycle. When such occurs, the F_A center axis (defined by the vacancy and the impurity ion) will have switched orientation from one to another of the crystal axes [100], [010], or [001]. Such reorientation must be taken into account in laser operation. First, in the F_A(II) center, the Li ion pins the center down to one small region of the crystal. Thus the center cannot wander, despite the constant flopping of the center axis that accompanies optical pumping. Second, both the direction of propagation and the polarization of the pump light must have the proper relation to the crystal axes in order to prevent orientational bleaching.

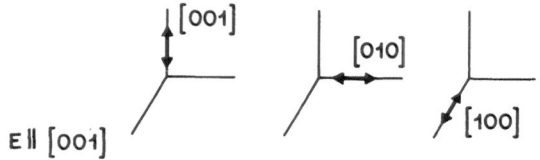

E ∥ [001]

For pumping in	Centers pumped?		
"green" band:	no	yes	yes
"red" band:	yes	no	no

The orientational bleaching [6.2] can be understood as follows. Let the two pump bands of the F_A center be known as the "red" and "green" bands, respectively. Note that for the pumping schemes shown in Fig. 6.9 (pumping with $E \| [001]$ and either in the red or green bands), one or more of the center orientations is *not* pumped. The probability that an optically excited F_A (II) center will reorient on return to the ground state is 2/3. Thus, in either of the two pumping schemes, after a few cycles, the center population will tend to accumulate in the unpumped orientation(s). Thus, neither of the two schemes of Fig. 6.9 would be suitable for the pumping of an F_A (II) center laser crystal.

Nevertheless, note that for pumping in the green band, bleaching can be avoided simply by rotating the pump polarization by 45° ($E \| [011]$), as then all three orientations are pumped. For pumping in the red band, bleaching can be avoided only if the E field has substantial components along all three major crystal axes. In practice, this is achieved in the following way: the laser crystals are slabs with [100] face normals at Brewster's angle to the beam, and with a [110] axis lying in the plane formed by the beam propagation direction and the E field. Thus there are equal components of the pump field along the [100] and [010] axes, and a smaller but substantial component along the [001] axis. Note that with this scheme for laser pumping in the red band, F_A (II) center emission is polarized along the various [110]-type axes; thus half the time they will (uselessly) emit with E field orthogonal to the field of the laser mode.

6.2.8 The F_B (II) Centers

The F_B centers [6.53] involve two nearest neighbor foreign metal ions and are obtained in substantial quantities when the dopant concentration is at least several percent. They are classified by their relaxation behavior in exact analogy to F_A centers: type-I F_B retain the vacancy configuration, while type-II relax to the double well. The normal configurations of F_B (I) and F_B (II) are shown in Fig. 6.10.

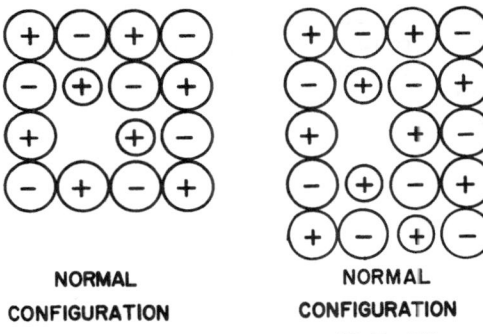

Fig. 6.10. Ionic configurations of $F_B(I)$ and $F_B(II)$ centers

NORMAL
CONFIGURATION
OF $F_B(II)$

NORMAL
CONFIGURATION
OF $F_B(I)$

Continuous wave cw laser action has been obtained with $F_B(II)$ centers [6.11,54] in KCl:Na (Table 6.1) and also in RbCl:Na. Since the tuning range $(2.5-2.9\ \mu m)$ with the RbCl:F_B center is nearly the same as that covered by the more efficient F_A centers (Fig. 6.7), the RbCl:F_B center has not been of commercial interest.

The $F_B(II)$ centers are always accompanied by a certain population of $F_A(I)$ and $F_B(I)$ centers, and the absorption bands of all three types tend to overlap. Thus, the efficiency of F_B center lasers is reduced. This problem can be circumvented, at least in part, by taking advantage of the orientational bleaching effects possible with $F_A(I)$ (see previous section), and of the similar effects possible with $F_B(I)$ centers. To accomplish this bleaching, the pump polarization is usually made to lie in a [100] plane. Nevertheless, the low slope efficiencies ($\sim 2\%$) observed with $F_B(II)$ center lasers may be due in part to a residue of unbleached $F_A(I)$ and $F_B(I)$ centers. That is, imperfect resolution of the pump bands makes the orientational bleaching effects incomplete, and optically pumped type-I centers most likely produce an absorption in the region of the type-II luminescence bands, just as the ordinary F center is known to do [6.55].

6.2.9 The F_2^+ Center

As already shown in Fig. 6.1, the F_2^+ center consists of a single electron trapped by a pair of adjacent anion vacancies. It has been successfully modeled [6.56,57] as an H_2^+ molecular ion embedded in a dielectric continuum, where the two vacancies play the role of the protons. In that model, energy levels are to be calculated as

$$E_{F_2^+} = k_0^{-2} E_{H_2^+}(r_{12})\ ,\qquad (6.5)$$

where the proton separation r_{12} is given by

$$r_{12} = k_0^{-1} R_{12}\ .\qquad (6.6)$$

Here k_0 is the dielectric constant and R_{12} is the vacancy pair separation. When all possible hosts are included, a large net tuning range is possible with F_2^+

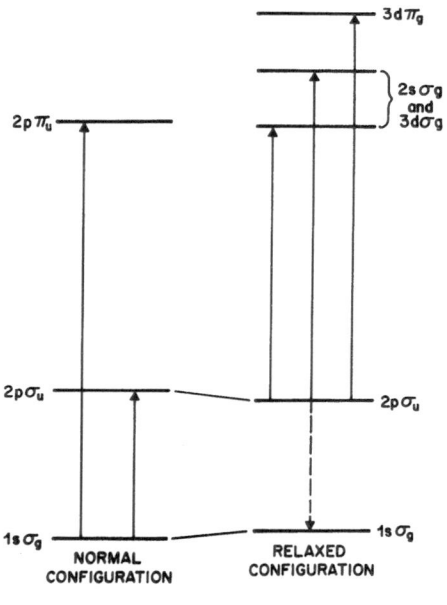

Fig. 6.11. Energy level diagram of the F_2^+ center; all of the indicated transitions have been observed directly

Fig. 6.12. The fundamental transition of the F_2^+ center in various hosts. *Above:* emission (laser tuning) bands; *below:* absorption (laser pump) bands

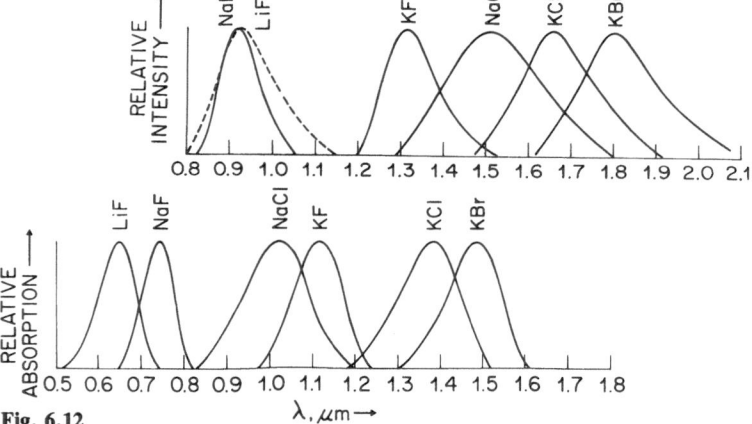

Fig. 6.12

centers or their derivatives. This fact is largely a consequence of (6.5) and of the large variation ($>2.5:1$) with host in k_0^2; the variation of R_{12} is helpful as well.

Figure 6.11 shows an empirically determined energy level diagram of the F_2^+ center [6.58]. There the levels are named after their molecular ion counterparts. The lasers use the lowest-energy ($1s\sigma_g \rightarrow 2p\sigma_u$) transition (Fig. 6.12). Its high oscillator strength ($f \sim 0.2$), temperature independent, 100% quantum efficiency, and small Stokes shift make it nearly ideal for laser action. Finally, there are no excited state absorptions at transition energies close to that of the emission.

The laser transition is completely polarized along the axis of the center (the line joining the vacancy pair). In Fig. 6.13, the six possible center axes (crystal type [110] axes) are represented by the face diagonals of a cube. Excitation of any of the higher-excited states (such as $2p\pi_u$, see Fig. 6.11) causes reorientation, with high quantum efficiency, among these possible directions [6.57]. To be sure, excitation of the fundamental (laser pump) band does not directly cause reorientation. Nevertheless, for the intensities ($\sim 10^6$ W/cm^2) encountered in laser pumping, multiphoton excitation of the higher levels is not completely negligible [6.5]. The resultant reorientation, although very much slower and less efficient than with direct excitation of the higher states, nevertheless can have devastating effects on laser operation.

To understand the reorientation and associated bleaching, consider the pump and laser mode beams to propagate and to be polarized as in Fig. 6.13. Suppose further that initially all the center axes are aligned in the direction marked 1. With multiphoton absorption of the laser beam, however, the center axes will gradually be flipped into the less-efficient orientations $3-6$ and finally all, or nearly all, will settle in the completely useless orientation 2. (That orientation is useless, of course, because its transition moment, which lies along the center axis, is orthogonal to the laser field.)

It is sometimes possible to flip the center back into the desired orientation with certain tricks of optical pumping. For example, consider a scheme that has had some success in extending the useful lifetime of F_2^+ centers in KF [6.5]. The laser was operated in a chopped mode, with the pump beam polarized as in Fig. 6.13 during the "laser on" part of the duty cycle. Nevertheless, in the "laser off" part of the cycle, the pump beam polarization was rotated by 90°, and the pump beam was joined by the beam (~ 3 mW) of a HeNe laser similarly polarized. The two light sources, acting together, created an efficient two-step excitation to the $2p\sigma_g$ state of the relaxed configuration (Fig. 6.11); this excitation was then able to promote a flipping of the centers back out of the 2 orientation and finally into the most desired orientation 1.

Figure 6.14 shows power output versus tuning for the KF, F_2^+ center laser [6.5]. The exceptionally high power and efficiency represented there serve to

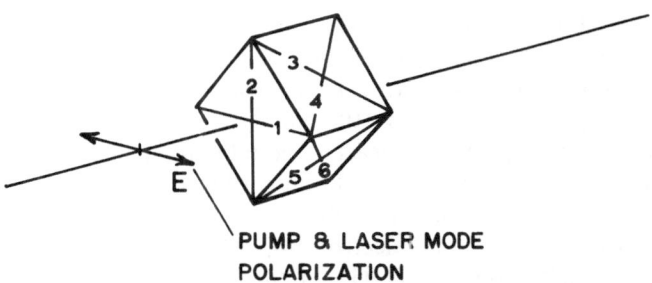

PUMP & LASER MODE
POLARIZATION

Fig. 6.13. The six possible orientations of F_2^+ center axes. The pump beam propagates along [001] and is polarized $\parallel$ [110]. (Courtesy of North-Holland [6.2])

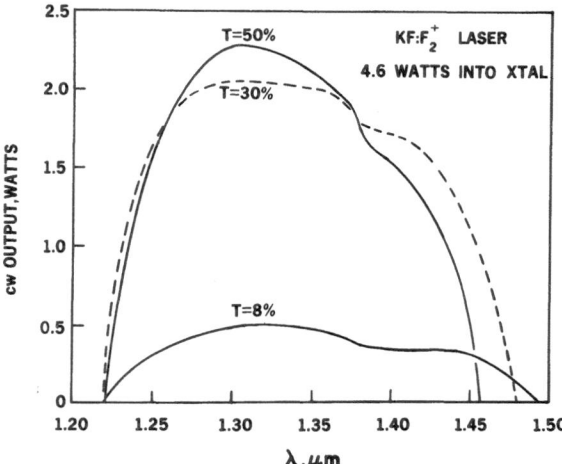

Fig. 6.14. Cw power output versus tuning for a KF, F_2^+ center laser

illustrate the convergence of positive factors that can occur in F_2^+ centers. First, the small Stokes shift minimizes the energy that must be dissipated as heat in the crystal, and thus maximizes the power that can be handled; it also has direct and obvious effect on the efficiency. Second, nearly all the centers are oriented for maximum contribution of their polarized emission into the (similarly polarized) laser mode. Finally, efficiency is enhanced by the high (small signal) gain, made possible in turn by the combination of large gain cross section, high quantum efficiency, and the center alignment. That is, high gain and the large output coupling it makes possible, allow both of the last two terms in (1.10) to be nearly unity.

For any given active spot on the laser crystal, however, the performance shown in Fig. 6.14 lasts for only a few minutes. Fading results from the constant reorientation of the F_2^+ centers and their subsequent random walk through the crystal. This leads to their eventual loss through aggregation or deionization. Thus, the only truly satisfactory cure to the orientational bleaching problem of F_2^+ centers lies in anchoring them to a definite place in the crystal. As will be detailed in the next section, this can be done by associating the center with an impurity or other defect.

As already mentioned in the section on radiation damage, F_2^+ centers are usually created by irradiation with an electron beam. Nevertheless, to create F_2^+ centers in densities high enough ($\sim 3 \times 10^{17}$ cm^{-3}) for the creation of efficient lasers, it is necessary to provide at least an equal density of extrinsic traps for the excess electrons. It has been found that certain transition metal ions (which always enter the alkali halides in the divalent state), such as Mn^{2+}, Cr^{2+}, Ni^{2+}, Co^{2+}, and Pb^{2+}, are all good electron traps [6.59]. It is also well known that debris from radiation damaged OH^- impurity ions makes a good trap [6.60–63]. None of these ions is universally the best, however, since those most easily incorporated vary with the particular alkali halide. Nevertheless, for any particular host, usually at least several ions can be found that will enter into solid solution at the required level of several hundred parts in 10^6.

The density of F_2^+ centers can often be greatly enhanced through laser excitation of F_2 centers in their fundamental absorption band. Such a source will then efficiently ionize the optically excited F_2 centers in a second step. The technique [6.59, 2] works so well that the conversion of F_2 to F_2^+ centers can be made virtually complete.

By the same token, it is easy to destroy F_2^+ centers through deionization. Such deionization represents an additional cause of fading during laser operation, and it accounts for the short (just a few hours) shelf life of most F_2^+ centers when kept in the dark at room temperature. Nevertheless, because association with defects produces such a great increase in the shelf life of F_2^+ centers (Sect. 6.2.10), it is unlikely that electron capture by *static* centers is of much importance. Therefore, deionization on the shelf probably requires thermally activated motion of the centers and tunneling from various nearby traps. Similarly, deionization during laser operation also probably involves motion of the centers.

To circumvent the shelf life problem, F_2^+ centers can be stored in the stable F_2 state, and then reionized when need. Either the efficient two-step process [6.59] described above, or a mild exposure to x-rays [6.64] can be used to reionize the centers; however, conversion with x-rays is not as complete as with the two-step optical process.

6.2.10 Defect Stabilized F_2^+ Centers: $(F_2^+)_A$, $(F_2^+)^*$, etc.

The F_2^+ centers can be associated with certain stabilizing defects or impurities. For example, if the defect is an alkali-metal impurity ion, then the center is called $(F_2^+)_A$. The $(F_2^+)^*$ and $(F_2^+)^{**}$ centers represent F_2^+ centers associated with various intrinsic defects created by radiation damage. Such association with impurities stabilizes the centers in two ways: it prevents orientational bleaching, and it greatly increases the room temperature shelf life of the centers, often to many months or more. It is also creates new and useful tuning ranges.

a) The $(F_2^+)_A$ Centers

Fundamental absorption and emission bands of those $(F_2^+)_A$ centers that have been used to date for successful laser action [6.7, 9, 10, 12, 65 – 68] are shown in Fig. 6.15. Note that in general the emission bands of the defect-perturbed centers are moved to longer wavelengths than the F_2^+ bands for the same host. One particularly noteworthy example is the $(F_2^+)_A$ center in KI : Li, whose emission band has allowed for laser tuning out to nearly 4 μm; this tuning band is of great importance to molecular spectroscopy, as it corresponds to various fundamental carbon-hydrogen bond-stretch frequencies.

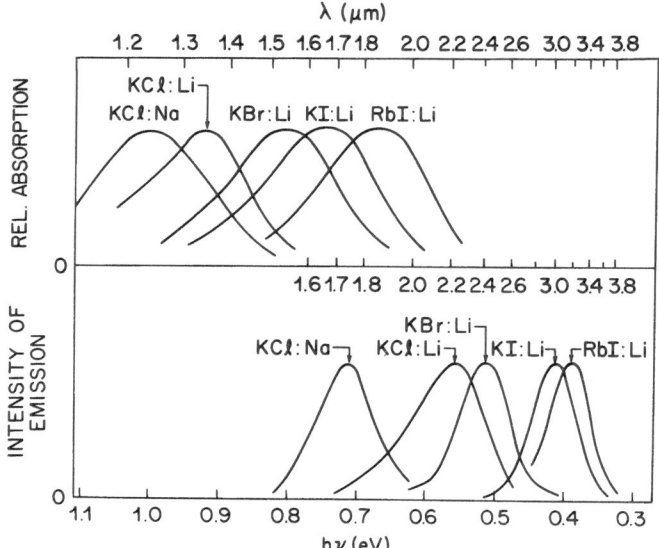

Fig. 6.15. The fundamental transition of $(F_2^+)_A$ centers in various hosts (adapted from [6.10, 65 – 67]). *Above:* absorption (laser pump) bands; *below:* emission (laser tuning) bands. (Courtesy of North-Holland [6.2])

Another noteworthy example is the recently discovered [6.7] stabilized (F_2^+) center in NaCl, whose exceptionally broad emission band is centered ($\lambda \sim 1.56\ \mu m$) on the most important wavelength region for fiber optics. Additionally, the NaCl (F_2^+) center laser is most efficiently pumped at 1.06 μm from a Nd:YAG laser, and has exhibited a non fading, cw output of up to 1 W at band center. Figure 6.16 shows power output versus tuning for this laser. Although K^+ was thought to be the stabilizing defect when [6.7] was written, the most recent evidence [6.69] is that OH^- or one of its components, but not K^+, is somehow involved. In particular, the highest densities of the center have been produced in crystals containing between 10 and 70 parts OH^- in 10^6. It is now thought that the center is actually stabilized by an O^{2-} ion derived from the OH^- ion. This idea is strongly supported by the fact that the centers can be created in crystals in which oxygen is the only significant impurity. Thus, the most logical name for the center is $F_2^+ : O^{2-}$.

The generation of $(F_2^+)_A$ centers is not a particularly easy task, as it involves the bringing together of three entities: two F centers, and the foreign alkali-metal ion. One scheme is first to create F_2^+ centers by radiation damage, and then to make the F_2^+ centers undergo a random walk (by a sequence of optically induced reorientations) until they meet and become attached to the foreign alkali-metal ion. Such was the basic scheme used [6.9] to create $(F_2^+)_A$ centers in the hosts KBr:Li, KI:Li, and RbI:Li. In particular, in KBr:Li, the random walk was excited by the 458 nm line of an Ar^+ laser, and carried out

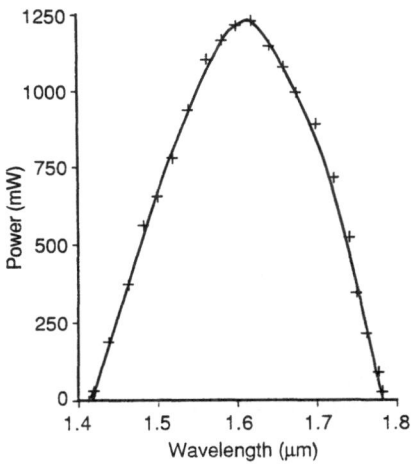

Fig. 6.16. cw power output versus tuning for a NaCl $F_2^+ : O^{2-}$ center laser

for about one hour at 158 K; in the other two crystals, the 514.5 nm line of the same laser was used, and the crystal temperature was 148 K.

The other scheme is that used [6.12] for the hosts KCl: Li and KCl: Na, and more recently for the host NaCl: O^{2-}. With the aid of light near 365 nm filtered from a mercury arc lamp, the F centers ($>10^{18}$ cm^{-3}) of an additively colored crystal were aggregated to form F_2, and also some $(F_2)_A$ centers. The crystals were then cooled to 77 K and reilluminated with the 365 nm source. This process simultaneously ionized the F_2 [and $(F_2^+)_A$] centers and caused the former to make a random walk (again through successive reorientations), converting them to $(F_2^+)_A$ centers. In additively colored crystals, F_A centers must be used as the electron traps for $(F_2)_A$ ionization. The use of F_A centers as traps is possible only because the F_A^- band edge (at least in KCl) lies higher in energy than the photon energy required for pumping of the $(F_2^+)_A$ centers. Finally, note that nonfading operation of most $(F_2^+)_A$ center lasers (including the NaCl: O^{2-} laser) requires the use of an auxiliary mercury arc lamp to provide continuous reionization or reorientation of the centers.

As one might well imagine, the final yield of $(F_2^+)_A$ centers from either of the techniques just described is not particularly high; the optical density in a several millimeters thick laser crystal is considerably less than unity. It would improve the efficiency of $(F_2^+)_A$ center lasers if the center densities could be increased.

b) The $(F_2^+)^*$ and $(F_2^+)^{**}$ Centers

The $(F_2^+)^*$ center was first observed [6.6] in crystals of NaF that had been heavily irradiated to produce a high density of F_2^+ centers. When these crystals were allowed to sit in the dark for a day, it was noticed that the F_2^+ band at 750 nm had largely disappeared and that another intense absorption band peaking at 870 nm had taken its place (Fig. 6.17). The new center was found to

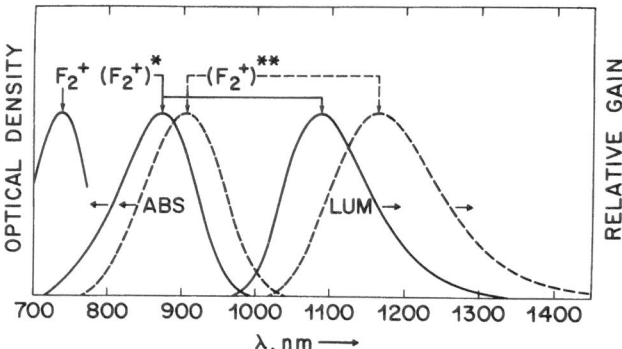

Fig. 6.17. Absorption and emission bands of the lowest energy transition of defect-stabilized F_2^+ centers in NaF at 77 K. For convenient reference, the absorption band peak of the ordinary F_2^+ center is also indicated

be amazingly stable, with a shelf life at room temperature measured in weeks or months.

Further investigation [6.6] revealed an emission band (again see Fig. 6.17), with quantum efficiency, band shape, Stokes shift, and emission decay time (40 ns) and polarization (along [110]), virtually the same as found in the ordinary F_2^+ center. Furthermore, it was discovered that the full ground- and excited-state absorption spectra of the two-center types were highly similar, in relative band strengths, energies, and polarizations. Thus, the center is almost certainly a modified F_2^+ center.

The precise identity of the modifying defect remains unknown. It soon became evident, however, that the divalent transition ions used as electron traps were *not* involved, since *exactly the same bands were obtained for all the ions used* (Mn^{2+}, Cr^{2+}, Ni^{2+}, and Pb^{2+}), and furthermore, substantial densities of $(F_2^+)^*$ centers were obtained in crystals doped only with OH^-. All other impurities were known to be at a negligible level (a few parts in 10^6 or less). Thus, it must be that the modifying defect is an intrinsic one created by the radiation damage.

The $(F_2^+)^{**}$ center was first discovered and shown to be laser active by workers in the USSR [6.70] who thought it was an F_3^- center. (Nevertheless, F_3^- centers, like most electron excess centers, are extremely volatile, i.e., they are easily ionized by optical pumping.) The center was later rediscovered in crystals doped with OH^-, and properly identified [6.71], much as described above for the $(F_2^+)^*$ center.

The $(F_2^+)^*$ and $(F_2^+)^{**}$ centers can be created easily and in large (nearly $10^{18}/cm^3$) densities, many times greater than the minimum required for efficient lasers. Figures 6.18, 19 show the history of samples of NaF: Mn^{2+} and NaF: OH^-, respectively, electron beam irradiated (1.5 MeV, 5 μA/cm², crystals at $-100°C$), for various total doses [6.71]. Shown there are the center concentrations obtained both immediately after irradiation, and following a

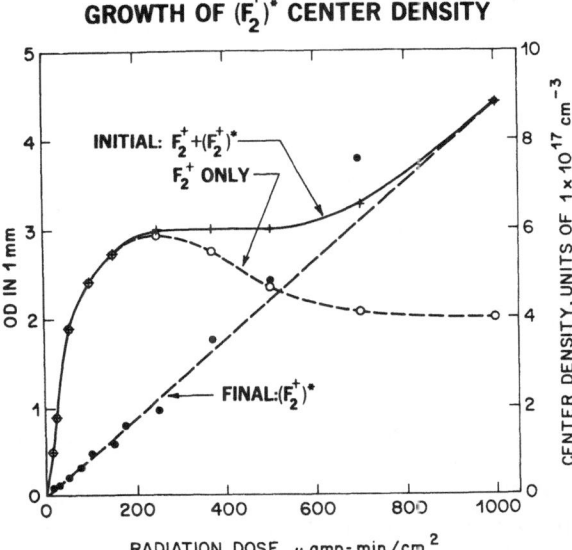

GROWTH OF $(F_2^+)^*$ CENTER DENSITY

INITIAL: $F_2^+ + (F_2^+)^*$

F_2^+ ONLY

FINAL: $(F_2^+)^*$

OD IN 1 mm

CENTER DENSITY, UNITS OF 1×10^{17} cm^{-3}

RADIATION DOSE, μ amp-min/cm^2

Fig. 6.18. Production of F_2^+ and $(F_2^+)^*$ centers in NaF:M_a^{2+} as a function of radiation dose and time. Following ~ 5 min at 300 K: $\times \times \times$, sum of F_2^+ and $(F_2^+)^*$ band heights; $\circ \circ \circ$, F_2^+ alone. After $12-24$ h at 300 K: $\bullet \bullet \bullet$, $(F_2^+)^*$ band height; F_2^+ band has disappeared

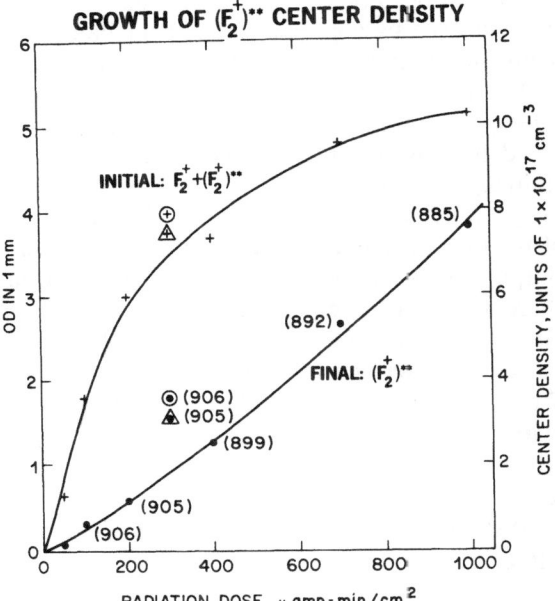

GROWTH OF $(F_2^+)^{**}$ CENTER DENSITY

INITIAL: $F_2^+ + (F_2^+)^{**}$

(885)

(892)

FINAL: $(F_2^+)^{**}$

(906)
(905)

(899)

(905)

(906)

OD IN 1 mm

CENTER DENSITY, UNITS OF 1×10^{17} cm^{-3}

RADIATION DOSE, μ amp-min/cm^2

Fig. 6.19. Production of F_2^+ and $(F_2^+)^{**}$ centers in NaF:OH$^-$ as a function of radiation dose and time. Following ~ 5 min at 300 K: $\times \times \times$, sum of F_2^+ and $(F_2^+)^{**}$ band heights. (Behavior of the F_2^+ band alone is quite similar to that shown in Fig. 6.32). After $12-24$ h at 300 K: $\bullet \bullet \bullet$, $(F_2^+)^{**}$ band height, F_2^+ band has disappeared. $\circ \circ \circ$, $\sim 2 \times 10^{-3}$ OH$^-$; triangle 1.5×10^{-3} OH$^-$; all others, $\sim 0.6 - 1 \times 10^{-3}$ OH$^-$. Numbers beside each point indicate band peak wavelength [nm]. Note that the pure $(F_2^+)^{**}$ center is formed only in the lower dosage range; see text

248 L. F. Mollenauer

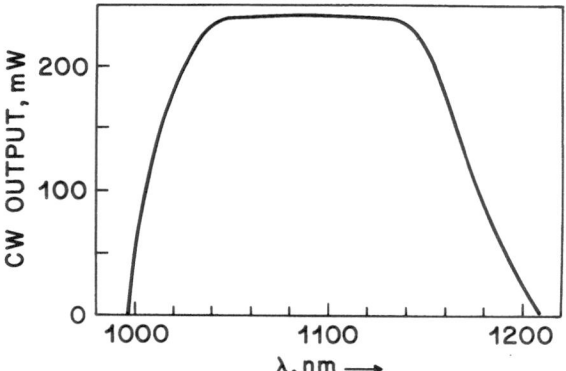

Fig. 6.20. Power output versus tuning of an undercoupled, NaF, $(F_2^+)^*$ center laser; input power $= 1$ W

period of $12-24$ h at room temperature and in the dark. Figure 6.18 shows unsaturating production of $(F_2^+)^*$ centers and a steadily increasingly efficiency of conversion (approaching 100%) from F_2^+ centers. As argued in detail in [6.71], this behavior represents yet more evidence that the * defect must be a product of the radiation damage itself.

On the other hand, Fig. 6.19 shows that production of the "pure" $(F_2^+)^{**}$ center occurs only in the lower dosage range, and also that such production is a function of the initial OH^- concentration. For higher dosages, one obtains a mixture of * and ** centers, as evidenced by a gradual shift of the band peak from 906 nm back toward 870 nm and by an increased bandwidth. This mixture is undoubtedly because of an exhaustion, or considerable reduction in number, of the ** defects as the higher dosages are achieved. Since one decay product of OH^- must be involved in the ** defect, the exhaustion must be due, of course, to the finite amount of OH^- initially present.

Laser operation with the $(F_2^+)^*$ and $(F_2^+)^{**}$ centers shows the same high efficiency and other effects of high gain evidenced by other F_2^+ center lasers. Figure 6.20 shows cw power output versus tuning for an undercoupled (output mirror $T = 3\%$), NaF, $(F_2^+)^*$ center laser, pumped with ~ 1 W [6.6]. The unusually flat curve is created by the undercoupling. In the same laser, but with $T = 30\%$, output power at band center nearly doubled.

In general, $(F_2^+)^*$ and $(F_2^+)^{**}$ center lasers show mild fading effects, but the fading decreases rapidly with increasing pump wavelength. For example, with the pump laser tuned to the long wavelength side of the absorption band (and with the crystal attached to a cold finger at 77 K), the cw operation described above and in Fig. 6.20 shows only $\sim 20\%$ output decline per hour for operation on a given spot [6.6].

6.2.11 The $Tl^0(1)$ Center

The $Tl^0(1)$ center was first discovered and analyzed [6.72, 73] by way of electron-spin resonance, and in those studies was shown to consist of a neutral

Tl atom perturbed by the field of a single, nearest-neighbor anion vacancy. Independently, laser-active centers were discovered in KCl: Tl, and were at first attributed to a Tl $-$ F$_A$ center [6.74]. Nevertheless, later optical [6.75 a] and magneto-optical [6.75 b] studies identified the new laser-active center as the Tl^0(1) center, and at the same time, considerably extended the model.

The electronic structure of neutral Tl consists of a single $6p$ electron outside various closed shells and subshells. In the free atom, the ground and first excited levels are $^2P_{1/2}$ and $^2P_{3/2}$, respectively, separated by a spin-orbit splitting of nearly $8000\ cm^{-1}$ (Fig. 6.21). Strong ($f \sim 0.15$) transitions exist between each of the those levels and the nearest-lying ($7s$) even parity state.

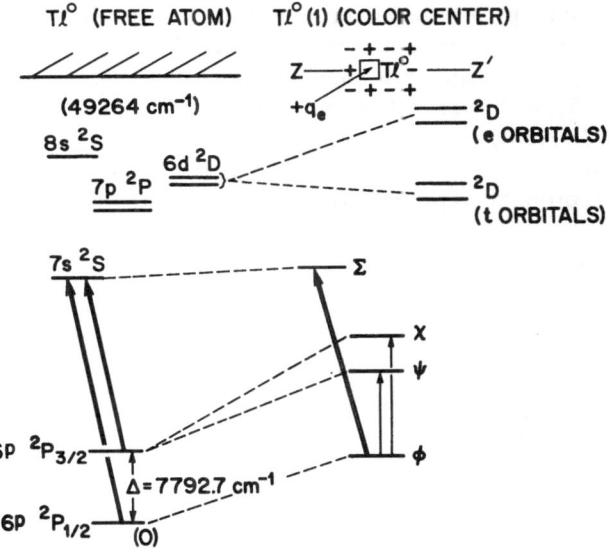

Fig. 6.21. *Left:* Energy levels of the free Tl atom. *Right:* Corresponding energy levels of the Tl^0(1) center

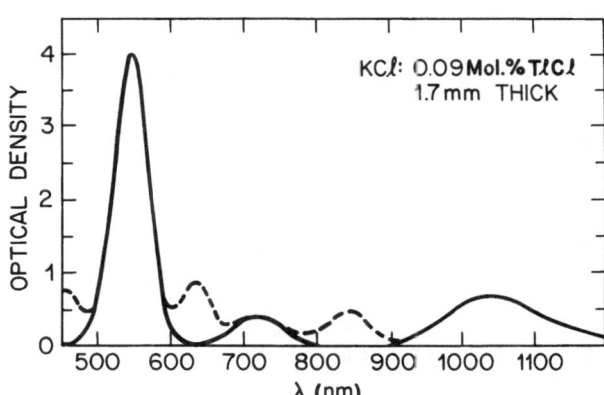

Fig. 6.22. Segment of the absorption spectrum of a radiation-damaged KCl:Tl crystal at 77 K. Only the bands shown in solid outline belong to the Tl^0(1) center

In the $Tl^0(1)$ center, the (odd) field of the vacancy does two important things: (1) it splits and mixes states of the $6p$ manifold, yielding the three Kramers doublets of Fig. 6.21 labeled Φ, Ψ, and χ, and (2) it mixes in higher-lying even parity states, and thus allows for electric dipole transitions of modest strength within the $6p$ manifold. Thus, the weaker long-wavelength bands (in KCl at 1040 and 720 nm; see Fig. 6.22) represent the crystal field induced absorptions to the Ψ and χ states, respectively, while the much stronger band occurring at shorter wavelength (in KCl, at 550 nm) is the transition to the nearest even parity state (Σ).

The lasers are based on the lowest energy ($\Phi \rightarrow \Psi$) transition, which is strongly polarized along the center axis. Although the $Tl^0(1)$ center has been created in several alkali halides [6.74], only the center in KCl has turned out to be of practical interest. (The potential loss here is not all that great, as the laser band energies change only little with changing host.) The laser pump band in KCl (peak at 1040 nm) has already been shown in Fig. 6.22; for the emission band (peak at 1510 nm), see Fig. 6.23.

For $T \lesssim 150\,K$, the luminescence decay time of 1.6 µs [6.75 a] is temperature independent and thus probably represents the true radiative decay time τ_r.

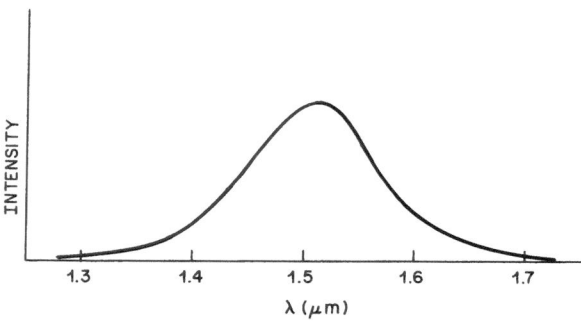

Fig. 6.23. Emission band of the $Tl^0(1)$ center in KCl:Tl (77 K)

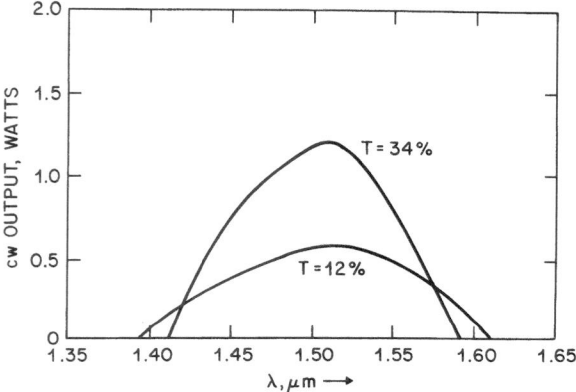

Fig. 6.24. Peak of chopped cw output power of a KCl, $Tl^0(1)$ center laser, as a function of wavelength Pump power at input is 6 W (peak)

From that decay time it can be shown that the oscillator strength of the emission is $f = 0.0075$. When combined with the emission bandwith of 670 cm^{-1}, this value for f yields a gain cross section $\sigma \sim 1 \times 10^{-17}$ cm^2. While this value is $\sim 30 \times$ smaller than the values quoted earlier for the other laser-active color centers, the $Tl^0(1)$ centers can be created in large enough density to compensate for the smaller σ. Also, the gain cross section of the $Tl^0(1)$ center is just large enough to allow mode locking via synchronous pumping, as will be shown in Sect. 6.4.1.

Figure 6.24 is a curve of cw power output versus tuning of a KCl, $Tl^0(1)$ center laser pumped with 6 W at 1.06 μm from a Nd:YAG laser [6.8]. The lower efficiency and peaked response stem from the smaller gain available from $Tl^0(1)$ centers, see (1.10).

Perhaps the most attractive feature of the KCl $Tl^0(1)$ center is its stability: it shows no detectable fading during laser action, and a given crystal can be made to last indefinitely. The stability of the $Tl^0(1)$ center is due in part to its refusal to reorient. The stability is also perhaps because of the coulomb binding between the Tl atom and the vacancy (with negative and positive effective charges, respectively). When stored at just a little below room temperature, the center in KCl has a shelf life in the dark of months. Thus, despite its lower efficiency, its exceptional stability and special tuning range have made the KCl, $Tl^0(1)$ center a most practical and important laser center.

The optimum Tl concentration for KCl $Tl^0(1)$ center laser crystals is about 0.2 mol% TlCl; this amount allows for the creation of just enough centers to make efficient laser crystals (O.D. $\sim 1 - 2$ mm^{-1} in the pump band). Higher Tl concentrations lead to the formation of significant amounts of Tl ion pairs and other aggregate centers; at least some of these seem to impair laser efficiency.

To create $Tl^0(1)$ centers, first large (several $\times 10^{18}$/cm^3) densities of F centers are created by electron-beam irradiation (1.5 MeV, $10 - 50$ μA min/cm^2) of crystals cooled to -100 °C or lower. Then the crystals at $T \sim -30$ °C are exposed for times of about 10 min to the light from a microscope lamp. Several processes then take place: (1) the Tl^+ ions capture electrons from ionized F centers, thereby becoming neutral Tl, or $Tl^0(0)$ centers, (the number in parentheses stands for the number of adjacent anion vacancies); and (2) the resultant anion vacancies move through the crystal until each encounters and becomes bound, through coulomb attraction, to a $Tl^0(0)$ center, thereby completing $Tl^0(1)$ center formation. The temperature of ~ -30 °C represents a compromise: it is high enough to allow for reasonable mobility of the anion vacancies, yet low enough to prevent destruction of the $Tl^0(0)$ centers that are required as an intermediate product and are unstable for significantly higher temperatures.

$Tl^0(1)$ centers cannot be created by the process of additive coloration, which merely turns the crystals a dirty brown color. Apparently the Tl^+ ions, because of their high electron affinity, are reduced to Tl^0 atoms which then aggregate to form a colloidal suspension in the crystal.

6.2.12 The F^+ Center in Alkaline-Earth Oxides

As a single electron trapped at an O^{2-} vacancy, the F^+ center in the alkaline-earth oxides is an analog of the ordinary F center in the alkali halides. Nevertheless, the deeper potential well of the F^+ center causes the electronic wave function to remain localized within the vacancy after relaxation. Thus the oscillator strength of the F^+ center is about the same ($f \sim 1$) in emission as it is an absorption, and the corresponding gain cross section is large: $\sigma \sim 2 \times 10^{-16}$ cm^2 [6.76]. Figure 6.25 shows the fundamental absorption and emission bands for the F^+ center in the host CaO [6.77]. Note that the transition energies tend to be considerably higher (once again because of the deeper potential well) than for the F center in the alkali halides.

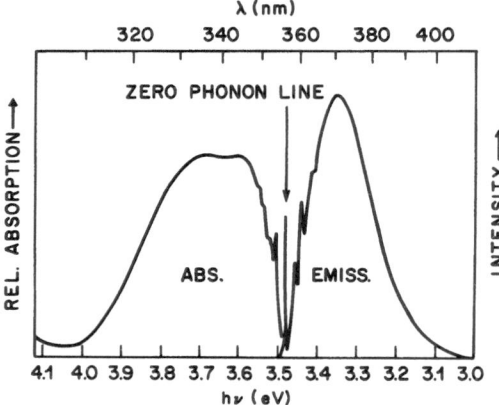

Fig. 6.25. Absorption and emission bands of the fundamental transition of F^+ centers in CaO. (From [6.77])

Nonfading cw laser action with low threshold pump powers has been reported [6.4] for the F^+ center in CaO, the only host explored so far (Table 6.1). There is no reorientation or motion of the centers to degrade laser operation, and the quantum efficiency of luminescence is thought to be nearly 100%.

Nevertheless, the F center (two electrons trapped in an O^{2-} vacancy) tends to absorb in the region of the F^+ emission band, and hence the F centers must be thoroughly bleached before F^+ laser action can begin. A tungsten lamp and filter to select out the F band (light of longer wavelengths releases too many electrons from shallow traps) are all that is required. The CaO, F^+ center laser could be operated at crystal temperatures up to ~ 200 K, where electrons released from shallow traps finally caused too much $F^+ \rightarrow F$ center conversion.

The F^+ centers can be produced in oxides by way of electron-beam irradiation. For the same center density, however, the required dose is much larger than in the alkali halides. For example, the dose required to obtain 2×10^{17}/cm^3 F^+ centers in CaO irradiated at room temperature with 2 MeV electrons was ~ 2600 μA min/cm^2. The electron traps required to maintain charge balance with F^+ centers are presumably provided by various transition metal ion impurities, typically found in the oxides at levels of some tens of parts in 10^6. Neutron irradiation of about 10^{19} n/cm^2 will also produce F$^+$ centers in the required density, but no F centers. Unfortunately, neutron irradiation also tends to produce various aggregate centers whose bands interfere with laser action.

The principal barrier thus far to further exploration of the F$^+$ center as a laser medium has been the relative difficulty in obtaining crystals (especially with high purity) of the highly refractory hosts. The problem is so severe that no others have so far been able to reproduce the laser action reported in [6.4], but if the materials problem can be solved, the F$^+$ center in the alkaline-earth oxides would seem to have a great potential for lasers in the visible as well as the near uv.

6.2.13 The H$_3$ Center in Diamond

Recently, stable and efficient laser action has been obtained at the emission peak of so-called H$_3$ centers [6.78] in a thin slab of diamond at room temperature. Nevertheless, with broad emission band peaking at ∼530 nm, and a computed gain cross section of ∼2×10^{-17} cm^2, the H$_3$ center should allow for laser action tunable from ∼500 to 600 nm. The ample gain cross section and high quantum efficiency should also allow for cw laser action, pumped for example at 488 nm (absorption band peak ∼475 nm) by an Ar$^+$ ion laser.

The H$_3$ center is thought to consist of a pair of nitrogen impurity atoms next to a carbon vacancy [6.79, 80]. The centers can be formed in abundance by irradiation of natural diamond with 1 − 2 MeV electrons to doses of $10^{18} - 10^{20}$ electrons/cm^2 and annealing at 1200 K for 20 h in vacuum [6.78, 81]. (Apparently the nitrogen itself is an abundant impurity in natural diamond.)

Because of the high temperatures involved in its formation, and because of the high thermal conductivity of diamond, it would seem likely that the H$_3$ center would stand up well to the rigors of cw pumping at room temperature. If so, the diamond color center laser would represent a most exciting development. It would mean that the superior frequency definition possible with color centers would at last be available for the visible. It would also mean an end to the paraphernalia and inconvenience associated with dye lasers, such as circulation pumps and the potential for exposure to carcinogens.

6.2.14 Laser-Active Color Centers in Sapphire

Recently, room temperature stable laser action has been reported in the Soviet literature [6.82, 83] for two different color centers in (initially) clear sapphire (α-corundum). Both center types were presumably produced by radiation damage. For one center type, the laser was continuously tunable over the range 540 − 620 nm with pulse pumping, and its pump band exhibited an absorption coefficient at band peak (460 nm) of $12\,\mathrm{cm}^{-1}$. The second type, when pumped with a ruby laser (693 nm), provided laser tuning from 750 to 900 nm. If these results prove easy to reproduce, and if the laser action is stable under cw pumping as well, then the laser-active color centers in sapphire are just as important and exciting a development, and for the same reasons, as are the diamond centers.

6.3 Some Examples of Color Center Lasers and Associated Hardware

6.3.1 A Typical cw Color Center Laser

The cw color center laser cavity [6.14, 5] shown in Fig. 6.2 embodies important design principles. The most rudimentary of these – the use of the folded, astigmatically compensated cavity, and the need for cryogenics and insulating vacuum – have already been treated briefly in the Sect. 6.1. Here we consider the design in greater detail.

a) Optical and Mechanical Design

In the cavity of Fig. 6.2, the radius of curvature of M_2 is 50 mm (hence $f = 25$ mm), and the folding angle 2θ (Fig. 1.3) is 20°. For those parameters, the thickness of crystal required for astigmatic compensation, see (1.18), is $t \cong 1.7$ mm when $n = 1.5$. (For other indices from 1.3 to beyond 2, the optimal thickness varies from that value by only a few tenths of a millimeter.) For a distance between M_0 and M_2 $(d_2) = 400$ mm and for $f = 25$ mm, $2S \sim 1.5$ mm, see (1.16). Thus, when the cavity is adjusted to the middle of its stability range, b_1 is also ~ 1.5 mm, a value comparable to the crystal thickness, as desired for optimum efficiency and low threshold. For the same adjustment and for $\lambda \sim 1$ μm, $w_{01} = 15.7$ μm, and the output beam diameter $2w_{02}$ is ~ 0.7 mm; for other wavelengths, those same two quantities scale as $\sqrt{\lambda}$, see (1.17c, d).

The radius of curvature of M_1 is the only purely arbitrary parameter for the cavity of Fig. 6.2; all other dimensions are closely interrelated. For example, if f is multiplied by a factor A, then d_2 must be multiplied by A^2, see (1.17c), to maintain constant beam waist at the crystal. Similarly, θ must be multiplied by $\sim A^{-1/2}$ if the crystal thickness is to remain constant as well (1.18).

Note from Fig. 6.2 that mirrors M_1 and M_2 are mounted *inside* the vacuum enclosure. Although this arrangement requires that most adjustments to M_1 and M_2 be made before the vacuum enclosure is sealed off, it avoids several serious problems posed by the alternatives. First, to mount the mirrors outside would require vacuum windows in the region of tight beam focus. This would make the astigmatic compensation difficult, as the total thickness of material at Brewster's angle (two windows plus crystal) would be too great. Because of the required tight spacing ($\leqslant 25$ mm) between the mirrors and the crystal, the arrangement would be mechanically awkward as well. (The Brewster's angle windows mounted in the side wall of the vacuum chamber produce negligible astigmatism, as they are in regions of nearly parallel light). The remaining alternative is to use the mirror blanks themselves as vacuum windows, mounted for adjustment on flexible metal bellows [6.84]. Nevertheless, in that case the mirror mounts become complex and expensive, and it is difficult to obtain tight mirror spacings. Also, that arrangement does not allow for the orthogonal and rapidly converging adjustment scheme (to be described shortly) that has been worked out for the cavity of Fig. 6.2.

Coaxial pumping is carried out through dichroic mirror M_1 ($T \sim 85\%$ or better at the pump wavelength). Lens L_3 and the substrate of M_1 (radius of curvature: 25 mm) form a thick lens of effective focal length ~ 33 mm, as measured from the surface of M_1. This compound lens provides the major focusing of the pump beam. Diverging lens L_1 ($f = -50$ mm) magnifies the beam and L_2 ($f = 150$ mm) then makes it gently converging, such that both final focusing and lateral adjustment of the focal spot can be achieved through adjustment of L_2. It is also possible to match the pump beam waist diameter to that of the laser cavity mode. To accomplish this, L_1 and L_2 are mounted on an optical rail and moved more or less as a unit along the pump beam axis; bringing the two closer to the cavity makes the beam spot at L_3 larger, thereby making the focal spot smaller.

b) Alignment

Alignment is carried out quickly and easily by using the pump beam to simulate the laser mode itself [6.2]. First, the external lenses L_1 and L_2 are temporarily removed and the beam made to propagate parallel to the optical rail and to be well-centered on M_1. The lenses are then replaced and adjusted until that part of the beam reflected by M_1 is sent right back along the incoming beam; this guarantees that the beam will be focused at the center of curvature of M_1. The necessary adjustment can be confirmed by observing the retro-reflected beam as it passes through a distant aperture centered on the original beam. One can often also observe a rise in power as M_1 begins to enhance the pump laser's output mirror; this effect is especially pronounced when the laser is operated near threshold.

The next step is to adjust M_2 such that the beam in the long leg of the cavity propagates along the desired path and is brought to a minimum spot size on a distant screen. The latter adjustment makes d_1 correspond approximately to one extreme edge ($\delta \cong 0$) of the stability range; d_1 is then increased by half the amount calculated from (1.16). The cavity is then in the middle of its stability range. To ease adjustment of d_1, M_2 is mounted on a miniature translation stage precisely oriented for motion along the pump beam axis and actuated by way of a mechanical feedthrough in the vacuum chamber wall.

Note that such motion does not upset the angular adjustment of M_2. All the above adjustments of M_2 are made with a dummy crystal (consisting of a slab of crystal or glass having the same thickness and about the same index as the amplifying crystal) temporarily mounted in place.

The final step involves the alignment of M_0. Usually, this adjustment can be completed only with the real laser crystal mounted in place (and cooled); with the aid of a detector and long-pass filter to eliminate all pump light, one then looks for the brightening of luminescence that presages laser action.

Once laser action has been achieved, mirrors M_1 and M_2 can be translated as a unit (once again with a miniature translation stage actuated by a mechanical feedthrough in the vacuum wall) along the pump beam axis, to obtain more perfect overlap of the pump and laser beams in the region of the beam waists; however, such adjustment must be tracked by corresponding focal adjustments in lens L_2. Considerable improvement in the laser output power can often be obtained in this way.

It should be emphasized that the alignment procedure described above can be accomplished quickly and efficiently; it is often possible to achieve laser action in a fraction of an hour, including time for crystal loading, chamber evacuation, and cool-down. This efficiency is due in no small way to the largely independent adjustments provided by the novel cavity design. It is also helpful that the large gain possible with color centers tends to make the lasers tolerant of minor imperfections in the cavities and their adjustment.

c) Cryogenic Details

The cavity vacuum chamber consists of a 76 mm high, ~200 mm diameter cylindrical wall, O-ring sealed to a flat base plate (Fig. 6.26). Its top is formed by the bottom plate of a completely detachable liquid N_2 storage Dewar, whose 4 l capacity yields a hold time of about 3 days. Prealignment is made, as described above, with the top of the chamber completely open (Dewar removed). To begin operation, first the crystal is mounted on the cold finger and the latter fastened in place. Then the Dewar (preevacuated and precooled) is set in place and the lower chamber evacuated to a pressure of ~10 μm Hg or less; at that point the external pump can be valved off, and the sliding valve between the lower chamber and the Dewar opened. The molecular sieve cryopump built into the Dewar will then take over and complete evacuation. With the aid of a tool thrust down through the throat of the Dewar, the retainer nut (again, see Fig. 6.26) is turned and the copper cone at the end of the flexible bellows is lowered until it settles into the mating cone of the cold finger. As the latter cools, it contracts tightly around the upper cone, yielding a contact of less than 1/2 K/W thermal resistance. To disconnect, the mating cones are jacked apart with aid of the central screw (acting through the smaller bellows). Then the larger bellows can be retracted, the sliding vacuum valve closed, the lower chamber repressurized to 1 atm, and the Dewar set aside. Note that this system allows the Dewar can to remain always evacuated

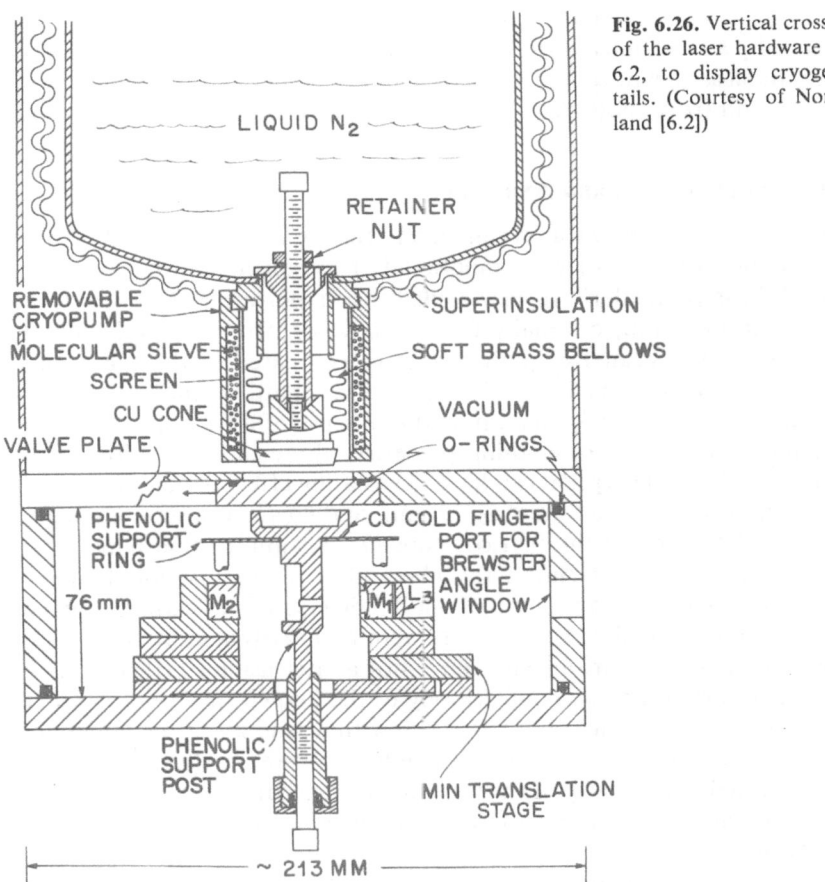

Fig. 6.26. Vertical cross section of the laser hardware of Fig. 6.2, to display cryogenic details. (Courtesy of North-Holland [6.2])

LIQUID N$_2$

RETAINER NUT

REMOVABLE CRYOPUMP

SUPERINSULATION

MOLECULAR SIEVE

SOFT BRASS BELLOWS

SCREEN

CU CONE

VACUUM

VALVE PLATE

O-RINGS

PHENOLIC SUPPORT RING

CU COLD FINGER

PORT FOR BREWSTER ANGLE WINDOW

76 mm

M$_2$

M$_1$ L$_3$

PHENOLIC SUPPORT POST

MIN TRANSLATION STAGE

~ 213 MM

and cooled, a great advantage. Evacuation of the Dewar itself, with its many layers of superinsulation and with its molecular sieve, is a slow process.

The crystals are held in place with a gentle spring clamp, a large slab face resting on the cold finger. Unfortunately, all known greases that might be used to aid thermal contact freeze at temperatures not far below room temperature, and the crystals tend to crack from differential contraction with respect to the cold finger. Thus, in general, thermally conductive grease can be used only on a small area under one far corner of the crystal slab. One exception is LiF, whose contraction (for temperatures below the freezing point of most greases) tracks that of copper almost perfectly.

Finally, the top of the cold finger is attached to the vacuum can with a "steering wheel" of 1.5 mm thick phenol-impregnated fiber material, such that motion in the horizontal plane is prohibited. Nevertheless, the bottom of the cold finger rests on a plastic post, and can be moved up and down a few millimeters by a mechanical feedthrough in the bottom of the vacuum can. In this way, the laser beam can be made to avoid scratches and other defects on the crystal, and fresh spots can be found on a partially "burned-out" crystal.

d) Water Vapor: Effects and Avoidance

The metal surfaces of the vacuum can tend to release water and possibly other trapped vapors for days or even months after establishment of a vacuum. (Anodized Al is particularly to be avoided, as its porous surface makes an efficient molecular trap, especially for water vapor.) In the high-insulating vacuum, molecular mean free paths are large compared to the can dimensions. Thus those released molecules on a direct path to the cooled laser crystal tend to become stuck there; note that the use of a high-speed external vacuum pump does little or nothing to deflect those molecules from their paths. The resultant absorption can have devastating effect on laser operation in the region of the water-ice band ($\lambda \sim 3$ μm), and it can even seriously affect operation in the region of the first overtone ($\lambda \sim 1.5$ μm) of that band.

For example, Fig. 6.27 shows the absorption of an uncolored, highly polished, and originally transparent alkali-halide crystal, following ~ 19 h at 77 K in a vacuum chamber whose walls were of heavily anodized Al [6.2]. (The measurements were made with the aid of a laser using $F_A(II)$ centers in RbCl.) The absorption corresponds closely to the known spectrum of water-ice, and is due entirely to ice layers, just a few thousand angstroms thick, on each of the two surfaces. To the eye, these layers were nearly invisible, and could be detected only by way of the weak interference fringes they produced. Yet at band center, absorption of the two surfaces summed to nearly 2.3 optical density.

Even in a chamber made of superior materials, there is always the possibility of slow buildup of such water-ice films. Such buildup can be avoided, however, by surrounding the cold finger with a metal shield, thermally anchored to the cold finger itself. The shield contains two holes, each just a bit

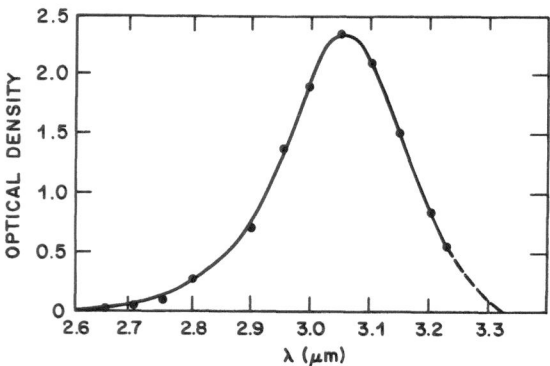

Fig. 6.27. Absorption band due to thin layers of water-ice on the surfaces of an (initially clear) alkali-halide crystal at 77 K (see text). (Courtesy of North-Holland [6.2])

larger in diameter than the laser beam, such that the crystal surfaces "see" only the opposing mirror surfaces (which do not emit water molecules). As a (perhaps not entirely necessary) refinement, the shield is provided with a shutter to close the holes when the laser is not in use. The use of such a shield has enabled the complete avoidance of serious long-term aging effects in lasers operating in the 1.5 μm region. The shield is strongly recommended for *any* color center laster using cooled crystals.

6.3.2 The Burleigh Laser

The first commercial color center laser, designed by K. German of Burleigh Instruments, Inc., is shown schematically in Fig. 6.28. Originally intended for ultrahigh precision molecular spectroscopy with F_A (II) and F_B (II) centers, it possesses several features worthy of discussion. The first of these is the novel arrangement used for pumping. Coaxial pumping is achieved by way of a special beam splitter, a plate of CaF_2 located at Brewster's angle to the mode beam [6.85]. Coated on one side for high ($\sim 95\%$) reflectivity in a band surrounding the major pump wavelength (647 nm), it nevertheless contributes losses of less than 0.2% for the lasing wavelengths ($2.2-3.3$ μm). By this scheme, dichroic mirror coatings (restricted in wavelength range) are not needed, and the two curved mirrors of the cavity have very broadband coatings of enhanced silver. Note also that the folding mirror serves as the focusing element for the pump beam. Nevertheless, the high reflectivity of the splitter is achieved only for *s* polarization, while the laser mode uses a *p* polarization. Fortunately, this arrangement is acceptable for use with F_A (II) or F_B (II) centers, as long as the pump electric field vector is oriented parallel to crystal [110] F_A (II) and [100] F_B (II) axes.

The Burleigh laser was designed for the achievement of stable single-frequency operation. The grating (457.8 lines/mm) is used in retroreflection, as described in Sect. 1.7.1 b, and is provided with a calibrated sine-bar drive. The intracavity etalon is required to eliminate spatial hole burning modes

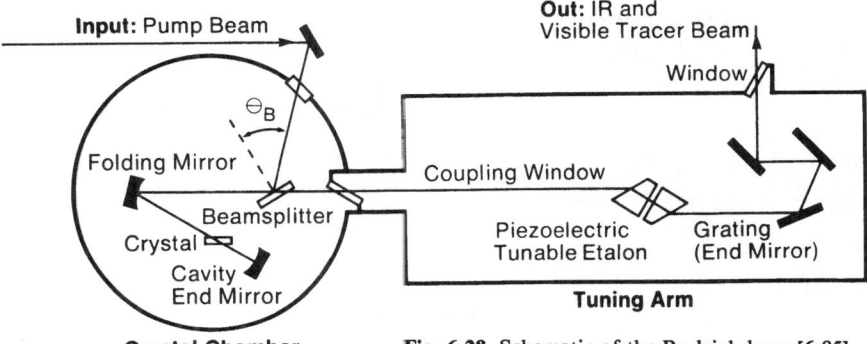

Fig. 6.28. Schematic of the Burleigh laser [6.85]

(Sect. 1.7.2). Both the etalon and one cavity mirror are piezoelectrically tuned, simultaneously and such that they track each other. All optical elements of the cavity are mounted on a base plate of Super-Invar. Measured long-term frequency drift of the laser's output is about 1 MHz/min. But the laser's frequency can also be locked with the piezoelectric controllers to an external cavity. When the latter, with Super-Invar spacers, is in a temperature and pressure stabilized enclosure, the overall drift can be much smaller, less than 1 MHz/h.

6.3.3 Single-Knob Tuning of a Single-Frequency, cw Color Center Laser

As shown theoretically in Sect. 1.7.2, single-frequency operation should be achievable with a grating as the sole tuning element, as long as (1.29) is satisfied. Such single-knob tuning is especially valuable when a large frequency range is to be scanned, as it avoids the awkwardness associated with simultaneous tuning of several elements. Figure 6.29 is the schematic of a cw color center laser cavity in which such single-frequency operation was achieved [6.86]. As shown there, a special lens-mirror element allows M_1 to be located as closely as possible to the crystal ($D = 3.7$ mm), and the beam diameter is magnified several times by an internal telescope (mirrors M_3 and M_4) before it is allowed to impinge on the Littrow grating (1 200 g/mm, incidence angle 64°).

Figure 6.30 shows the quantities $g(x)$, the grating response R, and their product γ, all as a function of the quantity $x = \Delta \nu / \Delta \nu_0$, (1.27, 28), for two cases: (a) R not adequate, (1.29) not satisfied, and (b) R adequate, (1.29) satisfied, for single-frequency operation; Fig. 6.31 shows the corresponding experimentally determined frequency spectra of the laser's output. Note that in Fig. 6.31a, the measured frequency spacing (5.6 GHz) is considerably less than the "spatial hole burning frequency" of ≈ 20 GHz, (1.28), just as predicted by the behavior of the quantity γ_a shown in Fig. 6.30.

Finally, note that the linewidths exhibited in Fig. 6.31 represent the long-term effects of low-frequency mirror vibration; observed short-term line-

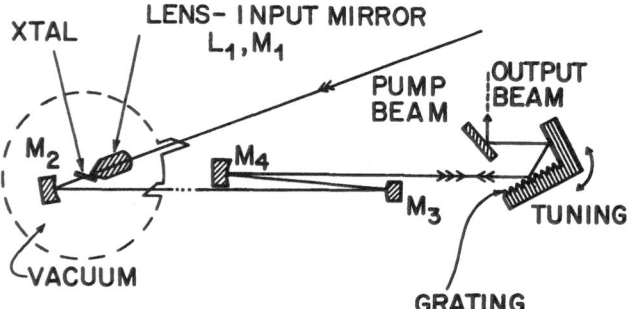

Fig. 6.29. Schematic of cw color center laser capable of single-frequency, single-knob tuning

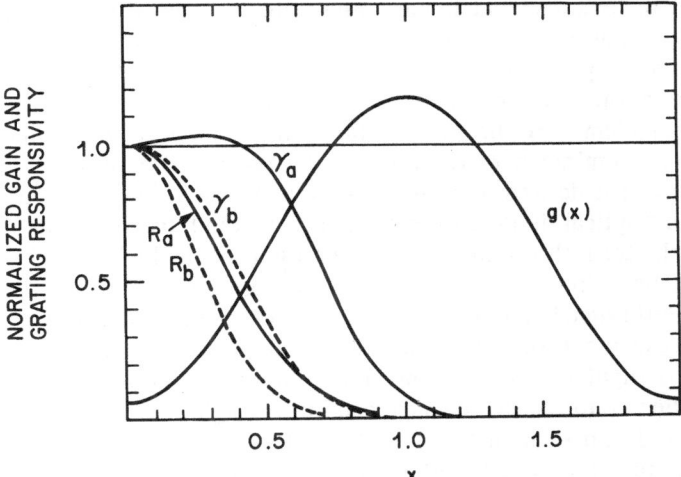

Fig. 6.30. The quantities $g(x)$, two grating response functions $R_a(x)$ and $R_b(x)$, and their products $\gamma_a(x)$ and $\gamma_b(x)$ (see text)

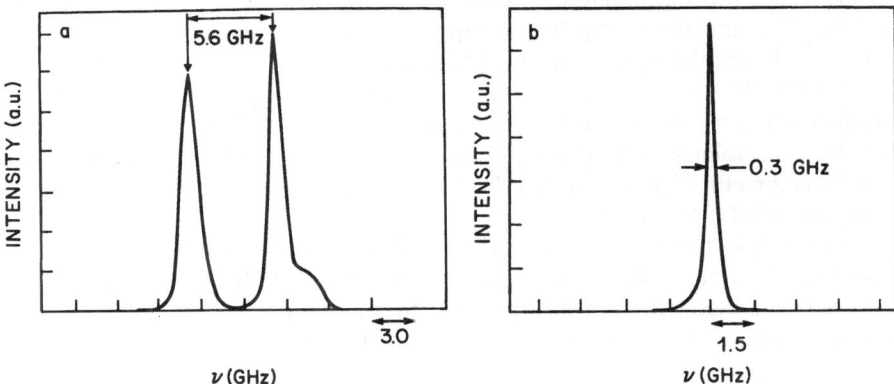

Fig. 6.31. (a) Spectrum of grating tuned, $\lambda \sim 1.5\,\mu m$, cw color center laser Beam dia. at grating = 4.2 mm. (b) The same, but grating beam dia. = 4.8 mm (see text)

widths were often instrument limited (less than 1/10 the widths shown). It should be possible to achieve such narrower linewidths all the time with the use of more stable mirror and grating mounts.

6.3.4 A Ring Cavity Color Center Laser

Ring cavities allow for unidirectional traveling-wave operation and the consequent avoidance of spatial hole burning (Sect. 1.7). Thus single-frequency operation is more easily achieved in ring cavities than in standing wave cavities. As an additional benefit, efficiency and power output for single-frequency operation can be somewhat higher. Figure 6.32 shows a ring color center laser whose design will serve to illustrate the major features of ring cavity construction and operation [6.87].

In single-frequency operation of the bare ring cavity (in Fig. 6.32, mirrors $M_1 - M_4$ and tuning elements) the propagation direction tends to switch randomly in response to minor perturbations. In this sense, the bare cavity is bistable. Nevertheless, the desired motion (counterclockwise in Fig. 6.32) can be enforced with an "optical diode", or less reliably, through use of the high reflector shown in the far right-hand corner of the figure. (The high reflector serves to feed light back into the desired direction from laser action in the wrong direction, should such begin to occur.) The optical diode [6.88] consists of an optically active element, such as a plate of crystal quartz, in series with a Faraday rotator. For light traveling to the right, the polarization rotations brought about by the two diode elements cancel, but for light traveling to the left, the rotations add and enough loss is induced to prevent lasing in that direction. The minimum required Faraday rotation is not large, a value of $\sim 2°$, corresponding to a differential loss of a few percent; this is more than sufficient to totally suppress lasing in the wrong direction.

The cavity of Fig. 6.32 can also be operated in a linear mode (the ring minus its bottom leg). Such operation considerably simplifies initial alignment of M_2, M_3, and the pump beam. Operation in the ring mode can then be achieved simply through rotation of M_1 and M_4.

Cavities like that of Fig. 6.32 have been used with $(F_2^+)^*$ centers in NaF, to produce intense tunable sources of narrow linewidth radiation near 1.08 µm for the spin polarization, through optical pumping, of He atoms. One such ring laser produced more than 300 mW cw of single frequency power when pumped with 1.5 W [6.89].

In principle, the cavity of Fig. 6.32 could be used with just about any other laser-active color center. The greatest difficulty usually arises in finding materials for the Faraday rotator that combine high Verdet coefficient with low absorption loss. One recently developed material, YIG (yttrium iron

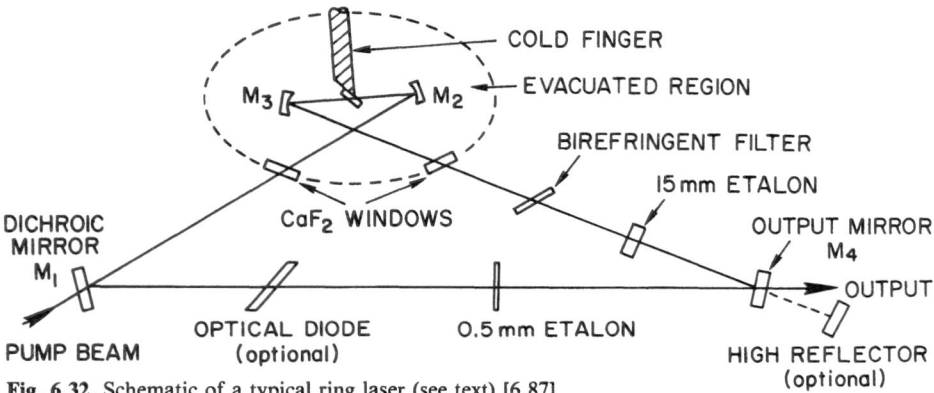

Fig. 6.32. Schematic of a typical ring laser (see text) [6.87]

garnet), makes an excellent Faraday rotator for the $1.15 - 6$ μm range [6.90]: in a saturating field (900 G, which can be provided by a Sm:Co permanent magnet), a crystal just 2.7 mm thick produces a rotation of 45° at 1.5 μm, and for that same wavelength the YIG shows a measured absorption loss between 0 and 0.2 dB. Thus the net loss of a YIG rotator depends primarily on the quality of its antireflection coatings.

6.4 Mode Locking and the Soliton Laser

The production of ultrashort pulses by mode locking has been described in general terms in Sect. 1.8. Here the mode locking of color center lasers is discussed, both that obtained in simple lasers by synchronous pumping, and the special operation, produced by pulse shaping effects in a fiber, of the "soliton laser".

6.4.1 Mode Locking by Synchronous Pumping

The large gain cross sections of most laser-active color centers make it easy to obtain mode locking by synchronous pumping (1.30). Thus, all of the color centers used so far-F_2^+ centers in LiF [6.5], KF [6.5], and NaCl [6.34], and the $Tl^0(1)$ center [6.8] in KCl- have mode locked well, producing pulse widths ranging from 4 to 10 ps. For all but the LiF center, the pump source has been a Nd:YAG laser, producing ~80 ps wide pulses at a pulse repetition rate of 100 MHz (corresponding cavity length: 1.5 m) and $\gtrsim 5$ W time averaged output power.

Figure 6.33 shows the laser of Fig. 6.2 as suitably modified [6.8, 34] for mode locking. Its cavity length has been stretched to 1.5 m to match the length

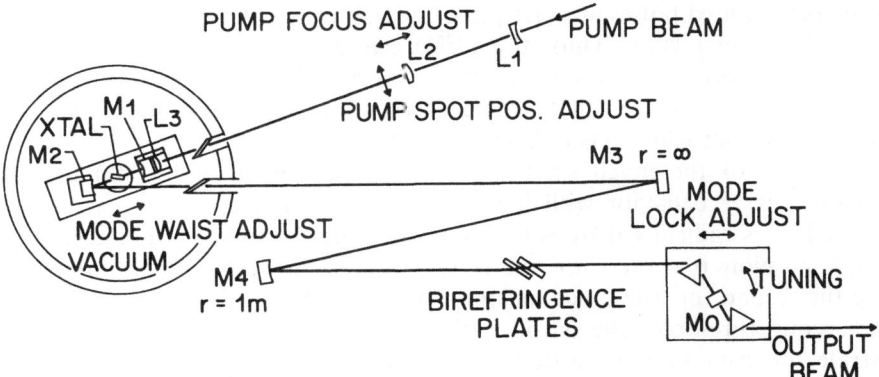

Fig. 6.33. Schematic of a mode-locked color center laser. The arrangement is the same as shown in Fig. 6.2, except that the cavity length has been stretched to 1.5 m. (Mirror M_4 relays the beam waist at M_0 to a point optically ~500 mm removed from M_1)

of the pump cavity. (Mirror M_4 relays the beam waist at M_0 to a point optically ~ 500 mm removed from M_1.) The cavity-length adjustment necessary to effect optimum mode locking is provided through mounting of the output mirror and tuning assembly on a precision translation stage.

The following behavior is common to all the ordinary mode-locked color center lasers studied to date, and is also remarkably similar to that reported for synchronously mode-locked dye lasers. As the cavity is shortened toward the optimum length, the pulse shape becomes narrower and taller until a maximum height is reached. The narrowest pulses are obtained with the cavity shortened just a few μm beyond the point of maximum height. Further shortening of the cavity often produces one or more satellite pulses a few tens of picoseconds removed from the main pulse.

Figure 6.34 is a scatter plot of the pulse widths obtained with the KF, F_2^+ laser for successive optimal adjustments of the system as a function of wavelength [6.5]. The ability of the laser to sustain short pulses over a good fraction of the entire tuning band reflects the large reservoir of gain available with the F_2^+ center. The open circles show the dramatic effect of the water-vapor band at ~ 1.38 μm on the mode locking when the cavity path was left open to the atmosphere. The effect is probably due less to the direct absorption, than to the associated anomalous dispersion and to its distortion of the laser cavity mode spacings.

An NaCl, F_2^+ center laser was used in the first experiments near 1.5 μm on pulse narrowing and solitons in optical fibers [6.34]. For that purpose the pulses are required to have no excess bandwidth. Indeed, the product $\Delta t \Delta f \cong 0.3$ of the pulses produced by that laser is the minimum allowed by the uncertainty principle, and the pulses are said to be "Fourier-transform-limited" (Fig. 6.35).

Unfortunately, the lasers using ordinary F_2^+ centers all show annoying fading effects, especially in mode-locked operation. Nevertheless, the $Tl^0(1)$ center laser shows no such effects, and it can also be made to produce transform limited pulses at ~ 1.5 μm of nearly identical shape to those of the NaCl, F_2^+ center laser. Thus, the $Tl^0(1)$ center laser has taken over as the source for fiber pulse studies. (Now, however, that laser is in turn being replaced by the NaCl, $F_2^+:O^{2-}$ center laser, which also shows no fading, see Sect. 6.2.10, but which has a much wider tuning range.)

Because of the much smaller gain cross section of the $Tl^0(1)$ center, however, laser behavior with increasing pump power and with decreasing cavity loss is radically different. The novel behavior [6.8] is summarized in Fig. 6.36. Note that there are two thresholds: the first for laser action per se and the second for true mode locking. That is to say, for input powers below the second threshold, the output pulses show a broad (>25 ps wide) peak astride an even broader pedestal, whereas for powers above the second threshold, the pulses have the normal shapes, like that shown in Fig. 6.35. Note the monotonic decrease in pulse width for increasing pump power, and the increase in pulse width with increasing output mirror transmission.

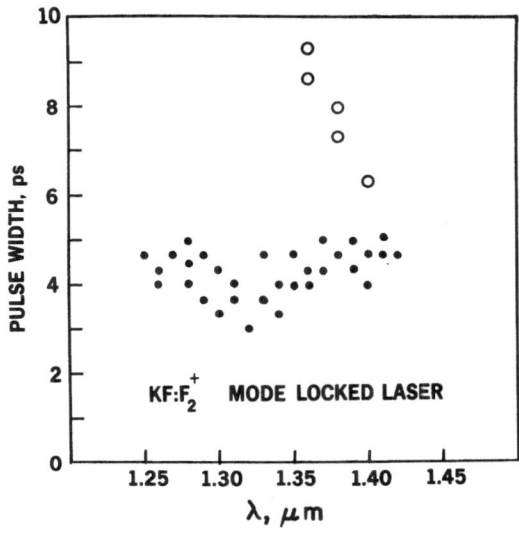

◀**Fig. 6.34.** Scatter plot of measured pulse widths of a mode-locked, KF, F_2^+ center laser. *Open circles* refer to operation without N_2 purging of the cavity beam path

Fig. 6.35. *Left:* Auto-correlation trace of the output of a mode-locked NaCl, F_2^+ center laser operating at $\lambda \sim 1.5$ μm. To obtain the true pulse width (7 ps), the auto-correlation width (10.8 ps) must be divided by the factor 1.55 appropriate for pulses having $\sim$ sech^2 shape. *Right:* Frequency spectrum of the same pulse. (Courtesy of North-Holland [6.2])

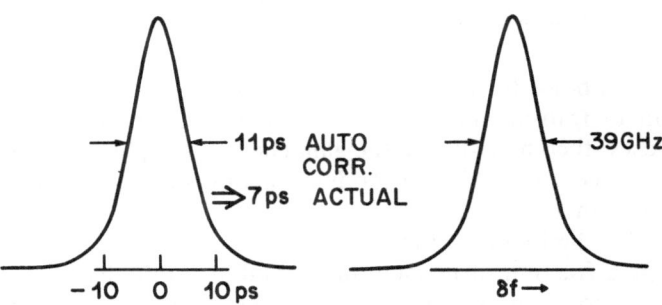

Fig. 6.35

The behavior just described can best be understood in the time domain. The width of the mode-locked pulses is determined by a balance between two factors: (1) the pulse broadening effect of the various dispersion elements in the cavity, and (2) the ability of rapidly varying gain to sharpen the pulse. For the $Tl^0(1)$ laser, the modulation of gain, and hence the ability to sustain narrow pulses, increases with increasing pump power, Fig. 6.37. The increased gain modulation can be understood as follows: At the lower-power levels [Fig. 6.37 c], the gain change ΔG brought about by a single pump pulse, (1.30), is small compared with the total gain required to overcome losses. Nevertheless, because the radiative decay time is long compared with the time between pulses, the gain level between pulses G_0 does not decay significantly. The accumulated effect of many pump pulses will then cause G_0 to rise until the peak gain crosses the loss line, as required to sustain laser action. By contrast, for higher pump power [Fig. 6.37 b], ΔG increases, G_0 decreases, and the fractional modulation of gain is higher.

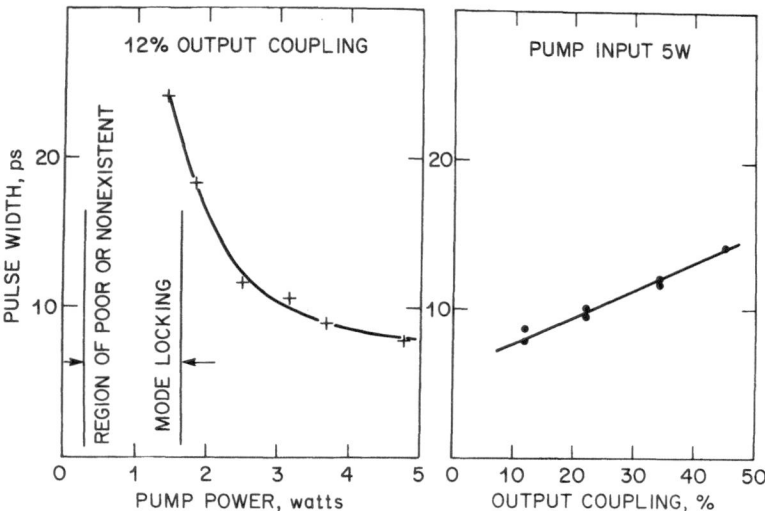

Fig. 6.36. Pulse widths (FWHM) of a mode-locked, KCl, $Tl^0(1)$ center laser as a function of pump power and output mirror transmission

A similar explanation holds for the observed increase of pulse width with increasing output mirror transmission: as the loss increases, the intracavity pulse intensity decreases, thereby reducing the gain change brought about by stimulated emission. Hence once again, the modulation of gain is reduced and the pulses are correspondingly broadened.

Finally, because of its limited gain the $Tl^0(1)$ center laser does not mode lock well over as large a fraction of its cw tuning band as do the F_2^+ center lasers. Because of its extraordinary stability, however, the $Tl^0(1)$ center laser is a most welcome addition to family of mode-locked lasers.

6.4.2 The Soliton Laser

The soliton laser [6.39 – 41] consists of a synch-pumped, mode-locked color center laser, tunable in the 1.5 μm region (the $Tl^0(1)$ center laser described in the previous section or its equivalent), coupled to a second cavity, containing a single-mode, polarization preserving optical fiber. Feedback from the fiber, where pulse compression and soliton formation take place, enables the color center laser to produce much shorter pulses than possible with ordinary mode locking. Transform limited pulses of ~ $sech^2$ intensity profile and of just about any desired pulse width, down to ~60 fs, can be obtained through choice of the control fiber length. Furthermore, tunability, limited only by power requirements for soliton production in the fiber, is in general much greater than in an ordinary mode-locked laser. Thus the soliton laser has become an extremely powerful and versatile tool. It has already been used in the measurement of the response times of ultrafast detectors and to study

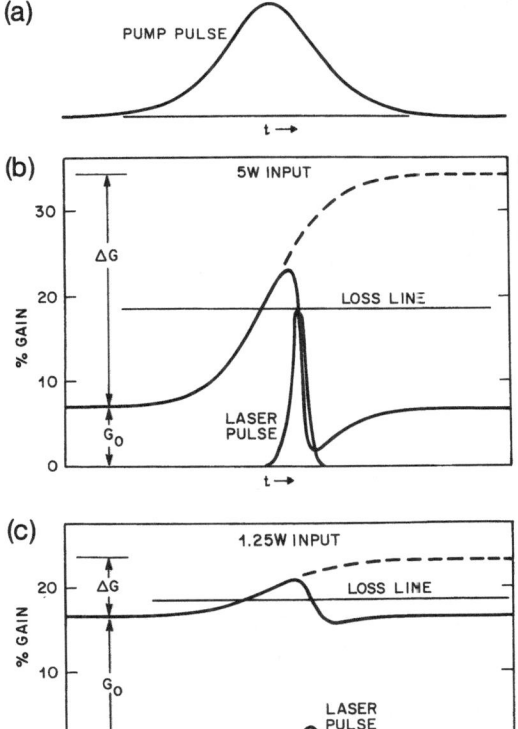

(a) PUMP PULSE

t →

(b) 5W INPUT

% GAIN

ΔG

30

20   LOSS LINE

10

G₀

LASER
PULSE

0

t →

(c) 1.25W INPUT

% GAIN

ΔG   LOSS LINE

20

G₀

10

LASER
PULSE

0

t →

Fig. 6.37a–c. Gain evolution in the mode-locked $Tl^0(1)$ center laser. (a) Pump pulse, ~ 80 ps wide (b) and (c) (– – –), effect of pump pulse alone; (————), net gain change, including effect of laser pulse. The loss line is based on $T = 12\%$ for the output mirror and on an additional 6% intracavity loss

fiber-soliton phenomena in the femtosecond regime. It is further expected to be of prime importance in the development of optical communications and in the study of ultrafast phenomena, especially in semiconductors.

Before we examine the soliton laser in detail, a bit of background on fibers and pulse formation may be helpful. The dispersive qualities of quartz glass and the loss per unit length of the best single-mode fibers now available are both shown in Fig. 6.38. As will be shown shortly, pulse compression and solitons are possible only in the region of "negative" group velocity dispersion $(\partial v_g / \partial \lambda < 0)$, which is, as shown by the figure, the region of wavelengths greater than ≈ 1.3 μm. (The net dispersion of a given fiber is also determined by the ratio of the core diameter to the wavelength. Nevertheless, such "modal" contribution can only push the zero of dispersion to longer wavelengths.)

Dispersion alone-regardless of sign-always causes the higher and lower frequency components of a pulse to separate, and thus always serves only to broaden the pulse. Pulse narrowing, and by extension, solitons, are made possible by nonlinearity of the medium. That is, the index of refraction is itself a function of the light intensity:

$$n = n_0 + n_2 I \ . \tag{6.7}$$

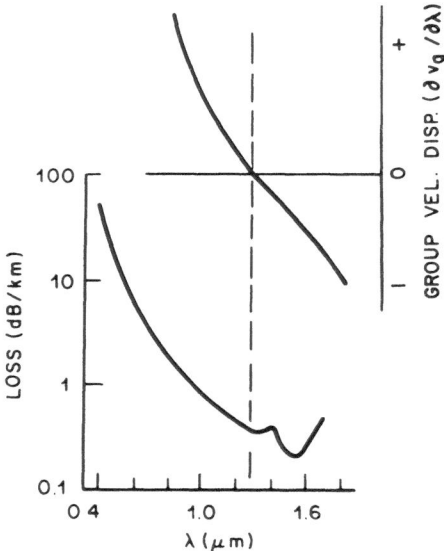

Fig. 6.38. Loss and group velocity dispersion as functions of wavelength for low loss single-mode fibers

Here n_2 has the numerical value 3.2×10^{-16} cm^2/W for quartz glass and I is the light intensity in compatible units. The nonlinearity produces "self-phase modulation", that is, a phase retardation in direct proportion to the intensity-induced change in index and to the length of fiber traversed:

$$\Delta \phi = \frac{2\pi}{\lambda} L n_2 I \ . \tag{6.8}$$

Nevertheless, in a pulse such as that shown in Fig. 6.39a, the rising and falling envelope intensity leads to a similar variation in the degree of phase retardation. Thus, the frequency "chirp" shown in Fig. 6.39b is generated. When such a chirped pulse is acted on by the fiber's negative group velocity dispersion, the leading half of the pulse, containing the lowered frequencies, will be retarded, while the trailing half, containing the higher frequencies, will be advanced, and the pulse will tend to collapse on itself (Fig. 6.39c). If the peak pulse intensity is high enough, such that the chirp is large enough, the degree of pulse narrowing can be substantial.

More generally, it can be shown [6.34] that the pulse envelope function $u(z, t)$, where z is distance along the fiber and t is time, is governed by the nonlinear Schrödinger equation:

$$i \frac{\partial v}{\partial \xi} = \frac{1}{2} \frac{\partial^2 v}{\partial s^2} + |v|^2 v \ , \tag{6.9}$$

where v, ξ, and s are dimensionless versions of u, z, and t, respectively. Although higher-order terms can be appended to (6.9), their elimination usually causes little or no error, except perhaps with the very shortest pulses.

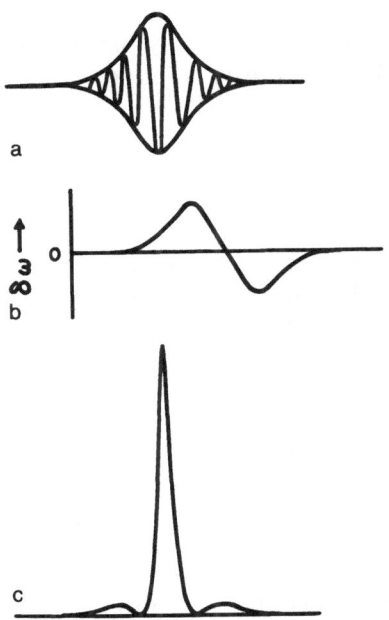

Fig. 6.39. (**a**) Optical pulse that has experienced self phase modulation. (**b**) Corresponding frequency chirp; frequencies in the leading half of the pulse are lowered, while those in the trailing half are raised. Because of the fiber's dispersion, the back of the pulse is advanced, and the leading half is retarded. Thus the pulse is compressed, as shown in (**c**) (see text)

An important set of solutions result from an input function of the form:

$$v(0,s) = N \mathrm{sech}(s) \ , \tag{6.10}$$

where N is an integer. The first two such solutions are shown graphically in Fig. 6.40. Corresponding to $N = 1$, one has the fundamental soliton, a pulse that never changes its (sech) amplitude shape as it propagates along the fiber. Physically, it represents a condition of exact balance between the pulse narrowing effect described earlier and the pulse broadening effect of dispersion

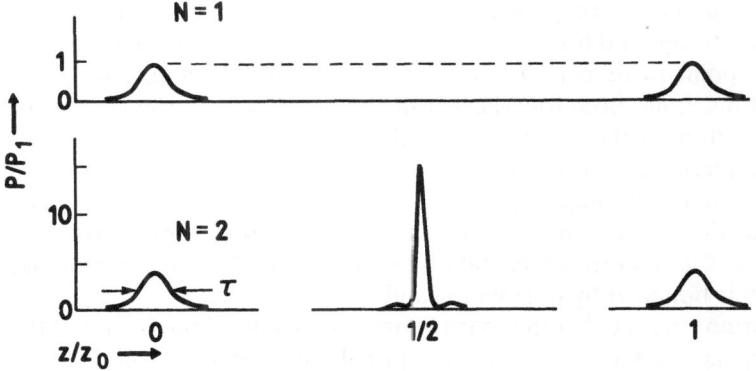

Fig. 6.40. Behavior of $N = 1$ and $N = 2$ solitons with propagation along a fiber

alone. Thus for $N < 1$, the purely dispersive effect dominates and the pulse broadens with propagation, while for $N > 1$ the pulse will always narrow, at least initially. In "real world" dimensions, the peak *power* corresponding to the fundamental soliton is given by the expression [6.34, 41]:

$$P_1 = 0.776 \frac{\lambda^3}{\pi^2 c n_2} \frac{|D|}{\tau^2} A_{\text{eff}} , \qquad (6.11)$$

where τ is the full width at half maximum (FWHM) of the input pulse, where n_2 has the numerical value given earlier, A_{eff} refers to the fiber core area, and where λ and c are the wavelength and speed of light, respectively, both as measured in vacuum. The dispersion parameter D reflects the change in pulse delay with change in wavelength, normalized to the fiber length.

For integers $N \geqslant 2$, sech input pulses always lead to pulse shaping that is periodic with period $\xi = \pi/2$. In real space, the period is [6.34, 41]

$$z_0 = 0.322 \frac{\pi^2 c}{\lambda^2} \frac{\tau^2}{|D|} \qquad (6.12)$$

where the various quantities on the right are defined as before for (6.11). The peak input powers, of course, are given by the expression

$$P_N = N^2 P_1 . \qquad (6.13)$$

For $N = 2$, the behavior is particularly simple: the pulse alternately narrows and broadens, achieving minimum width at the half period (again, see Fig. 6.40).

Although the $N = 1$ soliton shown in Fig. 6.40 is unique, the $N = 2$ soliton shown there represents but one member of a continuum. The $N = 2$ soliton can be looked on as a nonlinear superposition of two fundamental solitons; the continuum of solutions is obtained by varying the relative amplitudes and widths of the two components. The particular $N = 2$ soliton shown in Fig. 6.40 corresponds to components with amplitude and width ratios of 3:1 and 1:3, respectively, and it is the only $N = 2$ soliton to pass through the sech shape at any point in its period. Nevertheless, despite the potentially wide range of solitons, it has been found experimentally that the soliton laser tends to favor production of the "sech" $N = 2$ soliton.

The soliton laser is shown in Fig. 6.41. The mode-locked laser is coupled through beam splitter S and microscope objective L_1 to a length L of single-mode, polarization-preserving fiber; L_2 and M_3 form an efficient and stable cat's eye retroreflector at the other end of the fiber. Thus the fiber-containing, control cavity is bounded by mirrors M_0 and M_3.

It is important to note that the space between each end of the fiber and the corresponding lens surface is filled with a special index-matching oil. (The oil is a completely halogenated paraffin that has no detectable absorption in the

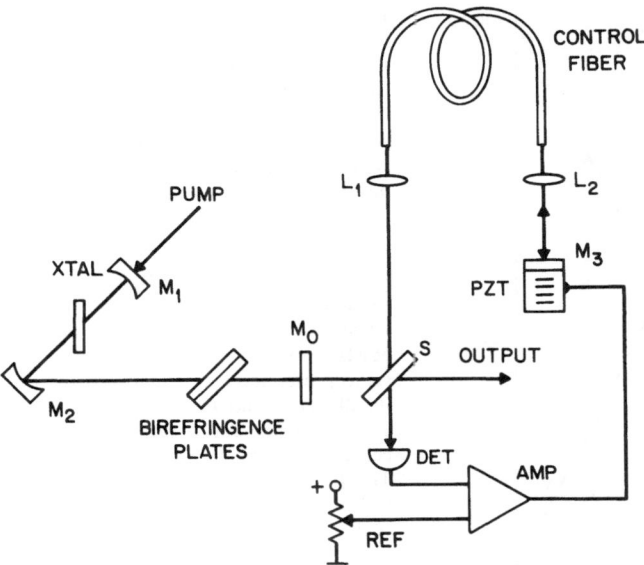

Fig. 6.41. Schematic of the color-center soliton laser. Typical reflectivities: $M_0 = 80\%$; $S = 30\%$

1.5 μm region.) Without this precaution, reflections from the fiber ends produce feedback that interferes with the soliton laser action [6.40].

Successful operation requires, of course, that pulses returned from the fiber be made coincident with those already present in the main cavity. Thus, the input end of the fiber and L_1 are mounted on a common translation stage, to enable final adjustment of the optical path length in the fiber arm to be an integral multiple of the main cavity length.

The device then operates as follows: as the laser action builds up from noise, the initially broad pulses are considerably narrowed by passage through the fiber. The narrowed pulses, reinjected back into the main cavity, force the laser itself to produce narrower pulses. This process builds on itself until the pulses in the fiber become solitons, at which point the laser operation reaches a stationary state.

Operation of the laser on $N \geqslant 2$ solitons is shown by the empirically determined dependence of the produced pulse width τ on the square root of fiber length (Fig. 6.42). Nevertheless, the values of peak power P implied by measurement of time-average powers in the fiber correspond, within experimental error, to values required for $N = 2$ solitons. Note that although for the shortest pulses, the peak powers in the fiber are large (nearly 10 kW), the corresponding time average powers remain modest (<100 mW), due to the long time (10 ns) between pulses.

The data points of Fig. 6.42 represent adjustment (through focusing or defocusing of L_1) of power in the fiber to obtain the narrowest possible pulses.

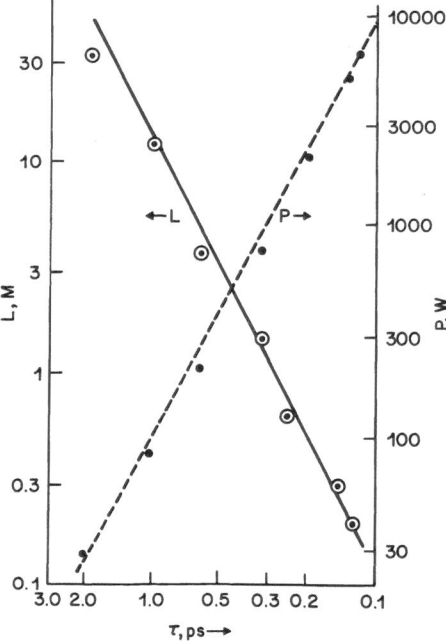

Fig. 6.42. Characteristics of the soliton laser as a function of produced pulse widths τ. *Circled dots:* control fiber lengths; *solid line:* $z_0/2$ calculated for "sech" $N = 2$ solitons; *dots:* experimentally determined peak powers at input to control fiber; *dashed line:* calculated peak power for "sech" $N = 2$ solitons

Fig. 6.43. FWHM of pulses emerging various lengths L' of test fiber (taken from the same spool as the 1.65 m control fiber) when sech²-shaped, laser output pulses of 560 fs FWHM were launched into it at the same power level as in the control fiber [6.92]. The *solid line* is the theoretically expected value for a sech $N = 2$ soliton calculated for $D = 14.5 \, \text{ps} \, \text{nm}^{-1} \, \text{km}^{-1}$. The power level was equal to the $N = 2$ soliton power as calculated from $n_2 = 3.2 \times 10^{-16} \, \text{cm}^2/\text{W}$ and $A_{\text{eff}} = 86 \, \mu\text{m}^2$ for 560 fs pulses. The datum point at $2L$ was found directly from the control fiber [6.92]

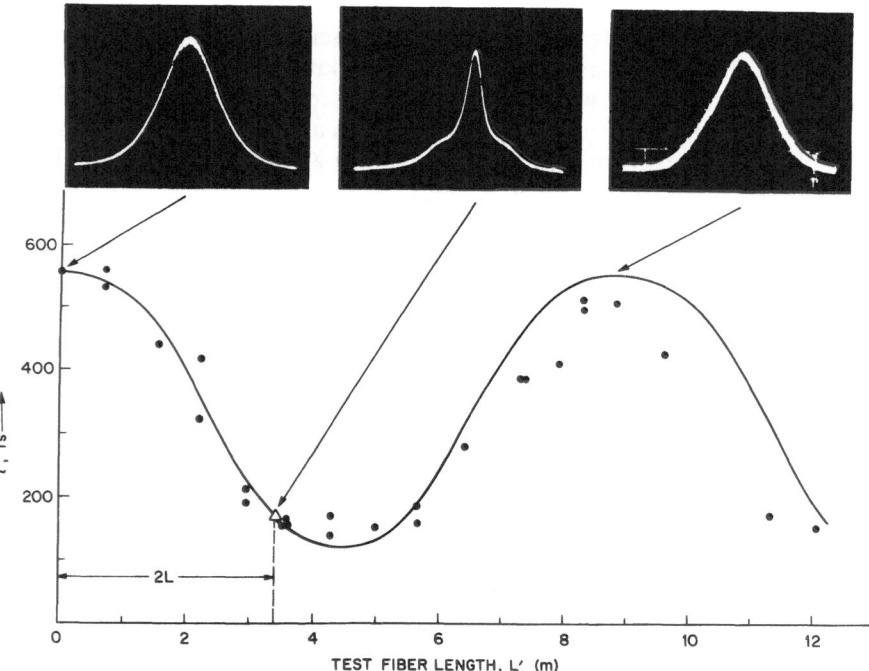

Fig. 6.43

That data also fits the relation $L \approx z_0/2$, with the implication that in the stationary state, pulses returned from the fiber have substantially the same width as those launched into it. Nevertheless, later experiments [6.92] have shown the existence of other stable modes of operation for which L is considerably less than $z_0/2$ (for example, see Fig. 6.43). Additionally, the *exact* condition $L = z_0/2$ probably cannot be reached in practice. That is, even in the stationary state, a certain degree of pulse narrowing is required of the pulses fed back from the fiber into the laser.

In the first experiments [6.39], it was observed that mirror vibration and drift caused the soliton laser action to flicker on and off at random. That is, laser action *per se* never stopped, but depending on mirror position, one had either the short, soliton laser pulses, or very broad ones. This instability results from dependence of the soliton laser action on the relative optical phase of pulses fed back from the fiber and those circulating in the main cavity. The electronic circuitry and piezotranslator shown in Fig. 6.41, a later development [6.92], form a servosystem that locks to the required phase and thus stabilizes the soliton laser action.

The scheme for operation of the servosystem is based on the fact that the power in the control cavity varies with ϕ, the round trip optical phase shift in the control cavity, about as shown in Fig. 6.44. It is also based on the observation that soliton laser action is correlated with a well defined level, lying somewhere within the middle of the range of cavity powers. The operation of the stabilization circuitry shown in Fig. 6.41 is now easily understood. The error signal for the control of ϕ is generated simply by taking the difference between the detector signal (a measure of the control cavity power, and hence a measure of ϕ) and a reference voltage corresponding to the soliton level. The op-amp magnifies that difference and drives the PZT translator of the end mirror M_3. Thus, assuming correct choice of signal polarity, the circuitry forms a closed negative feedback loop.

With stabilization, the laser exhibits uninterrupted soliton laser action, emitting a stream of pulses that are uniform both in width and height. The

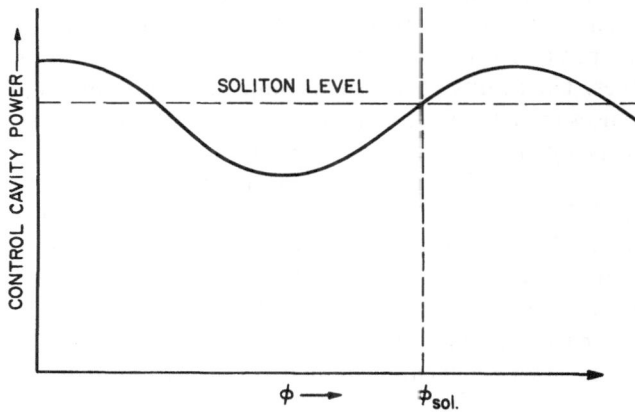

Fig. 6.44. Variation of control cavity power with round trip optical phase shift ϕ

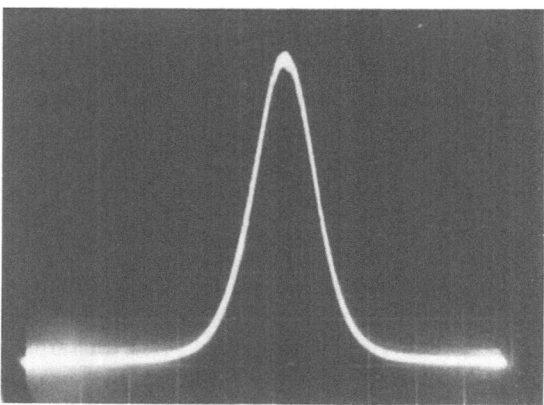

Fig. 6.45. A typical auto-correlation trace of the laser output pulses. In this example, the actual FWHM of the pulses is 580 fs ($L = 1.6$ m) [6.92]

stability and low noise of its output can be seen from the typical auto-correlation trace [6.92] shown in Fig. 6.45.

Although pulsing does not start by itself under cw pump conditions, it has been discovered [6.92] that once soliton laser action has begun, it will continue for a while under cw pumping (up to a minute has been observed) until some large perturbation interrupts the pulse stream. Thus, the soliton laser is passively mode locked. One consequence is that even in "synch-pumped" operation, the soliton laser's pulse repetition frequency is no longer determined by the repetition rate of the pump laser. To confirm that the soliton laser is mode locked at its own independent rate, a 100 MHz signal from the output of a fast photodiode in the soliton laser's output beam was mixed with a 100 MHz signal derived from the rf driver of the active-mode locker. The result was a fairly steady beat note, whose frequency could be tuned anywhere in a region ± a few Kilohertz about zero frequency, simply by adjusting the main cavity length. The existence of passive mode locking serves to illustrate in a dramatic way the fundamental idea in the invention of the soliton laser: that the pulse shaping of a fiber should be able to provide a much stronger drive for mode locking than that available from gain shaping by pump pulses.

Finally, we note that pulses from the soliton laser can be further compressed in an external fiber. The results of one such compression are shown in Fig. 6.46, where 150 fs pulses from the laser were reduced to ~50 fs [6.40, 41]. (Note that to obtain this result, it is mandatory to use index-matching oil at both ends of the compressor fiber, just as described above for the control fiber, and to use an isolator. Without those precautions, reflection from either or both of the fiber ends produces feedback that is destructive of the soliton laser action.) Recently, it has been possible to obtain stable 63 fs pulses directly from the soliton laser, and those pulses have been compressed in a "dispersion flattened" fiber to just 19 fs, or less than 4 optical cycles [6.92]. The bandwidths of such pulses ($\sim 500 \text{ cm}^{-1}$) match (or nearly match) the greatest known homogeneous line widths of various semiconductor, color

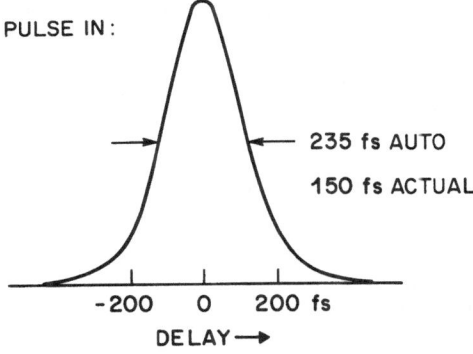

PULSE IN:

235 fs AUTO

150 fs ACTUAL

-200   0   200 fs

DELAY →

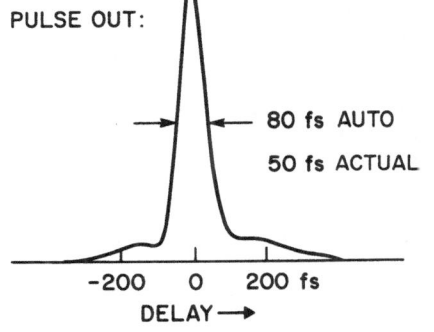

PULSE OUT:

80 fs AUTO

50 fs ACTUAL

-200   0   200 fs

DELAY →

Fig. 6.46. *Above:* Auto-correlation trace of pulses produced by the soliton laser with a 30 cm control fiber. *Below:* The same pulse following compression in a second, 35 cm fiber

center, or dye transitions. Thus they should allow for the measurement of relaxation phenomena in those systems to limits of resolution set only by the uncertainty principle itself.

A recently discovered [6.93] and analyzed [6.94] effect in optical fibers, the "soliton self frequency shift", should aid in the proposed measurements. In that effect, Raman gain causes a flow of energy from higher to lower frequency components, and thus the soliton's mean frequency is steadily lowered as it travels down the fiber. The effect scales as τ^{-4}, and for a ~20 fs pulse it is so strong that a wavelength increase of several hundred nm should occur in just a meter or two of fiber. Thus it should be possible to produce, from the same laser, synchronized femtosecond pulses of two distinctly different wavelengths for pump-probe experiments. Such a pulse pair could be used, for example, in a rise-of-gain measurement of the configurational relaxation time of the F_2^+ center in KBr, whose absorption and emission bands are centered at ~1.5 and ~1.8 μm, respectively.

References

6.1 A survey of color center physics as known in its prime can be found in: W. B. Fowler (ed.): *Physics of Color Centers* (Academic, New York 1968)

6.2 For an extensive survey of the physics of laser-active color centers, see L. F. Mollenauer: "Color Center Lasers" in *The Laser Handbook* ed. by M. Stitch, M. Bass (North Holland, Amsterdam 1985) Chap. 3

6.3 L. F. Mollenauer: "Color Center Lasers", in *Quantum Electronics,* Part B ed. by C. L. Tang (Academic, New York 1979) Chap. 6 (This is considerably earlier and less up-to-date than [6.2].)

6.4 B. Henderson: Opt. Lett. **6**, 437 (1981)

6.5 L. F. Mollenauer, D. M. Bloom: Opt. Lett. **4**, 247 (1979)

6.6 L. F. Mollenauer: Opt. Lett. **5**, 188 (1980)

6.7 J. F. Pinto, L. W. Stratton, C. R. Pollock: Opt. Lett. **10**, 384 (1985)

6.8 L. F. Mollenauer, N. D. Vieira, L. Szeto: Opt. Lett. **7**, 414 (1982)

6.9 I. Schneider, M. J. Marrone: Opt. Lett. **4**, 390 (1979)

6.10 I. Schneider, C. L. Marquardt: Opt. Lett. **5**, 214 (1980)

6.11 K. German: J. Opt. Soc. Am. B**3**, 149 (1986)

6.12 I. Schneider: Opt. Lett. **7**, 271 (1982)

6.13 L. F. Mollenauer, D. H. Olson: Appl. Phys. Lett. **24**, 386 (1974)

6.14 L. F. Mollenauer, D. H. Olson: J. Appl. Phys. **46**, 3109 (1976)

6.15 B. Fritz, E. Menke: Solid State Commun. **3**, 61 (1965)

6.16 C. R. Pollock, D. A. Jennings, F. R. Petersen, J. S. Wells, R. E. Drullinger, E. C. Beaty, K. M. Evenson: Opt. Lett. **8**, 133 (1983)

6.17 M. H. Begemann, R. J. Saykally: Opt. Commun. **40**, 277 (1982)

6.18 R. Beigang, J. J. Wynne: J. Opt. Soc. Am. **70**, 579 (1980)

6.19 C. Cheng, K. W. Giberson, A. R. Harrison, F. K. Tittel, F. B. Dunning, G. K. Walters: Rev. Sci. Instrum. **53**, 1434 (1982)

6.20 H. Gerhardt, T. W. Hänsch: Opt. Commun. **41**, 17 (1982)

6.21 K. W. Giberson, C. Cheng, M. Onellion, F. B. Dunning, G. K. Walters: Rev. Sci. Instrum. **53**, 1789 (1982)

6.22 L. G. Gray, K. W. Giberson, C. Cheng, R. S. Keiffer, F. B. Dunning, G. K. Walters: Rev. Sci. Instrum. **54**, 271 (1983)

6.23 D. J. Jackson, E. Arimondo, J. E. Lawler, T. W. Hänsch: Opt. Commun. **33**, 51 (1980)

6.24 D. J. Jackson, H. Gerhardt, T. W. Hänsch: Opt. Commun. **37**, 23 (1981)

6.25 P. J. Nacher, M. Leduc, G. Trenec, F. Laloe: J. Phys. Lett. **43**, L-525 (1982)

6.26 V. M. Baev, Gamalii, B. D. Lobanov, E. F. Martynovich, E. A. Sviridenkov, A. F. Suchkov, V. M. Khulugurov: Sov. J. Quantum Electron. **9**, 51 (1979)

6.27 V. M. Baev, H. Schröder, P. E. Toschek: Opt. Commun. **36**, 57 (1981)

6.28 R. L. Deleon, P. H. Jones, J. S. Muenter: Appl. Opt. **20**, 525 (1981)

6.29 T. E. Gough, D. Gravel, R. E. Miller, G. J. Scoles: J. Opt. Soc. Am. **70**, 665 (1980)

6.30 T. E. Gough, D. Gravel, R. E. Miller: Rev. Sci. Instrum. **52**, 802 (1981)

6.31 G. Litfin, C. R. Pollock, R. F. Curl, F. K. Tittel: J. Opt. Soc. Am. **70**, 664 (1980)

6.32 G. Litfin, C. R. Pollock, R. F. Curl, F. K. Tittel: J. Chem. Phys. **72**, 6602 (1980)

6.33 D. M. Bloom, L. F. Mollenauer, C. Lin, D. W. Taylor, A. M. Del Gaudio: Opt. Lett. **4**, 297 (1979)

6.34 L. F. Mollenauer, R. H. Stolen, J. P. Gordon: Phys. Rev. Lett. **45**, 1095 (1980)

6.35 L. F. Mollenauer, R. H. Stolen, J. P. Gordon, W. J. Tomlinson: Opt. Lett. **8**, 289 (1983)

6.36 R. H. Stolen, L. F. Mollenauer, W. J. Tomlinson: Opt. Lett. **8**, 186 (1980)

6.37 L. F. Mollenauer, R. H. Stolen, M. N. Islam: Opt. Lett. **10**, 229 (1985)

6.38 L. F. Mollenauer, J. P. Gordon, M. N. Islam: IEEE J. Quantum Electron. QE-**22**, 157 (1986)

6.39 L. F. Mollenauer, R. H. Stolen: Opt. Lett. **29**, 13 (1984)

6.40 L. F. Mollenauer, R. H. Stolen: In *Ultrafast Phenomena IV,* ed. by D. H. Auston, K. B. Eisenthal, Springer Ser. Chem. Phys., Vol. 38 (Springer, Berlin, Heidelberg 1984) pp. 2–6

6.41 L. F. Mollenauer: Phil. Trans. Roy. Soc. A**315**, 437 (1985)

6.42 W. B. Fowler (ed.): *Physics of Color Centers* (Academic, New York 1968) Chap. 2

6.43 R. S. Crandell, Mikkor: Phys. Rev. **138**, A1247 (1965)

6.44 H. Rogener: Ann. Phys. (Leipzig) **29**, 386 (1937)

6.45 A. B. Scott, W. A. Smith: Phys. Rev. **83**, 982 (1951)

6.46 L. F. Mollenauer: Rev. Sci. Instrum. **49**, 809 (1978)
6.47 J. N. Bradford, R. T. Williams, W. L. Faust: Phys. Rev. Lett. **35**, 300 (1975)
6.48 L. W. Hobbs, A. E. Hughes, D. Pooley: Proc. Roy. Soc. London A**335**, 167 (1973)
6.49 L. F. Mollenauer: Opt. Lett. **1**, 164 (1977)
6.50 H. Hartel, F. Luty: Z. Phys. **177**, 369 (1964)
6.51 B. Fritz, F. Luty, G. Rausch: Phys. Stat. Sol. **11**, 635 (1965)
6.52 F. Luty: In *Physics of Color Centers,* ed. by W. B. Fowler (Academic, New York 1968) Chap. 3
6.53 N. Nishimaki, Y. Matsusaka, Y. Doi: J. Phys. Soc. Jpn. **33**, 424 (1972)
6.54 G. Litfin, R. Beigang, H. Welling: Appl. Phys. Lett. **31**, 381 (1977)
6.55 K. Park, W. L. Faust: Phys. Rev. Lett. **17**, 137 (1966)
6.56 R. Herman, M. C. Wallis, R. F. Wallis: Phys. Rev. **103**, 87 (1956)
6.57 M. A. Aegerter, F. Luty: Phys. Stat. Sol. (b) **43**, 227 f, 245 f (1971)
6.58 L. F. Mollenauer: Phys. Rev. Lett. **43**, 1524 (1979)
6.59 L. F. Mollenauer, D. M. Bloom, H. Guggenheim: Appl. Phys. Lett. **33**, 506 (1978)
6.60 A. Chandra: J. Chem. Phys. **51**, 1499 (1969)
6.61 Yu. L. Gusev, S. N. Konoplin, S. I. Marennikov: Sov. J. Quantum Electron. **7**, 1157 (1977) [Kvantovaya Electronika **4**, 2024 (1977)]
6.62 W. Gellerman, F. Luty, K. T. Koch, G. Litfin: Phys. Stat. Sol. **57**, 111 (1980)
6.63 W. Gellerman, F. Luty, F. P. Koch, H. Welling: Opt. Commun. **35**, 430 (1980)
6.64 J. Nahum: Phys. Rev. **158**, 814 (1967)
6.65 I. Schneider: Solid State Commun. **34**, 865 (1980)
6.66 I. Schneider: Opt. Lett. **6**, 157 (1981)
6.67 I. Schneider, C. L. Marquardt: Opt. Lett. **6**, 627 (1981)
6.68 I. Schneider, S. C. Moss: Opt. Lett. **8**, 7 (1983)
6.69 J. Pinto, E. Georgiou, C. Pollock: Opt. Lett. **11**, 519 (1986)
6.70 Yu. L. Gusev, S. N. Konoplin, A. V. Kirpichnikov, S. I. Marenikov: Laser Spectroscopy Conf., Novosibirsk, USSR, 1979
6.71 L. F. Mollenauer: Opt. Lett. **6**, 342 (1981)
6.72 P. G. Baranov, V. A. Khramtsov: Phys. Stat. Sol. (b) **101**, 153 (1980)
6.73 E. Goovaerts, J. A. Andriessen, S. V. Nistor, D. Shoemaker: Phys. Rev. B**24**, 29 (1981)
6.74 W. Gellerman, F. Luty, C. R. Pollack: Opt. Commun. **39**, 391 (1981)
6.75a L. F. Mollenauer, N. D. Vieira, L. Szeto: Phys. Rev. B**27**, 5332 (1983)
6.75b F. J. Ahlers, F. Lohse, J. M. Spaeth, L. F. Mollenauer: Phys. Rev. B**28**, 1249 (1983)
6.76 J. Duran, P. Evesque, M. Billardon: Appl. Phys. Lett. **33**, 1004 (1978)
6.77 B. Henderson: "Anion Vacancy Centers in Alkaline Earth Oxides", in CRC *Critical Reviews in Solid State and Materials Sciences* (CRC Press, Boca Raton, FL 1980)
6.78 S. C. Rand, L. G. DeShazer: Opt. Lett. **10**, 481 (1985)
6.79 G. Davies: Diamond Res. 15–24 (1977)
6.80 M. D. Crossfield, G. Davies, A. T. Collins, E. C. Lightowlers: J. Phys. C**7**, 1909 (1974)
6.81 J. E. Field (ed.): *The Properties of Diamond* (Academic, New York 1979) pp. 23–27
6.82 E. F. Martynovich, V. I. Baryshnikov, V. A. Grigorov: Opt. Commun. **53**, 257 (1985) and Sov. Tech. Phys. Lett. **11**, 81 (1985)
6.83 E. F. Martynovich, A. G. Tokarev, V. A. Grigorov: Sov. Phys. Tech. Phys. **30**, 243 (1985)
6.84 G. Litfin, R. Beigang: J. Phys. E**11**, 984 (1978)
6.85 K. R. German: Opt. Lett. **4**, 68 (1979)
6.86 N. D. Vieira, L. F. Mollenauer: IEEE J. QE-**21**, 195 (1985)
6.87 K. W. Giberson, C. Cheng, F. B. Dunning, F. K. Tittel: Appl. Opt. **21**, 172 (1982)
6.88 T. F. Johnston, Jr., Wm. Proffitt: IEEE J. QE-**16**, 483 (1980)
6.89 G. P. Trenec, P. J. Nacher, M. Leduc: Opt. Commun. **43**, 37 (1982)
6.90 R. M. Jopson, G. Eisenstein, H. E. Earl, K. L. Hall: Electron. Lett. **21**, 783 (1985)
6.91 L. F. Mollenauer, D. Bloom, A. M. Del Gaudio: Opt. Lett. **3**, 48 (1978)
6.92 F. M. Mitschke, L. F. Mollenauer: IEEE J. Quantum Electron. QE-**22**, #12 (Dec. 1986)
6.93 F. M. Mitschke, L. F. Mollenauer: "Discovery of the Soliton Self Frequency Shift", Opt. Lett. **11**, 659 (1986)
6.94 J. P. Gordon: "Theory of the Soliton Self Frequency Shift", Opt. Lett. **11**, 662 (1986)

7. Fiber Raman Lasers

Chinlon Lin

With 20 Figures

By means of the stimulated Raman effect, laser-pumped optical fibers yield both cw and pulsed laser action from the near ultraviolet to the near infrared (from 0.3 to beyond 2 μm). A single Stokes order yields continuous tuning over several hundred cm^{-1}, and many Stokes orders are often produced from a single pump frequency. The basic principles and the performance of these efficient frequency down-converters are given, and the influence of fiber properties on performance is discussed.

7.1 Introduction

Nonlinear optical effects in glass fibers are known to present power transmission limitations for optical communication using glass fiber waveguides [7.1, 2]. Both the intensity and the spectral characteristics of the optical signal can be affected by various nonlinear processes [7.3], such as stimulated Raman scattering, stimulated Brillouin scattering, self-phase modulation, parametric mixing, etc. On the other hand, one can use nonlinear optical effects in fibers for efficient frequency conversion and generation of new frequencies. This has proven to be useful for various applications, including studies and measurements of multimode and single-mode fibers in the 1 − 1.7 μm spectral region [7.4 − 7].

Both single-mode and multimode fibers have been used for frequency conversion by stimulated Raman scattering, Raman oscillation, and phase-matched four-photon parametric mixing. Frequency conversion over a wide spectral range in the visible and the near-infrared region has been achieved, based on these nonlinear processes in fibers [7.8 − 14]. In this chapter we will describe frequency conversion in fiber Raman lasers, including recent results on uv fiber Raman lasers and optical Raman amplifiers. For other aspects of nonlinear optical effects in fibers, refer to previous reviews [7.2, 15 − 18].

7.2 Stimulated Raman Scattering in Optical Fibers

7.2.1 Raman Spectra of Optical Fibers

In Raman scattering, incident light is scattered by optical vibrational modes (optical phonons) of the material, resulting in a frequency-shifted Stokes

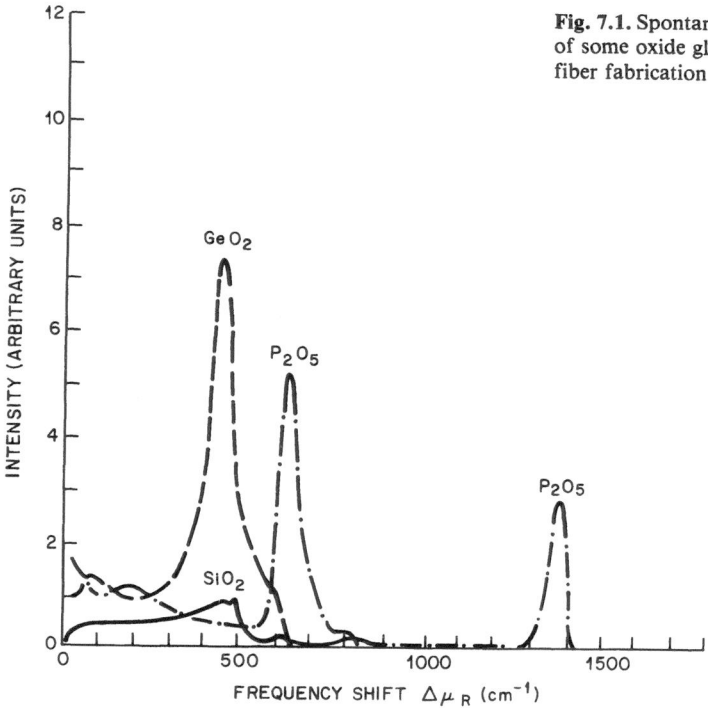

Fig. 7.1. Spontaneous Raman spectra of some oxide glasses used in optical fiber fabrication

RAMAN SPECTRA OF OXIDE GLASSES

light. The frequency shift is determined by the phonon frequency. The Raman spontaneous emission spectra for various silicate, germante, and phosphate glasses (Fig. 7.1) show a broad band of frequencies (molecular vibrations) associated with the amorphous nature of oxide glasses [7.19]. Note that the different glasses have different spectral features and varying Raman cross sections. These characteristics are important in broadband Raman amplifier and oscillator design. For example, the peak Raman scattering cross section of pure GeO_2 is about 10 times that of pure SiO_2; this is useful for obtaining higher Raman gain at lower pump powers. Note also that for these two glasses, the peaks occur at nearly the same shift frequency ($\Delta v_R \sim 450$ to $490 \, cm^{-1}$). Similarly, the additional Raman peak at $1330 \, cm^{-1}$ of P_2O_5 glasses [7.20, 21] is useful for obtaining a larger frequency shift in one-step Stokes conversion. Typically, in high-silica fibers, i.e., where the dopant glasses have relatively low molecular percentages compared with the silica glass, the Raman spectrum is mainly that of SiO_2.

Another way to modify the Raman spectra of optical fibers, other than using different dopant glasses, is to diffuse molecular gases, such as hydrogen or deuterium, into conventional solid silica glass optical fibers [7.22, 23]. With adequate concentration of H_2 or D_2 in the glass fiber, the Raman gain is comparable to that of silica glass. The most useful aspect of such gas-in-glass

Raman media is the large (respectively, ~4200 and ~3000 cm^{-1}) Stokes shifts of their Raman lines. Thus, for example, for a pump at 1.06 μm, the first Stokes will be at 1.56 μm for D$_2$ diffused silica fiber, whereas for the conventional silica glass fiber, the first Stokes would be peaked at 1.12 μm. However, the narrow Raman line of gases does not allow for frequency tuning.

7.2.2 Loss Characteristics of Optical Fibers

In addition to the Raman spectrum and Raman cross section, the other important material parameter is the loss characteristic. Depending on the spectral region of interest, the particular dopant oxide glass may have undesirable absorption losses for stimulated Raman scattering or Raman amplification. For example, high P$_2$O$_5$ doping in silica fibers contributes infrared absorption loss that rises faster in the long wavelength ($\lambda > 1.6$ μm) region, than for pure silica. For another example, to take advantage of the large Raman cross section of GeO$_2$, one would like to have a high germania doping, or even a pure germania fiber [7.24,25], but this raises the loss. While the silica-based glass fibers can have losses less than 1 dB/km in the $1 - 1.6$ μm spectral region, germania-based fibers typically have losses of about 50 dB/km in the same region [7.24 – 26], even though loss as low as 5 dB/km at 2 μm has been reported [7.26]. Thus, depending on the application, available pump power, and spectral region of interest, loss must be considered along with the other material parameters in selecting the fiber for optimum performance.

For nearly pure silica-core fibers with low OH$^-$ content, the loss can be low enough for fiber Raman laser applications from the uv to the near ir (0.3 – 2 μm). Figure 7.2 shows such a low OH$^-$, low-loss spectrum [7.27].

7.2.3 Raman Gain and Effective Interaction Length

For pump and signal fields of parallel polarization in a single-mode fiber, the coefficient of Raman gain is:

$$\alpha_g = R \frac{P_p}{A_{eff}} , \tag{7.1}$$

where P_p is the pump power, A_{eff} is the fiber core effective cross section area, and where α_g is defined as in Sect. 1.4. For a fused silica core at the Raman gain peak ($\Delta v_R \sim 490$ cm^{-1}), and for $\lambda_p = 1$ μm, the Raman gain coefficient $R \approx 1.0$ km^{-1}/MW/cm^2. For other pump wavelengths, the Raman gain scales in inverse proportion to λ_p. For other values of Δv_R, R scales in proportion to the Raman gain spectrum, which scales in turn like the Raman spontaneous emission spectrum at 0 K, times a factor λ_s^3. (The 0 K Raman emission spectrum can be inferred from measurements made at room temperature.) Figure 7.3 shows the gain spectrum for a pure SiO$_2$ fiber [7.2].

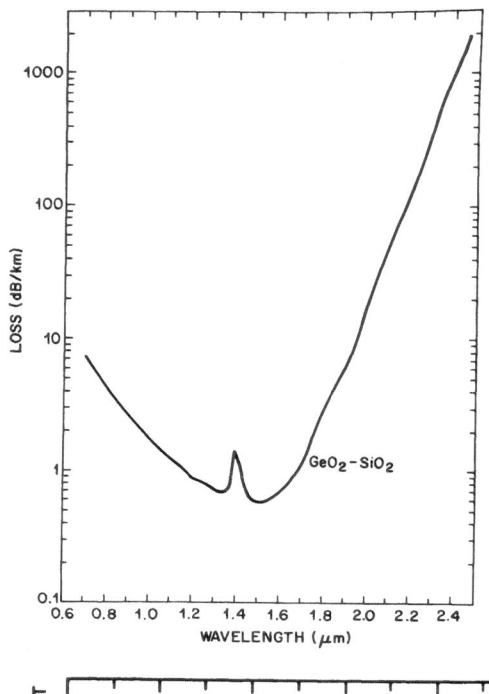

Fig. 7.2. Loss spectrum of a law OH⁻, low-loss multimode silica glass fiber

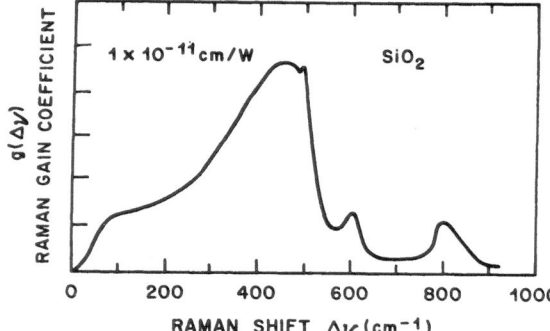

Fig. 7.3. Raman gain spectrum of a typical SiO_2 glass fiber

By virtue of a large, intentionally created birefringence, certain fibers are polarization preserving, i.e., they maintain the linear polarization of light launched with its E field parallel to one or the other of two orthogonal birefringent axes of the core. Most fibers are not polarization preserving, however, and for such ordinary fibers, the pump and Stokes polarizations vary rapidly with propagation and, because of dispersion, their relative polarizations also vary. There is no gain when the pump and Stokes waves are orthogonally polarized. Thus, for all but the shortest lengths of ordinary fiber, the effective value of R is reduced by one-half.

The principal advantage in using low-loss optical fibers for Raman amplification and other nonlinear interactions is that high intensity can be maintained over long interaction lengths. That is, A_{eff} can be as small as a few tens of μm^2, and the effective interaction length L_{eff} can be as long as kilometers if the fiber loss is low. In the small signal limit (negligible attenuation of pump power by stimulated emission), L_{eff} is, for a fiber of length L and a linear attenuation coefficient α_p (at λ_p),

$$L_{eff} = \int_0^L \exp(-\alpha_p x)\,dx$$

$$= \frac{1 - \exp(-\alpha_p L)}{\alpha_p}.$$

(7.2)

If the loss α_p is large or the fiber length is very long, so that $\alpha_p L \gg 1$, then

$$L_{eff} \simeq \frac{1}{\alpha_p}.$$

(7.3)

On the other hand, if the loss α_p is so small or L so short that $\alpha_p L \ll 1$, then

$$L_{eff} \sim \frac{1 - (1 - \alpha_p L)}{\alpha_p} = L,$$

(7.4)

i.e., the effective interaction length is approximately the true fiber length.

In practice, L_{eff} can be tens of kilometers long. Thus, even with the small Raman cross sections in glass fibers as compared with Raman cross sections in, for example, liquid CS_2, the low-loss characteristics and the resultant long interaction length lead to lower pump powers and/or higher gains by several orders of magnitude. This makes possible very efficient Raman amplification and Stokes conversion, even at low pump powers [7.1,2].

In the small signal limit, then, the overall Raman Stokes amplification is given by

$$\frac{I_{out}}{I_{in}} = \exp\left(R\,\frac{P_p}{A_{eff}}\,L_{eff} - \alpha_S L\right),$$

(7.5)

where the second factor accounts for ordinary fiber attenuation at the Stokes frequency. When the pump to Raman Stokes conversion is appreciable, however, such as in sequential multiple Stokes-Raman conversion, (7.5) no longer applies. Then, coupled wave equations must be solved to obtain relative pump and Stokes intensities as a function of interaction length, input pump intensity, etc. [7.28].

7.2.4 The Inhomogeneous Nature of Raman Gain

The Raman gain band in glasses is *inhomogeneously* broadened, since the various phonon frequencies are associated with different sites. Thus, in the small signal limit, laser operation at one frequency will not significantly affect such operation at another frequency. This fact makes it harder to obtain narrow linewidths with fiber Raman lasers than with lasers using homogeneously broadened gain media, such as dyes or color centers. Consequently, fiber Raman lasers have found application primarily for short pulse generation, or where the most precise frequency definition is not required.

Nevertheless, in a fiber Raman laser, the pump intensity plays a role similar to that of the excited state population in other systems. That is, gain saturation occurs when the stimulated Raman effect significantly reduces the pump intensity. Thus, in the large signal regime, operation on one laser frequency can reduce the gain available at other frequencies, and the Raman gain begins to behave more as if it were homogeneously broadened. One important consequence is that conversion of pump power to laser output power can be efficient in fiber Raman lasers when the gain is highly saturated.

7.2.5 Considerations for Choice of Fiber

For efficient Raman interaction, the general fiber requirements are low loss, small core (small A_{eff}), and that the glass has the desired Raman spectrum. For example, as indicated earlier, silica glass fibers have low loss and high damage threshold but small Raman cross section, while GeO_2-glass-based fibers have ~ 10 times larger Raman gain but higher losses. The exceptionally large Raman shift, also mentioned above, of high-phosphorous fibers are of particular interest for large frequency conversion in one step.

Fibers for efficient Raman amplification and frequency conversion are not always the same as those for high-bandwidth long-distance telecommunications (where low dispersion [7.29] and low-loss transmission is the primary concern). For example, large-core size is often desired for ease of coupling and splicing in communication fibers. In contrast, for high-intensity nonlinear interactions, small-core single-mode fibers are preferred. (Actually, such small-core fibers are beginning to take over in communications as well.) Nevertheless, multimode fibers find special application in, for example, higher overall Stokes energy generation, even though the conversion efficiency is lower because of the larger effective area.

In using single-mode fibers for efficient stimulated Raman scattering and Raman amplification, there are constraints on just how small the core can be. To discuss these constraints, we must introduce the V-number, or normalized frequency, a dimensionless parameter defined for step index fibers as

$$V = (n_1^2 - n_0^2)^{1/2} ka \ , \tag{7.6}$$

where n_1 and n_0 are the core and cladding indices, respectively, k is the propagation constant in vacuum, and a is the core radius. The physical significance of V is that it determines how many modes the fiber can support. In particular, if V is less than the cutoff value for the LP_{11}-mode ($V < 2.405$), only the LP_{01}-mode can exist and the fiber is single mode. Additionally, the fraction of the power carried in the core increases with increasing V number.

For example, if one wants the most efficient first Stokes conversion from $1.06 - 1.12$ µm, then a fiber with the largest allowable V for single mode ($V = 2.4$) should be used. However, if a wide spectral coverage by sequential, multiple-Stokes stimulated Raman scattering is desired, one has to keep in mind that $V = 2.4$ at 1.06 µm means smaller V, and hence a much less confined waveguiding mode at the longer wavelengths. [From (7.6) it can be seen that V is inversely proportional to λ in the absence of significant dispersion.] The conversion efficiency at the longer wavelength region could thus decrease so quickly that the sequential Stokes conversion stops at, say, 1.75 µm [7.9]. On the other hand, experiments in somewhat larger core multi-mode fibers show that conversion of 1.06 µm light is possible to 2 µm and beyond [7.9]. Therefore, if wide spectral range conversion is desired, rather than just efficient conversion to the first-order Stokes, one would use a fiber with, for example, $V = 3.2$ at 1.06 µm. The fiber would then operate in the two-mode regime at the pump wavelength, and the mode confinement at longer wavelength (for example, $V = 1.6$ at 2.1 µm) is still reasonable.

7.2.6 Picosecond Pulse Raman Interactions in Long Fibers: The Effect of Group Velocity Dispersion

In Raman conversion involving short pulses, the difference between the pump and generated Stokes pulse group velocities must be taken into account, since interaction ceases when the pump and the Stokes pulses no longer overlap. The velocity difference can be significantly large, even in the "low" dispersion (~ 1.3 µm) region. Thus, in general, the effective interaction length is smaller than that obtained with cw or long pulse pumping. In fact, the question of temporal overlap, as determined by pulse width and fiber dispersion, is as important as that of transverse spatial overlap between pump and Stokes modes in a multimode fiber. To obtain long interaction lengths, then, the group velocities must be matched or their difference at least kept small.

Two techniques have been proposed [7.15] for group velocity matching. The first is to use modal dispersion in low-mode-number fibers to compensate the material dispersion. For example, one can have the pump pulse in the fundamental mode and the Stokes pulse in a slower, higher-order mode [7.30]. Figure 7.4 illustrates the concept of group velocity matching in a two-mode fiber. Experimental realization was first obtained in a two-mode fiber Raman oscillator, pumped with a cw mode-locked (~ 140 ps) argon-ion laser [7.30]. This scheme is universally applicable over a wide spectral range.

The second technique applies only in the minimum chromatic dispersion region [7.29] of single-mode fibers. There one can achieve group velocity matching if the pump wavelength λ_p and the Stokes wavelength λ_s are approximately equally apart from the minimum chromatic dispersion wavelength λ_0 of the fiber [7.15] (Fig. 7.5). As long as λ_s is within the Raman gain bandwidth of λ_p, the group velocity matching will tend to make λ_s the actual Raman gain peak, because of the increased effective interaction length for λ_s.

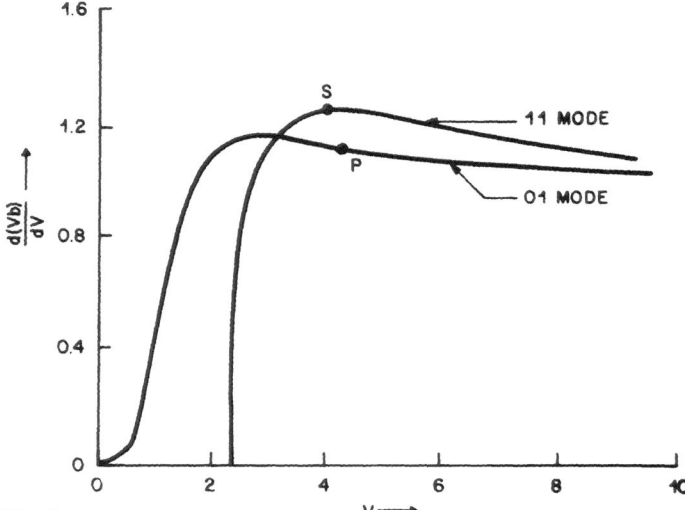

Fig. 7.4

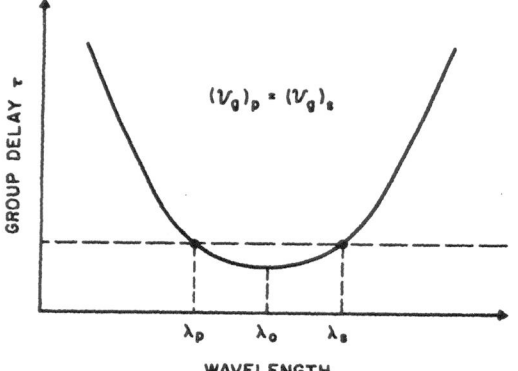

$(V_g)_p \approx (V_g)_s$

Fig. 7.4. Group velocity matching in a two-mode fiber. The pump pulse is in the lower-order mode and the Stokes pulse in the higher-order mode, if both are on the short wavelength side of minimum dispersion wavelength λ_0

Fig. 7.5. Group velocity matching for picosecond Raman interaction in a single-mode fiber based on equal group velocity on both sides of the minimum chromatic dispersion wavelength (λ_0)

This group velocity matching technique can be important to fiber Raman lasers operating in the region of λ_0. The technique is also useful for picosecond fiber Raman amplifiers involving direct optical amplification of high-speed injection laser signals. Such amplifiers may become important in high-bandwidth optical communication systems.

7.3 Single-Pass Fiber Raman Lasers

When the single-pass gain in a Raman amplifier becomes very large, say $\exp(16)$ or greater, Raman spontaneous emission from one end of the medium can be amplified to saturation at or before the output end. The frequency spectrum of the resultant output is greatly narrowed, as the high gain strongly emphasizes emission at the peak of the Raman band. Such single-pass laser action requires high pump intensities that can be supplied only from pulses. Thus, the single-pass laser is a traveling wave device, as only those waves traveling with the pump pulse will be significantly amplified. The process of light generation in such single-pass Raman lasers is often referred to simply as "stimulated Raman scattering", although as already indicated, spontaneous emission is also involved.

Generation of such Raman-frequency shifted, laser-like Stokes radiation by stimulated Raman scattering was first discovered in 1962 by *Woodbury* and *Ng* [7.31] and *Eckhardt* and coworkers [7.32]. Since then the effect has been seen in a large number of materials and used to generate new Stokes frequencies of coherent radiation [7.33].

The analogous effect in glass fibers was first observed and measured by *Stolen* et al. [7.34] in 1972. Figure 7.6 shows a typical experimental setup. In such single-pass fiber Raman lasers, the required pump pulse peak powers are in the 100 W – 1 kW range. The exceptionally wide bandwidth and low loss of silica glass fibers, mentioned earlier, makes them particularly attractive for single-pass lasers because of the long interaction length and the resultant efficient stimulated Raman scattering at low pump powers. Such low-loss fibers can be designed specifically for nonlinear optical interactions.

7.3.1 Stimulated Raman Scattering and Broadband Continuum Generation in the Visible Region

With a few kW of the second harmonic radiation ($\lambda_p = 0.53$ μm) of a Q-switched Nd:YAG laser, one can observe conversion to as many as 26 orders of Stokes components by sequential stimulated Raman scattering [7.34]. The resulting wavelengths extend from 0.53 to $0.8 - 0.9$ μm.

With nitrogen-laser-pumped dye lasers using dyes such as coumarins in a nonfrequency-selective cavity, one can have a blue laser ($\lambda_p \sim 440$ nm) with a large spectral width ($\Delta\lambda_p \sim 15$ nm) as the pump source. In an experiment using ~ 1 kW pulses from such a dye laser, coupled into a 19.5 m long silica core fiber with a core diameter of 7 μm, a wide band visible continuum covering 180 nm from the blue to the red ends of the visible spectrum was generated. The continuum resulted from a combination of sequential stimulated Raman scattering, self-phase modulation, and four-photon mixing [7.35]. As the pump laser light propagated in the fiber, because of the sequential Stokes conversion, the color of the scattered light changed from blue to green to red along the fiber from input to output ends. The continuum output appeared white before being dispersed by a prism.

The time-averaged continuum spectrum and the typical pulse characteristics of different spectral regions of the continuum are shown in Figs. 7.7a and b, respectively. Note the effect of depletion of the pump and lower-order

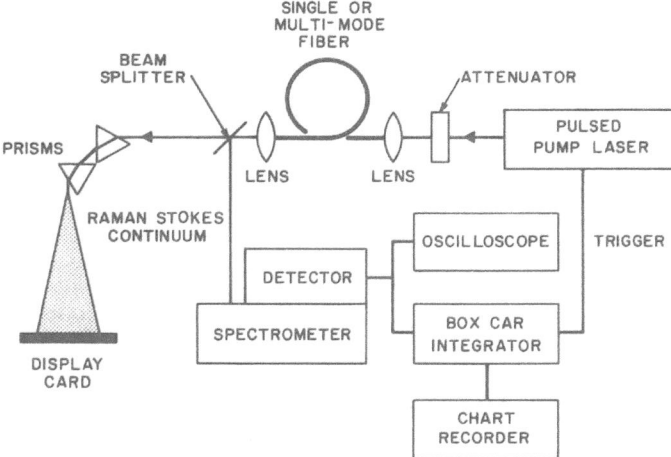

Fig. 7.6. Typical experimental setup for single-pass stimulated Raman scattering in optical fibers. Multiple-orders of Stokes lines or a wide-band continuum can be generated with appropriate pump lasers

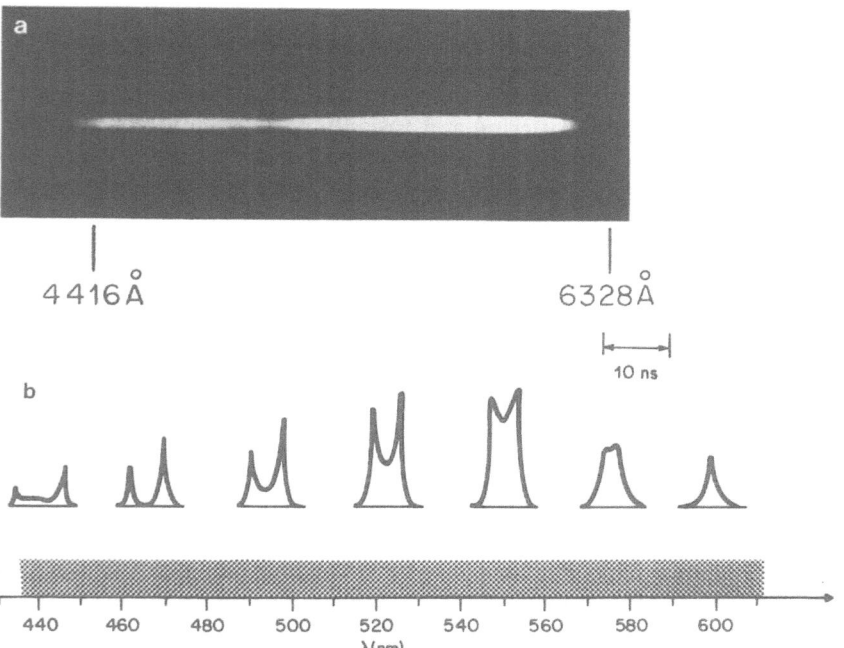

Fig. 7.7. (a) The time-averaged spectrum of a wide-band continuum generated in the visible, in a silica fiber pumped by a ~10 ns long pulses from a N_2-laser pumped broad-band ($\Delta\lambda_p \sim 150$ Å) coumarin blue dye laser ($\lambda_p \sim 4400$ Å). **(b)** Time characteristics of different spectral regions of the continuum, showing sequential Stokes conversion and pump depletion

Stokes and the conversion into higher-order Stokes in the longer wavelength region. The duration of the entire visible continuum is about 10 ns, and is therefore also ideal for nanosecond time-resolved spectroscopic applications.

7.3.2 Wide Band Near-IR Generation in Optical Fibers

From Fig. 7.2 it is obvious that for silica glass fibers, the near-ir spectral region has lower losses, and is thus ideal for long-length Raman interaction. Using a Q-switched Nd:YAG laser at $\lambda_p = 1.06\ \mu m$ as the pump source, Stokes conversion up to 2.1 μm has been demonstrated [7.9] in a low-loss multimode GeO_2-doped silica core fiber. Figure 7.8 shows the spectrum of the multi-Stokes, near-ir continuum.

In single-mode fibers the guiding loss is very large at wavelengths longer than 1.75 μm if the fiber is designed to have a cutoff wavelength $\lambda_c \leqslant 1.06\ \mu m$. Therefore, the spectral extent of the multiple-Stokes output generated in a single-mode fiber is typically limited to $\lambda \leqslant 1.75\ \mu m$ [7.36]. Of course for longer λ_p, for example $\lambda_p = 1.32\ \mu m$, one can design the single-mode fiber with a cut-off wavelength $\lambda_c \sim \lambda_p$ and generate Stokes to longer wavelengths.

7.3.3 Ultraviolet Stimulated Raman Scattering in UV Silica Fibers

In the uv spectral region, doped silica glass fibers usually have very high losses ($>500\ dB/km$). In specially selected pure silica core fibers, however, the loss was as low as $100 - 200\ dB/km$ [7.37]. Since the Raman gain is higher at

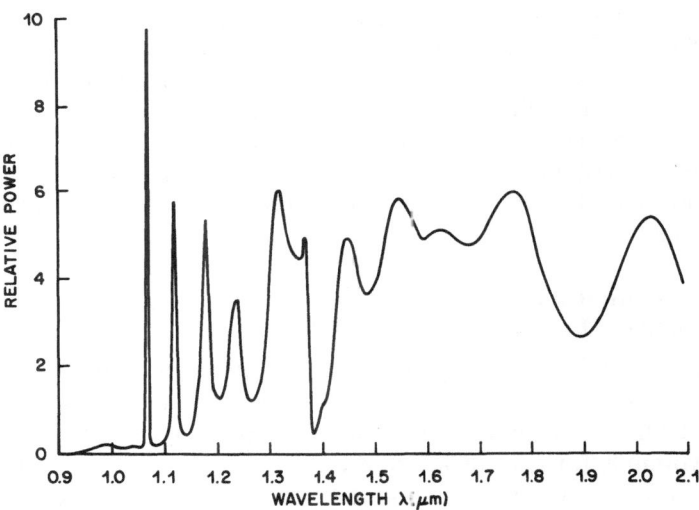

Fig. 7.8. A near ir wideband continuum generated in a multimode silica fiber pumped with a Q-switched Nd:YAG laser at 1.06 μm. The dip near 1.39 μm is due to OH⁻ absorption in the fiber. The spectral scan beyond 2.1 μm is limited by the spectrometer

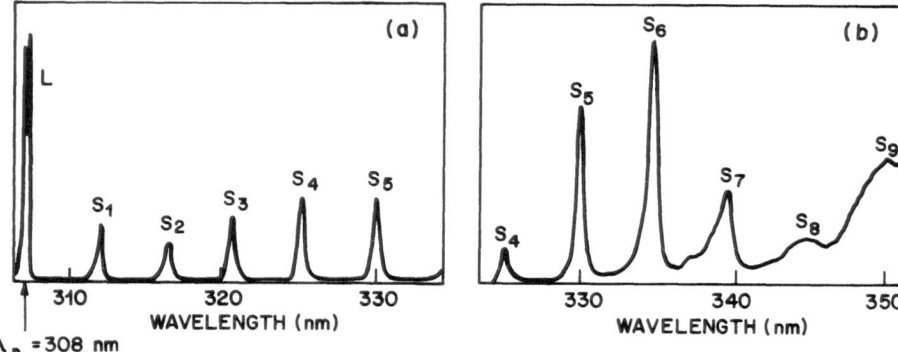

Fig. 7.9a, b. UV stimulated Raman scattering in a 180 m long, 200 μm core uv silica fiber (loss: 150 dB/km at λ_p = 308 nm). The XeCl excimer laser pump is at 308 nm. Spectra of Stokes light at (a) pump power = 550 kw, and (b) pump power = 650 kW are shown. S_i refers to ith Stokes order. The pump laser L works at 2 lines: 3079 and 3081 Å

shorter wavelengths ($g \propto 1/\lambda_p$), and high-power uv lasers (such as TEA excimer lasers) are readily available, it has been possible to demonstrate efficient uv stimulated Raman scattering in optical fibers [7.37–39]. Efficient stimulated Raman scattering in a 200 μm-core uv silica fiber has been obtained using the following excimer lasers as the pump: XeF (λ_p = 351 nm), XeCl (λ_p = 308 nm), and N_2 (λ_p = 337 nm). Figures 7.9a, b show the wideband uv spectra of the pump and many orders of Raman Stokes generated in 180 m of the 200 μm core fiber, using 550 and 650 kW, 7.5 ns pulses from the XeCl laser [7.38]. At the higher power, efficient stimulated Raman scattering in the uv fiber led to nine orders of Stokes generation, covering the spectral range of 308–350 nm. The Raman gain is $\sim 3 \times 10^{-11}$ cm/W in this wavelength range, while the fiber loss at λ_p = 308 nm was ~ 150 dB/km. Damage to the fiber end will eventually limit the amount of uv pump power which can be coupled in. In this experiment, the fiber end was submerged in water to raise the end damage threshold to 16 J/cm^2. In experiments of this kind, large-core multimode fibers produce the maximum overall Stokes power when high pump power is available.

7.3.4 Group Velocity Matching
in Picosecond Stimulated Raman Scattering in Fibers

As discussed in Sect. 7.2.6, group velocity dispersion is an important consideration in picosecond Raman interaction in long fibers. The technique of group velocity matching using modal dispersion to compensate for chromatic dispersion in a two-mode fiber, as depicted in Fig. 7.4, was first observed in an oscillator pumped by a picosecond mode-locked argon-ion laser [7.30]. A similar group velocity matching effect has been observed in single-pass stimulated Raman scattering in a 35 km long silica fiber pumped with a mode-locked Nd: YAG laser at 1.06 μm [7.40].

7.4 Tunable Fiber Raman Oscillators

7.4.1 Fiber Resonator for Raman Oscillation

When a resonator provides feedback of the Stokes component, oscillation threshold is reached when the gain equals the overall resonator loss. In a resonator with low round-trip loss (for example, 3 – 4 dB including fiber coupling loss), only modest (in our example, little more than 2 times) round-trip gain is required. The necessary pump power is then typically in the 1 W range, several orders of magnitude lower than the power required for single-pass Raman lasers. Thus, the pump power can often be supplied by a cw laser.

Figure 7.10 shows the schematic of a fiber Raman oscillator [7.41]. The input mirror is highly dichroic, with high transmission at the pump wavelength, and high reflectivity at the Stokes wavelength. The output mirror typically has a transmission of several tens percent or more at the Stokes wavelength. Without a frequency tuning element in the fiber resonator, the Stokes oscillation occurs at the peak of Raman gain spectrum.

7.4.2 Tunability in a Fiber Resonator with Frequency Selective Feedback

The broadband Raman spectra of glass fibers, as shown in Figs. 7.1,2, provide the possibility of frequency tuning in an oscillator with frequency selective elements such as prisms and etalons. For silica glass fibers the tuning range is ~ 500 cm^{-1} for one Stokes shift. If the pump-to-Stokes conversion is very efficient so that the Stokes depletes, for example, 70% – 80% of the input pump power, the one can achieve lasing at the second-order Stokes wavelength to obtain further frequency conversion. This cascade process

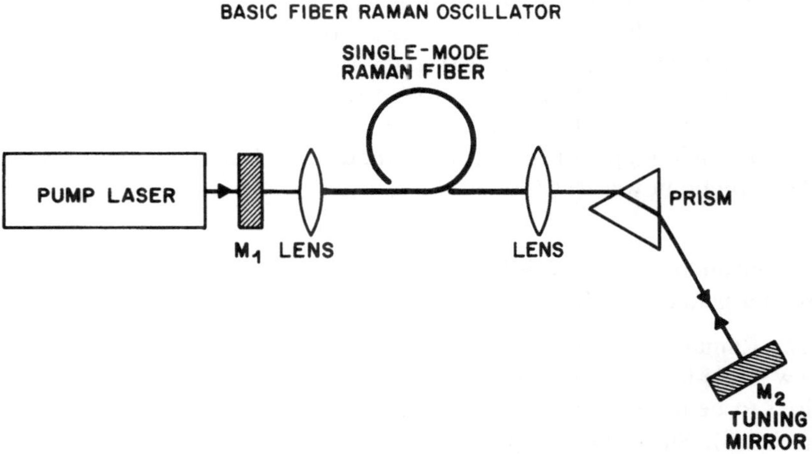

Fig. 7.10. Schematic of a fiber Raman oscillator. Mirrors M_1 and M_2 provide feedback at the Stokes wavelength

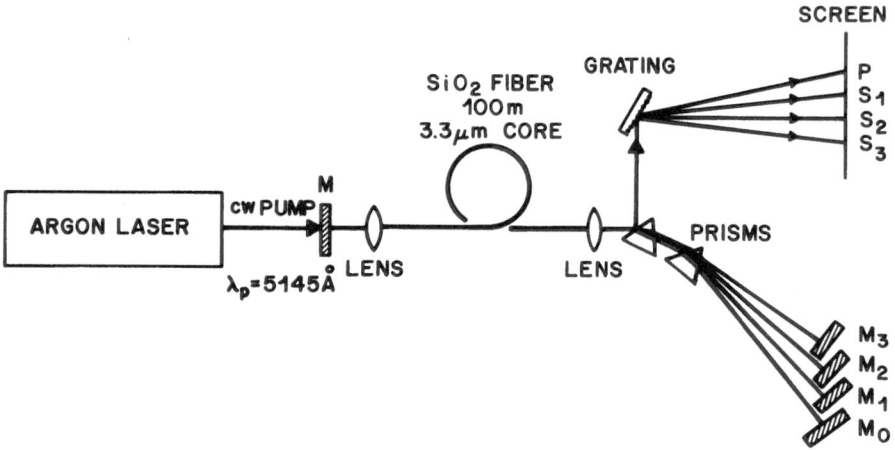

Fig. 7.11. A cw fiber Raman oscillator with multiple Stokes resonators, and pumped by a cw Argon laser at $\lambda_p = 0.53$ µm. Stokes tuning was obtained in the 5220 – 5660 Å range

often can be continued to many orders, but eventually it will be limited by the available pump power, fiber loss, and resonator loss for the higher-order Raman Stokes wavelengths.

Figure 7.11 shows a fiber Raman oscillator with multiple Stokes resonators pumped by a cw argon-ion laser at $\lambda_p = 514.5$ nm. The 100 m long single-mode fiber had a 3.3 µm core diameter and a loss of 17 dB/km at λ_p. With a coupled pump power of ~8 W in the fiber, four orders of Stokes oscillation have been obtained [7.42], allowing for tuning over the 520 – 566 nm spectral range. Output power versus tuning of the first Stokes oscillation was found to resemble the characteristic Raman spectrum of the silica glass fiber, and the laser linewidth was ~7 cm^{-1} [7.43].

Fiber Raman oscillation in the 1 µm region of near infrared has been obtained with a cw Nd:YAG laser at 1.064 µm as the pump [7.44]. The threshold for first Stokes oscillation was ~1 W of pump power in the fiber. In a similar experiment, over 70 nm of tuning for two orders of Stokes was observed in a 650 m long, prism-tuned, single-mode fiber Raman oscillator when the cw laser pump power was 5 W.

7.4.3 Synchronously-Pumped Fiber Raman Oscillators and Time-Dispersion Tuning

In a fiber Raman oscillator, synchronously pumped by picosecond pulses from a cw mode-locked laser, the cavity round-trip time for the Stokes pulses is required to be an exact integral multiple of the period between the pump pulses (see 1.8). Since the pump period is usually fixed, provision must be made for adjustment of the overall cavity length. Because of group velocity dispersion, the synchronous-pumping condition also allows for Stokes wavelength tuning by such adjustment of the resonator length. This scheme is called "time-dispersion tuning" [7.45] to distinguish it from the normal spatial dispersion tuning (as, for example, with prisms), and it relies on tuning the wavelength-dependent time delay difference. For two Stokes pulses at λ_1 and $\lambda_1 + \Delta\lambda$ in a single pass through the fiber, the time delay difference is

$$\Delta\tau = L D(\lambda) \Delta\lambda \ , \tag{7.7}$$

where $D(\lambda)$ is the group velocity dispersion in ps/nm-km, $\lambda = \lambda_1 + \Delta\lambda/2$, and L is the fiber length. By adjusting the resonator length by $\Delta l = C\Delta\tau$, one can tune the synchronized Stokes wavelength by $\Delta\lambda$ with a tuning rate of

$$\frac{\Delta\lambda}{\Delta l} = \frac{\Delta\lambda}{C\Delta\tau} = \frac{1}{C L D(\lambda)} \ . \tag{7.8}$$

In a time-dispersion tuned fiber Raman oscillator pumped by a cw, mode-locked argon laser at $\lambda = 0.53$ μm, the combination of time-dispersion and prism tuning has made possible first Stokes tuning over almost the entire Raman gain spectral range of ~ 500 cm^{-1} (from 5153 – 5287 Å) [7.45].

Figure 7.12 shows a near-ir tunable, multiple-Stokes fiber Raman oscillator pumped with a cw mode-locked Nd:YAG laser [7.45]. This multiple-Stokes resonator is similar to that of Fig. 7.11, except that here the position of each Stokes mirror is separately adjustable for time-dispersion tuning. Figure 7.13 shows the corresponding Stokes tuning curve, covering 1.07 – 1.32 μm. The change in mirror position can easily be translated into the equivalent delay difference. The chromatic dispersion $D(\lambda)$ of the fiber was determined [7.4] from the slope of this curve and (7.8).

Stokes tuning in the 1.32 – 1.41 μm spectral region has also been obtained in a 1 km long single-mode fiber with particularly low OH$^-$ content (1 dB/km near 1.39 μm OH$^-$ absorption peak). Figure 7.14 shows the loss spectrum of the low OH$^-$ single-mode fiber. The oscillator was pumped with cw mode-locked ~ 130 ps Nd:YAG laser pulses (100 MHz repetition rate and average power ~ 700 mW) at $\lambda_p = 1.32$ μm [7.47]. Tunable short Stokes pulses were obtained near 1.41 μm. Figure 7.15 shows the pump pulse and the Stokes pulses (~ 150 ps) at the output. Tunability into the 1.5 – 1.7 μm region should also be achievable with appropriate Stokes resonators.

Synchronous pumping of a 100 m long D_2 gas-in-glass fiber Raman oscillator has also been achieved by pumping with a mode-locked Nd:YAG laser at 1.06 μm. First Stokes pulses were generated at 1.56 μm [7.48].

In the synchronously pumped Raman laser action described above, picosecond Stokes pulses were obtained, but the linewidth was typically very large (a few tens of cm^{-1}) because of self-phase modulation [7.49]. [Editor's note: sub-picosecond pulses, tunable in the 1.5 μm wavelength region, and whose bandwidth was the Fourier transform of the pulse envelope, have recently been obtained in a synchronously pumped, fiber Raman soliton ring

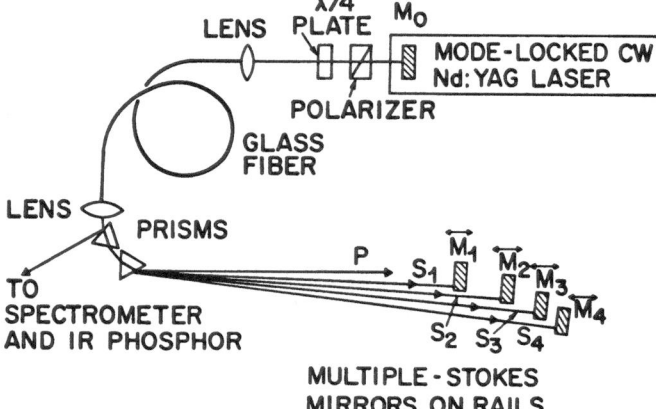

Fig. 7.12. A synchronously-pumped, near ir, fiber Raman oscillator with multiple-Stokes resonators. The pump is a train of 100 MHz, 130 ps pulses from a cw, mode-locked Nd:YAG laser operating at $\lambda_p = 1.06$ μm. For time synchronization each Stokes resonator length has to be adjusted individually

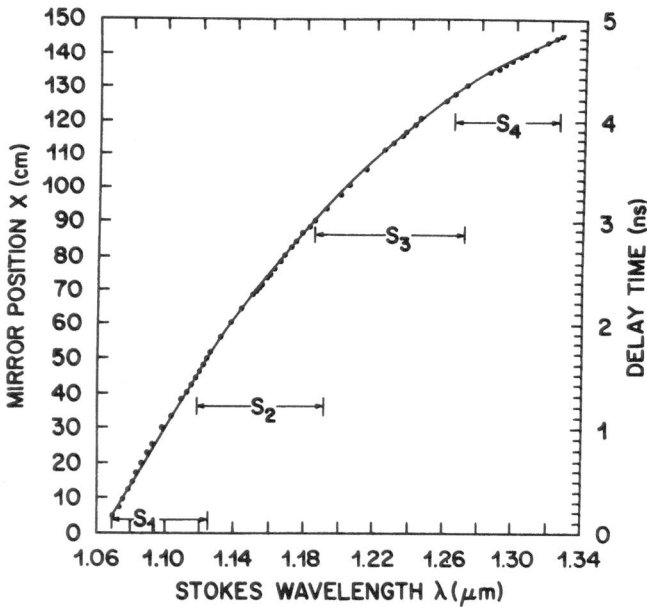

Fig. 7.13. The time-dispersion tuning curve for four orders of Stokes in the $1.07 - 1.32$ μm range obtained with the synchronously-pumped fiber Raman oscillator of Fig. 7.12

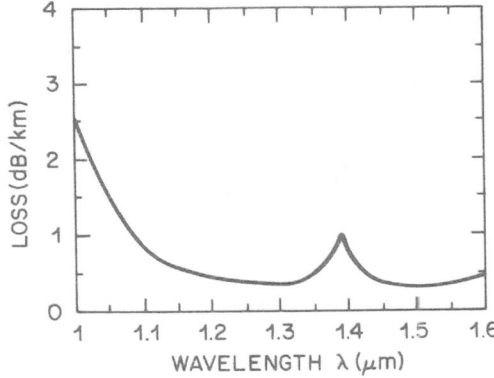

Fig. 7.14. Loss spectrum of a low OH⁻ single-mode fiber used in a fiber Raman oscillator pumped at 1.32 μm for tunable Stokes pulse generation up to 1.41 μm

Fig. 7.15. (a) Pump ($\lambda_p = 1.32$ μm), and (b) Stokes ($\lambda_s = 1.41$ μm) pulses at the output of a synchronously-pumped fiber Raman oscillator

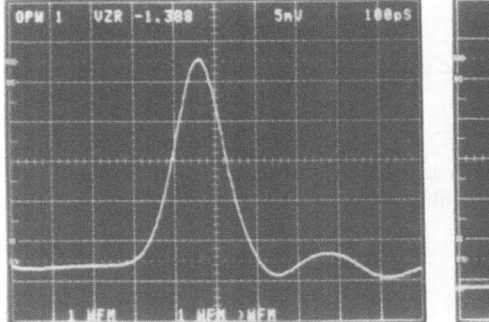
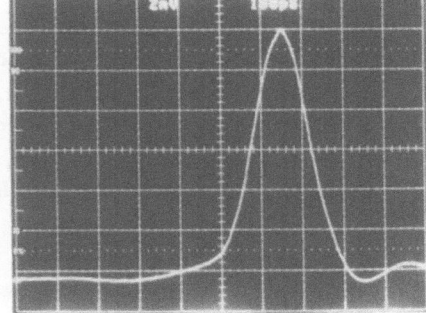

Fig. 7.15

laser [7.50]. As explained in Sect. 6.4.2 on the color center soliton laser, in the soliton regime ($\lambda > 1.3$ μm), self-phase modulation aids in pulse shaping, and does not necessarily add excess bandwidth.]

7.5 Applications of Fiber Raman Lasers

7.5.1 Dispersion and Bandwidth Studies in Multimode and Single-Mode Optical Fibers in the Near IR

Figure 7.16 shows a typical fiber dispersion measurement system based on a near-ir single-pass fiber Raman laser [7.36]. Subnanosecond (150 ps) optical pulses in the $1 - 1.7$ μm spectral range are obtained by pumping a single-mode Raman fiber (typically ~100 m long) with a Q-switched and mode-locked Nd:YAG laser at $\lambda_p = 1.06$ μm. Short pulses of various wavelengths are selected through a spectrometer and coupled into a test fiber. Both pulse delay

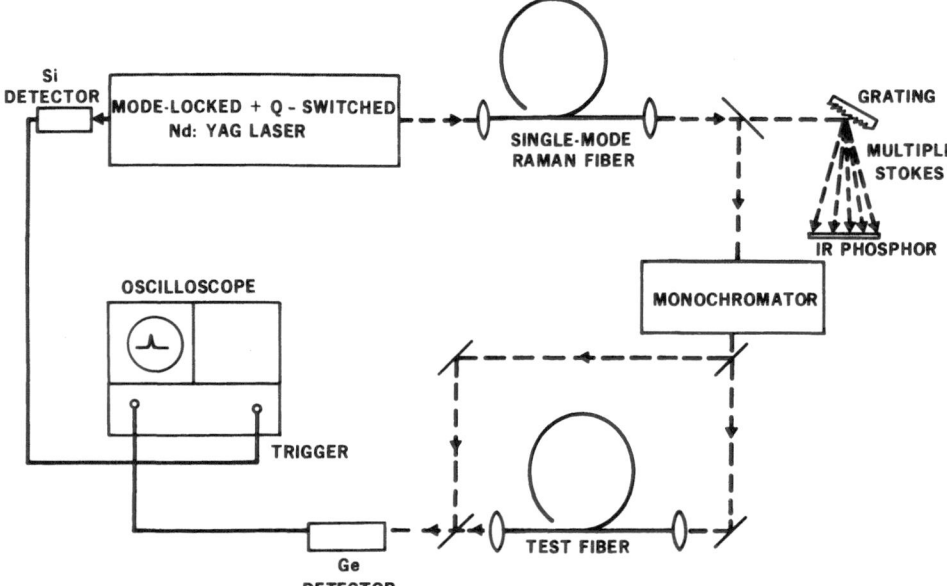

Fig. 7.16. Experimental setup for dispersion and bandwidth measurements in fibers in the $1-1.7\ \mu$m region using a near ir single-pass fiber Raman laser pumped by a Q-switched and mode-locked Nd: YAG laser

and pulse broadening in the test fiber and their wavelength dependencies are measured. This gives information about the fiber's dispersion characteristics and on its bandwidth spectrum [7.4 – 7,36].

The near-ir fiber Raman laser measurement setup of Fig. 7.16 has been widely adopted in the study of

a) chromatic dispersion properties of single-mode and multimode fibers as a function of dopant material, dopant concentration, and waveguide parameters [7.5, 51];
b) modal dispersion and bandwidth spectra of graded-index multimode fibers as a function of index profiles [7.6, 7]; and
c) the pulse propagation characteristics of two-mode fibers below and above mode cutoff [7.5, 7].

The test system is particularly useful for characterizing fibers of new design such as, for example, dispersion-shifted single-mode fibers [7.52 – 54] and broadband single-mode fibers [7.55], over the $1-1.7\ \mu$m spectral range. In fact it has become a standard facility for fiber dispersion measurement in many fiber-optics research laboratories worldwide. Figure 7.17 shows the results of typical dispersion measurement of a conventional single-mode fiber and of a dispersion-shifted single-mode fiber [7.4, 52]. The bandwidth

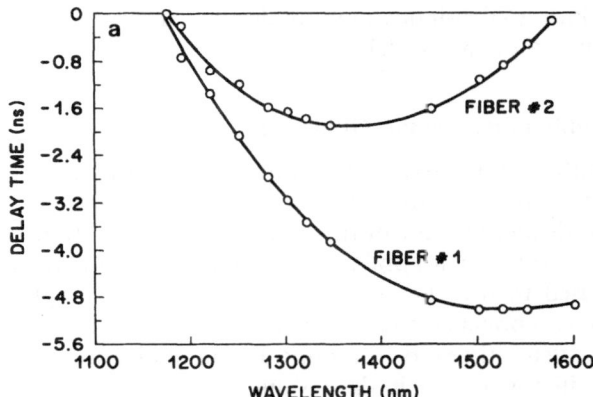

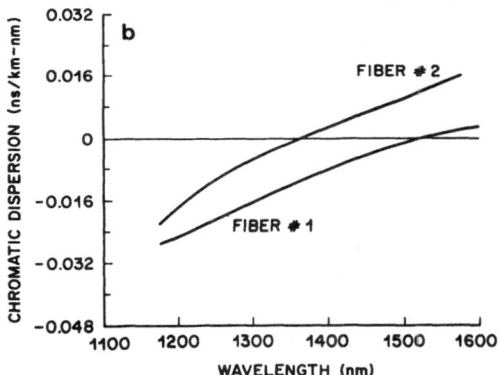

Fig. 7.17. Results of dispersion measurement using the setup of Fig. 7.16. Delay difference vs. wavelength was measured for a conventional single-mode fiber (#2) and a dispersion-shifted single-mode fiber (#1). Chromatic dispersion was obtained by taking the derivative of the delay curve with respect to wavelength

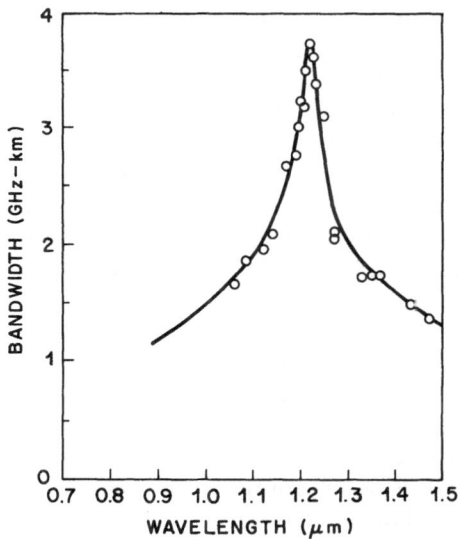

Fig. 7.18. Results of bandwidth spectrum measurement in a high-bandwidth graded-index multimode fiber using the fiber Raman laser measurement setup of Fig. 7.16

spectrum of a high-bandwidth graded-index multimode fiber measured with the fiber Raman laser is illustrated in Fig. 7.18 [7.7].

7.5.2 Applications of Tunable Fiber Raman Oscillators

A fiber Raman oscillator with continuously tunable wavelength output can be useful in a variety of spectroscopic measurements. Because tunable dye lasers are readily available in the visible, but not in the 1 μm region, fiber Raman oscillators are more attractive for near-ir applications. For example, the fiber Raman oscillator can be used to test the spectral response of wavelength-selective optical devices and components, such as filters, detectors, and wavelength-division multiplexers-demultiplexers for optical communication systems. A cw fiber Raman oscillator tunable in the 1.1 μm region has been used to measure the insertion loss and crosstalk of a GRIN-rod-grating wavelength-division multiplexer [7.56]. Presently, tunable fiber Raman oscillators are not widely used because of the experimental complexity involved for continuous tuning over wide spectral ranges. In future, it may be possible to simplify the fiber resonator design by utilizing micro-optic components being developed for optical fiber communication. It is likely that new applications will be found for such compact fiber Raman oscillators.

7.5.3 Optical Amplification and Pulse Shaping by Stimulated Raman Scattering

Figure 7.19 shows the basic concept of an optical fiber Raman amplifier [7.15]. One can have an optical amplifier with optical pulse shaping capability using stimulated Raman scattering in either the forward or the backward interaction. Figure 7.20 illustrates amplification, Stokes pulse steepening, and pump depletion in a backward fiber Raman amplifier in which the pump and Stokes pulses propagate in different directions in the fiber [7.57]. This experiment used two tunable dye laser pulses for providing the counter-propagating pump and Stokes signal pulses for signal amplification and pulse sharpening [7.57]. Similar optical amplification of high-speed injection laser signals (picosecond pulses) in the near infrared can be achieved with appropriate pump pulses and with appropriate consideration of group velocity matching between short pump and signal pulses. This may become more and more important as the technology of high-speed single-mode fiber transmission progresses.

Although there have been some demonstrations of linear optical Raman amplification of semiconductor laser pulses, where the amplifier was pumped with a Nd:YAG laser [7.23, 58, 59], the use of a high-power Nd:YAG laser is not compatible with the idea of a compact optical Raman amplifier for practical application in optical communications [7.15]. With future progress in high-power semiconductor lasers, and using appropriate fiber pigtail and coupling devices, a practical, compact semiconductor-laser-pumped optical

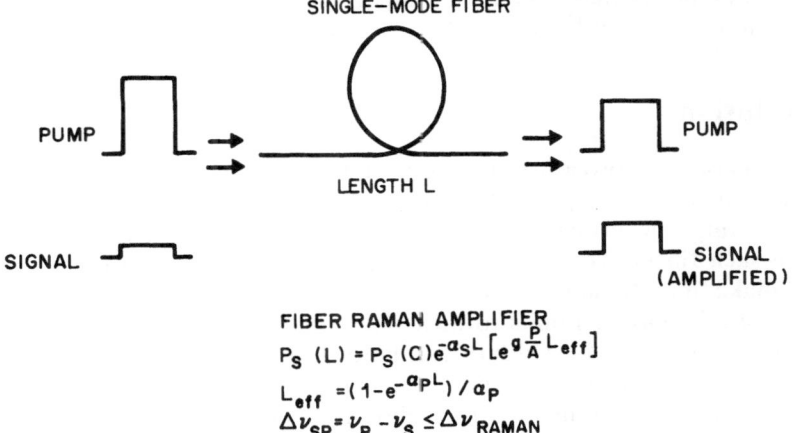

SINGLE-MODE FIBER

PUMP

LENGTH L

PUMP

SIGNAL

SIGNAL
(AMPLIFIED)

FIBER RAMAN AMPLIFIER

$$P_S (L) = P_S (C) \bar{e}^{\alpha_S L} \left[e^{g \frac{P}{A} L_{eff}} \right]$$

$$L_{eff} = (1 - e^{-\alpha_P L}) / \alpha_P$$

$$\Delta \nu_{SP} = \nu_P - \nu_S \leq \Delta \nu_{RAMAN}$$

Fig. 7.19. Schematic of an optical fiber Raman amplifier for linear amplification of optical signals at the Stokes wavelength

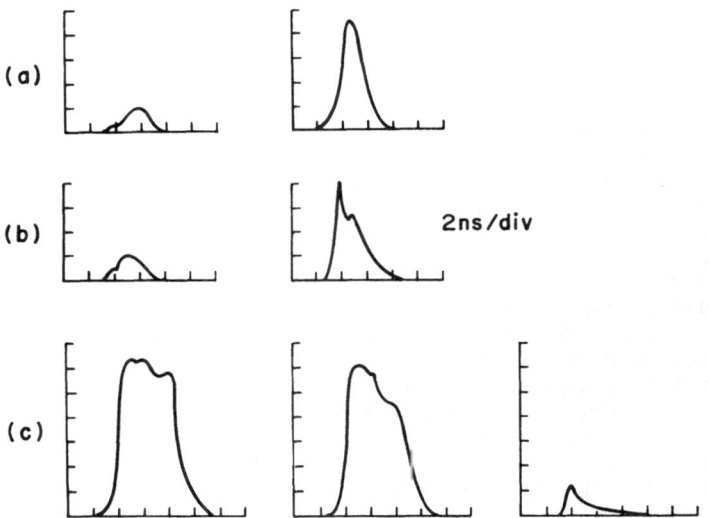

(a)

(b)   2ns/div

(c)

Fig. 7.20a – c. Optical amplification and Stokes pulse steepening in a backward fiber Raman amplifier. (a) small signal (10 mV/div, *left*) and amplified signal (100 mV/div, *right*); (b) large signal (*left*) and amplified, steepened signal (*right*), both 500 mV/div; (c) pump (*left*), pump with weak depletion (center) for the case (a), and pump after strong depletion for the case (b)

fiber Raman amplifier may become an important part of future high-speed optical communication technology.

7.6 Conclusion

Fiber Raman lasers are broadly tunable optical frequency converters based on stimulated Raman scattering and Raman oscillation in low-loss glass fibers. Their practicality and efficiency are based on the long interaction lengths possible in such fibers. Their relative simplicity and wide spectral region of operation make fiber Raman lasers attractive as widely tunable or wavelength-selectable optical sources. Limits to their operation [7.16] include fiber end damage, relatively broad linewidth of oscillation, and possible complications arising from various competing nonlinear processes (such as stimulated Brillouin scattering and stimulated four-photon mixing) [7.2]. Nevertheless, various forms of fiber Raman lasers have been obtained over a wide spectral range from the uv to the near ir (or beyond if liquid-core fibers [7.60] are included), using a variety of pump lasers with relatively low powers. Fiber Raman lasers have already found important and varied uses.

With future progress in low-loss fibers, fiber coupling and micro-optic wavelength-selective devices, high-power semiconductor lasers, etc., fiber Raman lasers and related devices will find more and more applications in opto-electronics technology.

References

7.1   R. G. Smith: Appl. Optics **11**, 2489 (1972)
7.2   R. H. Stolen: Proc. IEEE **68**, 1232 (1980)
7.3   L. F. Mollenauer, R. H. Stolen, J. P. Gordon: Phy. Rev. Lett. **45**, 1095 (1980)
7.4   L. G. Cohen, P. Kaiser, Chinlon Lin: Proc. IEEE **68**, 1203 (1980)
7.5   Chinlon Lin, L. G. Cohen, W. G. French, V. A. Foertmeyer: Electron. Lett. **14**, 170 (1978)
7.6   W. F. Love: Proc. 6th European Conf. on Optical Commun., York, U.K. (1980) p. 113
7.7   M. Horiguchi, Y. Ohmori, H. Takata: Appl. Optics **19**, 3159 (1980)
7.8   R. H. Stolen: IEEE J. QE-11, 100 (1975)
7.9   Chinlon Lin, V. T. Nguyen, W. G. French: Electron. Lett. **14**, 822 (1978)
7.10 K. O. Hill, D. C. Johnson, B. S. Kawasaki: Appl. Optics **20**, 1075 (1981)
7.11 Chinlon Lin, M. A. Bösch: Appl. Phys. Lett. **38**, 479 (1981)
7.12 E. M. Dianov, E. A. Zakhidov, A. Ya. Kayasik, P. V. Mamyshev, A. M. Prokhorou: JETP Lett. (English translation) **34**, 38 (1981)
7.13 R. H. Stolen, M. A. Bösch, Chinlon Lin: Optics Lett. **6**, 213 (1981)
7.14 Chinlon Lin, W. A. Reed, H. T. Shang, A. D. Pearson, P. F. Glodis: Electron Lett. **18**, 87 (1982)
7.15 Chinlon Lin: J. Opt. Commun. **4**, 1 (1983)
7.16 R. H. Stolen, Chinlon Lin: "Fiber Raman Lasers", in *CRC Handbook on Laser Technology,* ed. M. J. Weber (CRC Press, Boca Raton, 1983) Vol. 1, pp. 265 – 273
7.17 Chinlon Lin: "Nonlinear Optics in Fibers and Single Mode Optical Fiber Communications", in *Lightwave Technology,* ed. by W. T. Tsang, (Academic, New York to be published)
7.18 D. Cotter: J. Opt. Commun. **4**, 10 (1983)
7.19 F. L. Galeener, J. C. Mikkelsen, Jr., R. H. Geils, W. J. Mosby: Appl. Phys. Lett. **32**, 34 (1978)

7.20 V. I. Smirnov, Y. K. Chamorouski: Opt. & Quantum Electron. **9**, 351 (1977)
7.21 Chinlon Lin: unpublished
7.22 J. Stone, A. R. Chraplyvy, C. A. Burrus: Opt. Lett. **7**, 297 (1982)
7.23 A. R. Chraplyvy, J. Stone, C. A. Burrus: Opt. Lett. **7**, 415 (1983)
7.24 Chinlon Lin, L. G. Cohen, R. H. Stolen, G. W. Tasker, W. G. French: Optics Commun. **20**, 426 (1977)
7.25 H. Takahashi, I. Sugimoto, T. Sato, S. Yoshida: SPIE Proc. **320**, 320-16 (1982)
7.26 H. Takahashi, I. Sugimoto, M. Kimura, T. Sato: IOOC '83, Paper 30A1-5, Tokyo (1983)
7.27 Fiber literature form Fujikura Cable Co.
7.28 J. AuYeung, A. Yariv: IEEE J. QE-**14**, 347 (1978)
7.29 D. Marcuse, Chinlon Lin: IEEE J. QE-**17**, 869 (1981)
7.30 Chinlon Lin, R. H. Stolen, R. K. Jain: Optics Lett. **1**, 205 (1977)
7.31 E. J. Woodbury, W. K. Ng: Proc. IRE **50**, 2347 (1962)
7.32 E. J. Woodbury, G. M. Eckhardt: US Patent No. 3,371,265 (27 February 1968)
     G. Eckhardt, R. W. Hellwarth, F. J. McClung, S. E. Schwartz, D. Weiner, E. J. Woodbury: Phys. Rev. Lett. **9**, 455 (1962)
7.33 Y. R. Shen: In *Light Scattering in Solids*, 2nd. ed., ed. by M. Cardona, Topics Appl. Phys., Vol. 8 (Springer, Berlin, Heidelberg 1975) Chap. 7
7.34 R. H. Stolen, E. P. Ippen, A. R. Tynes: Appl. Phys. Lett. **20**, 62 (1972)
7.35 Chinlon Lin, R. H. Stolen: Appl. Phys. Lett. **28**, 216 (1976)
7.36 L. G. Cohen, Chinlon Lin: IEEE J. QE-**14**, 855 (1978)
7.37 R. Pini, M. Mazzoni, R. Salimbeni, M. Matera, Chinlon Lin: Appl. Phys. Lett. **43**, 6 (1983)
7.38 R. Pini, R. Salimbeni, M. Matera, Chinlon Lin: Appl. Phys. Lett. **43**, 517 (1983)
7.39 M. Rothschild, H. Abad: Opt. Lett. **8**, 653 (1983)
7.40 Y. Ohmori, Y. Sasaki, M. Kawachi, T. Edahiro: Electron. Lett. **17**, 595 (1981)
7.41 K. O. Hill, B. S. Kawasaki, D. C. Johnson: Appl. Phys. Lett. **29**, 181 (1976)
7.42 R. K. Jain, Chinlon Lin, R. H. Stolen, A. Ashkin: Appl. Phys. Lett. **31**, 89 (1977)
7.43 Chinlon Lin, R. K. Jain, R. H. Stolen: J. Opt. Soc. Am. **67**, 250A (1977); R. K. Jain, Chinlon Lin, R. H. Stolen, W. Pleibel, P. Kaiser: Appl. Phys. Lett. **30**, 162 (1977)
7.44 Chinlon Lin, R. H. Stolen, W. G. French, T. G. Melone: Opt. Lett. **1**, 96 (1977)
7.45 R. H. Stolen, Chinlon Lin, R. K. Jain: Appl. Phys. Lett. **30**, 340 (1977)
7.46 Chinlon Lin, W. G. French: Appl. Phys. Lett. **34**, 666 (1969)
7.47 Chinlon Lin, P. F. Glodis: Electron. Lett. **18**, 696 (1982)
7.48 A. R. Chraplyvy, J. Stone: to be published
7.49 R. H. Stolen, Chinlon Lin: Phys. Rev. **17**, 1448 (1978)
7.50 M. N. Islam, L. F. Mollenauer: Paper TUHH1, Digest of Tech. Papers, XIV, IQEC, 9 – 13 June, 1986 (Optical Society of America, 1986) p. 76
     M. N. Islam, L. F. Mollenauer, R. H. Stolen: "Fiber Raman Amplification Soliton Laser (FRASL) in *Ultrafast Phenomena V*, ed. by G. R. Fleming and A. E. Siegman, Springer Ser. Chem. Phys., Vol. 46 (Springer, Berlin, Heidelberg 1986) p. 46
7.51 K. I. White, B. P. Nelson, J. V. Wright, M. C. Brierly, A. Beaumont: Tech. Digest, Symposium on Optical Fiber Measurements (NBS Publ. No. 597, 89 (1980)
7.52 L. G. Cohen, Chinlon Lin, W. G. French: Electron. Lett. **15**, 334 (1979)
7.53 M. A. Saifi, S. J. Jang, L. G. Cohen, J. Stone: Opt. Lett. **7**, 43 (1982)
7.54 B. J. Ainslie, K. J. Beales, D. M. Cooper, C. R. Day, J. D. Rush: Electron. Lett. **18**, 842 (1982)
7.55 S. J. Jang, L. G. Cohen, W. L. Mammel, M. S. Saifi: Bell Syst. Tech. J. **61**, 385 (1982)
7.56 W. J. Tomlinson, Chinlon Lin: Electron. Lett. **14**, 345 (1978)
7.57 Chinlon Lin, R. H. Stolen: Appl. Phys. Lett. **29**, 428 (1976)
7.58 S. Kishida, Y. Aoki, H. Honmou, K. Washio, M. Sugimoto: IOOC '83, paper 2903-5, Tokyo (1983)
7.59 E. Desurvire, M. Papuchon, J. P. Pocholle, J. Raffy, D. B. Ostrowsky: Electron. Lett. **19**, 752 (1981)
7.60 J. Stone: Appl. Phys. Lett. **26**, 163 (1975)

8. Tunable High-Pressure Infrared Lasers

Tycho Jaeger and Gunnar Wang

With 15 Figures

The molecules CO_2, CO, CS_2, HF, DF, and N_2O collectively yield laser action on thousands of discrete lines in the wavelength range 2.5 to 20 µm. Tuning can be made continuous over segments of that range by using gas pressures of $3-10$ atm. Following review of the basic molecular processes and excitation schemes, experimental results are given for pulsed operation of high pressure CO_2, N_2O, and CS_2 lasers.

8.1 Background Material

During the 1960s several infrared lasers utilizing rotational-vibrational molecular transitions were demonstrated. Some of the most important laser-active molecules were CO_2, CO, HF, DF, and N_2O. A characteristic feature of these lasers is that they can oscillate at a great number of frequencies with frequency differences in the range of $20-60$ GHz. Among the numerous infrared lasers which were investigated, the CO_2 laser turned out to be one of the most efficient and scalable to high-output powers.

Several excitation schemes for molecular lasers, including excitation by electrical glow discharges, chemical reactions, optical pumping with laser radiation, and heating followed by gas dynamic expansion, have been used. The gas pressure was low, in the range from 0.001 to 1 atm, and the linewidth of the laser transitions was narrow compared to the spacing between the lines. It was possible to get laser radiation at thousands of lines in the range from 2.5 to 20 µm, using the numerous molecules, and including the different isotopes.

In 1971, it was suggested [8.1] that one could make a continuously tunable high-pressure CO_2 laser with a 600 GHz (20 cm^{-1}) broad-tuning range within the P- and R-branch of the laser bands. The basic idea was to take advantage of the pressure broadening of the laser transitions. If the laser could be operated at a high gas pressure ($3-10$ atm) depending on the gas mixture and the chosen isotopic species, the pressure-broadened lines would overlap and allow continuous tuning between the line centers. It was proposed to use a pulsed laser with a transverse electrical discharge (TE laser). The main problem was to obtain a uniform glow discharge without arcing at pressures necessary to obtain overlap between the lines. Continuous frequency tuning was demonstrated by several groups using uv preionized self-sustained discharges [8.2] and electron-beam controlled discharges [8.3,4]. Later, continuous frequency tuning was demonstrated with a radio frequency excited waveguide laser [8.5]. It has been demonstrated that the pressure requirement can be reduced by using a mixture of different CO_2 isotopes [8.6]. Frequency tuning has also been obtained with N_2O and CS_2 [8.7].

Direct optical excitation of high-pressure CO_2 and N_2O lasers has been investigated using the output from a HBr laser [8.8,9]. Experiments have also been made with a combination of optical excitation and collisional energy transfer to CO_2 and N_2O. The transfer molecules have been CO [8.10], DF [8.11], and CO_2 [8.9,12,13], which have been excited with radiation from a frequency-doubled CO_2 laser, a DF laser, or a HBr laser. There has also been an interest in utilizing the broad gain band width to generate short pulses, and subpicosecond pulses have recently been demonstrated [8.14].

The basic molecular processes which are important for the understanding of tunable high-pressure lasers, and which are common for the different types of lasers, will be reviewed in Sect. 8.2. Lasers with the $^{12}C\ ^{16}O_2$ isotope have been used in most experiments. Such a laser will be used as a typical example for the features which are general for the molecules investigated. Following a brief description of the different excitation schemes, a survey will be given of the frequency tuning ranges which have been obtained.

Experiments with tunable ir lasers will be reviewed in Sect. 8.3. A considerable amount of work has been done on high-pressure CO_2 lasers, but only the results which are relevant for continuous frequency tuning will be covered in this chapter. There have been few mode-locking experiments demonstrating picosecond pulses, but recent experiments indicate that high-pressure lasers can be developed into a powerful source for subpicosecond pulses (Sect. 8.4).

A great number of application experiments have been made with line-tunable ir lasers, and in some of these advantage has been taken of the limited tuning range which can be achieved at low pressures. For many of these applications a continuously tunable laser would be of great importance, but few experiments have so far been made. In Sect. 8.5 we will therefore include a discussion on potential application areas of tunable high-pressure ir lasers.

A great deal of development work has still to be done before this type of tunable laser can become a practical tool. In Sect. 8.6 the problem areas, and the need for further investigations and technical development will be discussed.

8.2 Amplification in High-Pressure Gases

In this section we will review in some detail the spectroscopic and dynamic properties of the laser-active molecules. We will emphasize the features which are important for the understanding of high-pressure gas lasers and which are common for the different excitation schemes. We will outline the calculation of the gain for a given gas mixture and for a given degree of excitation.

The energy transfer from the excitation source to the laser-active molecules will be briefly reviewed, in order to get a background for the more detailed description of the tunable lasers.

Most of the experimental and theoretical investigations have been based on $^{12}C\ ^{16}O_2$, which is the most abundant isotope of CO_2. We will therefore use this molecule as an example, but the general characteristics including the normal modes will also be valid for the other molecules which have been used, viz., CS_2 and N_2O. A survey of these molecules, including different isotopes, will be given at the end of this section. Several review papers and books discuss the physics and technical aspects of gas lasers working at low and high pressures [8.15 – 23].

8.2.1 Gain and Molecular Kinetics in Excited High-Pressure Gases

The infrared spectrum of CO_2 corresponds to transitions between vibrational-rotational energy levels in the molecule [8.15, 16]. The normal modes of vibration of the CO_2 molecule are shown in Fig. 8.1a. The three modes which are characteristic for linear three-atomic molecules, are the symmetric stretch mode, the bending mode, and the asymmetric stretch mode. The modes are designated with the vibrational frequencies v_1, v_2, and v_3. To the first approximation each vibrational mode behaves like a harmonic oscillator. The energy levels for each mode are shown in the energy diagram (Fig. 8.1b). The frequency differences between the levels within each "ladder" decreases with increasing quantum number because of the anharmonic potential. The vibrational states are designated by $(v_1 v_2\ ^l v_3)$ where v_1, v_2, and v_3 are the quantum

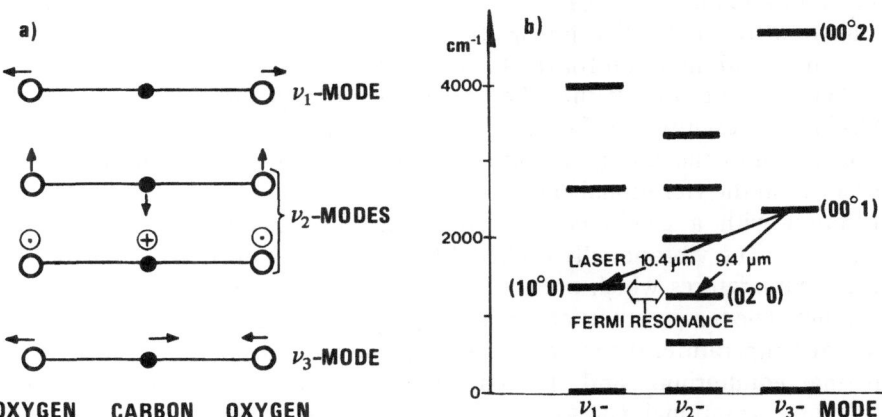

Fig. 8.1. The normal modes of vibration of the CO_2 molecule (a) and the corresponding energy diagram for each mode (b)

numbers of the v_1, v_2, and v_3 vibrational modes. In the v_2-mode the molecule can oscillate in two orthogonal directions. The rotational quantum number l can have the values (v_2, $v_2 - 2, \ldots, - v_2$).

The actual molecular vibrational states for CO_2 are somewhat more complicated than described above. Because of nonlinear coupling, mixing occurs among the set of states designated by the values of v_1 and v_2, where ($2v_1 + v_2$) is constant and the symmetry character is the same. This effect is called Fermi resonance. The mixing of states is important for the gain of the laser transitions, and gives significant differences in gain characteristics for the different CO_2 isotopes [8.24, 25].

The states (10^00) and (02^00) mix to form two states of the form $A(10^00) - B(02^00)$ and $B(10^00) + A(02^00)$, which are the lower levels for the laser transitions originating from the (00^01) level. The two states are designated by $[(10^00), (02^00)]_I$ and $[(10^00), (02^00)]_{II}$. We will, however, as shown in Fig. 8.1 b, designate the lower laser levels with (10^00) and (02^00) instead of the correct notation. This simplified presentation is frequently used in the literature. The degree of mixing determines the relative strength of the two vibrational transitions, and is given by the energy difference between the two states (10^00) and (02^00). The amplitudes A and B are equal at zero energy difference, but A increases and B decreases with increasing energy difference.

The N_2O and CS_2 molecules have energy level diagrams similar to CO_2, but laser action has only been observed on the vibrational transitions from the (00^01) level to the (10^00) level [8.19].

Due to the strong line broadening at high gas pressure, we also have to take into account transitions from the higher v_3 vibrational levels [8.26 – 28] shown in Fig. 8.2. The vibrational-rotational bands associated with each of these transitions are called sequence bands and hot bands. The vibrational-rotational transitions in the 10.4 μm band are indicated in the energy diagram in Fig. 8.3. The rotational energy levels with quantum number J are added to the vibrational levels in the diagram. The levels are missing for every second value of J for the $^{12}C^{16}O_2$ isotope [8.15]. The band is divided into the P- and R-branch, with notation for the laser lines as indicated in the figure.

Gain on the lines within the regular bands and the sequence bands is obtained by exciting the CO_2 molecules to the higher levels in the v_3-mode. Calculation of the gain is simplified by the fact that the population on each level within the vibrational and rotational modes is given by a Boltzmann distribution with a characteristic temperature for each degree of freedom [8.23, 29]. The vibrationally excited CO_2 molecules can be characterized by three temperatures which are the variables determining the gain at a given pressure. The temperatures are the rotational temperature T_1 which is equal to the gas temperature, the temperature T_2 of the v_1- and v_2-modes which are in thermal equilibrium, and the temperature T_3 of the v_3-mode. Through collisions, rotational energy of the molecules is rapidly exchanged with translational energy ($R - T$ transfer) in the laser-gas mixture. The rotational temperature is therefore equal to the gas kinetic temperature. The relaxation

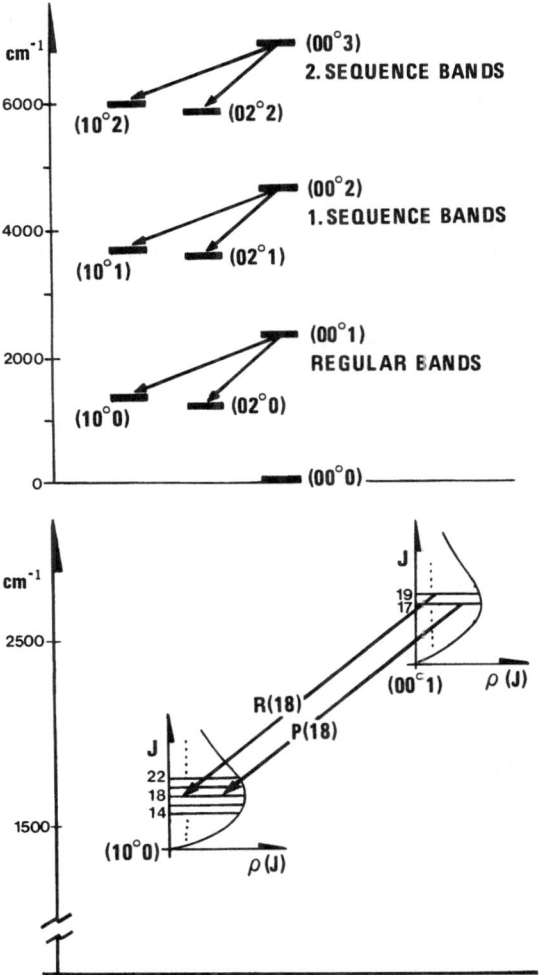

Fig. 8.2. Simplified vibrational level diagram of CO_2, showing regular bands and sequence bands

Fig. 8.3. Rotational sublevel manifolds of the (00^01) and (10^00) CO_2 vibrational levels. Two laser transitions from the P- and R-branch are shown. The relative population of the rotational levels, $\varrho(J)$, is indicated

time constant is of the order of 1 ns at 1 atm. The relaxation within each vibrational mode is rapid because of resonant exchange of vibrational energy quanta ($V - V$ transfer). The energy exchange between the ν_1- and ν_2-mode is rapid, because of the Fermi resonance.

The gain is continuous within the P- and R-branches of a high-pressure gas laser. This results from the overlap of all the pressure-broadened transitions in both the regular band and the sequence bands. We consider the gain in a gas mixture when the pressure is increased, but with constant vibrational temperatures. The gain at line center for each individual line is then independent of the pressure, but the linewidth will increase proportionally with the pressure. At sufficiently high pressure we get substantial overlap between the lines in the regular band, and a transition to a continuous gain is achieved. The gain will now increase with pressure due to the increasing number of amplifying molecules per unit volume.

The frequencies of the lines in the sequence bands are located arbitrarily between the lines in the regular band. At low pressure they therefore have negligible influence on the gain at the regular lines, except for some coincidences [8.30, 31]. In the high-pressure region the sequence bands will, however, give a significant addition to the total gain. It is interesting to note that the average gain per molecule in the high-pressure laser, therefore, is approximately proportional to the energy in the v_3-mode.

In a CO_2 laser, the gain $\alpha_{ul}(v)$ at frequency v corresponding to a transition from a upper level u to a lower level l in the regular band is given by

$$\alpha_{ul}(v) = \psi_{CO_2} N \sigma_{ul} \left(n_u - \frac{g_u}{g_l} n_l \right) g(v - v_{ul}) . \tag{8.1}$$

The terms in (8.1) are defined as follows:

σ_{ul}: cross section for stimulated emission. Dependent on the rotational quantum number of the upper and lower level,
N: molecular number density [cm^{-3}],
ψ_{CO_2}: relative partial pressure of CO_2 in the gas mixture referred to the total pressure,
$g(v - v_{ul})$: Lorenzian lineshape,
n_u, n_l: relative number of molecules in upper and lower level,
g_u, g_l: degeneracy of upper and lower level.

The lineshape function due to pressure broadening is given by

$$g(v - v_{ul}) = \frac{1}{\pi \Delta v} \frac{\Delta v^2}{(v - v_{ul})^2 + \Delta v^2} . \tag{8.2}$$

The halfwidth Δv is given by

$$\Delta v = p \left(\psi_{CO_2} \cdot \gamma_{CO_2} + \sum_M \psi_M \gamma_M \right) \left(\frac{T_1}{T_0} \right)^\alpha$$

$$= p \gamma_0 \tag{8.3}$$

where

α: approximately 0.5,
γ_{CO_2}: self-broadening coefficient for CO_2,
γ_M: pressure-broadening coefficient of CO_2 due to collisions with molecule M,
ψ_M: relative partial pressure of molecule M,
γ_0: linewidth at 1 atm for the gas mixture,

p: total gas pressure,
T_0, T_1:   273 K and the gas temperature, respectively.

The relative number of molecules in the upper and lower level is given by

$$n_{\mathrm{u}} = Q_{\mathrm{v}}^{-1} \exp(- h v_3/k T_3) \varrho(J_{\mathrm{u}})$$

$$n_{\mathrm{l}} = Q_{\mathrm{v}}^{-1} \exp(- E_{\mathrm{l}}/k T_2) \varrho(J_{\mathrm{l}})$$

(8.4)

where

E_{l}:   energy of the lower laser level,
Q_{v}:   partition function for the vibrational modes, which is dependent
            on T_2 and T_3, and

$\varrho(J_{\mathrm{u}})$ and $\varrho(J_{\mathrm{l}})$ are given by

$$\varrho(J_i) = \left(\frac{2 h c B_i}{k T_1}\right) g_i \exp\left[- B_i J_i(J_i+1) \frac{h c}{k T_1}\right] , \quad i = \mathrm{u}, \mathrm{l}$$

(8.5)

where B_i is the rotational constant.

Before considering the resulting gain added from all lines, we will discuss the gain due to a single line when the pressure p is increased while having a constant excitation (constant T_3). Combining (8.1) and (8.2) we get

$$\alpha_{\mathrm{ul}}(v) = \psi_{\mathrm{CO_2}} \frac{N}{p} \sigma_{\mathrm{ul}} \left(n_{\mathrm{u}} - \frac{g_{\mathrm{u}}}{g_{\mathrm{l}}} n_{\mathrm{l}}\right) \frac{(\gamma_0 p)^2}{\pi \gamma_0 [(v - v_{\mathrm{ul}})^2 + (\gamma_0 p)^2]} .$$

(8.6)

We see from (8.6) that the gain at the line center is constant and independent of the pressure, and from (8.3) that the linewidth increases proportionally with pressure. Both are dependent on the gas mixture, see (8.3).

The sequence bands give significant contributions to the gain in lasers with strongly excited molecules. The influence of these bands increases with increasing vibrational temperature T_3. In particular, optical excitation can result in high T_3 values. In this context it should be pointed out that the validity of the harmonic oscillator model for the vibrational modes is important for the interaction between the laser radiation and the molecules. The cross section for a transition between two vibrational levels with quantum number v and $v-1$ is proportional to v. The gain at a sequence band with upper vibrational level v_3 is therefore given by

$$\frac{\text{gain of sequence band}}{\text{gain of regular band}} = v_3 \exp[-(v_3-1) h v_3/k T_3] .$$

(8.7)

We have a corresponding relation for the hot bands. The frequencies of the sequence bands decrease with increasing vibrational quantum number

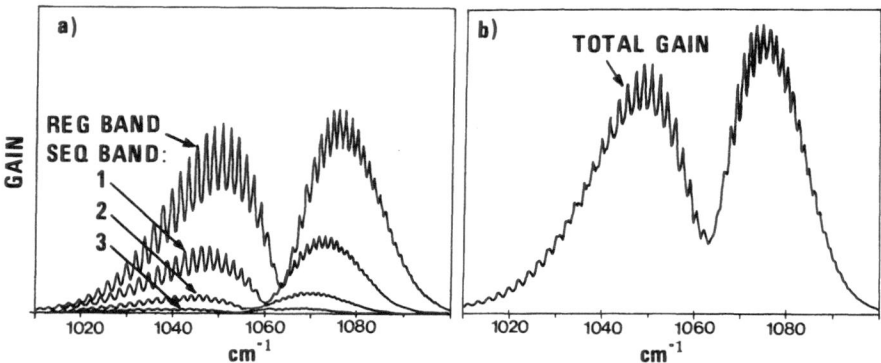

Fig. 8.4a, b. Calculated gain in a 10 atm CO_2 : N_2 : He/2 : 2 : 96 gas mixture with vibrational temperature $T_3 = 2000$ K at 9.4 µm. (a) Relative contributions to the gain from regular and sequence bands; (b) Sum of contributions shown in (a)

compared to the regular bands, but there is still overlap between them. The total gain resulting from all the bands is therefore approximately proportional to the average energy in the v_3-mode per molecule or the average number of vibrational quanta.

The gain in $^{12}C\,^{16}O_2$ at 10 atm is shown in Fig. 8.4a for the regular band and the three first sequence bands. The total gain is shown in Fig. 8.4b. Experimental results with high-pressure lasers (up to 20 atm) have given results in good agreement with the theory which has been outlined. We have assumed pressure-broadened lines with a Lorenzian lineshape. A more complete theory of pressure broadening predicts a non-Lorenzian lineshape with an increase of the gain at the line centers [8.26, 32 – 34], but this effect does not change the general picture given above.

All high-pressure molecular lasers have so far been operated in a pulsed, gain-switched mode, and the excitation pulse (10 – 500 ns) has, with a few exceptions, been short compared to the build-up time of the laser pulse and the decay time of the upper laser level. It is therefore convenient to present measurements and calculations of gain as a function of excitation energy density. In order to compare results at different pressures, the energy density is most frequently given in units of Joules per liter · atmosphere.

8.2.2 Excitation Techniques

We will now give an overview of the excitation techniques which have been investigated. The energy flow from the excitation sources to the CO_2 v_3-vibrational mode is illustrated in Fig. 8.5 for the different excitation techniques used so far.

The *direct optical excitation* technique requires a laser with frequency matching a vibrational-rotational transition. Because of the pressure broad-

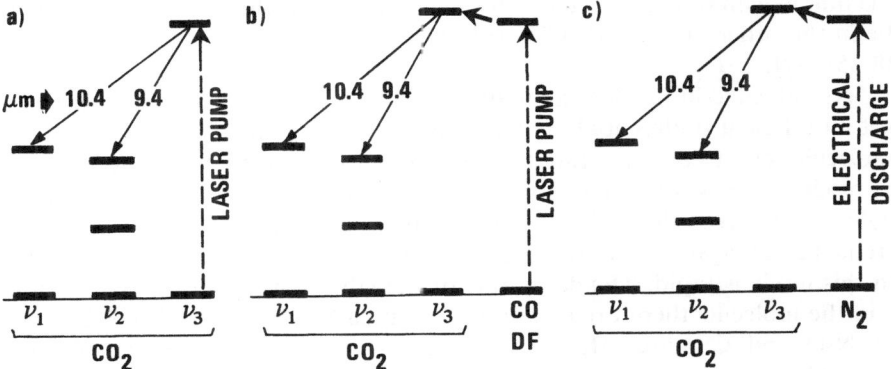

Fig. 8.5a–c. Schematic diagrams of the excitation processes, (a) direct optical pumping; (b) optical pumping using transfer molecule; (c) excitation by electrical discharge

ening, the tolerance on the laser frequency is not as critical as for optical excitation of low-pressure lasers. The vibrational mode can be strongly excited by optical pumping, corresponding to a high vibrational temperature T_3. A high density of laser molecules and a high vibrational temperature produces a high-gain coefficient. It is therefore possible to get high gain with a gas cell only a few millimeters long. This is advantageous in the design of frequency tunable lasers.

Excitation can also be achieved by *optical pumping of a transfer molecule*. The energy is then transferred by collisions from this molecule to the laser molecule (Fig. 8.5). In order to get a high efficiency, the transfer rate should be rapid compared to the de-excitation rate of the laser molecule. This can, however, be obtained without having accurate coincidence between the energy levels of the two molecules.

Compared to excitation by an electrical discharge, the optical excitation technique has several advantages. In addition to the higher gain, dissociation of the laser molecules is avoided. Such dissociation is a problem with glow discharge excitation. The optical excitation technique is especially useful for molecules which cannot be excited by electrical discharges, and when using expensive isotopes.

In the majority of experiments a *pulsed glow discharge* has been used. Self-sustained discharges – both uv preionized and rf discharges – and electron-beam controlled discharges have been used at high pressure. We will refer to the lasers as uv preionized TE lasers (transversely excited), rf excited lasers, and e-beam controlled TE lasers.

The excitation rates of the molecular vibrations are strongly dependent on the velocity distribution of the electrons, whose distribution is determined by the electric field per unit number density (E/N ratio) for the gas mixture [8.23]. Maximum vibrational temperature, which can be obtained for the ν_3-mode, is limited by de-excitations due to collisions with electrons. The maximum gain is determined by this temperature T_3, which has been shown to be in the order of $2000 - 2500$ K for pulsed and cw self-sustained discharges [8.35 – 37].

As indicated in Fig. 8.5, a mixture of CO_2, N_2, and He is used in order to get an efficient preferential excitation of the vibrational ν_3-mode of the CO_2 molecule, and to obtain a rapid de-excitation of the strongly coupled ν_1- and ν_2-modes. The vibrational modes of CO_2 and N_2 are excited by inelastic collisions between the molecules and electrons in the plasma. The vibrational frequency of N_2 is close to the ν_3-mode of CO_2, and therefore energy transfer to this mode is rapid. The decay rate of the vibrational levels of N_2 are slow and the molecule therefore also serves as an energy reservoir. The molecules of N_2O and CS_2 are efficiently excited by using CO or N_2 as transfer molecules.

An important problem has been to achieve arc-free discharges at the high gas pressures. In TE lasers an arc-free glow discharge is obtained by a uniform preionization of the gas with uv radiation. Then a self-sustained glow discharge is generated by discharging a high-voltage capacitor.

In the rf excited laser the glow discharge is obtained in an alternating electrical field. High-pressure experiments have been made with 40 and 100 MHz excitation [8.5, 38]. A uniform glow discharge can be initiated without preionization.

The e-beam controlled laser uses a high-energy electron beam (about 250 kV) to ionize the gas and control the low-energy electron number density [8.23, 39]. An electrical field across the gas accelerates the electrons and provides the electrical excitation of the molecules. The main advantage is that one can choose the voltage giving the optimum value of the E/N-ratio for efficient excitation of the laser [8.23]. Since the voltage is lower than necessary for a self-sustained discharge, it is possible to attain longer arc-free discharge pulses at the high pressures required for continuous frequency tuning.

8.2.3 Survey of Molecules for Tunable Lasers

Continuous frequency tuning has so far been observed using CO_2, N_2O, and CS_2. Spectroscopic data showing the overall tuning range will be summarized in this section.

The most abundant and easily available isotopes have been used in the majority of experiments with high-pressure lasers. The reported tuning ranges are shown in Fig. 8.6. The overall tuning range of high-pressure lasers can be increased by using various isotopes [8.41].

The vibrational frequencies are determined by the atomic weights of the atoms in the molecules. Carbon and nitrogen have two isotopes each, oxygen and sulphur have three and four, respectively. Combinations like $^{12}C^{16}O^{18}O$ have also been used in high-pressure experiments.

The pressure threshold for continuous tuning is determined by the distance between line centers and the pressure-broadening coefficient. The pressure-

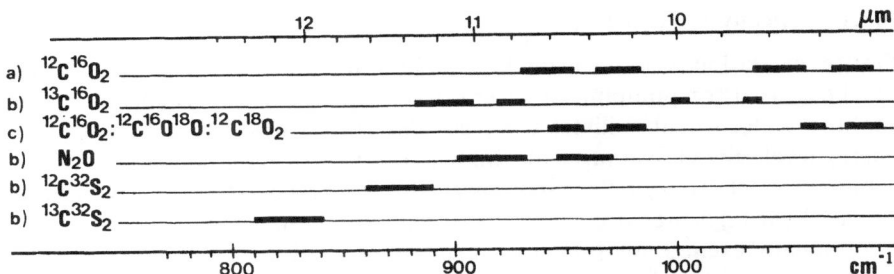

Fig. 8.6a–c. Frequency tuning ranges observed with different CO_2, N_2O, and CS_2^- isotopes. (a) Typical tuning range obtainable with e-beam controlled and uv preionized TE lasers. (b) Tuning ranges of the lasers described in [8.40], (c) Tuning ranges of the laser described in [8.6]

Table 8.1. Pressure broadening of the CO_2 laser transitions*

Collision broadening by	CO_2	N_2	He
Broadening coefficient γ [GHz/atm]	2.9	2.1	1.8

* Broadening coefficient dependent on J.
  Data given for the $P(20)$ and $R(20)$ lines [8.31].

broadening coefficients do not differ very much for the three molecules. The pressure-broadening coefficients for CO_2 is given in Table 8.1. The frequency difference when all the rotational levels are allowed to be occupied, are about 25, 25, and 10 GHz for CO_2, N_2O, and CS_2, respectively [8.19]. In $^{12}C^{16}O_2$ every second level is occupied, and the resulting frequency difference requires about 10 atm pressure. This is also the case for the other CO_2 molecules with $^{16}O_2$ or $^{18}O_2$. The threshold pressure can be reduced to 5 atm by using a molecule where all levels are occupied, for instance $^{12}C^{16}O^{18}O$, which has two different oxygen isotopes, or molecules with $^{17}O_2$ [8.15]. In N_2O, which has the structure $N-N-O$, all the levels are occupied [8.15], and the resulting pressure threshold for continuous tuning is about 5 atm. In CS_2 molecules with $^{32}S_2$ and $^{34}S_2$, every second level is occupied, while all levels are occupied for molecules with $^{33}S_2$ [8.15]. The most abundant sulphur isotope is ^{32}S.

8.3 Experimental Investigations of Tunable Lasers

This section is divided according to excitation technique. The various techniques are very different with respect to basic physical problems, apparatus used in the experiments, performance, and potential for practical applications.

8.3.1 Optically Excited Lasers

Optical excitation of lasers using laser radiation is a versatile technique which has been used for pumping of several laser types covering frequencies from the ultraviolet to submillimeter waves. The resultant selective excitation of energy levels gives advantages like high quantum efficiency and laser action at frequencies selectively determined by the excited levels. High-energy density, essential for high-pressure gas lasers, is easy to obtain.

Optically excited low-pressure gas lasers oscillating at a great number of frequencies in the infrared spectral range from $2.5-21$ μm have been demonstrated. The following gases have been used: CO_2, N_2O, CS_2, HF, NH_3, COF, OCS, C_2H_4, SiH_4, C_2H_4, CF_4, NOCl, and CF_3I [8.42]. Some of the molecules were excited using CO, DF, HF, HBr, and CO_2 as a transfer molecule. The excitation was performed with HF, DF, HBr, and CO_2 lasers, and frequency-doubled CO_2 lasers. Laser action at pressures adequate for continuous frequency tuning has been achieved with CO_2 and N_2O. A summary of the experimental results is given in Table 8.2.

Most experimental investigations of high-pressure optically pumped lasers have been carried out using a longitudinal pumping geometry, as shown in Fig. 8.7. The laser gas is excited by radiation from the pump laser, which is transmitted into the gas cell through a dichroic mirror. In the directly optically pumped laser, the radiation is absorbed by the laser molecule. In the optically pumped transfer laser, the radiation is absorbed by a transfer molecule and the energy is subsequently transferred to the laser molecule by a $V-V$ transfer (Sect. 8.2.2). In order to get high efficiency, the transfer rate should be high compared to the de-excitation rates of the two molecules. Helium is well suited as a buffer gas and is therefore used to optimize the performance of optically pumped lasers. For example, helium pressure broadens CO_2 about

Table 8.2. Optically pumped high-pressure laser experiments

Pumping source	Laser molecule	Transfer molecule	Maximum gas pressure [atm]	Maximum output energy [mJ]	Peak putput power [kW]	Pulse length [ns]	Quantum efficiency [%]	Reference
HBr	CO_2	–	112*	–	–	–	–	[8.13]
laser	CO_2	–	33	–	1	2.2	<2.2	[8.8]
	N_2O	–	7.5	–	–	–		[8.9]
	N_2O	CO_2	42	–	0.6	7–15	<1	[8.9]
Frequency doubled CO_2 laser radiation	CO_2	CO	18*	0.25	–	50	9	[8.10]
DF laser	CO_2	DF	19*	6		100	20	[8.11]

* Denotes that He has been used as a buffer gas.

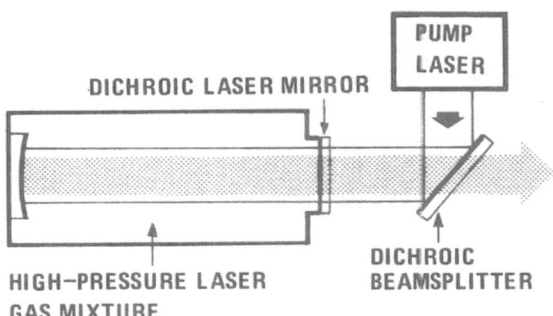

Fig. 8.7. Optically pumped laser with longitudinal pumping geometry

2/3 as much as the CO_2 self-broadening, but deactivates the CO_2 v_3-mode only 1/4 as rapidly. It can therefore be used to control the partial pressure of the active molecules and the pressure broadening independently.

In order to realize the inherent high quantum efficiency of optical pumping, several factors have to be taken into account in the design of the laser. Close overlap between the excited gas volume and the mode volume of the laser resonator is essential. When the laser is operated in a gain-switched mode, the build-up time of the laser pulse should be short compared to the time constant for de-excitation of the v_3-mode of the laser molecule.

The duration of the output pulse and the overall efficiency may be reduced by refractive index distortions in the laser gas due to heating during pumping and laser action. The laser transitions excite the v_1- and v_2-modes, and this energy is transferred to translational energy during the laser pulse. The thermal gradients generate density variations, and the time constant for these processes limits the duration of the laser pulse. The absorption length of the pump radiation can be increased, and the heating reduced, by using He as a buffer gas.

In direct optical pumping at high gas pressure, the molecules are excited to higher vibrational levels in the v_3 mode because of overlapping absorption bands. The absorption of pump radiation will therefore not saturate. This is due to the fact that the molecules behave like harmonic oscillators, and the cross section for absorption and stimulated emission between level $(00^0 v_3)$ and $(00^0 v_3 - 1)$ is therefore proportional to v_3.

Direct optical pumping of high-pressure lasers was first demonstrated in CO_2 and N_2O using a pulsed HBr laser oscillating on 18 wavelengths between 4 and 4.6 μm. The observed efficiencies were very low. The primary reason was probably that the pump lasers did not have optimum spatial and temporal behavior.

We shall briefly review three experiments. The common experimental setup is shown in Fig. 8.8 [8.8,9]. In the first experiment 0.5 μs long output pulses from a multiline, multimode HBr laser were focused into the gas cell filled with high-pressure, pure CO_2. Laser action was demonstrated at gas pressures ranging from 10 to 33 atm, with threshold pump energy varying

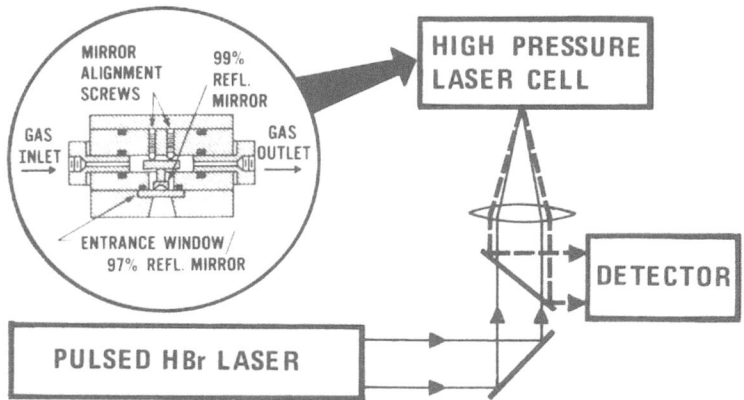

Fig. 8.8. Schematic diagram of an optically pumped high-pressure laser system. Gas cell constructed for short cavity lengths in the millimeter range. Adapted from [8.9,13]

from 5 to 30 mJ. The output pulse length was approximately 2 ns at 20 atm. Absorption measurements in a 1 mm cell showed that most of the radiation at the HBr lines is completely absorbed at 4 atm pressure. With the same experimental setup, laser action with pure N_2O was observed at gas pressures up to 7.5 atm with 20 mJ threshold input power. N_2O does not absorb the radiation from the HBr laser as strongly as CO_2.

In the third experiment, laser action at gas pressures up to 112 atm [8.13] was achieved using a CO_2: He mixture and a transverse pumping geometry. The experimental arrangement is nearly the same as in Fig. 8.8. By adding He, it is possible to get an absorption length which matches the diameter of the mode volume. In order to get lasing at increasing pressure it was necessary to increase the amount of He. At the highest pressure (112 atm with 12% CO_2), the threshold was about 20 mJ.

The optically excited transfer laser scheme can be used to excite several different molecules with a single combination of pump laser and transfer molecule. This is because efficient $V - V$ energy transfer to the laser molecule can be obtained without close coincidence between the energy levels. Applied to high-pressure lasers, this technique should make it possible to cover a broad-frequency range with a single pump laser. The transfer scheme has been demonstrated at high pressure for the following laser and transfer molecule combinations; N_2O/CO_2, CO_2/CO, and CO_2/DF. Continuous frequency tuning has been observed, and it has also been verified that high quantum efficiency can be obtained.

In the first experiment [8.9] using the N_2O/CO_2 combination, CO_2 was excited with a HBr laser in the experimental arrangement shown in Fig. 8.8. The rate constant for $V - V$ energy transfer from the (00^01) level of CO_2 to the 125 cm^{-1} lower level in N_2O is high (8×10^7 s^{-1} atm^{-1} [8.9]). Laser oscillation at the CO_2 laser transitions was avoided by using a small fraction of CO_2

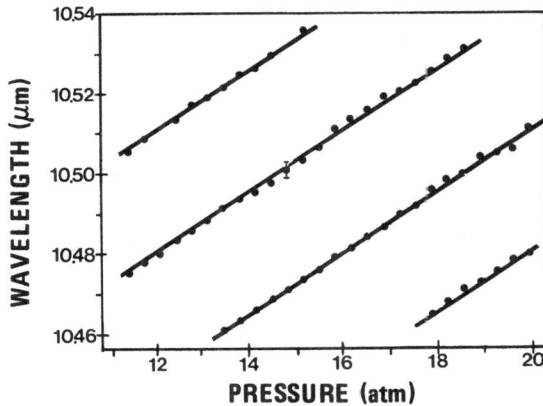

(about 6%). It was found that the optimum resonator length is dependent on the total pressure, and is 4 mm at 10 atm, 2 mm at 20 atm, and 1 mm at 40 atm. Threshold pump energy was about 10 mJ at 10 atm, and 20 mJ at 42 atm which was the maximum pressure. The output pulse appears as a single, short pulse of approximately 600 W peak power. The pulse width varies from 10 ns at 15 atm to 7 ns at 25 atm.

The frequency of this laser could be tuned continuously over 5 cm^{-1} at 10.5 µm by varying the resonator length (Fig. 8.9) [8.12]. The resonator was 1.82 mm long and included a 0.4 mm iris to eliminate higher-order transverse modes. The gas cell was operated at 22 atm with a 90:10 mixture of $N_2O:CO_2$. Under these conditions the peak output power was about 150 W in a pulse of 2.2 – 3 ns duration at a pulse repetition rate of 1 pps.

A high-pressure CO_2 laser excited by energy transfer from CO pumped with radiation from a frequency-doubled CO_2 laser has been demonstrated [8.10]. This source/transfer-molecule combination has also been used to excite other molecules at low pressure. The second harmonic of the CO_2 $P(24)$ line in the 9.4 µm band falls within 0.003 cm^{-1} of the CO 0-1 $P(14)$ transition. The CO molecule has a low de-excitation rate, and is therefore well suited for storing vibrational energy.

The experiment was performed with a $CO_2:CO:He$ gas mixture in a 4.3 cm long cell with internal mirrors. The laser was excited longitudinally with a 50 ns long pulse. For a 13:2:85 mixture at 16 atm the measured threshold pump energy was 2.1 mJ, slope quantum efficiency was 13%, and the maximum output energy was 0.25 mJ. The output pulse length was about 50 ns.

The third energy transfer scheme which has been demonstrated at high gas pressures, utilizes radiation from a pulsed DF laser to pump DF molecules, with subsequent energy transfer to CO_2 [8.11]. The vibrational level of DF is 550 cm^{-1} higher than the ν_3-level of CO_2, and the transfer rate is high, 1.5×10^8 s^{-1} atm^{-1}. Practically all the absorbed energy will therefore be

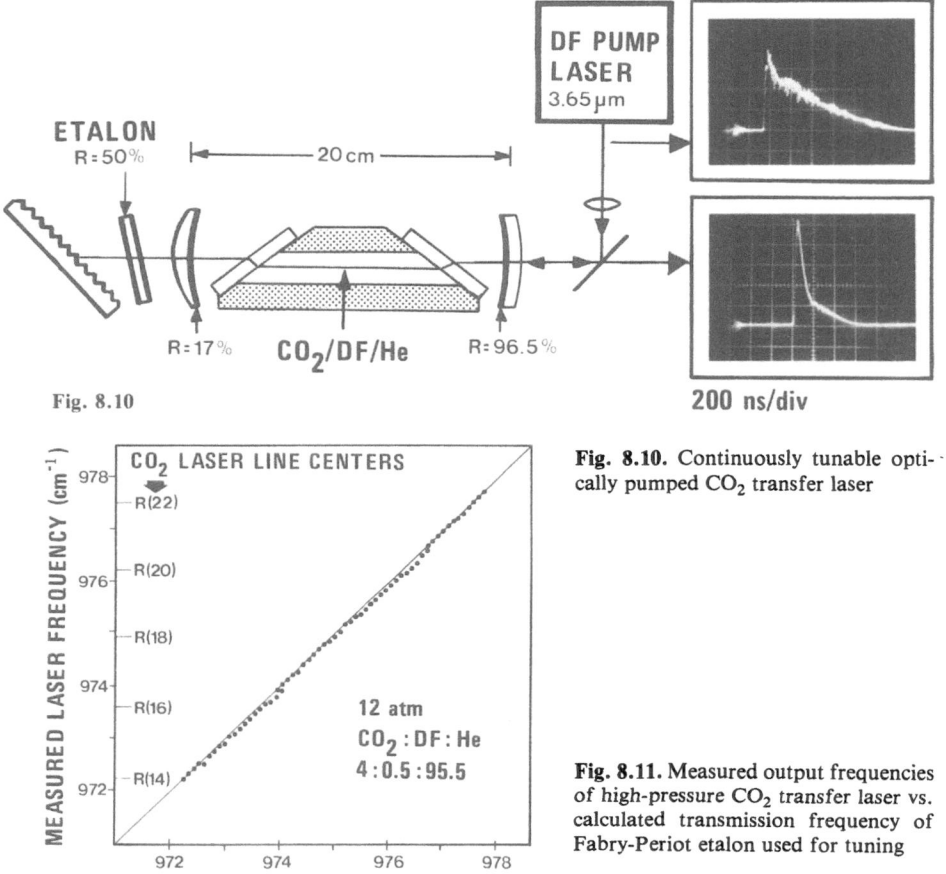

Fig. 8.10

200 ns/div

Fig. 8.10. Continuously tunable opti- · cally pumped CO_2 transfer laser

Fig. 8.11. Measured output frequencies of high-pressure CO_2 transfer laser vs. calculated transmission frequency of Fabry-Periot etalon used for tuning

stored in the CO_2 molecules, giving rise to high gain and reduced saturation of the absorbing DF transition.

Continuous frequency tuning was obtained using the low-loss tunable resonator shown in Fig. 8.10. The frequency selective elements are a diffraction grating and a solid ZnSe Fabry-Perot etalon with 80 GHz free spectral range for fine tuning. The extra mirror is used to reduce the resonator losses caused by the tuning elements.

Figure 8.11 shows the measured CO_2 laser frequencies when the laser was tuned over a 5.3 cm^{-1} wide spectral range in the R-branch of the 10.4 μm band. Deviations from the calculated transmission frequency of the FP etalon is less than the 2 GHz experimental uncertainty. The laser frequency linewidth was less than 3 GHz. The threshold pump energy was about 50 mJ, and 0.8 mJ output pulses could be obtained with the maximum pump energy, 85 mJ.

High quantum efficiency has been obtained in experiments performed at 10 atm with a two-mirror CO_2 laser resonator, where a 100% mirror was used instead of the extra mirror shown in the figure. The optimum CO_2: DF: He mixing ratio was found to be 0.5:5:94.5 at 10 atm, but this ratio is not critical. The threshold pump energy and slope quantum efficiency were 35 mJ and 35%, respectively. Maximum output energy was 6 mJ. About 25% single-pass peak gain was also measured. The experimental results were reasonably well reproduced by computer simulations.

8.3.2 Ultraviolet Preionized TE Lasers

Continuous frequency tuning was first demonstrated with a high-pressure uv preionized self-sustained TE CO_2 laser [8.2]. This work was followed by several investigations of gain characteristics, frequency tuning, reduction in size, and other features important for the development of a tunable laser for practical applications [8.43 – 49]. There are also high-pressure lasers on the market based on this scheme [8.50]. A tunable high-pressure CO_2 laser with electron-beam (instead of uv) preionization of a self-sustained discharge has also been reported [8.51].

The main elements of the uv preionized TE CO_2 laser are shown schematically in Fig. 8.12. The laser is mounted inside a high-pressure chamber, and consists of a pair of electrodes connected to a high-voltage pulse generator, a uv radiation source, and the laser resonator with mirrors and frequency tuning elements. The CO_2: N_2: He gas mixture in the discharge volume is preionized with uv radiation prior to the firing of the pulse generator which supplies the electrical energy to the discharge. Most experiments have been made with a discharge which is $20 - 50$ cm long and about 1×1 cm^2 in cross section, but efficient high-pressure operation of smaller lasers has also been

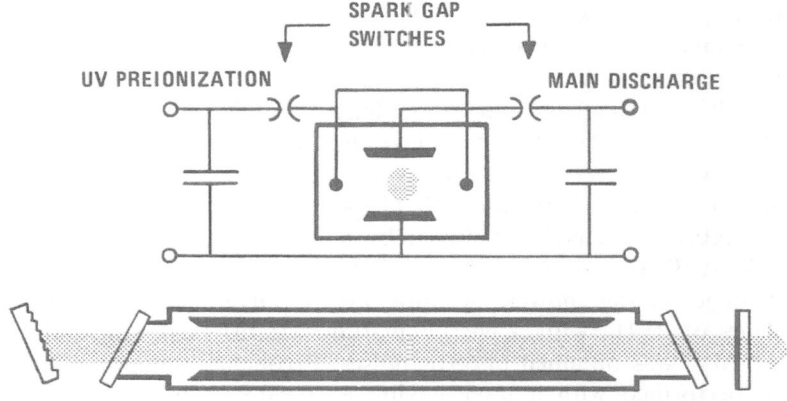

Fig. 8.12. Schematic diagram of uv preionized TE laser. The main discharge is delayed with respect to the uv preionization

demonstrated. The voltages used in the experimental work are roughly of the order of 100 kV.

The electron density is maintained by a Townsend avalanche process. This is an unstable situation where density fluctuations easily lead to formation of arcs. It becomes more difficult to get an arc-free discharge with increasing pressure, and this has been a major problem in the development of tunable lasers [8.23].

In order to get an arc-free uniform glow discharge with the required energy density, several factors are important. Before the high-voltage pulse generator is fired, the gas mixture must be ionized and the electron number density must exceed a minimum value in the discharge volume [8.52]. The discharge should be driven by a uniform electrical field which is obtained with specially shaped electrodes. The duration of the current pulses has typically been as short as $2-100$ ns, to avoid arc formation. The formation of arcs can be reduced by using a gas mixture with a low CO_2 and N_2 content, typically $2\% - 5\%$.

Continuous frequency tuning was first demonstrated with a laser having a $0.7 \times 0.7 \times 26$ cm^3 discharge volume and 10 atm gas pressure [8.2]; the $CO_2 : N_2 : He$ mixing ratio was $10 : 10 : 80$. The 130 cm long tunable resonator consists of a spherical mirror, a grating, and a NaCl prism for intracavity beam expansion, in order to get higher spectral resolution and reduce the energy density on the grating. About $5\% - 10\%$ output coupling was provided by reflection from one surface of the prism. The frequency tuning range was 20 cm^{-1} in both of the P-branches and slightly less in the R-branches at 9.4 and 10.4 μm. The output energy was of the order of $40-80$ mJ. In a later experiment [8.43], a spectral bandwidth of 3 GHz was measured. With a two-mirror resonator, maximum output of ~1 J in 30 ns was obtained.

A careful investigation of the small signal gain in the pressure range from 4 to 19 atm has been made using the laser described above (Fig. 8.13a) [8.26]. The gain at different pressures was measured with constant energy density per unit pressure (Fig. 8.13b). Great care was taken to ensure that the gas was excited with approximately constant E/N-ratio. It was also shown that it is important to take the sequence bands and hot bands into account when calculating the gain.

A smaller laser system, with discharge length $16-20$ cm, and electrode spacing from 0.4 to 0.7 cm, has been investigated with respect to voltage requirement and electrical characteristics of the discharge, gain, output energy, and closed cycle operation [8.44 – 48]. In one experiment, the discharge, $20 \times 0.7 \times 0.7$ cm^3, was driven by a single spark gap switching circuit [8.47]. At 10 atm with a $2.5 : 2.5 : 95$ $CO_2 : N_2 : He$ mixture and 43.5 kV discharge voltage, average powers of 2.6 W were obtained at a pulse repetition frequency of 25 Hz with a gas circulation system. Peak single-mode power of 1.1 MW was obtained in 65 ns pulses. Frequency tuning measurements were performed with a laser having a $16 \times 0.9 \times 0.6$ cm^3 discharge volume and a 65 cm long resonator with a grating and a 20% transmitting plane output mirror [8.46]. A 6 mm aperture restricted the laser to single-

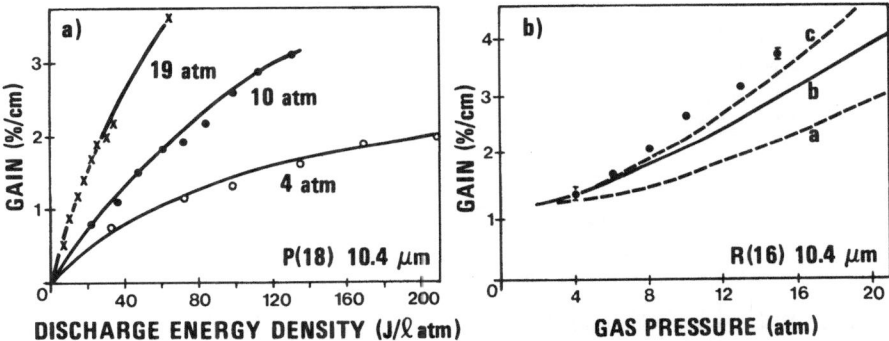

Fig. 8.13a, b. Observed peak small signal gain in uv preionized TE laser with a $5:5:90/CO_2:N_2$: He gas mixture. **(a)** Gain vs. energy deposited into the discharge. **(b)** Gain vs. gas pressure for constant energy density, 90 J/1 atm. Comparison with model predictions for (*a*) gain including regular plus hot bands, (*b*) additional contribution from sequence bands, (*c*) corrections for non-Lorentzian line overlap effects. Vibrational temperature $T_3 = 1630$ K is assumed. After [8.26]

mode operation. At 10 atm, frequency tuning over 17 cm^{-1} in the *P*-branch and 12 cm^{-1} in the *R* branch of the 10.4 μm band was obtained at output energy levels of 40 – 80 mJ in 50 ns pulses. The power reduction between line centers was about 20%, and this was reduced to less than 15% by increasing the pressure to 12 atm. No frequency pulling was observed between line centers. Amplitude modulation of the output pulse showed that the laser was oscillating at several longitudinal modes. In a later work, [8.48] single-frequency operation was demonstrated at arbitrary, fixed frequencies by using a second partially reflecting (17%) mirror between the grating and the discharge. In this case the laser was operated close to threshold.

In view of the high-voltage requirement, we note an experiment with a compact waveguide laser [8.53], operated at 12 atm and voltage as low as 18 kV. Tuning could easily have been demonstrated in this experiment.

The pressure required for continuous tuning with CO_2 can be reduced by using mixtures of isotopes. This has been experimentally verified using $^{12}C^{16}O_2$: $^{12}C^{16}O^{18}O$: $^{12}C^{18}O_2$ in a mixing ratio of $1:2:1$ [8.6]. The $^{12}C^{16}O^{18}O$ molecule has transitions for both odd and even rotational quantum number *J*. At a gas pressure of 4.5 atm, frequency tuning was achieved within the *P*- and *R*-branch of both the 9.4 and 10.4 μm bands; see Fig. 8.6. The results were obtained with a sealed off, uv preionized laser system with a catalyst [8.54] which reversed the dissociation of CO_2.

8.3.3 Electron-Beam Controlled TE Lasers

The best controlled arc-free discharges have been obtained by using electron-beam ionization of the laser gas. This technique has also been successful in exciting high-pressure N_2O and CS_2.

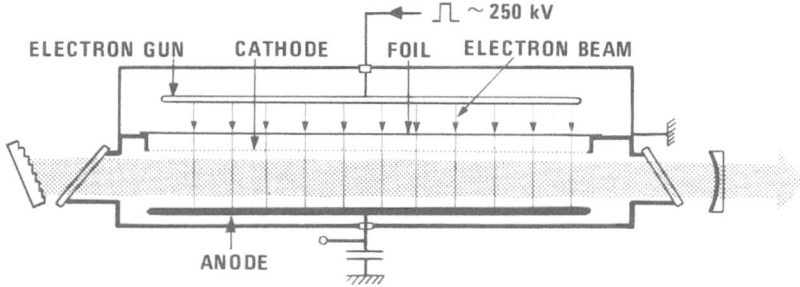

Fig. 8.14. Electron-beam controlled TE laser

The laser consists of a high-pressure discharge chamber, a cold cathode electron gun and discharge circuitry (Fig. 8.14). The electron beam is transmitted through a thin foil, for instance 25 μm thick titanium, into the high-pressure gas mixture. The foil is mounted close to a grid which forms the cathode of the sustainer discharge circuit. The high-energy electrons, at about 250 keV, generate low-energy secondary electrons through ionization of the gas molecules. The energy loss per collision is about 30 eV. The resulting electron number density is determined by equilibrium between the generation rate, given by the electron-beam current density, and the electron loss rate. The electron gun can generate current pulses with approximately constant amplitude from a fraction of a microsecond to several microseconds. The E/N ratio is determined by the voltage of the sustainer capacitor and is independent of the gas mixture.

Several laser experiments with continuous frequency tuning at gas pressures from 5 to 15 atm have been reported [8.4, 40, 55 – 61]. The highest pressure so far has been 50 atm [8.62]. The results from an extensive investigation of a 15 atm e-beam sustained TE laser [8.4, 55, 58] are representative for this laser type and will be discussed in some detail.

The discharge volume of the laser is $20 \times 1 \times 1$ cm^3 and the gas is preionized with a 0.5 μs long electron-beam pulse. High gain, 0.052 cm^{-1}, was obtained with 115 J/l · atm excitation energy density. The optimum values of CO_2:N_2:He gas mixture and E/N-ratio for high gain were 25:5:70 and 1.3×10^{-16} V cm^2, respectively. The gain is higher than for uv preionized TE lasers at the same energy density, and no saturation was observed.

In this experiment, the tunable resonator is designed with an intracavity absorbing gas cell for selection of the 9.4 or 10.4 μm bands and a Fabry-Perot etalon with 65 cm^{-1} free spectral range for frequency tuning. Tuning is obtained by tilting the FP etalon. A second intracavity FP etalon with 2.3 cm^{-1} free spectral range is used for spectral narrowing of the laser radiation [8.58]. With this system 100 ns long laser pulses with 1 GHz linewidth were observed. The pulse energy in this setup was limited to 100 mJ by damage of optical components. The overall tuning range was 70 cm^{-1} for $^{12}C^{16}O_2$.

A similar laser with a 40 cm long discharge chamber has been used to investigate the overall tuning range for $^{13}C^{16}O_2$, N_2O, $^{12}CS_2$, and $^{13}CS_2$ [8.40]. The results are shown in Fig. 8.6. The N_2O and CS_2 molecules were excited in a gas mixture with CO and He. CO was used as transfer gas instead of N_2, because the vibrational levels have a closer coincidence with the asymmetric mode of N_2O and CS_2. Because of the small frequency difference between the line centers in these molecules, 20 – 30 GHz, continuous tuning can be obtained at about 5 atm. In the experiments with CS_2, the gas mixture had to be cooled to 270 K in order to reduce the lower laser level absorption.

In several investigations [8.59 – 61] with $^{12}C^{16}O_2$, it has been shown that frequency tuning between line centers can be obtained with about 5 atm pressure by taking advantage of the high gain in e-beam controlled lasers.

8.3.4 Radio Frequency Excited Waveguide Lasers

Excitation of waveguide CO_2 lasers by means of radio frequency electrical fields was first reported in 1978 [8.63]. The gas was excited by a transverse electrical field and a stable arc-free glow discharge was maintained, because the alternating electrical field counteracts arc formation which requires a minimum build-up time.

High-pressure operation has been reported with pulsed excitation at 1 atm [8.64], 2 atm [8.65], 3 atm [8.66], and 10 atm [8.67] gas pressure and continuous frequency tuning has been demonstrated [8.5]. Recent work has indicated that continuous operation at pressures adequate for continuous frequency tuning should be possible [8.68].

The laser used in the tuning experiments is shown in Fig. 8.15. The laser has an open waveguide geometry with walls made of dielectric ribbons between the electrodes. With this design it is possible to have a rapid transverse gas flow, which is necessary for high prf operation. Electrical breakdown of the laser gas is obtained by means of the overvoltage which is generated due to impedance mismatch at the leading edge of each rf pulse. To avoid arcing at high gas pressure, dielectric waveguide walls were used. The

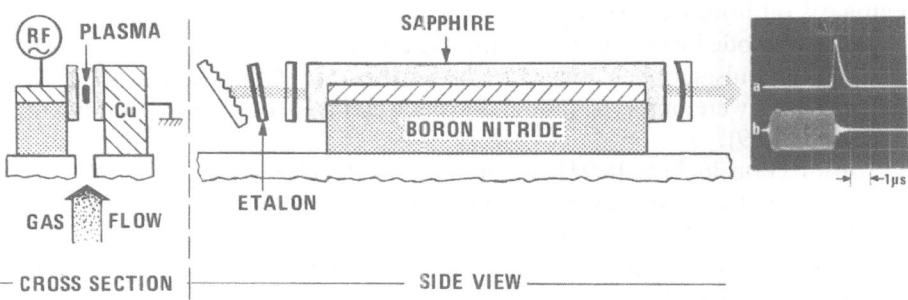

Fig. 8.15. Tunable high-pressure rf CO_2 waveguide laser and gain-switched output pulse. (*a*) Output pulse from the laser operated at 10 atm; (*b*) Input 40 MHz rf voltage waveform

pulse duration before arcing depends on several parameters; the dimensions of the waveguide-electrode configuration and the dielectric material, the electrical power density, the gas pressure and mixture, the rf frequency, and the gas flow. It is assumed that there is a waveguide mode distribution in the horizontal plane and a free space Gaussian beam distribution in the vertical plane when a spherical mirror is used.

High-pressure operation and continuous frequency tuning at 10 atm pressure have been demonstrated using the laser in the gain-switched mode. The electrode is 2×130 mm^2, and the waveguide walls have a spacing of 1.2 mm. The excitation frequency is 40 MHz and the maximum available power 7 kW. It was necessary to use a helium-rich gas mixture, with 2% CO_2 and N_2, in order to get $6 - 10$ μs arc-free discharge pulses at maximum power. A typical gain-switched output pulse is shown in Fig. 8.15.

Continuous frequency tuning over a 10 cm^{-1} wide frequency range in the *R*-branch of the 10.4 μm band has been obtained using the tunable resonator shown in Fig. 8.15. The peak output power was about 0.25 kW. Single-frequency operation was obtained by proper adjustment of the resonator length. The excitation energy density was well below the level giving gain saturation, and it should therefore be possible to attain continuous frequency tuning in both the *P*- and *R*-branch in the 9.4 and 10.4 μm bands, with increased rf power.

Using two mirrors with 98% and 100% reflectivity, the threshold power was approximately 2 kW, and the peak output power increased linearly to 1.4 kW at the maximum input power. In order to get maximum output power the gas in the discharge has to be replaced between each pulse.

8.4 Short Pulse Generation

High-pressure gas lasers are well suited for short pulse generation, due to the large bandwidth, but few experiments have been published. We will briefly describe some mode-locking experiments which have been reported, with emphasis on injection locking.

Passive mode locking using a saturable absorber of *p*-type germanium has given 80 ps pulses, much longer than the theoretical limit [8.69, 70]. Even longer pulses were observed (about 800 ps) when an acousto-optic modulator was used [8.69].

In injection locking [8.71], low-power short pulses are injected into the high-pressure CO_2 laser cavity. The time interval between the pulses has to be exactly equal to the cavity roundtrip time. Since the high-pressure CO_2 laser primarily acts as an amplifier, the output pulse length is to first order given by the input pulse length.

In both injection-locking experiments described below, a dye laser controlled, semiconductor reflection switch [8.72] was used as an optical switch to generate 10 μm input pulses. Mode-locked dye laser pulses produce a dense plasma on the surface of a semiconductor, which turns the surface into a good reflector; CdTe, Ge, and Si have been used. The risetime of this switch can be made less than one picosecond. Using two switches in series, the first in reflection mode and the next in transmission mode, picosecond pulses can be generated from a cw CO_2 laser.

In one experiment [8.73], one Ge switch was used to produce subnanosecond pulses from a 1 W cw CO_2 laser. The output mode-locked pulses from the 7 atm CO_2 laser were less than 200 ps wide with a pulse energy of more than 10 mJ.

The two-switch technique was applied in another experiment [8.14] using a 10 atm CO_2 laser. A single, 100 kW, 2 ps pulse was generated by gating a CO_2 TEA laser. The pulse in the high-pressure laser cavity reached an energy of 15 mJ, corresponding to 10^{12} W/cm^2 power density. In this experiment, pulse compression was observed after the laser pulse had reached maximum energy in the resonator. A plasma-breakdown wave, traveling with the amplified pulse, together with the dispersion in the NaCl Brewster windows, resulted in compression of the pulses to less than 1 ps.

These few experiments confirm that high-pressure gas lasers and amplifiers have good potential for short pulse generation.

8.5 Applications

Most publications involving tunable gas lasers concern aspects of the laser itself: physics of the gas molecules, excitation techniques, tuning techniques, etc. Rather few applications of tunable gas lasers have been published so far. This is in fact not surprising, since a lot of further experimental and development work has to be done before the sources are easily accessible and technically mature. At the end of this section we will describe some of the few papers in which tunable lasers have been used in spectroscopy and as a source for optical pumping of ir lasers. We will, however, first point out a few of the potential application areas, rather than list specific applications in detail.

Optical remote sensing and atmospheric spectroscopy has been an active research area for many years. Most pollution monitoring at infrared frequencies has been restricted to measuring species having absorption lines which strictly coincide with one of the many lines from the available infrared gas lasers, primarily low-pressure CO_2 lasers. Lead-salt laser diodes have been used, but the power is low and the diodes are not easily operated.

Continuously tunable gas lasers are ideal for DIAL (**D**ifferential **A**bsorption LIDAR, where LIDAR: **L**ight **D**etection **A**nd **R**anging experiments because one or more absorption lines can be scanned. With pulse powers of kilowatts and with control of the inherent frequency chirp such lasers should make accurate long-range measurements possible. Both power and heterodyne detection can be used. A large number of important pollutants have absorption lines in the 9 – 13 μm atmospheric window, which is well covered by the known high-pressure lasers. In the ever increasing measurement programs of atmospheric transmission, tunable ir lasers have the potential of being an important new tool. It should be possible to measure, with any desired spectral resolution, time as well as space-resolved absorption spectra.

In *spectroscopy*, the potential for very high spectral resolution at high powers, as well as time-resolved spectroscopy, are the most important advantages of introducing tunable lasers. As an example, real-time spectroscopy in manufacturing is just one of the many specific applications.

A few papers have been published where a tunable CO_2 laser was used for optical excitation of ir lasers. In two experiments, efficient optical pumping of NH_3 and C_2D_2 has been demonstrated, by using a tunable 10 atm CO_2 TE laser. The NH_3 molecules were pumped at otherwise inaccessible absorption lines. Strong laser action in the 11 – 13 μm region on transitions formerly only obtained in buffered NH_3 mixtures was observed [8.74]. In C_2D_2 a total of 16 new laser transitions from 17.5 – 20.3 μm were observed [8.75]. Continuous tuning of a CH_3F Raman far-ir laser has been demonstrated by pumping with a tunable 10 atm CO_2 TE laser [8.76]. Continuous tuning in three spectral bands was reported within the 220 – 400 μm wavelength region.

A few experiments have been made which demonstrate the application of tunable lasers in spectroscopy. In these experiments ethylene and ammonia absorption spectra [8.12, 3, 56], and resonance excitation of luminescence in ethylene [8.57] were measured in the 10 μm range. The resolution was a few hundredths cm^{-1}.

A certain level of technical perfection is required for general utilization of tunable high-pressure gas lasers. The amplifying medium probably meets the minimum requirements, whereas substantial development must be put into the tuning technique and frequency control.

8.6 Recent Progress

The CO_2 laser has important industrial and scientific applications. The physics and technology of this laser therefore is an active research area including work on high-pressure CO_2 lasers. Some results are summarized below.

X-ray preionization of high-pressure TE CO_2 lasers has been demonstrated [8.77 – 79]. This technique has several advantages. The x-ray beam can be transmitted through an aluminum window with thickness in the order of 1 mm. The x-ray radiation has a large penetration depth in the laser gas, and an axial preionization geometry has been demonstrated [8.77]. Furthermore, the absence of uv sparks within the laser inclosure may significantly reduce CO_2 dissociation and the build-up of impurities which appear to limit the lifetime of sealed off high-pressure gas lasers.

Narrow band (~ 1 GHz) [8.80], and single-frequency [8.81 – 83] operation and tuning have been reported. A tunable laser resonator for continuous single-frequency tuning over a broad frequency range has, however, not yet been developed. The tuning range of the radio-frequency excited waveguide laser has been increased [8.83] and continuous tuning in both the *P*- and *R*-branch in the 9.4 and 10.4 µm bands has been attained. Direct optical pumping of high-pressure CO_2 and N_2O lasers with a pulsed HF laser has been demonstrated [8.84]. Frequency measurements of the lasing transitions in nine CO_2 isotopic species [8.85] give information on the potential tuning range for high-pressure lasers. Amplification of subpicosecond pulses in multiatmosphere CO_2 lasers has been investigated [8.86].

8.7 Status and Future Technical Development

Continuous frequency tuning of high-pressure infrared gas lasers has only been reported for the three molecules discussed, CO_2, N_2O, and CS_2. In a study of optical pumping of high-pressure gases no other molecule was found to give laser operation [8.10]. No complete analysis of all possible molecules has been published, to our knowledge. The only likely candidate molecules should be linear, with few vibrational modes which can deactivate the upper laser level. The use of isotopes and mixtures of isotopes of three gases has been explored in only two publications [8.6, 7]. The potential for extending the tuning range and for reducing the required gas pressure by the use of isotopes is great enough to be of technical interest.

It is of interest to compare the excitation techniques with respect to technical and economical aspects and limitations.

- *Ultraviolet preionized transverse excitation* will most likely be limited to short pulse operation and modest to high pulse energies. The high voltages required – tens of kilovolts – is to some extent a technical disadvantage.
- *Electron-beam controlled transverse excitation* is an expensive technique, giving a very homogeneous excitation. It is probably scalable to the highest excitation energies of the four techniques, but is only economically feasible for special purposes.
- *Radio frequency* excitation has so far been limited to low energies in pulsed operation, ~ 1 mJ. The technique gives reasonable arcing thresholds, and cw operation is most probably possible. High-efficiency, high-power rf sources, especially for pulsed operation, have to be developed. The rf excitation may turn out to be the most reliable and cost efficient technique.
- *Optical pumping* is in principle a very efficient and reliable technique, and represents several advantages: no dissociation of the laser molecules – very high quantum efficiency – potential for sealed-off and cw operation. The basic and severe limitation is the lack of efficient laser pump sources. It is

therefore not very likely that optical pumping has a bright technical future, but several interesting physics experiments are forseen using this technique.

So far only pulsed operation of high-pressure lasers has been demonstrated. Much work is forseen before the pulse shapes are sufficiently controlled. In compact TE-waveguide lasers, cw operation seems to be within reach. Quasi continuous operation − 5 ms pulses at 8 atm − has been demonstrated with thresholds in the range of 0.5 kW rf power [8.68].

For a tunable laser to be useful, its frequency control and stability have to be in accordance with the application. These problems have been given little attention up to now. No measurements of linewidth and frequency chirp have been included in any of the high-pressure experiments published. The majority of the lasers have run simultaneously on several longitudinal modes. In a few experiments single-frequency operation has been demonstrated. Frequency chirp phenomena, like those observed at 1 atm, must also be expected at higher pressures.

Tuning experiments and measurements have been carried out, but the resonator and tuning techniques have been far from optimal. Combinations of gratings, prisms, and Fabry-Perot etalons have been used as tuning elements, and beam expanders have been introduced into the cavities to increase the frequency stability and tuning resolution. The frequency tuning therefore has been a tedious process, and would probably be unacceptable for many applications. Development of technically acceptable tuning techniques many well represent an even greater challenge than the development of the high-pressure laser media.

References

8.1 V. N. Bagratashvili, I. N. Knyazev, V. S. Letokhov: Opt. Commun. 4, 154−156 (1971)
8.2 A. J. Alcock, K. Leopold, M. C. Richardson: Appl. Phys. Lett. 23, 562−564 (1973)
8.3 V. N. Bagratashvili, I. N. Knyazev, Yu. A. Kudryavtsev, V. S. Letokhov: Opt. Commun. 9, 135−138 (1973)
8.4 F. O'Neill, W. T. Whitney: Appl. Phys. Lett. 26, 454−456 (1975)
8.5 S. Løvold, G. Wang: IEEE J. QE-20, 182−185 (1984)
8.6 R. B. Gibson, K. Boyer, A. Javan: IEEE J. QE-15, 1224−1228 (1979)
8.7 F. O'Neill, W. T. Whitney: Appl. Phys. Lett. 28, 539−541 (1976)
8.8 T. Y. Chang, O. R. Wood: Appl. Phys. Lett. 23, 370−372 (1973)
8.9 T. Y. Chang, O. R. Wood: Appl. Phys. Lett. 24, 182−183 (1974)
8.10 H. Kildal, T. F. Deutsch: "Optically Pumped Gas Lasers", in *Tunable Lasers and Applications,* ed. by A. Mooradian, T. Jaeger, P. Stokseth, Springer Ser. Opt. Sci., Vol. 3 (Springer, Berlin, Heidelberg 1976)
8.11 K. Steneresen, G. Wang: IEEE J. QE-19, 1414−1426 (1983)
8.12 T. Y. Chang, J. D. McGee, O. R. Wood II: Opt. Commun. 18, 279−281 (1976)
8.13 T. Y. Chang, O. R. Wood II: IEEE J. QE-13, 907−915 (1977)
8.14 P. B. Corkum: Opt. Lett. 8, 514−516 (1983)
8.15 G. Herzberg: *Infrared and Raman Spectra of Polyatomic Molecules* (Van Nostrand, New York 1945)

8.16 P. K. Cheo: "CO$_2$ lasers", in *Lasers*, Vol. 3, ed. by A. K. Levine, A. J. DeMaria (Dekker, New York 1971) Chap. 2, pp. 111−267

8.17 G. Bekefi (ed.): *Principles of Laser Plasmas* (Wiley, New York 1976)

8.18 T. Y. Chang: "Optical Pumping in Gases", in *Nonlinear Infrared Generation,* ed. by Y. R. Shen, Topics Appl. Phys., Vol. 16 (Springer, Berlin, Heidelberg 1977) Chap. 6, pp. 215 to 272

8.19 M. J. Weber (ed.): *Gas Lasers,* CRC Handbook of Laser Science and Technology, Vol. II (CRC Press, Boca Raton, FL 1982)

8.20 E. W. McDaniel, W. L. Nighan (eds.): *Gas Lasers*, Applied Atomic Collision Physics, Vol. 3 (Academic, New York 1982)

8.21 K. J. Button (ed.): *Coherent Sources and Applications, Part II,* Infrared and Millimeter Waves, Vol. 7 (Academic, New York 1983)

8.22 N. G. Basov, E. M. Belenov, V. A. Danilychev, A. F. Suchkov: Sov. Phys.-Usp. **17**, 705−721 (1975)

8.23 O. R. Wood II: Proc. IEEE **62**, 355−397 (1974)

8.24 M. Silver, T. S. Hartwick, M. J. Posakony: J. Appl. Phys. **41**, 4566−4568 (1970)

8.25 L. E. Freed, C. Freed, R. G. O'Donnell: IEEE J. QE-**18**, 1229−1236 (1982)

8.26 R. S. Taylor, A. J. Alcock, W. J. Sarjeant, K. E. Leopold: IEEE J. QE-**15**, 1131−1140 (1979)

8.27 B. G. Whitford, K. J. Siemsen, J. Reid: Opt. Commun. **22**, 261−264 (1977)

8.28 J. Reid, K. J. Siemsen: IEEE J. QE-**14**, 217−220 (1978)

8.29 K. Smith, R. M. Thomson: *Computer Modeling of Gas Lasers* (Plenum, New York 1978)

8.30 R. K. Brimacombe, J. Reid: IEEE J. QE-**19**, 1674−1679 (1983)

8.31 R. K. Brimacombe, J. Reid: IEEE J. QE-**19**, 1668−1673 (1983)

8.32 J. L. Miller, E. V. George: Appl. Phys. Lett. **27**, 665−667 (1975)

8.33 J. L. Miller, A. H. M. Ross, E. V. George: Appl. Phys. Lett. **26**, 523−526 (1975)

8.34 J. L. Miller: J. Appl. Phys. **49**, 3076−3083 (1978)

8.35 C. Dang, J. Reid, B. K. Garside: IEEE J. QE-**16**, 1097−1103 (1980)

8.36 C. Dang, J. Reid, B. K. Garside: Appl. Phys. B **27**, 145−151 (1982)

8.37 C. Dang, J. Reid, B. K. Garside: IEEE J. QE-**19**, 755−764 (1983)

8.38 S. Løvold, N. Menyuk: Unpublished data, MIT Lincoln Laboratory, Lexington MA 02173

8.39 J. D. Daugherty: "Electron beam ionized lasers", in *Principles of laser plasmas*, ed. by Georg Bekefi (Wiley, New York 1976) Chap. 9, 369−419

8.40 F. O'Neill, W. T. Whitney: Appl. Phys. Lett. **31**, 270−272 (1977)

8.41 C. Freed: IEEE J. QE-**18**, 1220−1228 (1982)

8.42 C. R. Jones: Laser Focus **14**, 68−74 (August 1978)

8.43 D. Rollin, A. J. Alcock: Opt. Commun. **23**, 11−14 (1977)

8.44 T. Carman, P. E. Dyer: J. Appl. Phys. **49**, 3742−3746 (1978)

8.45 T. W. Carman, P. E. Dyer: Appl. Phys. **17**, 27−30 (1978)

8.46 T. W. Carman, P. E. Dyer: Opt. Commun. **29**, 218−222 (1979)

8.47 P. E. Dyer, B. L. Tait: Appl. Phys. Lett. **37**, 356−358 (1980)

8.48 B. K. Deka, P. E. Dyer, R. J. Winfield: Opt. Commun. **39**, 255−258 (1981)

8.49 Yu. T. Mazurenko, Yu. A. Rubinov, P. A.Shakhverdov: Sov. J. Opt. Technol. **46**, 341−344 (1979)

8.50 Lumonics Inc, Kanata (Ottawa), Ontario, Canada
   Laser Applications Ltd, Hull, North Humberside, England

8.51 W. Chong-Yi, C. Schwab, W. Fuss, K. L. Kompa: Opt. Commun. **46**, 311−314 (1983)

8.52 J. I. Levatter, S.-C. Lin: J. Appl. Phys. **51**, 210−222 (1980)

8.53 D. J. Brink, V. Hasson: J. Appl. Phys. **49**, 2250−2253 (1978)

8.54 R. B. Gibson, A. Javan, K. Boyer: Appl. Phys. Lett. **32**, 726−727 (1978)

8.55 N. W. Harris, F. O'Neill, W. T. Whitney: Appl. Phys. Lett. **25**, 148−151 (1974)

8.56 V. N. Bagratashvili, I. N. Knyazev, V. V. Lobko: Sov. J. Quantum Electron. **5**, 857−859 (1975)

8.57 V. N. Bagratashvili, I. N. Knyazev, V. S. Letokhov, V. V. Lobko: Opt. Commun. **14**, 426−430 (1975)

8.58 N. W. Harris, F. O'Neill, W. T. Whitney: Opt. Commun. **16**, 57 – 62 (1976)
8.59 V. N. Bagratashvili, I. N. Knyazev, V. S. Letokhov, V. V. Lobko: Sov. J. Quantum Electron. **6**, 541 – 549 (1976)
8.60 Yu. I. Bychkov, G. A. Mesyats, V. M. Orlovskii, V. V. Osipov, V. V. Savin: Sov. J. Quantum Electron. **8**, 870 – 872 (1978)
8.61 S. S. Alimpiev, Yu. I. Bychkov, N. V. Karlov, E. K. Karlova, G. A. Mesyats, Sh. Sh. Nabiev, S. M. Nikiforov, V. M. Orlovskii, V. V. Osipov, A. M. Prokhorov, E. M. Khokhlov: Sov. Tech. Phys. Lett. **5**, 336 – 338 (1979)
8.62 N. G. Basov, E. M. Belenov, V. A. Danilychev, O. M. Kerimov, I. B. Kovsh, A. S. Podsosonnyi, A. F. Suchkov: Sov. Phys.-JETP **37**, 58 – 64 (1973)
8.63 K. Laakmann: "Transverse rf excitation for waveguide lasers", Proc. Int. Conf. on Lasers '78, pp. 741 – 743
8.64 J. L. Lachambre, J. MacFarlane, G. Otis, P. Lavigne: Appl. Phys. Lett. **32**, 652 – 653 (1978)
8.65 C. P. Christensen, F. X. Powell, N. Djeu: IEEE J. QE-**16**, 949 – 954 (1980)
8.66 A. E. Bakarev, L. S. Vasilenko, V. G. Gol'dort, A. É. Om, O. M. Skhimnikov: Sov. J. Quantum Electron. **10**, 243 – 245 (1980)
8.67 S. Løvold, G. Wang: Appl. Phys. Lett. **40**, 13 – 15 (1982)
8.68 S. Landrø, G. Wang: "High-pressure cw RF-excited CO_2 waveguide laser", CLEO '84, paper THL 1
8.69 A. J. Alcock, A. C. Walker: Appl. Phys. Lett. **25**, 299 – 301 (1974)
8.70 A. C. Walker, A. J. Alcock: Opt. Commun. **12**, 430 – 432 (1974)
8.71 P. B. Corkum: Laser Focus **15**, 80 – 84 (June 1979)
8.72 A. J. Alcock, P. B. Corkum: Can. J. Phys. **57**, 1280 – 1290 (1979)
8.73 P. B. Corkum, A. J. Alcock, D. F. Rollin, H. D. Morrison: Appl. Phys. Lett. **32**, 27 – 29 (1978)
8.74 B. K. Deka, P. E. Dyer, R. J. Winfield: Opt. Commun. **33**, 206 – 208 (1980)
8.75 B. K. Deka, P. E. Dyer, R. J. Winfield: Opt. Lett. **5**, 194 – 195 (1980)
8.76 P. Mathieu, J. R. Izatt: Opt. Lett. **6**, 369 – 371 (1981)
8.77 P. E. Dyer, D. N. Rauf: Opt. Commun. **53**, 36 – 38 (1985)
8.78 Krishnaswamy Jayaram, A. J. Alcock: J. Appl. Phys. **58**, 1719 – 1726 (1985)
8.79 K. Midorikawa, H. Tashiro, S. Namba, M. Okada, H. Kubomura: "Continuously tunable CO_2 laser system with an x-ray preionized amplifier", CLEO '86, paper WS 3
8.80 P. E. Dyer, D. N. Rauf: Appl. Opt. **24**, 3152 – 3154 (1985)
8.81 I. N. Knyazev, A. A. Sarkisian: Opt. Commun. **52**, 421 – 424 (1985)
8.82 B. K. Deka, M. A. Rob, J. R. Izatt: Opt. Commun. **57**, 111 – 116 (1986)
8.83 S. Løvold, G. Wang, K. Stenersen: "Single mode continuously tunable pulsed rf-excited CO_2 and N_2O waveguide lasers", CLEO '85, paper FN 5
8.84 K. Stenersen, G. Wang: "Direct optical pumping of high-pressure CO_2 and N_2O lasers with a pulsed HF laser", CLEO '85, paper FN 2 (to be published in IEEE J. Quantum Electron.)
8.85 L. C. Bradley, K. L. Soohoo, C. Freed: IEEE J. QE-**22**, 234 – 267 (1986)
8.86 P. B. Corkum: IEEE J. QE-**21**, 216 – 232 (1985)

9. Tunable Paramagnetic-Ion Solid-State Lasers

John C. Walling

With 49 Figures

Tunable paramagnetic ion lasers, particularly those based on transition metals, offer the potential for high energy storage and high pulse output power. This potential derives from the low oscillator strength of their laser emission bands and the consequent slow rate of spontaneous emission. It also derives from the fact that the host crystals are often hard, durable gemstones with high thermal conductivity. Thus, some broadly tunable paramagnetic ion lasers offer the pulse energy and power of the traditional fixed wavelength, lamp-pumped, solid-state lasers, such as Nd:YAG and ruby.

On the other hand, the low oscillator strength is a mixed blessing, as, for example, it often leads to low gain and the need for highly efficient cavities. The advantages of lasers using high oscillator strength transitions, such as color center and dye lasers (which do not have the potential for significant energy storage), have already been discussed in Chaps. 1 and 6.

The first tunable transition metal ion lasers were discovered in the early 1960s at Bell Laboratories, within only three years of the discovery of the ruby laser, the first solid state laser. Interest in the tunable transition metal ion lasers was eclipsed for a time by the discoveries of Nd:YAG and liquid dye lasers, but was renewed in the late 1970s when the tunable, flash lamp pumped, alexandrite laser was discovered at Allied Corporation. Flash lamp pumped alexandrite could be tuned over a significant wavelength range, 710 to 820 nm at room temperature, and had good energy storage for Q-switching. At about the same time, a group at Lincoln Laboratories began to reexamine a number of the materials explored earlier at Bell Laboratories, this time using Nd:YAG as a pump source, which greatly improved performance. Some of the materials studied at Lincoln Laboratories demonstrated especially wide tuning ranges. Particularly interesting is Ti-doped sapphire, which not only has an exceptionally broad tuning range, $660-1060$ nm, but operates well at room temperature. Ti:sapphire, however, lacks significant energy storage and can not be as effectively Q-switched in flash lamp pumped operation. Many other tunable transition metal ion lasers have now been discovered, which together extend wavelength coverage from 660 nm to 2280 nm (except for two small gaps), with varying degrees of performance.

Herein, after a brief outline of fundamentals and a discussion of vibronic laser kinetics, the transition metal ion lasers are reviewed. This general review is followed by a more extensive discussion of the alexandrite laser. At present, alexandrite is the most fully developed of the transition metal ion lasers, providing tunable emission with performance in other respects much like that of ruby and Nd : YAG.

9.1 Overview of Fundamentals

This section treats the fundamental concepts of vibronic lasers, with emphasis on the phonon interactions and optical spectroscopy, elaborating in part on the discussion in Chap. 1.

9.1.1 Crystal Field States

Optical transitions within the paramagnetic shell of transition-metal ($3d$ shell) and rare-earth ($4f$ shell) ions have wavelengths and oscillator strengths well suited for laser applications, particularly where energy storage is desired. Optical spectra of these ions in crystalline solids exhibit both narrow (purely electronic) transitions and, particularly in the $3d$ case, broad bands associated with vibrational interactions. These broad emission bands provide an opportunity for making broadly tunable lasers.

The strength of interaction with the lattice is critical to practical attainment of tunable laser action. Of particular importance is the difference in the way the transition-metal $3d$ shell and the rare-earth $4f$ shell interact with the lattice. In the $4f$ case, shielding of the crystal field is provided by higher-lying $5s$ and $5d$ electrons, largely isolating the $4f$ shell from the lattice. Thus, interaction with the lattice is weak, and rare-earth $4f$-$4f$ transitions usually appear as sharp, "no-phonon" lines with weak or nearly undetectable vibronic sidebands. By contrast, the $3d$ shell of transition metal ions participates to some degree in lattice bonding. Thus, crystal field terms and interaction with the lattice are much stronger. This in turn makes the vibronic sidebands much stronger; the major fraction of the oscillator strength of a given transition often appears in the sidebands, rather than in the no-phonon line. This high relative strength of the vibronic sidebands makes tunable laser action possible.

An exception to the isolation of the rare earths is the second member of the rare-earth series, the Ce^{3+} ion, where the $4f$ shell contains only one electron. In this case, the positive nuclear charge is insufficient to fully contract the $4f$ electron deep within the $5s$ and $5p$ charge distribution, and the $4f$ electron interacts more strongly with the lattice. As a result, Ce^{3+} plays a unique, intermediate role in tunable paramagnetic ion solid-state lasers.

Transition strengths are strongly influenced by certain symmetry properties. Electric dipole transitions within the $3d$ shell are strictly parity forbidden between pure $3d$ states, whereas magnetic dipole states are parity allowed. Nevertheless, except in cases where the crystal field point symmetry at the active ion includes inversion, the $3d$ transitions are dominantly electric dipole. This occurs because the crystal field mixes into the $3d$ states minor contributions from shells with parity opposite to that of the $3d$. This "configurational mixing" accounts for the electric dipole oscillator strength, but by no means offers a direct and simple way to derive the associated matrix elements or to fully understand the polarization properties of these transitions. This mixing, which is highly dependent on the specific ion and host, largely accounts for the great variance in oscillator strengths among the several transition-metal-ion – host combinations. Symmetry properties of the crystal field also have a major influence on transition strengths and polarization properties. If the site of the active ion has inversion symmetry, electric dipole transitions are disallowed. As a result the oscillator strength depends on the magnetic dipole interaction, which is usually weaker by at least an order of magnitude.

The term "vibronic" refers to simultaneous transitions between interacting electronic and vibrational states that involve quantum number changes in each. The electronic component of the transition carries the dominant part of the transition energy (on the order of $10\,000\;\mathrm{cm}^{-1}$, whereas the vibrational component is on the order of $1000\;\mathrm{cm}^{-1}$). Consequently, the vibronic transitions appear in the optical spectra as sidebands to the no-phonon (purely electronic) transition with which they are associated. If the vibronic interaction is strong, the sideband carries the greater share of the oscillator strength to the point that the no-phonon line disappears entirely.

Two sidebands occur, situated quite symmetrically on either side of the no-phonon line. At low temperatures, only the sideband that involves the emission of phonons, the Stokes sideband, is present; the other, the anti-Stokes sideband, is suppressed. The associated phonons are simply not available for absorption. In optical absorption, the Stokes sideband appears on the high photon energy side of the no-phonon line, whereas the anti-Stokes sideband appears on the low-energy side. In emission, these position are reversed. At higher temperatures (generally above room temperature), both sidebands appear in both the absorption and emission spectra.

In order to obtain continuous broad band emission, the $3d$ transition-metal ions have been widely used. Certain of these ions have been more successful than others by reason of a particularly well adapted electronic structure and stability of the ionic charge in the host crystal. In particular, Cr^{3+}, which is an exceptionally stable ion, has been used profitably in a number of hosts, including chrysoberyl, the pure host of alexandrite. The Tanabe-Sugano energy level diagram for the $3d^3$ configuration, which represents the energy level structures of both Cr^{3+} and V^{2+} in octahedral coordination, is presented in Fig. 9.1 [9.1]. This diagram depicts the state energies as a function of the cubic crystal field strength; co-valence effects, higher-order crystal field terms and the spin-orbit coupling were omitted in its derivation. Also neglected was the Jahn-Teller interaction, which is a frequently invoked contribution to the splitting of the $^4T_2$ multiplet. Where present, it reduces the lower (metastable) $^4T_2$ level energy relative to that of ^2E

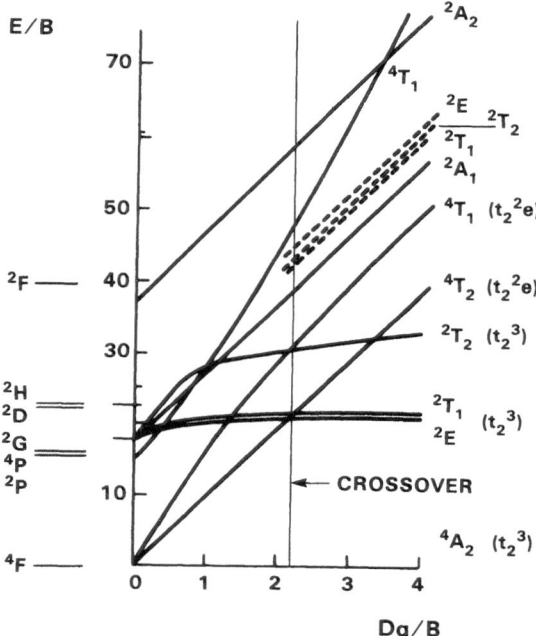

E/B

70

2A₂

⁴T₁

2E   2T₂
2T₁
2A₁

50

⁴T₁ (t₂²e)

2F ——   40

⁴T₂ (t₂²e)

2T₂ (t₂³)

30

2H
2D
2G
4P
2P

2T₁ (t₂³)
2E

10

◄— CROSSOVER

4F ——

⁴A₂ (t₂³)

0   1   2   3   4

Dq/B

Fig. 9.1. Tanabe-Sugano diagram of the $3d^3$ configuration in an octahedral crystal field corresponding to the ions Cr^{3+} and V^{2+}. The crossover is between "weak" and "strong" fields

and $^2T_1$. The diagram is valuable for a qualitative interpretation of spectral features. The cross-over line illustrated in the figure is discussed in Sect. 9.3.

Transitions between states of the same spin, the "spin-allowed" transitions, $^4A_2$ to $^4T_2$ for example, generally have greater oscillator strength than those which involve a spin change (said to be "spin-forbidden"). Also, if a spin-allowed transition also involves a substantial change in orbital configuration, a strong vibrational sideband generally occurs. In direct contrast are the "spin only" transitions, exemplified by ^2E to $^4A_2$, where the transition is basically a spin flip with only a minor accompanying orbital change. Such transitions are typified by weak vibronic sidebands. A strong vibrational interaction can be anticipated if the slopes of the energy levels vs crystal field, as represented in Fig. 9.1, are significantly different, as is the case for the $^4T_2$ to $^4A_2$ transition. By contrast, the ^2E curve is nearly flat and the $^4A_2$ to ^2E (spin-forbidden) transition has a comparatively weak crystal field interaction. From the standpoint of optically pumped laser operation, pump bands and broadly tunable gain bands derive from spin-allowed transitions that involve a significant orbital change. These would include the $^4A_2$ to $^4T_2$ and $^4T_1$ transitions for pumping in alexandrite (and ruby) and the $^4T_2$ to $^4A_2$ transition for broadly tunable lasing.

Figure 9.2 illustrates examples of transitions arising from the $^4A_2$ state for the mirror-symmetry Cr^{3+} ion in alexandrite at both room temperature and 4 K. The broad bands are the vibronic sidebands of the $^4A_2$ to $^4T_2$ transitions in question. The displacement in the peaks of this broad band in different

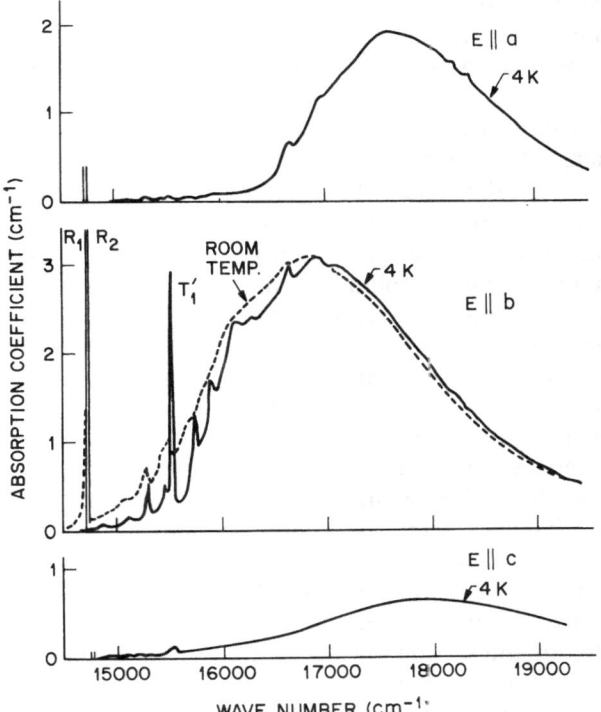

Fig. 9.2. Alexandrite absorption spectra at 4 K and room temperature. The major broad peaks are vibronic sidebands resulting from spin-allowed transitions ($^4A_2$ to $^4T_2$); the narrow peaks $R_{1,2}$ are no-phonon spin-forbidden transitions ($^4A_2$ to ^2E). T_1' is a no-phonon spin-allowed transition ($^4A_2$ to $^4T_2$)

polarizations is attributed to the splitting of the $^4T_2$ states. The splitting arises from a combination of spin-orbit, low-order crystal field, and possibly Jahn-Teller interactions. The only confirmed no-phonon $^4T_2$ line is indicated as T_1', which has diminished and disappeared into the background at room temperature. The structure immediately to the left of T_1' are no-phonon lines of the $^2T_2$ multiplet. The structure on the immediate right of the T_1' line includes resolved vibrational modes and other $^4T_2$ no-phonon transitions not yet unambiguously identified. The R_1 and R_2 lines from the mirror-site Cr^{3+} ion are also indicated.

9.1.2 Phonon Modes

In the formation of a solid, the vibrational modes of molecules become bands of vibrational states. The states in each band are very closely spaced in energy and are not generally resolved in measurements. In a three-dimensional lattice, there are three bands for each atom in the unit cell, corresponding to the number of degrees of freedom. The three lowest energy bands contain acoustic modes and are dominantly characterized by uniform motion of atoms over broad spatial regions, as in a pressure or sound wave. Above the highest-energy acoustic mode there is usually a band gap above which lies the remainder of the bands, usually also separated by gaps, which are the optical phonon bands. Optical modes are characterized by the opposing relative motion of adjacent atoms or ions. A discussion of phonon modes may be found in [9.2].

An impurity, (e.g., Cr^{3+} substituted for Al^{3+} in alexandrite or ruby) causes a discontinuity in the regular lattice array. This becomes a unique point of symmetry where the lattice is perturbed locally by a difference in mass and/or effective spring constant. The pure lattice modes are slightly shifted in energy by this impurity (by an amount less than the separation between adjacent unperturbed modes), but retain their finite amplitude throughout the lattice. On the other hand, a local mode is a strongly perturbed lattice mode that has been pulled from a local energy extremum of the band into the band gap. This mode loses its delocalized character, and becomes centralized about the impurity. A discussion of local modes and their characteristics can be found in [9.3].

The vibrational component of a vibronic transition may be a wave packet composed of many lattice modes derived from a position of near-constant transitional energy within the dispersion manifold. This would include local extrema of a given phonon band as occur at symmetry points within the Brillouin zone. At the moment of their creation, these packets have large, vibrational amplitudes positioned at the active ion, which then disperse as their constituent phonon eigenstates propagate.

Because the photon−phonon interaction in vibronic transitions is coupled through the electronic states, the interaction is concentrated in the vicinity of the active ion in either case. Consequently, conservation of crystal momentum ($\hbar k$), which plays an essential role in the Raman effect, is irrelevant or much less relevant in this case. Lattice phonons may be properly represented as stationary waves having zero real momentum. The small photon momentum, which is on the order of 10^{-4} times that of the highest (real) momentum traveling phonon mode, is conserved by the translational momentum of the lattice as a whole.

9.1.3 Vibronic Interaction

The vibronic interaction [9.4] is key to understanding the spectral characteristics portrayed in Fig. 9.2. We present here only a brief heuristic discussion.

First of all, the interaction of the radiation field with the phonons is via the electronic states of the impurity ion. The susceptibility of the electronic states to the radiation field is felt by the vibrational states as both a comparatively short-range multipole interaction involving both exchange and Coulomb terms, and a dipolar interaction of exchange and also comparatively long-range Coulomb (electric dipole field) terms. The strength, linearity, and the range of the interaction all influence the line shape of the emission spectra.

The phonon lifetime at low temperatures, as discussed below, is too long relative to the time of one phonon cycle to allow the interaction dynamics of the electronic and vibrational states during the transition simply to be ignored. Nevertheless, it is common practice to assume an instantaneous electronic state change, which provides a step function drive input to the vibrational Hamiltonian. Such an approximation is often used to describe the broad spin-allowed transitions (in connection with coordination diagrams), as discussed briefly in Chap. 1.

The appearance of the observed optical spectrum at low temperatures depends intimately on the character of the interaction. As a general rule, a strong, localized interaction that involves significant local displacements in atomic equilibrium positions will lead to a broad, essentially featureless sideband, typified, in fact, by color centers. In this case the coupling is to a broad spectrum of optical and acoustic phonon modes in multiple order. At the other extreme, a weak interaction, particularly one involving little orbital wave function change and having perhaps a greater long-range component, provides a narrow no-phonon line and a finely structured sideband that reflects the density of states of the lattice phonon modes. Evidence for many "in between" conditions are seen in various spectra. This evidence generally appears as peaks or oscillations on the side of the broad spectra with period or displacement from a no-phonon line which corresponds to some dominant phonon energy.

A commonly cited model for coupling intermediate between the two above-mentioned extremes is illustrated in Fig. 9.3 [9.2]. Here the structure of such bands are expressed in terms of a no-phonon line combined with convolved multiple orders of a fundamental wave packet of phonons (including local mode vibrations) characteristic of the interaction. However, the interaction assumed by this interpretation lacks the diverse properties present in

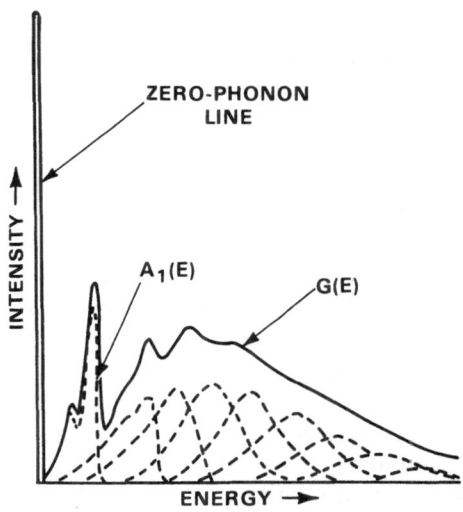

ZERO-PHONON
LINE

$A_1(E)$

$G(E)$

INTENSITY →

ENERGY →

Fig. 9.3. Decomposition of a vibronic spectrum into its no-phonon line, one-phonon wave packet spectrum $[A_1(E)]$, and multiple orders of same $[G(E)]$. (From [9.2])

most real systems. This, together with the added complexity of multiple electronic levels, generally makes this picture inadequate for interpreting many real systems, as in alexandrite (Fig. 9.3).

9.1.4 Linewidths and Lifetimes

Other broadening mechanisms play a significant role for both fixed frequency and tunable laser performance. The sharpest spectral features are produced by transitions from metastable states, particularly those that involve weak phonon interaction, such as the *R*-lines of ruby. For these, the dominant broadening mechanisms are crystal strain and a weak spin-orbit splitting. At higher temperatures these are broadened further by phonon vibrations via the crystal field [9.5]. No-phonon lines and vibronic sideband peaks arising from states above metastable states are "lifetime broadened" (by zero-point vibrationally induced transitions to lower states at low temperatures and further by phonons at higher temperatures). Finally, additional broadening at still higher temperatures derives from "fill in" transitions originating from thermally excited vibronic states.

Strain broadening implies an inhomogeneous crystal field environment, i.e., one that differs slightly for each ion. Phonon-induced broadening, by contrast, is homogeneous, i.e., the same for each ion. Homogeneous broadening is generally desired for lasers because the energy stored in all ions is accessible even for very narrow linewidths, although the effects of inhomogeneous broadening can be relieved to a limited extent by rapid excitation transfer among ions, as occurs in Nd: glass lasers. Similarly, rapid relaxation among the vibronic states of the excited multiplets is essential if homogeneous broadening is to have its desired effect for lasers.

As mentioned above, the weak sidebands of spin-only transitions are often highly structured at low temperatures. The more finely detailed features of these weak bands, in some cases less than 4 cm^{-1} wide, can be identified with Raman wavelengths, indicating that their origins are closely associated with lattice optical phonon modes. Local modes may or may not be distinguished as features in these spectra; it is difficult to make their identification unambiguous given the spectral complexity.

The lower limit to the lifetime of the phonon associated with a 4 cm^{-1} spectral linewidth is placed at 3.8 ps by use of the uncertainty principle derived from Gaussian wave packets (where $\Delta t[\text{ps}] = 14.64/\Delta E[\text{cm}^{-1}]$). This lifetime is long compared the typical time for one optical phonon cycle, 0.166 ps for a 200 cm^{-1} phonon. Phonon lifetimes vary widely, however, and other parts of the spectrum exhibit broader features. At room temperature, almost all vibrational features broaden to the point that their implied lifetime is short compared to one phonon cycle.

The nonradiative relaxation of the electronic levels by phonon interaction is discussed in several texts [9.2,6]. A summary and guide to the theory of paramagnetic ion vibronic lasers, including nonradiative relaxation, has also been presented [9.7]. Nonradiative relaxation competes with radiative rates and, even at low temperatures, may dominate to the point that no fluorescence is observed, particularly if the level separation falls within one Debye frequency. When the transition gap is large and must be spanned by several phonons, as for the laser transition, strong phonon-induced perturbations generally are needed to precipitate the nonradiative relaxation. Nonradiative relaxation is an exceedingly varied and complex phenomenon. One mechanism frequently invoked is as follows: As the temperature rises, higher-lying vibrational states, particularly optical phonon states, become occupied. These then perturb the local crystal field to such an extent that the energies of the excited and ground electronic levels, each being differently affected by the perturbation, are momentarily brought in close proximity. The probability of this occurring increases exponentially with temperature, roughly according to a Boltzmann/activation energy description. Once the transition occurs, the new vibrationally excited state relaxes with the creation of a number of additional phonons.

Nonradiative transitions are extremely important in the laser context, first, because they relax the states in the excited multiplets among themselves, as they do also for the ground multiplets, thus repopulating a laser depleted upper state and depopulating a filled lower state (Sect. 9.2). Spectral hole burning is thereby suppressed or eliminated. Of equal importance, however, nonradiative transitions are also very deleterious to lasers when they cause relaxation between the metastable and ground multiplets in competition with the radiative laser transition, thereby reducing the storage time. The impact of nonradiative decay on specific tunable transition metal ion lasers will be discussed in Sect. 9.3.

9.2 Vibronic Laser Kinetics

While the kinetics of paramagnetic ion lasers are very similar to those of color center lasers, they are often significantly more complex. First of all, multiple electronic levels are frequently involved in both the upper and lower transition levels, which may or may not be in equilibrium during laser action. Moreover, the Stokes shift is much less in the paramagnetic ion case, because the vibronic interaction is weaker, leading to significant ground state absorption at normal operating temperatures. Excited state absorption also plays a much greater role in the paramagnetic ion case, and some means is required to differentiate it from the emission cross section in order to properly understand and characterize the laser. For these reasons, it is necessary to develop the theory of vibronic laser kinetics beyond the treatment given in Chap. 1.

In this section, we first summarize the theory of vibronic laser gain, emphasizing the common case in which the wavelength-dependent emission cross section can be computed from the observed fluorescence emission. The standard rate equations for a three-level laser are then rewritten in terms of the vibronic laser gain derived from this theory to provide a bridge to conventional laser resonator kinetics theory. Finally, the various cross sections are discussed in light of this theory for real materials, alexandrite in particular.

9.2.1 Vibronic Laser Gain

The principal theoretical work expressly devoted to vibronic laser kinetics is by *McCumber* [9.8] and is based on the Einstein detailed balance relations. McCumber shows that the vibronic laser gain can be related to the observed fluorescence spectrum under certain special conditions, in spite of the fact that several nondegenerate vibronic levels contribute.

McCumber divides the vibronic states into independent manifolds, internally coupled thermodynamically at temperature T through the lattice phonon interactions, but uncoupled with each other. The manifolds are then divided into two classes. The "j" manifolds contain upper states of the laser transition and the "i" manifolds contain the lower states, as illustrated schematically in Fig. 9.4. The vibronic levels associated with each electronic level represent the many phonon energies in the system in single and multiple orders, only a few of which are illustrated. Even though the coupling to different vibrational states will vary, and for some be extremely weak, each state should be represented. We associate each electronic state, having energy E_j (or E_i), with its own separate vibronic manifold, whether the state be degenerate or not. The distribution of vibronic energy levels in each manifold is nearly the same, although not exactly the same. The laser gain per unit length $g_\lambda(k, E)$, for radiation with energy E, unit wave vector k, and polarization λ is given by

$$g_\lambda(k, E) = \sum_{ji} \{n_j - n_i \exp[(E - \hbar\mu_{ji})/kT]\} F_\lambda(k, E)_{ji} [c^2 h^3/E^2 n_\lambda^2(k, E)] , \tag{9.1}$$

where n_j (or n_i) is the average number of ions per unit volume that occupy the specific electronic state, $\hbar\mu_{ji}$ is a temperature-dependent excitation potential from i to j, $n_\lambda(k, E)$ is the index of refraction, k is Boltzmann's constant, and T is temperature. McCumber describes $\hbar\mu_{ji}$ as the "net free energy" required to excite one impurity ion from multiplet i to multiplet j while maintaining the initial lattice at temperature T. The function $F_\lambda(k, E)_{ji}$ is the rate at which photons of polarization λ are emitted per unit solid angle about k, per unit energy about E, as a result of a vibronic transition from j to i. This function includes contributions from both phonon-assisted and no-phonon transitions.

The gain expression may be simplified if, in addition to thermodynamic equilibrium within the vibrational manifold, we assume that both the excited and ground manifold groups are in equilibrium internally but that the two groups remain thermodynamically uncoupled with each other. The validity of this assumption (the quasi-thermodynamic-equilibrium approximation)

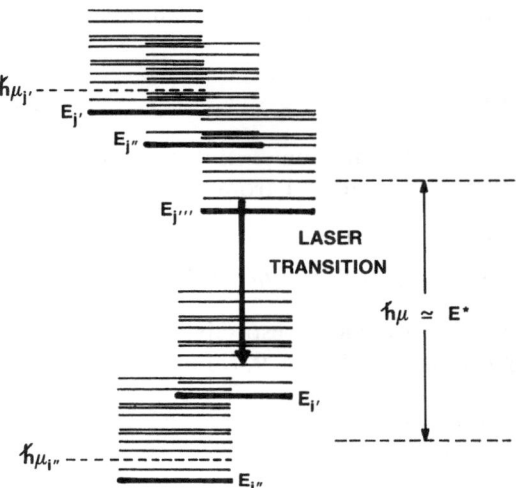

Fig. 9.4. Schematic illustration of manifolds of vibronic states associated with each electronic state. E^* is defined by (9.11) below

depends on the extent to which the *intragroup* transition rates are fast compared to the *intergroup* rates even during laser operation. In alexandrite, these (intragroup) relaxation times are on the order of a few picoseconds or less, short compared to the laser Q-switched pulse duration.[1]

The transition rate is related to the stimulated emission cross section $\sigma_{e\lambda}(k, E)_{ji}$:

$$\sigma_{e\lambda}(k, E)_{ji} = F_\lambda(k, E)_{ji} \frac{h^3 c^2}{E^2 n_\lambda^2(E)} \ . \tag{9.2}$$

Making the quasi-thermodynamic-equilibrium approximation, the population of the electronic states j and i are then given by

$$n_j = \frac{n_2 \exp(-\hbar\mu_j/kT)}{\sum_j \exp(-\hbar\mu_j/kT)} \quad \text{and} \tag{9.3}$$

$$n_i = \frac{n_1 \exp(-\hbar\mu_i/kT)}{\sum_i \exp(-\hbar\mu_i/kT)} \ , \quad \text{where} \tag{9.4}$$

$$\mu_j - \mu_i = \mu_{ji} \ . \tag{9.5}$$

Here n_2 is the total concentration of ions in the excited multiplets (the j group) and n_1 is the total concentration of the ions in the ground multiplet(s) (the i group).

---

[1] In mode-locked operation it is the pulse train duration that is relevant, not the individual pulse duration. In this case, the level depopulation during a single passage of a pulse is small (only the accumulated effect is large), and the system relaxes between occurrences.

By carrying out the summations indicated in (9.1), the gain assumes the simplified form

$$g_\lambda(k, E) = \{n_2 - n_1 \exp[(E - \hbar\mu)/kT]\} \sigma_{e\lambda}(k, E) \ . \tag{9.6}$$

In this equation, $\sigma_{e\lambda}(k, E)$ is the "effective" emission cross section of the total system of states, and $\hbar\mu$ is the excitation potential from the ground to the excited manifolds given by

$$\hbar\mu = -kT\,[\ln \textstyle\sum_j \exp(-\hbar\mu_j/kT) - \ln \textstyle\sum_i \exp(-\hbar\mu_i/kT)] \ . \tag{9.7}$$

The first and second terms of (9.7) are represented, respectively, by the upper and lower dotted levels appearing on the right in Fig. 9.4. Finally,

$$F_\lambda(k, E) = \frac{\sum_{ji} F_\lambda(k, E)_{ji} \exp(-\hbar\mu_j/kT)}{\sum_j \exp(-\hbar\mu_j/kT)} \ . \tag{9.8}$$

Referring to (9.3), $F_\lambda(k, E)$ is, therefore, the total emission rate function. The effective emission cross section is given by

$$\sigma_{e\lambda}(k, E) = F_\lambda(k, E)\,\frac{h^3 c^2}{E^2 n_\lambda^2(k, E)} \ . \tag{9.9}$$

The laser gain from (9.6) consists of two terms, the second of which is a loss term corresponding to ground state absorption. The effective ground state absorption cross section $\sigma_{a\lambda}$ is related to the effective emission cross section by

$$\sigma_{a\lambda} = \exp[(E - \hbar\mu)/kT]\,\sigma_{e\lambda} \ . \tag{9.10}$$

Since the states are partitioned such that each vibronic multiplet contains one electronic state, each ij specifies a single no-phonon transition of energy E_{ij}.

One may approximate $\hbar\mu_{ji}$ by the no-phonon transition energies E_{ji}. According to (9.7), the error introduced by this approximation depends on the relative energies of the vibrational states before and after the transition [as they impact $\mu_i(\mu_j)$]. At 0 K, the error caused by this approximation vanishes. As discussed in Sect. 9.1, the presence of the impurity ion itself only slightly perturbs the vibrational states except for a very few local mode states; the change in electronic state energy must produce an even smaller effect. The approximation is therefore expected to be a very good one even at room temperature. Making this approximation, (9.7) becomes

$$E^* = -kT\,[\ln \textstyle\sum_j \exp(-E_j/kT) - \ln \textstyle\sum_i \exp(-E_i/kT)] \ , \tag{9.11}$$

where E_j and E_i are energies of the electronic states. This approximation provides a direct and convenient means to estimate the excitation potential from the optical spectra provided the lower-energy no-phonon lines are resolved. Because of the Boltzmann factors in (9.11), only the lowest electronic levels in each group contribute significantly to the sum. This approximation has been verified for alexandrite [9.9] by comparing E^* with a more direct measurement of $\hbar\mu$. The latter is obtained using (9.10) to relate the measured emission to the absorption cross sections in the same spectral region (see Fig. 9.26).[2]

It now becomes possible to relate the absolute spectral emission rate to the laser gain. Using E^* in lieu of $\hbar\mu$ from here on for convenience, E^* regulates the ground state absorption for all wavelengths according to temperature. However, experimentally it is more expedient to measure the decay rate, and to use this to normalize the relative fluorescence spectra in order to obtain the absolute spectral emission rate, than to obtain the absolute rate directly. The fluorescence quantum yield (η), required for this approach, may be determined with satisfactory accuracy by photoacoustic spectroscopy [9.11]. Whereupon, the observed decay time τ is related to the fluorescence rate by

$$\tau^{-1} = (8\pi/3\,h\eta) \sum_\lambda \int_0^\infty dE\, F_\lambda(k',E) \;, \tag{9.12}$$

where k' is perpendicular to the λ component of the electric vector. This expression is derived by integrating the dipole emission over the solid angle and is valid as written for pure electric dipole emission, the usual case of interest.

Self-Consistency of Emission Parameters

Vibronic lasers are "resonator tuned" over a wide band of fluorescence emission. The strength and breadth of this band is inversely proportional to, and thus "trades off" with, storage time. This inherent constraint puts all such lasers at a certain disadvantage relative to their fixed frequency counterparts and may determine their eventual competitive relationship with these lasers in applications where tunability is not an issue. This constraint, derived from (9.9 and 12) assuming the emission spectrum is Gaussian, is expressed in (9.13), where the parameter SC (self-consistency) is approximately unity for all materials. We have

$$SC = 3.74 \times 10^{-20} \frac{\eta\, G(\lambda_p\,[\text{nm}])^2}{n^2\,\tau[\mu s]\,\sigma_p[\text{cm}^2]\,\text{FWHM}_F[\text{cm}^{-1}]} \simeq 1 \;, \tag{9.13}$$

where G is the total emission divided by the emission in the laser polarization, n the refractive index, σ_p the peak emission cross section, FWHM_F the full width at half maximum of the fluorescence emission band, and λ_p the peak emission wavelength. [Editor's note: (9.13) is equivalent to (1.8). ($G = \frac{1}{3}$ for an isotropic system.)]

---

[2] The assumption of ground state degeneracy made in [9.10] is not necessary provided E^* is defined by (9.11) above.

9.2.2 Laser Rate Equations

The expressions for vibronic laser emission and absorption cross sections can replace the corresponding quantities in conventional three-level laser rate equations (as developed, for example, in [9.12]). Using the notation developed in this reference, the basic rate equations for a vibronic laser can now be written

$$dn_1/dt = \{n_2 - \exp[(E-E^*)/kT] n_1\} c \Phi \sigma_{e\lambda} + n_2/\tau - W_p n_1 \ , \tag{9.14}$$

$$n_{tot} = n_1 + n_2 = \text{const.} \ , \quad \text{and} \tag{9.15}$$

$$d\Phi/dt = c \Phi(\sigma_{e\lambda} - \sigma_{2a\lambda}) - \Phi/\tau_c + S \ , \tag{9.16}$$

where Φ is the photon density, τ is the fluorescence storage time, W_p is the pumping rate, and S is the rate that spontaneous emission adds to the resonator mode. The excited state absorption cross section $\sigma_{2a\lambda}$ has also been introduced as discussed below. It remains, at this point, an experimentally determined parameter that is a function of wavelength, polarization, and temperature. The lower and upper state degeneracies g_1 and g_2, normally present in the excitation equation (9.14), have been replaced by a term in E^* as follows:

Standard theory   *Vibronic theory*

$$(\gamma - 1) = g_2/g_1 \quad \equiv \quad \exp[(E-E^*)/kT] \ . \tag{9.17}$$

a) Ground-State Absorption

An interesting aspect of vibronic laser kinetics is the combination of three- and four-level laser operation in a single material (a property of alexandrite and emerald). The foregoing analysis shows that, in the quasi-thermo-dynamic-equilibrium approximation, the ground state absorption is controlled by a wavelength-independent but temperature-dependent parameter, E^*, which contains all effects of state splittings and degeneracies in both excited and ground manifolds. The rate equations for a vibronic laser provide an understanding of how three- and four-level laser kinetics are unified in this system.

For a "pure" four-level laser, γ, defined by (9.17), is unity, and the ground state absorption is zero. The relaxation of the terminal level to the ground state must be fast compared to the pulse duration, the essence of the quasi-thermodynamic assumption. For alexandrite, this relaxation time is on the order of 10^{-13} s, see Sect. 9.2. However, as the wavelength approaches the R-line, E approaches E^*. When E becomes equal to E^*, γ is 2 and the kinetics are that of a pure three-level laser with equal degeneracies in the upper and lower laser levels.

The vibronic gain for alexandrite, from (9.6), is plotted in Fig. 9.5 [9.10]. (A similar plot for the case of MgF_2:Ni^{2+} was made by *McCumber* [9.8].)

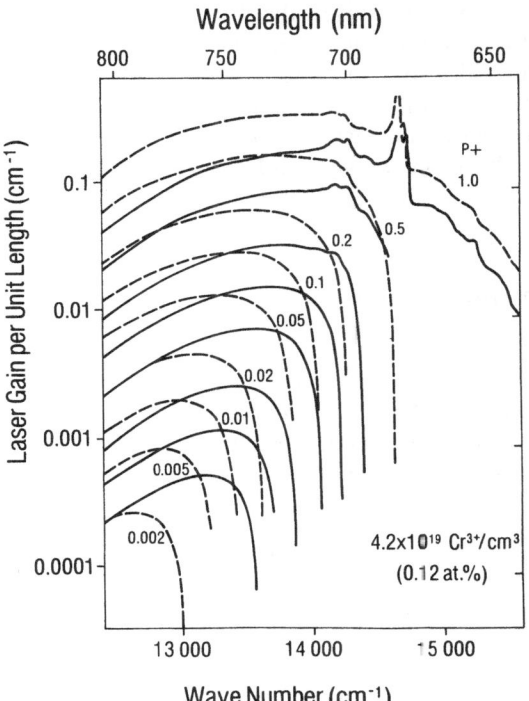

Fig. 9.5. Alexandrite laser gain obtained from the emission spectrum of Fig. 9.24 and from (9.6, 9); excited state absorption is neglected. The numbers next to each pair of curves indicate $p+$, the excitation level. (——) are for 25 °C; (– – –) are for 75 °C

This plot, which excludes excited state absorption, illustrates the continuous transformation between three- and four-level laser action and the role played by the strong R-line. The plot is made with excitation level as a parameter, where the excitation level, p^+, is defined as the fraction of active ions excited. When the excitation level reaches 50%, the R-line gain clearly becomes dominant. With $E = E^*$, the lasing is formally three level (with equal degeneracy in the upper and lower states), irrespective of whether or not E^* falls precisely on the R-line.

b) Effect of Excited State Absorption

Excited state absorption does not enter the excitation rate equation (9.14) because the decay path of the doubly excited state is nonradiative to the excited state with high yield and with a fast (picosecond) relaxation rate. Excited state absorption is, however, a loss mechanism for resonator photons, and it does enter the flux equation (9.16). If the excited state absorption cross section exceeds the emission cross section, then no gain will occur regardless of the strength of excitation. This property limits the lasing band in alexandrite on the long-wavelength side.

Excited state absorption, as a loss mechanism, differs from ground state absorption by reducing the slope efficiency in addition to raising the laser threshold. Ground state absorption only raises the threshold. However, neither mechanism operates entirely independently of the other. Neglecting the ground state absorption, we find that excited state absorption increases the threshold by the factor $\sigma_e/(\sigma_e - \sigma_{2a})$ in Q-switched operation. However, so long as the initial excitation level is increased by the same factor as the threshold, the Q-switched pulse duration, power, and energy are the same. The only change is in the efficiency, which is reduced by $(\sigma_e - \sigma_{2a})/\sigma_e$.

9.3 Tunable Paramagnetic Ion Lasers: Classes and Characteristics

In this section, the more important classes of magnetic ion lasers are examined. Emphasis is placed on their material and electronic characteristics and the impact of these on laser properties. Table 9.1 lists popular names and

Table 9.1. Popular names and chemical formulas of paramagnetic ion laser materials

Name or acronym	Chemical formula
Alexandrite	$BeAl_2O_4 : Cr^{3+}$
Emerald	$Be_3Al_2(SiO_3)_6 : Cr^{3+}$
Cr : GSGG	$Gd_3(Sc,Ga)_2Ga_3O_{12} : Cr^{3+}$
Cr : GSAG	$Gd_3Sc_2Al_3O_{12} : Cr^{3+}$
Cr : GGG	$Gd_3Ga_5O_{12} : Cr^{3+}$
Cr : LLGG	$(La,Lu)_3(Lu,Ga)_2Ga_3O_{12} : Cr^{3+}$
Cr : YSGG	$Y_3Sc_2Ga_3O_{12} : Cr^{3+}$
Cr : YGG	$Y_3Ga_3O_{12} : Cr^{3+}$
Cr : KZnF$_3$	$KZnF_3 : Cr^{3+}$
Cr : ZnWO$_4$	$ZnWO_4 : Cr^{3+}$
Cr : SrAlF$_5$	$SrAlF_5 : Cr^{3+}$
V : MgF$_2$	$MgF_2 : V^{2+}$
V : CsCaF$_3$	$CsCaF_3 : V^{2+}$
Ti : sapphire	$Al_2O_3 : Ti^{3+}$
Co : MgF$_2$	$MgF_2 : Co^{2+}$
Co : KZnF$_3$	$KZnF_3 : Co^{2+}$
Co : KMgF$_3$	$KMgF_3 : Co^{2+}$
Co : ZnF$_2$	$ZnF_2 : Co^{2+}$
Ni : MgF$_2$	$MgF_2 : Ni^{2+}$
Ni : KMgF$_3$	$KMgF_3 : Ni^{2+}$
Ni : MgO	$MgO : Ni^{2+}$
Ni : CAMGAR	$CaY_2Mg_2Ge_3O_{12} : Ni^{2+}$
Ce : YLF	$LiYF_4 : Ce^{3+}$
Ce : LaF$_3$	$LaF_3 : Ce^{3+}$
Sm : CaF$_3$	$CaF_3 : Ce^{3+}$
Ho : YAG	$Y_3Al_5O_{12} : Ho^{3+}$
Ho : YLF	$LiYF_4 : Ho^{3+}$
Ho : BaY$_2$F$_8$	$BaY_2F_8 : Ho^{3+}$

Table 9.2. Cr^{3+} and V^{2+} ion doped materials in which tunable laser operation has been demonstrated

Name or acronym	Peak wavelength [nm]	Tuning range [nm]	Operating temperature	Storage time [μs]	Pumping means	Slope effic.[a] [%]	Peak emiss. cross section [10^{-20} cm²]	Refs.
Alexandrite	752	710–820	22–300°C	260–60	Flash lamp	0.5–5	0.7–2	[9.10,13–16]
Alexandrite	756		RT	262	Kr+ laser	51	0.7	[9.17]
Emerald		720–842			Kr+ laser	34	3.1	[9.17,18]
Emerald	684.8	658–720		65	630 nm[b]	1.7[c]		[9.19]
Emerald	765	729–809	RT		Kr+ laser (cw)			[9.20]
Cr:ZnWO4		980–1090	77 K	8.6	Kr+ laser	13	43	[9.21]
Cr:ZnWO4	1030		RT	5.4[d]	Pulsed dye			[9.22]
Cr:GSAG	784	735–820[e]	RT	150	Kr+ laser	18.5		[9.23,24]
Cr:GGG	745		RT	159	Kr+ laser	10	0.6	[9.23]
Cr:LLGG	850		RT	68	Kr+ laser	3[e]	1.6	[9.23]
Cr:YSGG	750		RT	139	Kr+ laser		0.6	[9.23]
Cr:YGG	730		RT	241	Kr+ laser		0.36	[9.23]
Cr:GSGG	770		RT	115	Kr+ laser		0.8	[9.23]
Cr:GSGG	777	742–842	RT		Kr+ laser	28		[9.24,25]
Cr:GSGG		766–820	RT		Flash lamp	0.05		[9.26]
Cr:KZnF3	825	785–865	RT			14		[9.27–29]
Cr:KZnF3		758–845	80 K	270				[9.22]
V:MgF2	1120	1070–1150	80 K	2300	Kr+ laser		0.087[f]	[9.30–32]
V:CsCaF3	1282	1240–1340	80 K	2500	Kr+ laser	0.06[c]		[9.29]
Cr:SrAlF5	925	825–1010	RT	30	Kr+ laser	3.6	2	[9.33,34]

[a] In laser-pumped cases, number is in relation to energy absorbed by crystal.
[b] 8 ns pulsed laser.
[c] Low because of apparent color center formation or crystal defects.
[d] Two Cr sites may exist; other site has value of 0.5 μs.
[e] Reported in [9.7] in reference to unpublished work by Struve and Huber.
[f] Peak laser gain cross section (emission cross section about 5×10^{-21} cm² is reduced by excited state absorption).

chemical formulas of most of these laser materials. A summary of their laser properties is presented in Tables 9.2 and 3.

Most of the success in vibronic lasers has derived from the divalent $3d$ ions, primarily Ni, Co, and V, and from trivalent Cr and Ti. Because Cr^{3+} and V^{2+} both have three electrons in the $3d$ shell, they are very similar in spectroscopic and laser properties. Divalent Ni, Co, and Mn have lased in a number of hosts, but only at cryogenic temperatures. The ions Ti^{3+} and Cu^{2+}, which have only one electron (or one hole) in the $3d$ shell also have similar properties, notably the absence of excited state absorption and a potential tuning range that is consequentially broad. Nevertheless, only Ti^{3+} has been successfully lased to date, as problems of crystal growth and stability have prevented lasing with Cu^{2+}. Finally, Ce^{3+}, with its unique electronic properties mentioned briefly in Sect. 9.1, has been lased on broadly tunable

Table 9.3. Ti^{3+}, Co^{2+}, Ni^{2+}, Ce^{3+}, Sm^{2+} and Ho^{3+} doped materials in which tunable laser operation has been demonstrated

Name or acronym	Peak wavelength [nm]	Tuning range [nm]	Operating temperature	Storage time	Pumping means	Slope effic.[a] [%]	Peak emiss. cross section $[10^{-20}$ $cm^2]$	Refs.
Ti : sapphire	750	660 – 986	RT	3.2 μs	Ar^{+}[b]	53	10 – 20	[9.35, 36]
Co : MgF_2	1850	1510 – 2280	80 K	1.2 ms	1.32 YAG	40	0.15	[9.6, 37]
Co : $KZnF_3$		1650 – 2070	80 K		cw 1.32 μm[c]			[9.38]
Co : $KZnF_3$	1970	1700 – 2260			Ar^{+}	8		[9.39]
Co : $KMgF_3$	1821		77 K	3.1 ms	Xe flash lamp			[9.40, 41]
Co : ZnF_2	2165		77 K	0.4 ms	Xe flash lamp			[9.40, 41]
Ni : MgF_2	1668	1600 – 1740	77 K	12.8 ms	1.32 YAG	28		[9.42]
Ni : $KMgF_3$	1591		77 K	11.4 ms	Xe flash lamp			[9.43]
Ni : MgO	1.32, 1.41		>80 K	3.6 ms	1.06 YAG	57		[9.6, 44]
Ni : CAMGAR			80 K			0.7[d]		[9.42]
Ce : YLF	308, 325			40 ns	KF-excimer		800	[9.45]
Ce : LaF_3				18 ns	KF-excimer		700	[9.46]
Sm : CaF_2		708.5,	<210 K	~2 μs	QS ruby			[9.47]
Ho : YAG[e]	2.05	745	77 K		cw tungsten	5[d]		[9.41, 48]
Ho : YLF[e]			77 K		cw tungsten	5[d]		[9.49]
Ho : YLF[e]	2.05, 2.063		RT	12 ms	Flash lamp			[9.50]
Ho : BaY_2F_8[e]	2.171		RT					[9.51]

[a] In laser-pumped cases, number is in relation to energy absorbed by crystal.
[b] Argon ion laser.
[c] Nd : YAG laser.
[d] Overall efficiency.
[e] Sensitized with Er^{3+} and Tm^{3+} to improve lamp pumped efficiency.

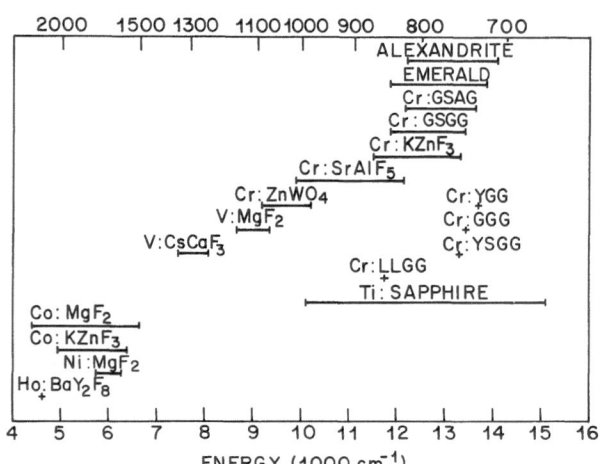

Fig. 9.6. Wavelength coverage of tunable transition metal ion lasers, according to data from Tables 9.2 and 3

$5d$ to $4f$ transitions in selected hosts. Combined, the vibronic lasers offer an extensive range of wavelength coverage, as illustrated in Fig. 9.6.

9.3.1 Cr^{3+} and V^{2+}

Chromium, used in ruby for the first visible laser [9.52], has since provided a class of tunable lasers when incorporated in several hosts, including chrysoberyl, beryl, several garnets, $ZnWO_4$, and more recently $SrAlF_5$. The laser materials of this series and many of their properties are outlined in Table 9.2. Undoubtedly, this list is not complete, as work in this area continues vigorously.

The spectroscopic properties of this series of materials present a definite pattern when considered relative to the strength of the crystal field interaction on the Cr^{3+} or V^{2+} ion. These properties may be illustrated with the aid of the Tanabe-Sugano diagram for the $3d^3$ configuration in octahedral coordination presented in Fig. 9.1. (For tetrahedral sites, the energy level structure is entirely different [9.1].) Here the diagram illustrates the crossover between the ^2E and $^4T_2$ state multiplets with increasing cubic field interaction, delineating thereby the so-called strong and weak field lasers. It also identifies the $^4T_2$ and $^4T_1$ multiplets responsible for the double-humped laser-pumping bands characteristic of these systems, and the general position of the $^2T_2$ and $^2A_1$ multiplets, potential contributors to excited state absorption as discussed below.

The energy gap between the $^4A_2$ and $^4T_2$ multiplets is denoted in the diagram by $10Dq$, which is a measure of the strength of the cubic crystal field interaction. The Racah parameter, B, is an attribute of the free ion, and in effect specifies the energy of the $^4T_1$ multiplet relative to that of the $^4T_2$. In the weak field case, the value of $10Dq/B$ lies below the cross-over point between the $^4T_2$ and ^2E multiplets; for the strong field case, it lies above. Examples of strong field lasers are ruby, $YAG:Cr^{3+}$, and alexandrite. The weakest field Cr laser thus far discovered is $ZnWO_4:Cr^{3+}$, although $CsCaF_3:V^{2+}$ has the weakest crystal field interaction in this series overall, and exhibits the longest wavelength range. Near the crossover are emerald and $Cr:GSGG$.

The relative position of these levels profoundly affects the storage time of the excited ion. Figure 9.7 illustrates schematically the temperature dependence of storage time for materials with progressively stronger fields. The drop in lifetime below 400 K results from thermal excitation of the $^4T_2$ multiplet that has a much higher fluorescence rate. Because this reduction in lifetime is caused by increased radiative decay, there is a corresponding increase in the effective emission cross section, which generally implies a higher gain (as discussed in Sect. 9.2). The initial increase in the storage time for curve (d) is the result of thermal population of the ^2E level. At much higher temperatures, nonradiative relaxation of $^4T_2$ to $^4A_2$ occurs, and is accompanied by a corresponding reduction in quantum yield and no increase in emission cross section. For curves (e) and especially (f) the falloff due to nonradiative relaxation

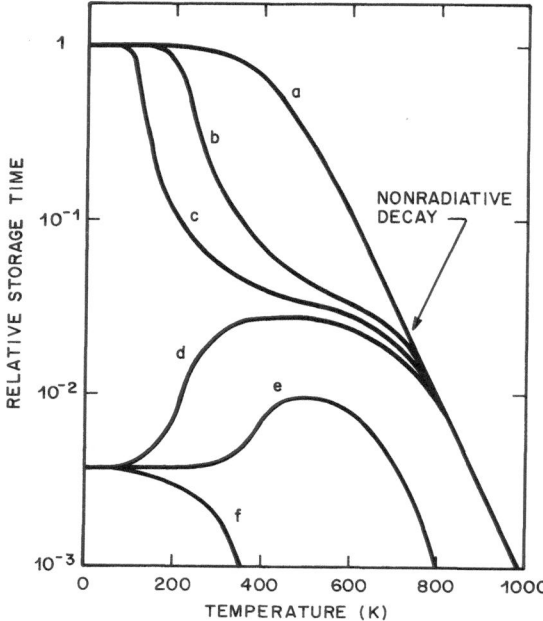

Fig. 9.7. Schematic illustration of the temperature dependence of storage time for $3d^3$ ion vibronic lasers: (a) based on actual ruby data [9.53,54]; (b) low temperature component based on alexandrite data (Fig. 9.23); (c) based on emerald data [9.53,54]; (d and e) for low field examples where in (d) the initial rise at low temperature is due to the thermal population of ^2E and in (e) the early falloff is characteristic of low field materials; (f) specifically related to the behavior of Cr:CsCaF$_3$ [9.29]

occurs earlier, consistent with what is usually observed with the lower field materials. Curve (f) specifically relates to Cr:CsCaF$_3$ having a low temperature lifetime of about 3 ms [9.29]. The storage times at zero temperature are normalized to two different values, one for the high and one for the low field materials, in order to represent the disparity between the radiative lifetimes of the ^2E and $^4T_2$ multiplets. Figure 9.7 also illustrates how the distinction between the high and low field cases becomes discrete at low temperatures.

a) Ruby, Cr:YAG, Alexandrite, and Emerald

All four of these strong field lasers exhibit three-level (nonvibronic) laser operation on the ^2E to $^4A_2$ transition – the R-line. Since Maiman's early work on ruby, R-line lasing has been demonstrated in Cr:YAG [9.55], in alexandrite [9.56,57], and in emerald [9.19]. Only in alexandrite and emerald [9.58] has vibronic lasing also been observed. [The author has made a cursory (unsuccessful) attempt to lase both ruby and Cr:YAG at an elevated temperature.] Figure 9.7 also illustrates how the distinction between the high and low temperature.

The energy level diagram for alexandrite is presented in Fig. 9.8. From this diagram and the alexandrite fluorescence lifetime data of Fig. 9.23, a five-level model of the alexandrite laser kinetics has been constructed, as shown in Fig. 9.9. The introduction of the storage level provides a convenient way to describe the temperature-dependent gain of alexandrite and emerald.

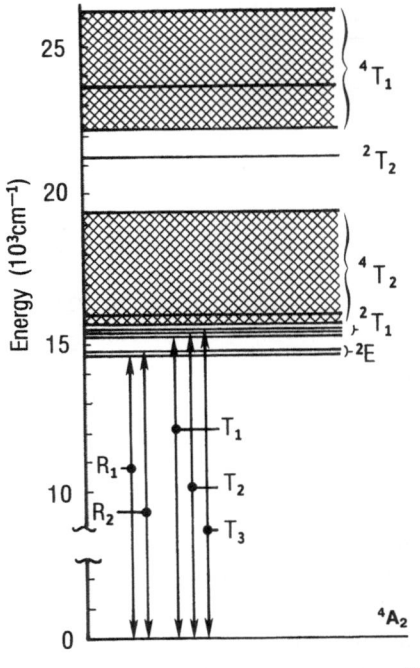

◄ **Fig. 9.8.** Alexandrite mirror-site energy levels. Crosshatched broad bands are vibronic sidebands of absorption to $^4T_2$ and $^4T_1$ [9.10]

Fig. 9.9. Schematic illustration of the "five-level" laser kinetics scheme of alexandrite [9.10]. Parameters are derived from the theoretical fit to the lifetime data in Fig. 9.23

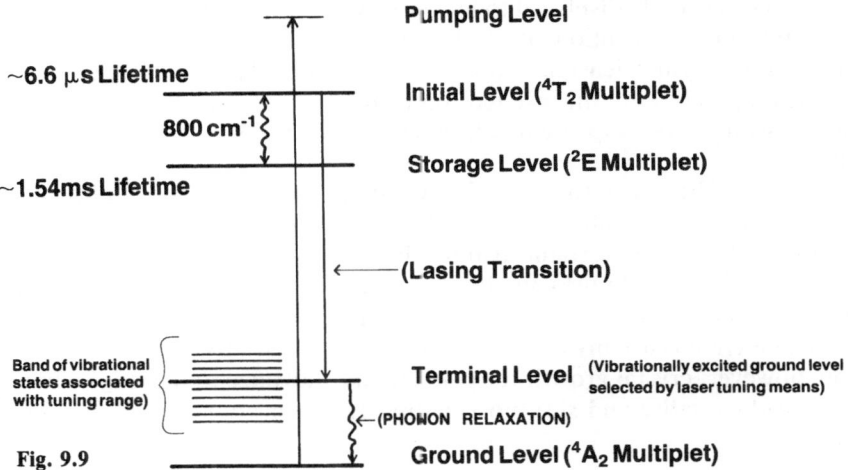

Fig. 9.9

Many of the salient characteristics of emerald and alexandrite are compared in Table 9.4. This table illustrates correlation between emission parameters discussed in Sect. 9.2. For alexandrite, G is comparatively large, approximately 80%, because the emission is strongly polarized (Fig. 9.24). The emerald parameters are obtained from [9.59].

The comparison between alexandrite and emerald is an interesting one, with both materials having attractive properties. The comparatively poor

Table 9.4. Comparison of properties of alexandrite and emerald

Property	Alexandrite	Emerald
Hardness [Mohs]	8 – 8.5	7.5 – 8
Thermal conductivity [W/cm K]	0.23	0.05
Thermal expansion coefficient [$10^{-6} K^{-1}$]	~6	~1.2
(Rod) Thermal load to fracture [kW/cm]	0.6	0.57
Fraction of emission in lasing polarization, G	0.8	0.5
Refractive index, n	1.74	1.57
Quantum yield, η	~1	~1
Fluorescence bandwidth [cm^{-1}]	~1800	~1720
Fluorescence emission peak wavelength [nm]	704	720
Peak emission cross section, σ_p [10^{-21} cm]	10	31
Peak excited state absorption cross section [10^{-21} cm]	~1	3 – 17
Storage time [µs]	260	65
Self consistency parameter, SC	1.05	1.13

thermal conductivity of emerald is compensated by its low coefficient of thermal expansion, which makes the thermal load to fracture roughly comparable for the two materials. In the balance, the principal difference between the materials is the fact that emerald operates optimally near room temperature, where it is comparable to alexandrite operating near 170 °C. The emerald band is slightly displaced to longer wavelengths, and emerald exhibits less ground state absorption at its lower optimal operating temperature. However, the thermal lensing properties of emerald lasers have not been determined as yet, and the excited state absorption may be significant. Nevertheless, a laser-pumped emerald laser has yielded 34% slope efficiency (Table 9.2).

The greatest drawback to emerald thus far encountered is the quality and size of the crystals produced. Nevertheless, recently, an attenuation in emerald of only 0.4%/cm round trip has been reported [9.18]. Emerald can not be grown by Czochralski methods because emerald undergoes a solid phase transition during cool down that inevitably cracks the crystal. Consequently, emerald is usually grown hydrothermally, a relatively slow process. In time, however, crystal growth for emerald may advance to the stage where material in the quality and size most desired for laser application is readily available.

b) Garnets

The garnets comprise a wide range of compositions and allow for many substitutions because of their open crystal structure. This same property is largely responsible for their use in magnetic bubble domain films where bubble characteristics can be compositionally tuned to the application. For lasers, as we have seen, the garnets offer substantial variation in crystal field strength. Thus an opportunity exists for obtaining extended wavelength coverage by variations in composition.

The crystal field, nevertheless, remains cubic with only slight distortions induced by the substituted ions. The cubic symmetry of garnets is a disadvantage for vibronic lasers for several important reasons. First of all, the emission is nonpolarized, and the unwanted polarization component merely detracts from storage time, and provides no compensating gain. Cubic crystals have no appreciable natural birefringence that would otherwise suppress the negative effects of thermally induced birefringence on laser emission, as occurs, for example, in alexandrite. Compounding this effect is the relatively poor thermal conductivity of the garnets, an inherent property associated with the high concentration of heavy ions. The inversion site symmetry of the garnets does increase the storage time of the excitations (albeit at the expense of gain cross section) and thereby makes possible Q-switched operation under flash lamp excitation. By contrast, those weak field materials that do not have inversion-site symmetry have exceptionally short lifetimes (a few microseconds), largely preempting this important mode of operation.

In GSGG (Gd_3 (Sc, Ga)$_2$ Ga_3O_{12}) and GSAG ($Gd_3Sc_2Al_3O_{12}$), the crystal field is essentially at the crossover. The R-lines still do not appear at room temperature, because the inversion-site symmetry does not allow for a strong electric dipole transition. Also, lifetime broadening occurs from thermally induced transitions with the $^4T_2$ levels that are in close proximity. The $^2T_1$ multiplet, lying just above the ^2E, is not important in laser operation, although it may be involved in excited state absorption as discussed in Sect. 9.4.2. The quantum efficiency η is near unity for Cr:GSGG and Cr:GGG, at least for concentrations up to 3×10^{20} ions/cm^3 [9.60].

Both Cr:GSGG and Cr:GSAG have proved to be suitable tunable laser materials. Of the two, Cr:GSAG in particular is attractive because of its favorable crystal growth characteristics [9.22]. In fact, this material grows extremely well by Czochralski methods. Al_2O_3 is less aggressive to the iridium crucible than is Ga_2O_3 and has much lower vapor pressure [9.22]. Hence, the composition of Cr:GSAG remains stoichiometric and uniform over the entire pulling period. Moreover, the distribution coefficient is near unity, ensuring uniform doping throughout the crystal and no difficulty in introducing adequate dopant for absorption of pump light. No concentration quenching of the fluorescence rate is observed up to a concentration of 5×10^{20} cm^{-3} [9.22].

Flash lamp pumping of Cr:GSGG has also been reported [9.26,61]. A slope efficiency of 0.33% was obtained using a $NaNO_2$ uv filter, resulting in 14 mJ output from a 76 mm long laser rod [9.61].

The garnet vibronic lasers with weaker crystal fields have only begun to be studied and relatively little has been reported on their properties. Highlights of the available information are presented in Table 9.2.

c) Other Materials

Zinc tungstenate ($ZnWO_4$) has the monoclinic wolframite structure wherein Cr^{3+} substitutes for divalent Zn. The growth and laser properties of $ZnWO_4$

have been reported [9.21]. The ionic radius of Cr^{3+} is small compared to Zn^{2+}, which accounts for the particularly weak crystal field in this case. However, substituting an ion of another charge into a lattice site creates a space charge that must be compensated. Crystal growth is significantly complicated when a compensating dopant must be added; frequently, such a dopant will not have the same probability of entering the crystal as the primary dopant and a heterogeneous crystal having inferior properties will result. In the case of $ZnWO_4$, charge compensation may be achieved by Zn vacancies induced during growth by the entry of the Cr^{3+} impurity − one vacancy for every two Cr ions. The result is the creation of Cr^{3+} species with different lifetimes at room temperature, one of which is apparently paired with the Zn vacancy.

The fluorides tend to have compensating characteristics. The thermal conductivity is comparatively low because of the weaker fluoride bond strength. But the refractive index is also low, which aids in extending the fluorescence lifetime for a given gain − bandwidth product. Also, the second index of refraction is typically low for fluorides, which is beneficial in reducing nonlinear self-focusing and resulting optical damage. As a rule, fluoride laser hosts are chemically less stable than oxides and must be handled more carefully.

The first tunable fluoride laser to operate at room temperature was $KZnF_3:Cr^{3+}$ [9.25, 27 − 29]. It provides a respectable 14% slope efficiency when laser pumped by a Kr^+ ion laser at room temperature. It is grown by the Bridgman method as a naturally charge compensated, singly doped material. Three Cr^{3+} sites are produced with 90% of the total chromium in the laser effective site. The quantum efficiency remains high (80%) at room temperature. The tuning range covers a large fraction of the fluorescence emission spectrum, as shown in Fig. 9.10. The full role of excited state absorption has not yet been evaluated.

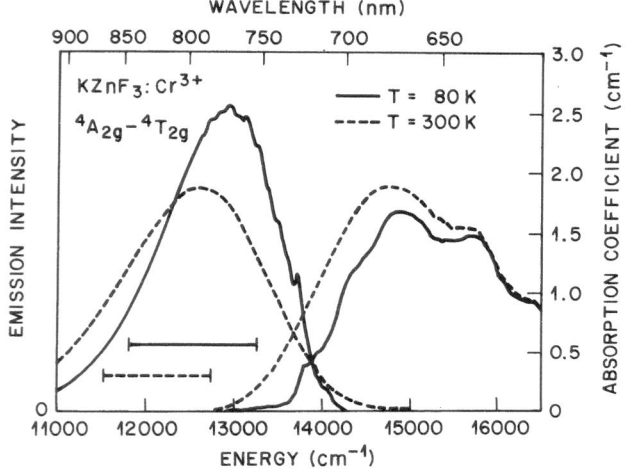

Fig. 9.10. Emission and absorption spectra of $KZnF_3:Cr^{3+}$ at 80 and 300 K (⊢——⊣), (⊢ − − ⊣) indicate the corresponding tuning ranges [9.25]

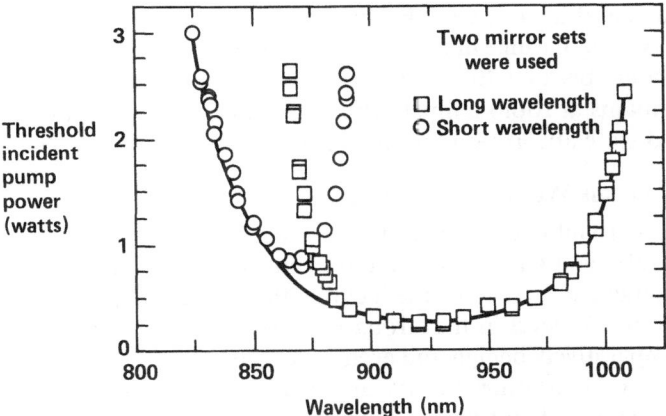

Fig. 9.11. Laser threshold curve for $SrAlF_5$ [9.33]. (Courtesy of John Caird, Lawrence Livermore National Laboratory)

One of the most recently reported Cr-doped fluoride vibronic lasers is $SrAlF_5$ [9.33, 34], which also offers one of the broadest tuning ranges (Fig. 9.6). It is uniaxial with little fluorescence polarization dependence. The fluorescence lifetime does not change appreciably between 4 and 300 K, which would ordinarily indicate a high quantum yield at room temperature. There is evidence, however, that certain of the four nonequivalent Cr sites may absorb, but not fluoresce. If true, this could be a significant source of loss, greatly reducing the material's utility. A laser of this material has been tuned at room temperature using a birefringent tuner (Fig. 9.11). In this case, the $SrAlF_5$ crystal was 0.193 cm thick, doped with about 2×10^{20} Cr^{3+} ions/cm³ and located in the center of a nearly concentric laser resonator. The pump source was a 647 nm, cw Kr ion laser, chopped to reduce the duty factor to 2% in order to minimize thermal lensing.

d) $V : MgF_2$

The ion V^{2+} behaves very similarly to Cr^{3+}, but generally has a weaker crystal field interaction, causing operation at longer wavelengths. *Johnson* and *Guggenheim* [9.62] were the first to explore $MgF_2 : V^{2+}$. More extensive later work [9.30] sparked interest in this material in connection with the laser fusion program at Lawrence Livermore National Laboratories. It was recognized that the placement and width of its absorption bands, together with its long storage time, made for particularly efficient and long-lived flash lamp operation [9.32]. Moreover, its low nonlinear index and other properties imbue the material with a high damage threshold (<100 J/cm²) that compensates to a degree its high saturation fluence. However, it was found that excited state absorption in this material mitigates strongly against efficient operation and prevents lasing at room temperature.

Further, V^{2+} has lased also in $CsCaF_3$ at 80 K [9.28], where it has produced the longest wavelength range in the series. This material may potentially lase at room temperature because 30% of its near-unity quantum efficiency at low temperatures remains at room temperature. How much the performance is affected by excited state absorption has not yet been reported, however.

e) Merits of Strong Versus Weak Field Materials

Several factors conspire to inhibit lasing on the vibronics in strong field materials, particularly for ruby and Cr: YAG where no vibronic lasing has been reported. Major difficulties can arise from the fact that the crystal must be heated in order to populate the $^4T_2$ level in the strong field case. (In alexandrite, the consequences are comparatively benign and even have some operational advantages.) In general, however, heating the crystal increases the ground state absorption, truncating the laser band on the short-wavelength side. This results from the temperature dependence of the factor $\exp[(E-E^*)/kT]$ in the ground state absorption, as discussed in the previous section. From a practical standpoint, vibronic lasing does not occur when $E > E^*$ because of the very high ground state absorption. With increasing crystal field, the increasing separation of the $^4T_2$ and ^2E levels moves E^* (E^* approximately equals the R-line energy) across the fluorescence band of the $^4T_2$ level from the shorter to longer wavelengths. In essence, this erases that portion of the fluorescence band from the lasing band. In addition, the populated ^2E and $^2T_1$ states provide the opportunity for spin allowed excited state absorption to the $^2T_2$ and $^2A_1$ multiplets, whereas such absorption from $^4T_2$ is spin forbidden. Thus, the chance for excited state absorption is increased, while at the same time, the important $^4T_2$ multiplet is depopulated.

For the low-field case the situation is simpler since $^4T_2$ is the only multiplet occupied. Excited state absorption occurs from this quartet state to nonquartet states ($^2T_2$ and $^2A_1$) and must be weak, but also broad, because these states are of different orbital composition. Excited state transitions to $^4T_1$ may be strong, but also may be relatively narrow because of the similar orbital composition. On the other hand, there is no guarantee that $^4T_1$ and $^4T_2$ are coupled to the same lattice distortion mode, and, if not, the transition could again be broad. Nevertheless, the position of the quartet levels are readily obtained from the known absorption spectra. Thus it becomes fairly straightforward to determine if the $^4T_1$ levels are in a position to absorb laser photons from the excited $^4T_2$ free energy storage level.

Exploring this further, the emission and absorption spectra for several $3d^3$ materials are presented in Figs. 9.12 and 13. In these figures, the emission band has been reconstructed in a position displaced to higher energy by the apparent energy of the lowest energy no-phonon lines of the $^4T_2$ multiplet. Even though this $^4T_2$ no-phonon line may not be directly observed, its position is near the long-wavelength base of the vibronic absorption band, symmetrically located, or nearly so, between the emission and absorption cross-section peaks. In those cases where the $^4T_2$ absorption band is clearly split, the long wavelength peak must be used.

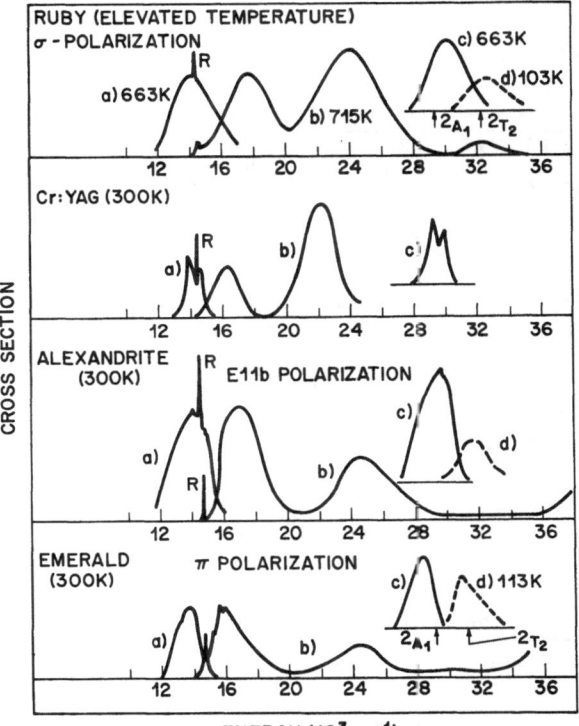

Fig. 9.12. Optical spectra of selected high-field $3d^3$ series members: (*a*) emission cross section; (*b*) absorption cross section; (*c*) emission cross section displaced by approximate $^4T_2$ lower no-phonon line energy; (*d*) excited state absorption displaced by *R*-line energy. References: ruby [9.53, 63], Cr: YAG [9.64]; alexandrite [9.10, 65]; emerald [9.19, 63]

Universal in this comparison is the fact that the excited state absorption falls in the broad gap beyond the $^4T_1$ multiplet. (Care must be taken in interpreting absorption spectra because the base line is often displaced by surface reflection. This is particularly true in the uv region where the problem is more accentuated.) However, for Cr: $SrAlF_5$ and V: $CsCaF_3$, the position of the displaced fluorescence is on the high-energy shoulder of this absorption band. The fact that excited state absorption has not prevented lasing in these materials suggests that the vibronic band of the $^4T_2$ to $^4T_1$ transition may indeed be substantially narrower than that of the $^4A_2$ to $^4T_1$ and confined to the left of the displaced peak. The levels of the $^2A_1$ multiplet fall within this gap, however, and excited state absorption involving them could contribute, particularly for the high-field materials as discussed above. The position of these higher levels and some excited state absorption data are presented for certain materials in Fig. 9.12 in connection with the displaced spectra. The source of this excited state absorption is the ^2E multiplet.

9.3.2 Ti^{3+} and Cu^{2+}

While no Cu^{2+} lasing has been reported thus far, Ti^{3+} has lased in sapphire and yttrium orthoaluminate, YALO. Ti^{3+} substitutes for aluminum in both

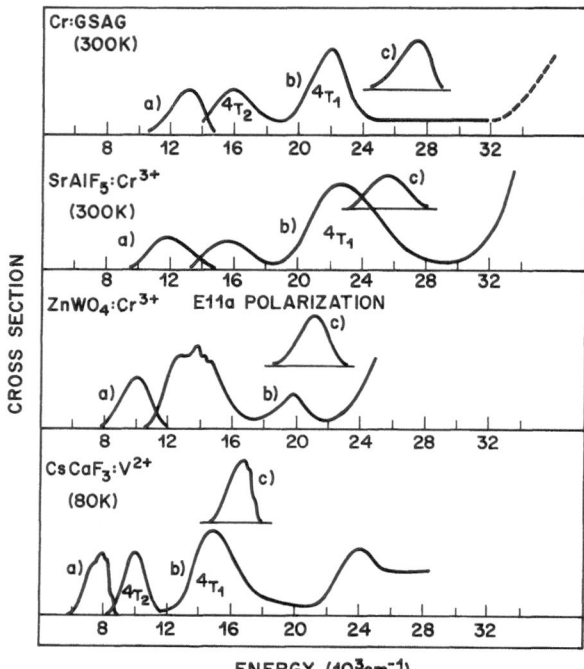

Fig. 9.13. Optical spectra of selected low-field $3d^3$ series members: (*a*) emission cross section; (*b*) absorption cross section; (*c*) emission cross section displaced by approximate $^4T_2$ lower no-phonon line energy. References: Cr:GSAG [9.22]; Cr:SrALF$_5$ [9.49]; Cr:ZnWO$_4$ [9.21]; V:CsCaF$_3$ [9.29]

materials. The absorption and fluorescence spectra of Ti^{3+}-doped sapphire appears in Fig. 9.14. There is considerable interest in this material because of sapphire's excellent laser host properties and the exceptionally broad tuning range made possible by the absence of excited state absorption. Laser emission has been achieved in this material over a continuous band from 660 nm to 986 nm, comprising most of the fluorescence band in Fig. 9.14. Excited state absorption is absent because the extremely simple state structure of the single-electron d shell presents no levels above the broad pump band in a range where excited state absorption could be a factor. Important also is the high peak emission cross section, $\sigma_e = 3 - 4 \times 10^{-19}\,\mathrm{cm}^2$, which greatly facilitates energy extraction. Ti: sapphire does exhibit a rapid reduction in η (quantum yield) near and above room temperature [9.35]. However, the principal intrinsic drawback of this material is its very short storage time of 3.8 μs. This property is a direct consequence of the combined large cross section and large bandwidth, as discussed above. A storage time this short makes Q-switched operation under flash lamp pumping nearly impractical. (Lasing in this material by flash lamp pumping has been demonstrated, however [9.66].) Consequently, Ti^{3+}:Al$_2$O$_3$ will likely find its greatest application when pumped by a pulsed laser such as frequency-doubled, Q-switched Nd:YAG. The material Ti^{3+}:YALO also has promise for practical application, largely because of a somewhat longer, 10 μs, lifetime and the possibility of lasing

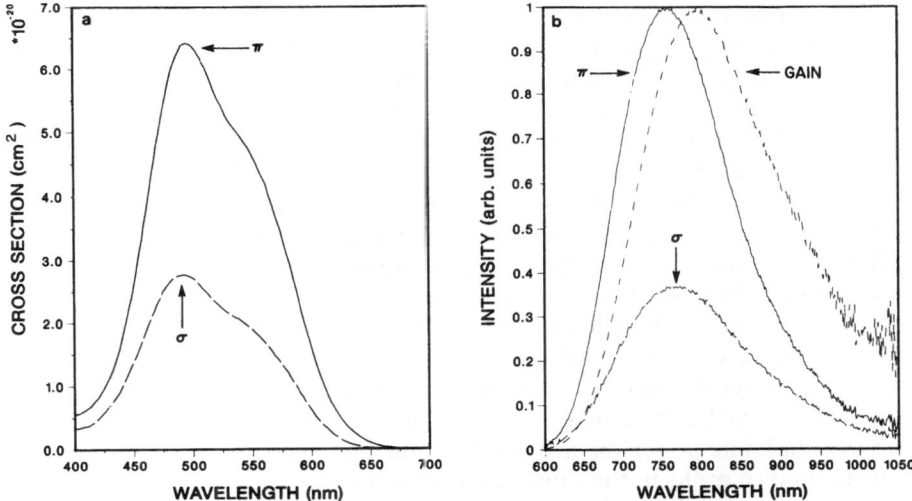

Fig. 9.14. Absorbance (**a**) and fluorescence (**b**) of Ti:sapphire for the $E \perp c$ polarization at room temperature. (After *Moulton* [9.7], courtesy of North-Holland)

down to 600 nm [9.31]. Traditionally, YALO has been a difficult material to grow, however.

Currently, Ti:sapphire has not been grown with consistently high material quality. In fact, substantial loss in the laser band is reported [9.35], which is believed to be due to impurities induced during growth. Annealing the crystals greatly reduces the absorption, but loss coefficients on the order of 1%/cm remain. Defect-induced optical damage may limit Ti:sapphire rod life. While sapphire has excellent thermomechanical properties for a laser host, it is also known for its propensity to harbor point defects. These may account for ruby's low (compared to alexandrite and Nd:YAG) tolerance to repeated pulses at high fluence (Fig. 9.48). The low saturation fluence of Ti:sapphire may well compensate for any such effect.

9.3.3 Ni²⁺ and Co²⁺

The ions Ni^{2+} and Co^{2+} were among the first explored for tunable solid-state lasers [9.40, 41, 62, 67]. Their performance was substantially improved later when cw Nd:YAG lasers were used for excitation [9.68, 69]. All of the materials that utilize these ions require cryogenic cooling. Also, the thermal conductivity of insulating crystals increases dramatically when cooled, which greatly reduces thermally induced optical distortion. The cryogenic laser-pumped resonators that have been generally employed are similar to those developed for color center lasers as illustrated in Fig. 6.2.

In particular, Ni: MgF_2 has been studied extensively [9.7, 42, 44] using a cw Nd: YAG laser pump at 1.32 μm. Nearly 2 W of cw output power and continuous tuning were obtained with a 28% slope efficiency (relative to the power absorbed by the crystal) for a crystal at 77 K. The efficiency, while high, is about half that predicted from resonator loss, and the difference is attributed to excited state absorption.

One of the most ideally suited materials for a laser host, but one very difficult to produce, is MgO. One major advantage of Ni: MgO over Ni: MgF_2 is that the absorption band is shifted to be accessible by 1.06 μm Nd: YAG, which is much more efficient than Nd: YAG at 1.32 μm. Moreover, with the crystal cooled to 80 K, 10 W average power has been obtained [9.36]. The exceptionally high thermal conductivity of MgO, particularly at 80 K, is largely responsible for this high average power performance.

Unfortunately, the tuning range is separated (presumably by excited state absorption) into a region around 1.32 μm and one around 1.41 μm that correspond to peaks in the fluorescence spectrum.

Of the compositions in this category, $MgF_2: Co^{2+}$ has shown particular promise [9.44]. Its tuning range covers virtually the entire fluorescence spectrum. The fluorescence spectra and laser performance vs wavelength obtained from pulsed and cw operation are shown in Figs. 9.15, 16, and 17, respectively. In order for the tuning range to be so broad, excited state absorption must be very low. Nevertheless, nonradiative transitions affect the quantum yield, which steadily deteriorates by two orders of magnitude from 4 K to room temperature. This degradation prevents practical use in cw

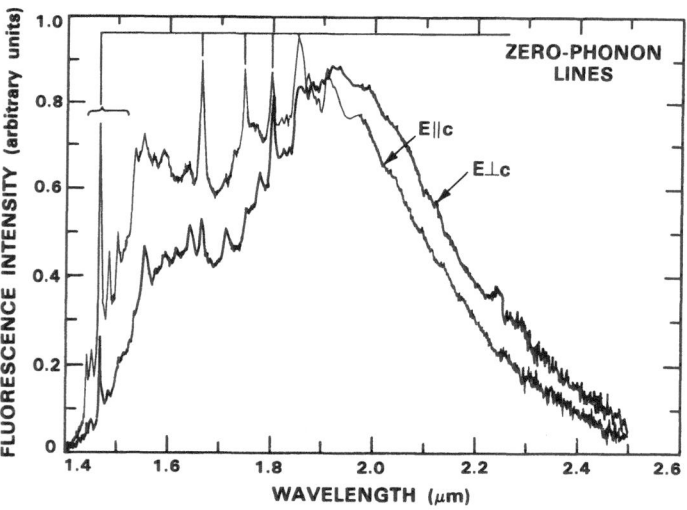

Fig. 9.15. Polarized fluorescence spectra for Co: MgF_2 at 77 K, with no-phonon lines indicated. (After *Moulton* [9.7], courtesy of North-Holland)

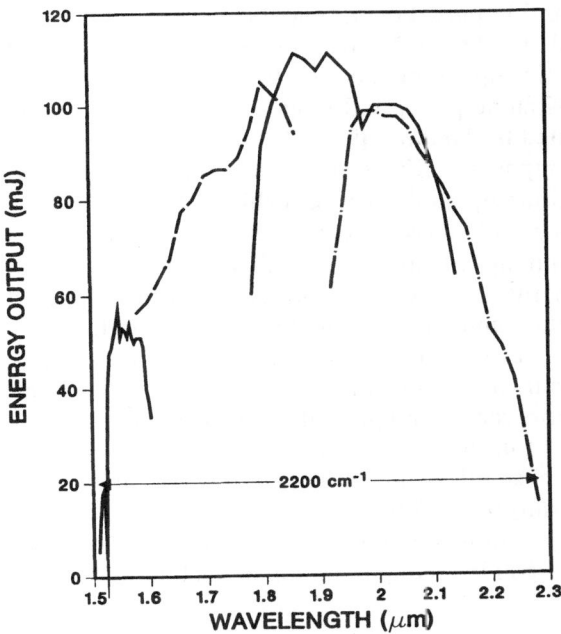

Fig. 9.16. Energy output vs wavelength for a pulsed Co: MgF$_2$ laser operating at 80 K, with input of 500 mJ at 1.34 μm from a Nd: YALO laser. Different curves are associated with different sets of cavity mirrors. The total tuning span is indicated. (After *Moulton* [9.7], courtesy of North-Holland)

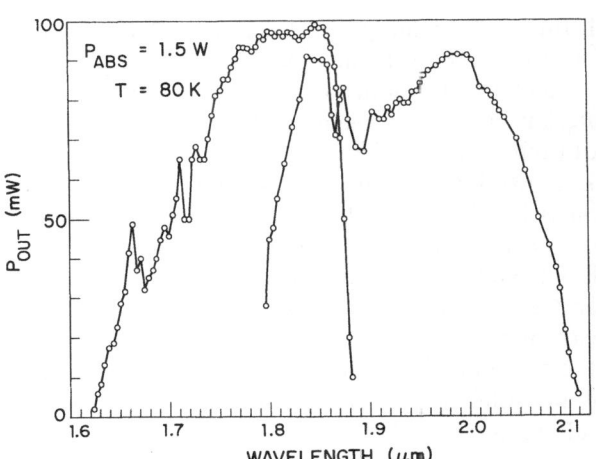

Fig. 9.17. Continuous-wave power output vs wavelength for a three-mirror-cavity, Co: MgF$_2$ laser. Crystal temperature was 80 K and the absorbed 1.33 μm pump power from a cw Nd: YAG laser was 1.5 W. The two curves resulted from different sets of cavity mirrors. (After *Moulton* [9.7], courtesy of North-Holland)

operation above about 80 K and in pulsed operation above about 225 K. The latter temperature is accessible by thermoelectric coolers, obviating the need for cryogenic liquids. Alternative operational modes have also been explored in $Co:MgF_2$, including Q-switching [9.70] and mode-locking [9.71]. Extensive reviews have been presented by *Moulton* [9.7, 37].

Laser properties of the composition $KZnF_3:Co^{2+}$ have also been studied [9.39]. Cryogenic cooling is required, as this material exhibits the same rapidly decreasing quantum efficiency with increasing temperature as $Co:MgF_2$. Thus, the resonator and pumping scheme were similar to that shown in Fig. 6.2. The pump beam in this case was double passed, which enabled absorption of up to 80% of the pump energy. The tuning bandwidth is very broad, as in the $Co:MgF_2$ case, as once again, the excited state absorption is low. As a result, dielectric mirrors could not be manufactured with high enough reflectivity over the entire range, and pure silver mirrors had too much loss for the low gain. Mirrors that were silver coated but enhanced with a dielectric overcoat provided a solution, achieving an overall cavity loss, excluding output coupling, of only 0.7%. With this device, over 40 mW power was obtained in single mode operation and up to 55 mW multimode while the crystal absorbed about 1.7 W argon ion pump radiation in the blue and green.

9.3.4 Ce^{3+}

As mentioned in Sect. 9.1.1, the Ce^{3+} ion has properties intermediate between typical rare-earth and transition-metal ions. Lasing on transitions between the $5d$ and $4f$ shells are possible in certain Ce-doped compositions, but there is a wide variation among materials in the wavelength at which fluorescence occurs and in the amount of excited state absorption present.

Both $Ce:YLF$ and $Ce:LaF_3$ exhibit lasing from a $5d$ to $4f$ transition when pumped by a KrF excimer laser [9.45, 46] (Fig. 9.18). Ce^{3+} lasers have much higher emission cross sections ($7-8 \times 10^{-18} cm^3$), because the intershell laser transition is parity allowed, but the storage time is correspondingly short, 40 and 18 ns, respectively, for $Ce:YLF$ and $Ce:LaF_3$. In this respect, they behave more like a dye laser or color center laser than like other tunable paramagnetic lasers. Nevertheless, they are paramagnetic ion lasers having the shortest laser wavelength of any solid-state laser discovered to date, emitting as short as 285.5 nm ($Ce:LaF_3$) [9.46].

9.3.5 Sm^{2+} and Ho^{2+}

On the whole, the rare-earth spectra are dominated by much sharper lines with comparatively weak vibronic sidebands. Thus, the emission and excited state absorption cross sections for vibronic lasing are low. The material $CaF_2:Sm^{2+}$ was first lased on no-phonon lines [9.72]. Later, vibronic lasing in this material was achieved at cryogenic temperatures at wavelengths between 708.5 and 745 nm [9.47]. The first *room temperature* vibronic laser, rare earth or

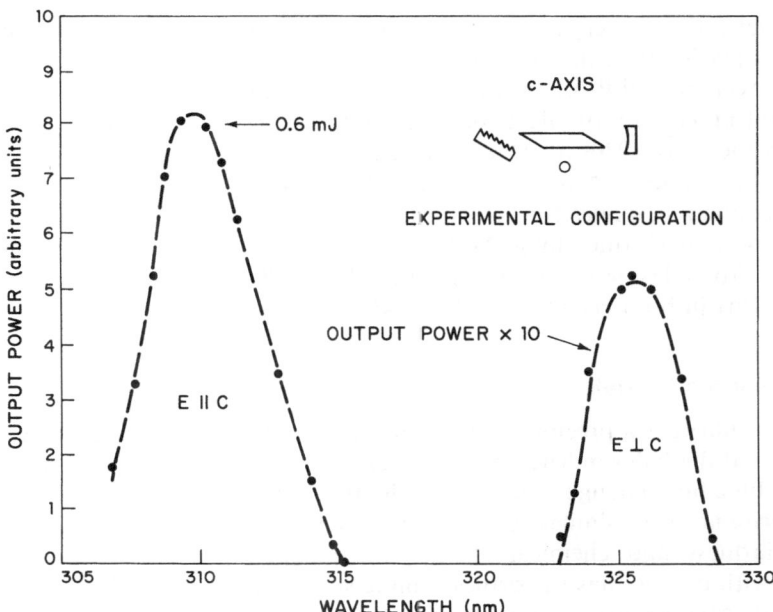

Fig. 9.18. Energy output vs wavelength for a grating-tuned Ce: YLF laser. The two tuning regions had different output polarization. (After *Ehrlich* et al. [9.45, 46])

transition metal, was $BaY_2F_8: Ho^{3+}$ [9.51]. Lasing was obtained in this case using pulsed xenon lamp excitation. In both of these materials, the host had been sensitized with Er and Tm to increase the absorption of lamp radiation.

9.4 Alexandrite Material Characteristics

We have outlined in the previous section a number of ion – host combinations (including alexandrite) that are proven tunable lasers. Alexandrite is advanced among these, however, and we therefore now wish to discuss its material properties in greater detail. Achieving an operating laser in the laboratory is only the first step toward developing a laser for practical application. It is important to be aware, at the onset, of the important role host properties play in making a practical laser. The crystalline host has two primary functions: (1) it provides the electronic and vibrational environment that imbue the active ion with required kinetics properties, and (2) it provides the mechanical, thermal, and optical properties required for high-performance laser operation. The latter becomes a determining factor for practical laser devices. It is, therefore, not surprising that the most successful solid state lasers are based on sapphire (ruby) and garnets (Nd: YAG), both extremely hard gem-like materials. Chrysoberyl, the pure alexandrite host, has laser-relevant physical properties intermediate between those properties of sapphire and YAG, that recommend it for broadly based applications.

Czochralski growth of alexandrite crystals for laser applications was initiated in the early 1970s at Allied Chemical (now Allied-Signal) Corporation by C. Cline and R. Morris as an integral part of a program to develop beryllium-containing oxides for laser hosts. Preliminary laser experiments on alexandrite were performed by R. Webb also at Allied Corporation. The discovery of vibronic lasing in alexandrite in the late 1970s [9.73] greatly stimulated the effort in both crystal growth [9.74] and laser development.

9.4.1 Alexandrite Crystals

Natural alexandrite is a precious gem, known particularly for its pronounced trichroism and the "alexandrite" effect, where the intensities of its deep red, green, and blue hues change according to lighting conditions. The properties of alexandrite are very similar to ruby, both being Cr^{3+} doped oxides with extreme hardness and chemical stability. Chrysoberyl, $BeAl_2O_4$, is isomorphous with olivine, has the orthorhombic space group Pnma (D_{2h}^{16}), and has the lattice constants $a = 0.9404$, $b = 0.5476$, and $c = 0.4427$ nm.[3] In chrysoberyl, the oxygen atoms assume an approximate hexagonal-close-packed structure with the aluminum atoms at the octahedral interstices, whereas the beryllium atoms occupy tetrahedral sites (Fig. 9.19). *Farrell* et al. [9.76] first produced synthetic chrysoberyl by flux growth, performed a detailed x-ray crystallographic analysis of chrysoberyl, and suggested the suitability of the host for laser application.

In chrysoberyl, there are two crystallographically nonequivalent Al^{3+} sites that occur in equal number upon which Cr^{3+} substitutes: one with mirror, C_s, and one with inversion, C_i, site symmetry. The mirror site, which is preferentially occupied by Cr^{3+}, plays a nearly exclusive role in laser operation. The combined density of aluminum sites is 3.509×10^{22} sites per cubic centimeter. Many of the physical and laser-related materials properties of alexandrite are summarized in Table 9.5.

a) Chromium Concentration

The presence of two chromium sites in alexandrite significantly complicates the measurement of chromium concentration in either site. A suitable approach involves the use of electron paramagnetic resonance (EPR) to determine the occupancy ratio and neutron activation analysis (NAA) to determine the total Cr content. An EPR analysis on alexandrite [9.77] has determined that $78\% \pm 3\%$ of the Cr ions fall on mirror sites. This study also

---

[3] The axis labels appear in all permutations in the early literature on chrysoberyl. The usage herein complies with the convention established in [9.75].

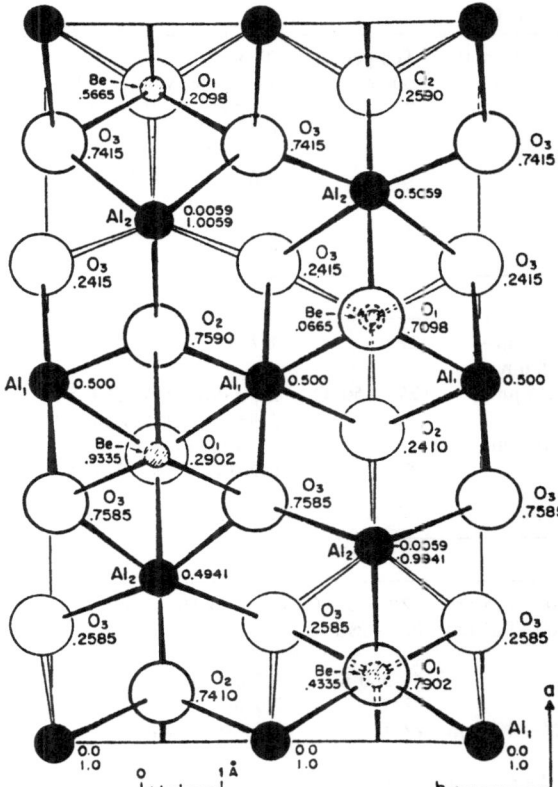

Fig. 9.19. Crystal structure of chrysoberyl. (After *Farrell* et al. [9.76])

shows that the occupancy ratio remains constant for a variety of Cr concentrations and Czochralski growth techniques. Routine nondestructive monitoring of the Cr concentration of laser rods, using the mirror-site R-line absorbance that has been normalized to the EPR and NAA results, provides a measure of the average concentration down the rod length to within 2% relative and 5% absolute accuracy for laser rods used in the experiments described here.

b) Birefringence

Losses caused by depolarization arising from thermally induced birefringence can be severe in high average power slab or rod lasers employing optically isotropic gain elements, such as Nd:YAG or Nd:glass, unless special resonator designs are used to counter it. A naturally occurring birefringence, present in anisotropic materials, is usually strong compared to that thermally induced. This is true of alexandrite. Measurements of the depolarization loss during alexandrite oscillator operation at 10 Hz indicate that the depolarization is less than 1%, and is clearly not a factor.

Table 9.5. Summary of alexandrite properties[a]

Physical properties	Symbol	Wavelength range	Temperature range	Typical parameter values
Thermal conductivity	k		22 °C	0.23 W/cm^2
Breaking stress	σ_B			$4 - 9 \times 10^8$ Pa
Cr concentration	C			$0.05 - 0.23$ at.%
Fraction Cr on mirror site	f_m			0.78 ± 0.03
Specific heat	C_p			?
Optical properties				
Refractive index	n	250 nm – 2.6 μm	22 °C	1.74
dn/dT		1.15 μm	25° – 50 °C	$E \parallel a$ $(9.4 \pm 0.5) \times 10^{-6}$ K^{-1} $E \parallel b$ $(8.3 \pm 0.5) \times 10^{-6}$ K^{-1}
Scattering loss	ε	1.06 μm	22 °C	<5 db/km
Nonlinear refractive index coefficient	γ		22 °C	$(2 \pm 0.03) \times 10^{-20}$ m^2/W
Stress optic				?
Laser properties				
Emission cross section	σ_e	700 – 825 nm	22° – 290 °C	$0.7 - 3 \times 10^{-20}$ cm^2
Excited state absorption cross section	σ_{2a}	700 – 825 nm	22° – 290 °C	$2 - 7 \times 10^{-21}$ cm^2
Storage time	τ		4 – 500 K 22 °C	1.5 ms – 70 μs 262 μs
Ground state absorption in pump bands	σ_{ap}	350 – 800 nm	22 °C	0.5×10^{-19} cm^2
Excited state absorption in pump bands	σ_{2ap}	425 – 670 nm	22 °C	0.3×10^{-19} cm^2
Thermal lens		730 – 790 nm	90 °C	0.56 diopters/kW
Absorbed power in rod to fracture				0.6 kW/cm
Optical damage threshold bulk and surface		750 nm	22 °C	>30 GW/cm^2

[a] References given in [9.13].

c) Thermal Lensing

Alexandrite's thermal lens, induced by optical pumping, is generally less than that experienced by Nd:YAG. There is a weak astigmatism that is correctable, in principle, by pump chamber design. A total power of 1 kW delivered to flash lamps in a dual ellipse pump chamber will produce a thermal lens in alexandrite of about 3 m focal length.

9.4.2 Optical Cross Sections and Their Role in Laser Performance
a) Pump Band Absorption

The combined mirror- and inversion-site absorption spectra for alexandrite in the three electric field polarizations [9.10, 57] are given in Fig. 9.20. The in-

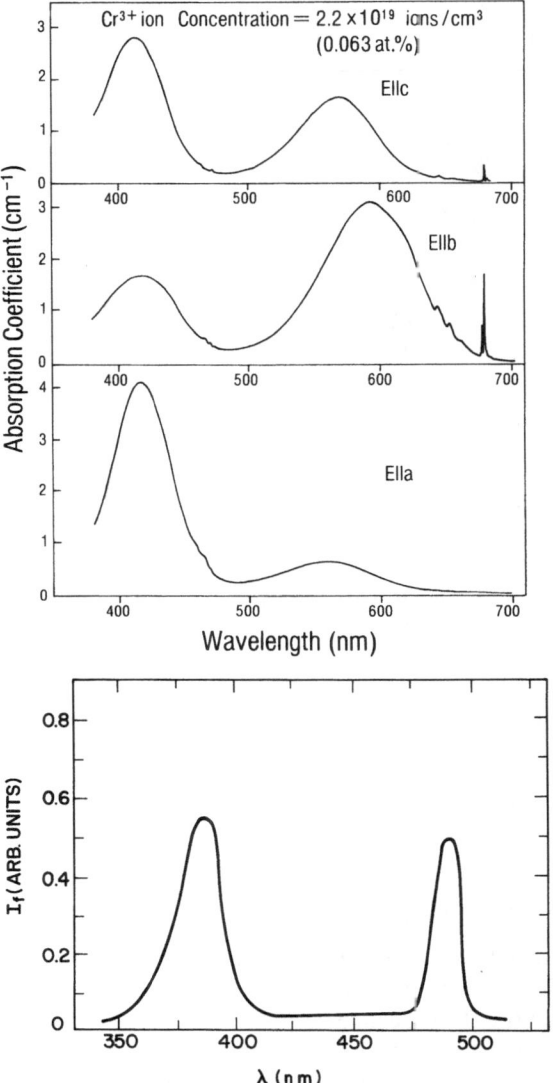

Fig. 9.20. Absorption spectra of alexandrite [9.10]

Fig. 9.21. Selective excitation spectrum showing the absorption of the inversion site in alexandrite. (After *Powell* et al. [9.78])

version-site component is very small and has been obtained independently by excitation spectroscopy [9.78] as presented in Fig. 9.21. The peak absorption cross section of the inversion site at 480 nm contributes about half of the absorption in the trough in Fig. 9.20 [9.79], which partially explains its minimal role in laser emission. Also important is the very large separation between the inversion site ^2E and the $^4T_2$ multiplets, which effectively eliminates the $^4T_2$ fluorescence sideband and its contribution to laser gain at normal temperatures of laser operation.

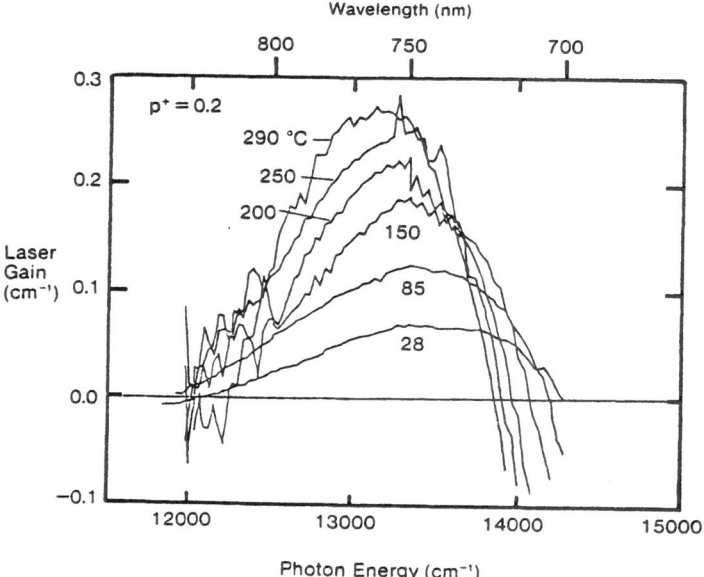

Fig. 9.22. Temperature-dependent gain in alexandrite. (After *Shand* and *Jenssen* [9.14])

b) Laser Gain Cross Section

The laser gain cross section, which is the difference between the emission and the excited state absorption cross sections, has been measured by determining the single pass gain using a pulsed dye laser probe beam [9.14] and is given in Fig. 9.22. The alexandrite sample was a laser rod, excited in a laser pump chamber.

c) Emission and Excited State Absorption Cross Sections

In order to obtain independent emission and excited state absorption cross sections from the laser gain cross section given above, the emission cross section is obtained directly from the fluorescence rate spectrum using (9.9). As stated in Sect. 9.2, calibration of the observed spectra to the emission rate is achieved by independent measurements of the quantum yield (η) and the radiative lifetime. The alexandrite quantum yield at room temperature is 95% ± 5%, determined by photoacoustic spectroscopy [9.11].

With η near unity, the radiative lifetime is essentially the observed fluorescence decay time. The decay time has been measured using the focused beam from a pulsed dye laser to excite a crystal near to its surface in order to minimize effects of fluorescence reabsorption and reemission [9.10]. Also, samples of several concentrations between ≪0.005 at.% and 0.5 at.% were examined. The results obtained are given in Fig. 9.23 as a function of temperature for a sample containing less than 0.005 at.% Cr. No significant

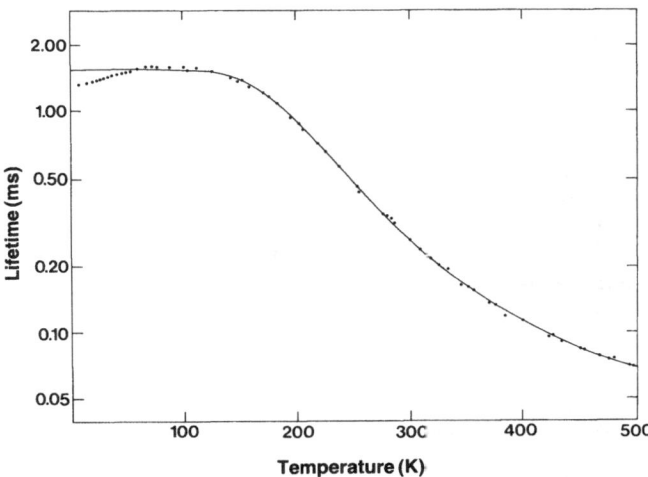

Fig. 9.23. Fluorescence lifetime of alexandrite. The theoretical curve is based on two occupied upper levels as illustrated in Fig. 9.9, and on the parameters given there [9.10]. Cr^{3+} concentration: <0.001 at.%. Data, (● ● ●); theory using two-level model, (————)

concentration quenching was observed, at least for materials with less than 0.5 at.% Cr, well within the range of interest for most laser applications.

The room temperature fluorescence rate spectra are given in Fig. 9.24. No detectable magnetic dipole effect is seen in the $E \parallel b$ spectra, although some effect is seen in the weaker $E \parallel a$ and $E \parallel c$ components of the mirror-site R-lines. The strong polarization of these spectra, 10 to 1 in favor of $E \parallel b$ at room temperature, is of major benefit to laser performance. The emission and excited state cross sections can now be differentiated, as shown in Fig. 9.25.

d) Ground State Absorption in the Laser Band

The ground state absorption cross section, throughout the laser wavelength band, can be obtained directly from the emission cross section using (9.17). It can also be measured directly using absorption spectroscopy at the shorter wavelengths and then extended to longer wavelengths by excitation spectroscopy, where the absorption is otherwise too weak to be measured directly. The ground state absorption cross section obtained by the two methods for the important $E \parallel b$ polarization is presented in Fig. 9.26. The theoretically predicted ground state absorption at elevated temperature has not been confirmed experimentally.

e) Mirror-Site Excited State Absorption

If the excited state absorption were to arise from the ^2E levels and not the $^4T_2$ levels, one would expect to reduce σ_{2a} by increasing the temperature, thus populating the $^4T_2$ levels and enhancing the emission cross section, while at the

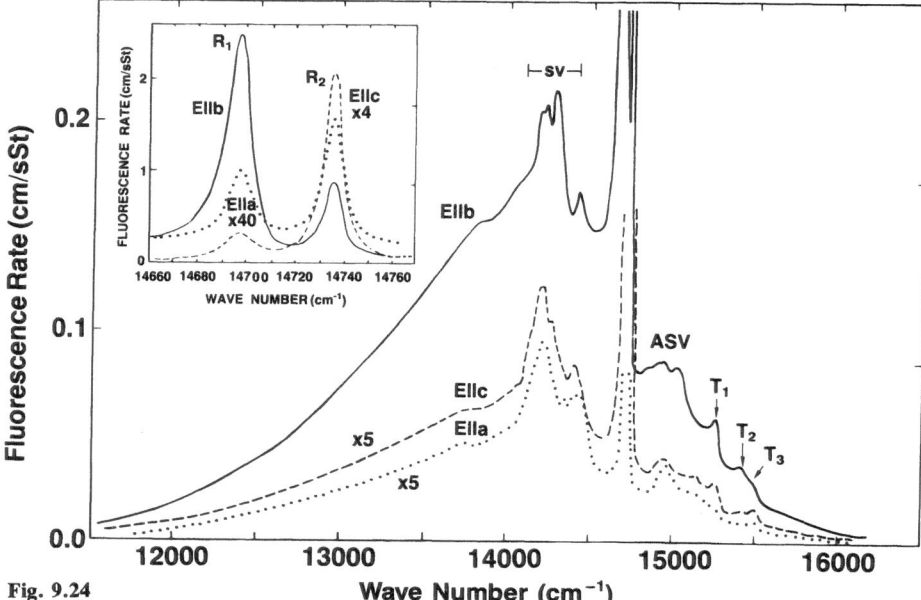

Fig. 9.24

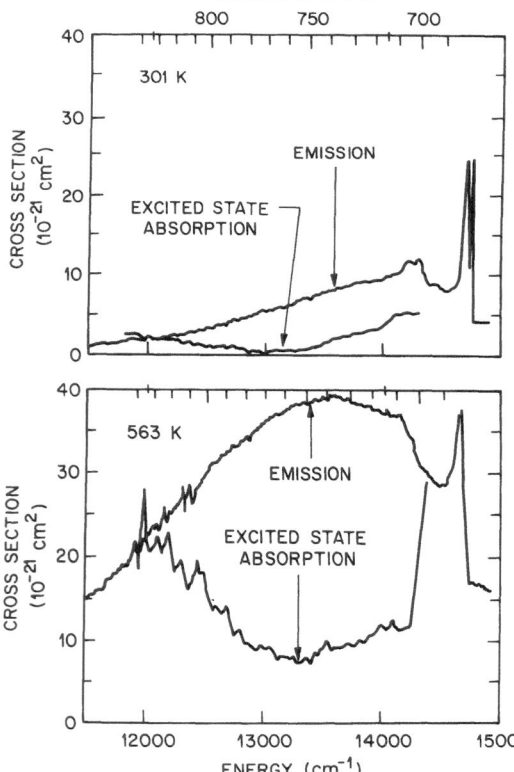

Fig. 9.24. Fluorescence spectra of alexandrite at room temperature [9.10]. "SV" denotes Stokes sideband attributed to lower member of ^2E and "ASV" denotes the anti-Stokes component

Fig. 9.25. Emission and excited state absorption cross sections of alexandrite. (After *Shand* and *Jenssen* [9.14])

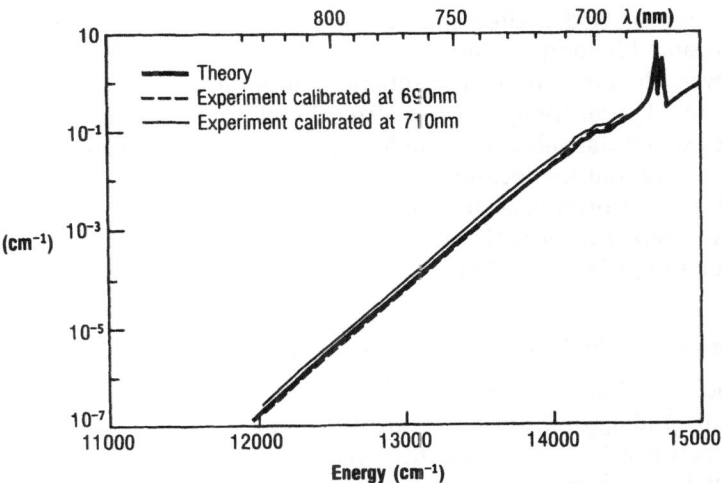

Fig. 9.26. Absorption coefficient of alexandrite at room temperature [9.9]. Experimental curves are excitation spectra calibrated by measured absorption at wavelengths indicated; theory curve is obtained from fluorescence spectrum

same time suppressing the excited state absorption. However, at the longer wavelengths, where the laser bandwidth is limited by excited state absorption, the opposite is observed. Figure 9.27 illustrates the measured ratio of the excited state absorption to the emission cross section at 825 nm (data points) as a function of temperature, plotted in relation to the calculated ratio of occupancy for several multiplets relative to ^2E. All are normalized to their value at room temperature. The fact that the data points fall about the $^4T_2$

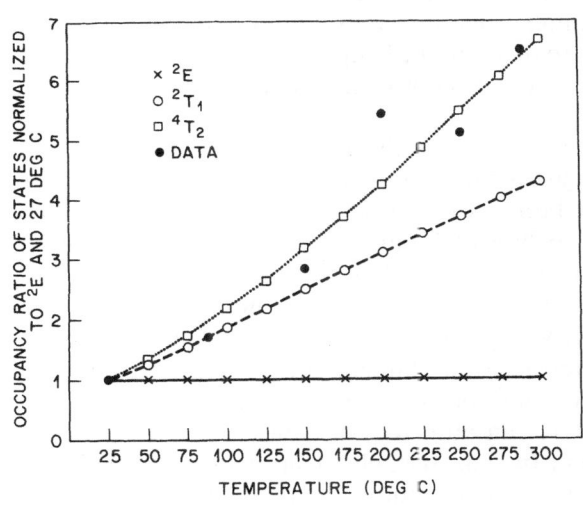

Fig. 9.27. Source of excited state absorption in alexandrite. Points are from temperature-dependent gain data in [9.14]; curves are derived assuming that the indicated lower state is the source, and are based on its Boltzmann occupation probability

curve strongly suggests that excited state absorption arises from $^4T_2$ at this wavelength, presumably terminating on $^4T_1$ where the transition is spin allowed. However, the precision of measurement does not allow a contribution from a $^2T_1$ to $^2A_1$ transition, which is also spin allowed, to be ruled out. (Note, also, the excited state absorption in low temperature ruby and emerald presented in Fig. 9.12 and the location of the $^2T_2$ and $^2A_1$ states.) It is clear that the excited state absorption at the longer wavelengths does not arise from ^2E, however. Additional information on excited state absorption in alexandrite can be found in [9.14, 65, 80, 81].

f) Energy Transfer and the Role of the Inversion Site in Laser Operation

In principle, the inversion site might add to the total vibronic gain, and/or reduce the gain by adding to the excited state absorption. However, as stated above, the gap between $^4T_2$ and ^2E for the inversion site is very large (larger than in ruby), and there is no broad band emission to contribute to the gain. The suppressed electric dipole oscillator strength of the inversion site ions reduces both the pump band absorption and the excited state absorption of these ions.

A strong argument can be made that in the pulsed laser the effect of inversion-site excited state absorption is insignificant. The evidence comes from the fact that threshold is reached throughout the gain region by *cw-pumped* alexandrite lasers. In the cw case, the equilibrium population is proportional to storage time, and the excited state absorption for the cw laser operation would be heightened by the ratio of the mirror-site to inversion-site storage times, i.e., by a factor of 48 ms/0.262 ms = 184. Consequently, in pulsed operation, the inversion-site excited state absorption must be less than $1/184 = 0.54\%$ of the gain in pulsed operation. The threshold for the cw laser is, however, approximately the value predicted by the same cross sections as for the pulsed case, which implies that the contribution in the pulsed case is lower yet. For this argument to be valid, the lamp spectra must be comparable for both pulsed and cw cases; this is approximately true for the cw Xe arc lamps (Sect. 9.5.2).

The inversion site may be excited indirectly via the mirror site by excitation transfer. However, long-range excitation transfer between mirror and inversion sites in alexandrite has been shown to be absent by four-wave-mixing experiments [9.82]. Further analysis [9.78] indicates that short-range mirror-to inversion-site excitation transfer occurs but with no back transfer. The author has seen evidence that back transfer may occur, not to the mirror site, but to a third species (possibly Cr-Cr pairs which have intermediate, metastable-state energy) in the transfer chain linking mirror and inversion sites. Inversion-site trapping, active only for nearby mirror- and inversion-site Cr ions, is probably responsible for an observed slight increase in storage time with chromium concentration. In sum, the role of inversion sites in alexandrite lasers is minimal, at least for temperatures less than 90°C.

9.5 Alexandrite Laser Operation and Performance

Alexandrite lasers operate well under flash lamp and cw arc lamp excitation as well as laser pumping. Their performance, in many ways, is comparable to that of Nd: YAG, ruby, and dye lasers. Rods range in size from 2 mm to 1 cm in diameter and up to 12 cm long, and have less than 1 fringe of optical distortion at 1.15 µm. Standard rods are 6.4 mm in diameter and 11 cm long and contain 0.14 at.% Cr (5×10^{19} Cr atoms/cm^3).

Most conventional operating modes have been explored with alexandrite. Many of these are reviewed in this section along with some novel resonator designs effective in achieving improved beam quality or otherwise refined performance. Further information on alexandrite's laser properties, including mode locking, is presented in a recent review [9.13]. Table 9.6 summarizes many of the results obtained. This section closes with a discussion of optical damage as experienced with alexandrite laser operation.

9.5.1 Flash Lamp Pumped (Pulsed) Alexandrite Lasers

Flash lamp pumped solid-state lasers available commercially often utilize an oscillator stage, to establish a beam of refined characteristics, followed by one or more stages of amplification, as required for power. For the oscillator, both stable and unstable configurations are used depending on the gain of the system. Our discussion here will first focus on the oscillator, where we examine the various types used with alexandrite and review the performance obtained. Included in this is a discussion of preliminary work with zigzag alexandrite slabs.

a) Basic Oscillator

The characterization and optimization of alexandrite oscillators is made more complex and time consuming by the dependence of optical cross sections on both temperature and wavelength, dependencies of little or no consequence in ruby or Nd: YAG. The desirability of elevated temperature operation for alexandrite becomes immediately apparent the moment one begins to operate the laser. The output pulse energy increases with each successive shot as the rod is heated by the flash lamp. This ability of alexandrite to work optimally above room temperature facilitates heat removal, leading to smaller, more-effective cooling systems. In spite of the fact that the laser performance of alexandrite is generally optimized above the boiling point of water, the convenience and superiority of a water coolant has led to its widespread use with alexandrite, with rod temperatures around 100 °C.

Figure 9.28 illustrates the performance of a simple stable alexandrite oscillator in normal mode (non-Q-switched) operation, where the effect of temperature on performance is readily apparent. Simple normal mode oscillators utilize low output coupling in the 10% – 20% range and have produced

Table 9.6. Alexandrite laser performance summary

Operational mode	Rod size Diameter [mm]/ Length [mm]	Cr^{3+} conc. [at.%]	Wave-length [nm]	Band-width [nm]	Pulse energy [J]	Pulse duration [ns]	Pulse repeti-tion [Hz]	Average power [W]	Mode character	Diffrac-tion limit	Effi-ciency [%]
Normal	6.3/10	0.28	755	1.0	7	200 μs	5	35	Multimode	–	2.5
Normal	6.3/10	0.14	750	1.0	4.5	150 μs	20	90	Multimode	30×	1.2
Normal	6.3/11	0.14	750	1.0	0.6	150 μs	20	12	Multimode	5×	0.3
Normal	6.3/11	0.14	750	1.0	1.2	150 μs	125	150	Multimode	15×	–
Q-switched	6.3/2×11	0.14	790	0.01	0.4	<1 μs	125	50	Multimode	15×	0.4
Q-switched	6.3/11	0.14	750	0.1	1.5	28	20	30	Multimode	30×	0.55
Q-switched	6.3/11	0.14	750	0.1	2.0	40	20	40	Multimode	30×	0.19
Q-switched	6.3/11	0.14	740	0.1	0.55	25	10	5.5	Multimode	10×	0.19
Q-switched	6.3/11	0.14	750	0.1	0.55	16	10	5.5	Multimode	12×	0.21
Q-switched	6.3/11	0.14	765	0.1	0.55	26	10	5.5	Multimode	10×	0.19
Passive mode-locked	6.3/11	0.14	750	0.025	0.5 mJ	38 ps	10	–	TEM(00) transform limited	1.5×	–
Active mode-locked and Q-switched	6.3/11	0.14	750	0.02	2 mJ	150 ps	10	–	TEM(00)	–	–
cw pumped	3/10	0.2	755	1	–	–	–	60	Multimode	–	1.21
cw pumped	5/10	0.09	755	1	–	–	–	2	TEM(00)	1.3×	0.31
Oscillator/ Amplifier	6.3/11	0.14	755	1	3	30	10	30	Multimode	7×	–
Oscillator/ Amplifier	6.3/11	0.14	755	1	0.5	30	10	5	Near TEM(00)	1.2×	–
Oscillator/ Amplifier	6.3/11	0.14	755	1	1	10	10	10	Multimode	7×	–

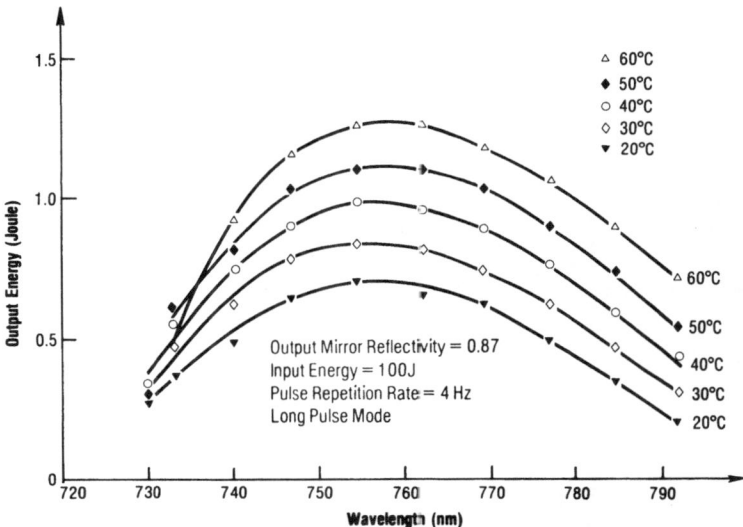

Fig. 9.28. Alexandrite oscillator output energy vs tuning, for various temperatures [9.83]

pulse energies of 7.5 J from standard rods in multimode emission. Overall efficiencies exceeding 2.5% have been demonstrated [9.10].

At higher repetition rates, thermal lensing must be compensated in order to obtain good rod fill and good efficiency from a multimode, nontuned oscillator. Figure 9.29 illustrates such a resonator with thermal lensing compensation, provided by a convex output mirror curvature, and the performance obtained. Normal mode pulse durations are on the order of a few hundred microseconds and track the duration of the pumping pulse once threshold is reached. The temporal emission behavior of normal mode pulses is characteristically spiked as shown in Fig. 9.30 in comparison with those from ruby. These spikes are related to multiple transverse and longitudinal mode buildup and decay, and are more accentuated than in either Nd:YAG or ruby because of the lower emission cross section. Individual spikes are typically on the order of 1 μs in duration.

Tuning has been accomplished using a birefringent tuner almost exclusively because of its convenience, very low loss, and high damage resistance (see Chap. 1). With tuners of this type, linewidths of 0.2 – 0.4 nm are typical, and tuning can be achieved over nearly the entire gain band provided the tuner is properly designed and adjusted.

b) Q-Switching

Q-switched operation offers the greater opportunity for application and the greater challenge to the design engineer. Typical Q-switched pulse durations are between 30 and 200 ns in alexandrite lasers, though 10 ns has been

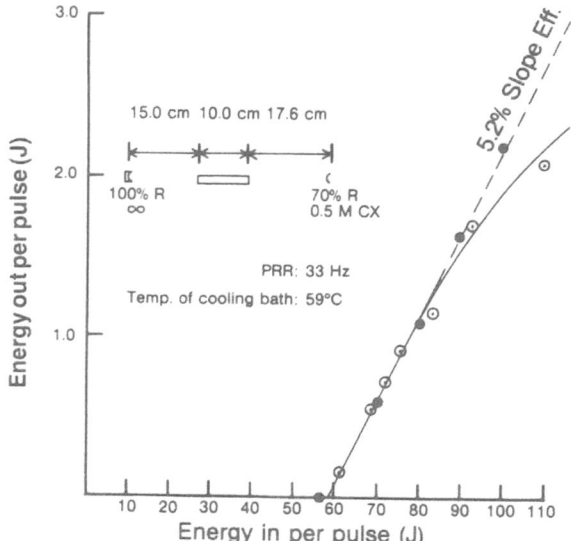

Fig. 9.29. Energy out vs flash lamp energy for an alexandrite multimode oscillator in a thermally compensated resonator [9.15]

ALEXANDRITE

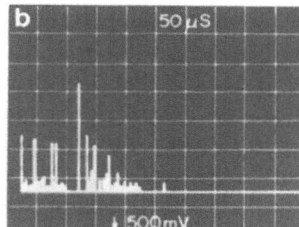

PUMP LEVEL: 2.1 × Threshold
OUTPUT ENERGY: 2.9J
PULSE DURATION: <200µs

RUBY

PUMP LEVEL: 2.0 × Threshold
OUTPUT ENERGY: 1.3J
PULSE DURATION: >300µs

Fig. 9.30a, b. Comparison of temporal characteristics of long pulse emission of ruby (**a**) with alexandrite (**b**). (Courtesy of D. R. Siebert, Allied-Signal Corporation)

achieved, and can be tailored both with temperature and wavelength. Optical damage, discussed below, can be controlled to a large degree by resonator design and by careful operation. In designing resonators, it is extremely important to control reflections from surfaces, so they do not focus upon any of the optical elements, and to utilize high quality, protective apertures to control stray emissions.

In multimode, Q-switched performance, 2 J per pulse at 20 Hz for an average power of 40 W has been demonstrated with a standard rod in a stable oscillator. The pulse duration was 30 ns and the beam divergence was 20 × diffraction limited. (For alexandrite the diffraction limit corresponds to a full angle beam-divergence-aperture product of approximately 1 mm-mrad). At 5 × the diffraction limit, the energy is reduced to 0.5 J at the same repetition rate and pulse duration. As such, the performance of the stable alexandrite oscillator is nearly comparable to Nd : YAG in an unstable resonator with about half the peak power and twice the energy. The primary difference is in beam divergence, in which Nd : YAG is superior. Superior alexandrite performance is obtained from other types of resonators including the unstable resonator as discussed below.

The spatial beam characteristics of a multimode Q-switched and normal mode alexandrite oscillator, obtained with a TV camera, are distinctly different, as illustrated in Fig. 9.31. The complex mode pattern of the normal mode emission becomes curiously smooth when the Q-switch is activated. Using a Pockel cell to gate and sample the output temporally, this uniformity has been shown [9.84] to result from an integration of several spatially structured modes that appear sequentially during the period of the Q-switched pulse (Fig. 9.32). Initially, the TEM_{00} mode appears. This is subsequently replaced by irregular modes of increasing order. Restricting the number of

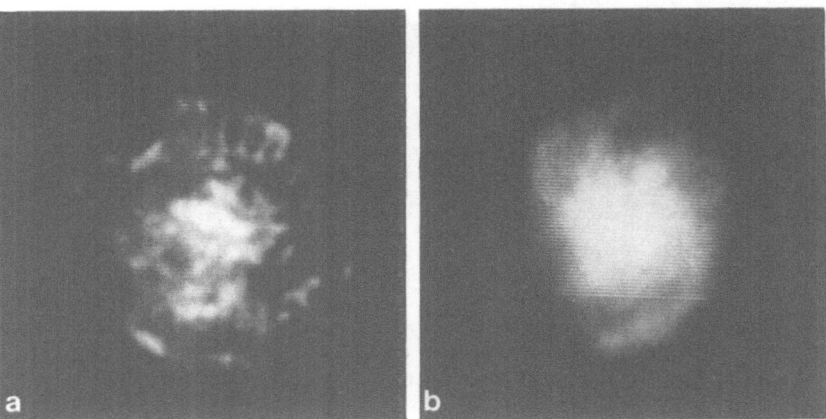

Fig. 9.31a, b. Spatial mode pattern of an alexandrite multimode oscillator. (a) Normal mode; (b) Q-switched

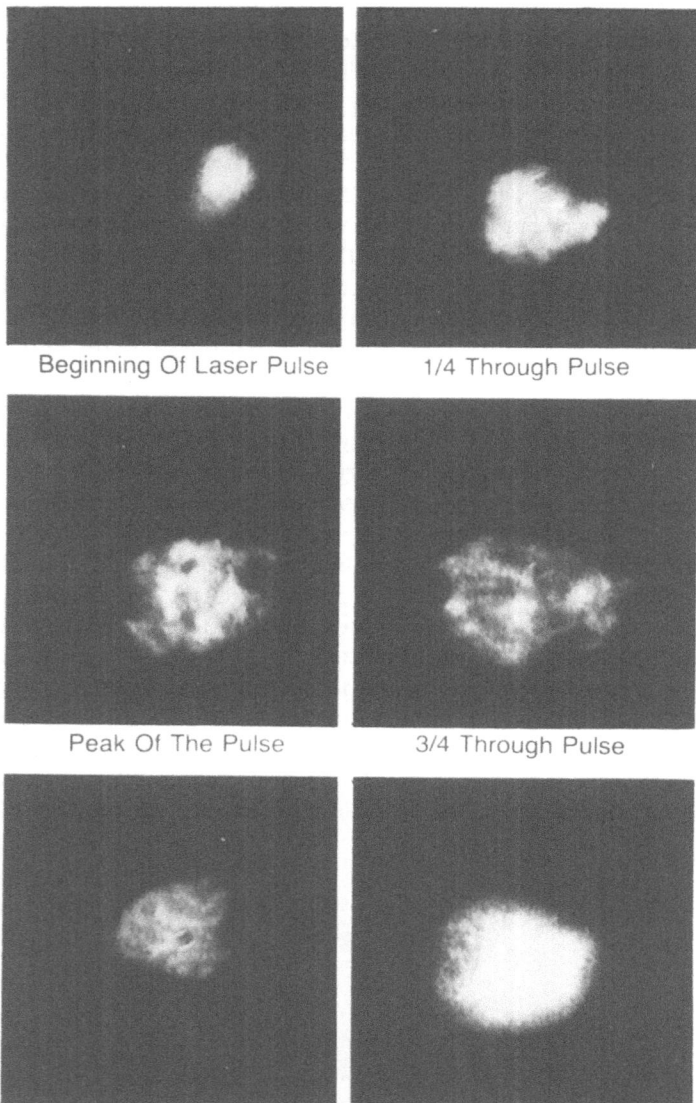

Beginning Of Laser Pulse   1/4 Through Pulse

Peak Of The Pulse   3/4 Through Pulse

Spatial Profile Integrated Over Entire Laser Pulse

Fig. 9.32. Time-resolved spatial mode pattern of an alexandrite Q-switched oscillator: bottom two pictures are with different absorbers in front of vidicon. Dark spot in lower-left picture is diffraction due to dust. (Figure supplied courtesy of D. J. Harter, Allied Corporation)

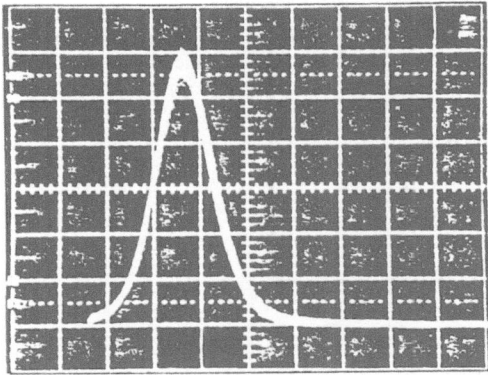

Fig. 9.33. Twenty overlapping pulses to illustrate pulse stability of a Q-switched multimode alexandrite oscillator [9.13]

modes that can oscillate, by using a birefringent tuner or by operating close to threshold, causes the Q-switched pulse to be more structured. Surprisingly, the temporal profile of the Q-switched pulse is very regular and repeatedly stable, in spite of the complex modal development (Fig. 9.33).

c) Q-Switching on the R-Line

For lasing alexandrite on the R-line a lower concentration is preferred, because it is necessary to achieve 50% inversion to reach threshold, and standard rods are too opaque to achieve this. Operation on the R-line has been achieved with rods having a Cr concentration ranging near 0.05 at.%, see Fig. 9.34. In this mode, the Q-switched operation is similar to that of ruby except that the pulse duration is shorter, roughly 10 ns, indicative of a high

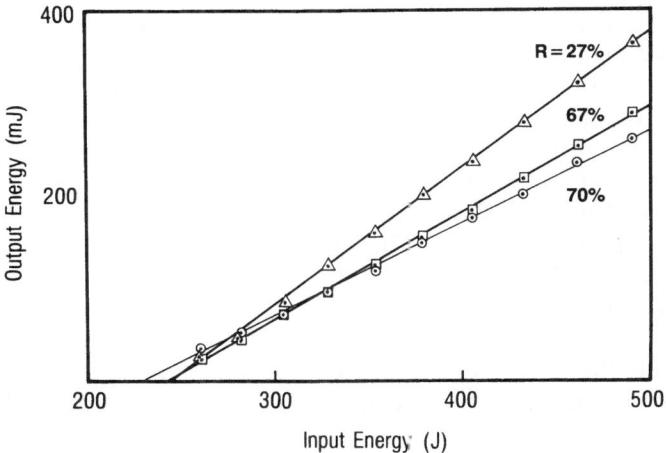

Fig. 9.34. Output vs flash lamp energy for a Q-switched alexandrite laser operating on the R-line [9.85]

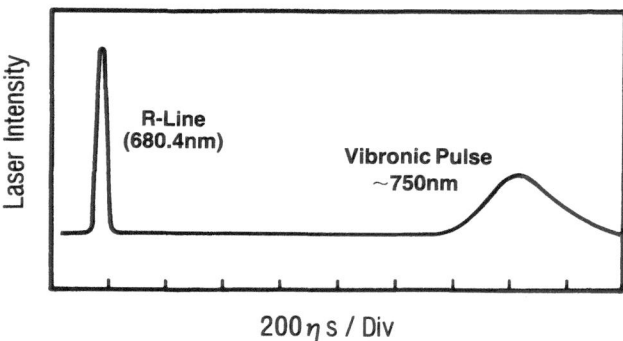

Laser Intensity

R-Line
(680.4nm)

Vibronic Pulse
~750nm

200 η s / Div

Fig. 9.35. Q-switched alexandrite *R*-line and vibronic pulses in tandem [9.85]

emission cross section more comparable to Nd:YAG. If low output coupling is used, it is possible to obtain an initial *R*-line pulse followed by a vibronic pulse, as illustrated in Fig. 9.35. The *R*-line gain ceases once the excitation level drops below 50%, but the vibronic gain remains finite, accounting for this effect. The difference in pulse widths is accounted for by the difference in emission cross sections, approximately 3×10^{-19} cm^2 and 7×10^{-21} cm^2 for the *R*-line and vibronic pulses, respectively.

d) Alexandrite Unstable Resonator Oscillators

Resonators can be designed by selecting the mirror curvature and placement to be geometrically unstable and yet still support the development of a coherent beam [9.86]. In such unstable resonators, the power in the developing beam migrates radially outward and is extracted more from the outer region of the rod cross section than from the central region as in a conventional stable oscillator. The unstable resonator combines the favorable properties of the oscillator and the amplifier in a single oscillator device. For Nd:YAG, unstable resonators are widely used to obtain nearly diffraction-limited beams with good rod fill and high extraction efficiency.

Of the various methods for achieving an unstable resonator, one is particularly well suited for low gain lasers. This method uses for output coupling a device which simulates a mirror whose reflectivity decreases parabolically with distance from the center. This device consists of birefringent flats and lenses (the radial birefringent elements, RBEs) and, when combined with a polarizer, is called a radial birefringent device [9.87]. The advantage of this device, schematically represented in Fig. 9.36, stems from the fact that the condition of stability vs instability is continuously variable. Furthermore, the reflectivity parameters can be changed by rotating the birefringent lenses about the central axis even while the laser is operating. The stability condition of the resonator may be easily adjusted during laser operation, via thermal lensing of the laser rod by varying the repetition rate. The convenience derived from these "in vivo" parameter adjustments for experimentation cannot be overemphasized. In practical applications, such adjustment becomes problematic,

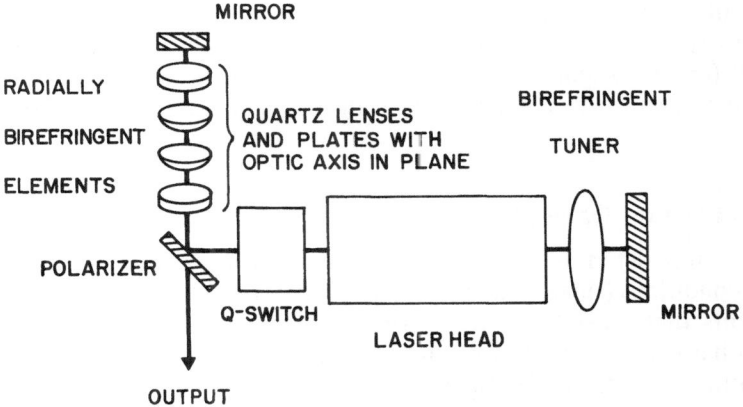

Fig. 9.36. Radial birefringent element (RBE) unstable resonator for alexandrite [9.88]

because the performance is optimized only at one operating design point. However, this is true even of commercial Nd:YAG unstable resonators.

The RBE resonator has been applied to alexandrite [9.88], achieving 400 mJ pulses, 40 ns in duration and $2.5 \times$ diffraction limited, at an 8 Hz repetition rate (Fig. 9.37). The emission was nearly diffraction-limited with

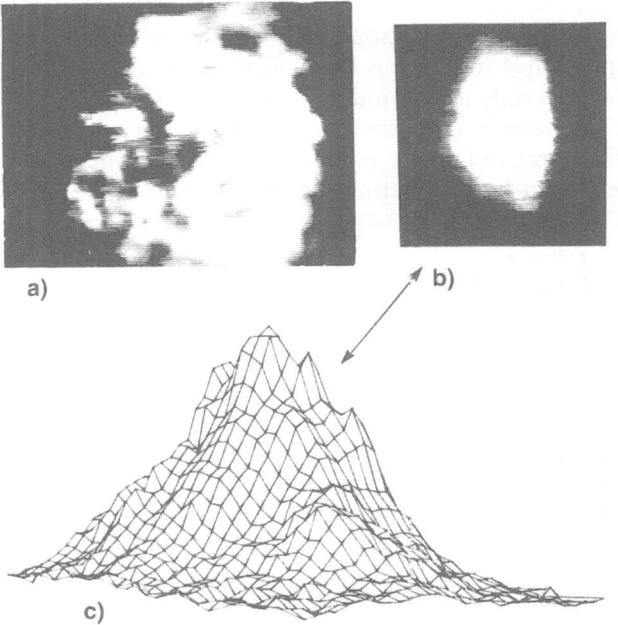

Fig. 9.37 a – c. Spatial profile of alexandrite Q-switched laser outputs for (**a**) stable resonator and (**b, c**) RBE unstable resonator

250 mJ per pulse. Optical damage for this resonator was limited to the polarizer beam splitter used for output coupling, which was essentially the "optical fuse" for the system. Improved polarizer beam splitters and rods of improved optical quality would probably lead to yet higher levels of performance.

e) Achievement of Low Order Modes

One of the greatest deterrents to high performance from oscillators under refined beam conditions (single axial mode, etc.) is that the many beam-conditioning elements that must be used contribute a substantial loss. A novel approach that has been developed to counter this problem is the self-injection locked oscillator or SILO [9.89]. In this technique, shown schematically in Fig. 9.38, the frequency control elements are placed in one branch of a Y-shaped resonator where the beam is allowed to build up under near threshold conditions. Once a quality beam is established, the beam path is electro-optically switched to the other (component-free) branch for power development and output coupling. In the SILO approach, a 100 mJ tunable single longitudinal and transverse mode output has been achieved with alexandrite (Fig. 9.39). This performance compares closely to that obtained in conventional injection locking, where two independent oscillators are used, but with considerable reduction in the size and complexity of the device. In order to successfully injection lock, the injected signal must dominate spontaneous fluorescence. Nevertheless, for low emission cross section materials, the required signal is correspondingly low. It has been possible to injection lock and alexandrite oscillator with only a nanojoule of injected energy [9.90].

Unidirectional, traveling wave, ring resonators have been used extensively with dye lasers to obtain improved single mode performance, where they significantly increase the output by eliminating spatial hole burning. Spatial

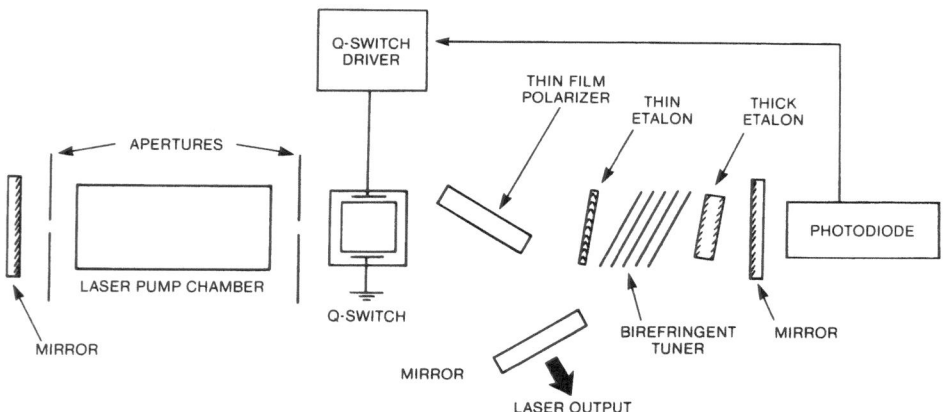

Fig. 9.38. Schematic of self-injection locked oscillator (SILO). (After *Harter* et al. [9.89])

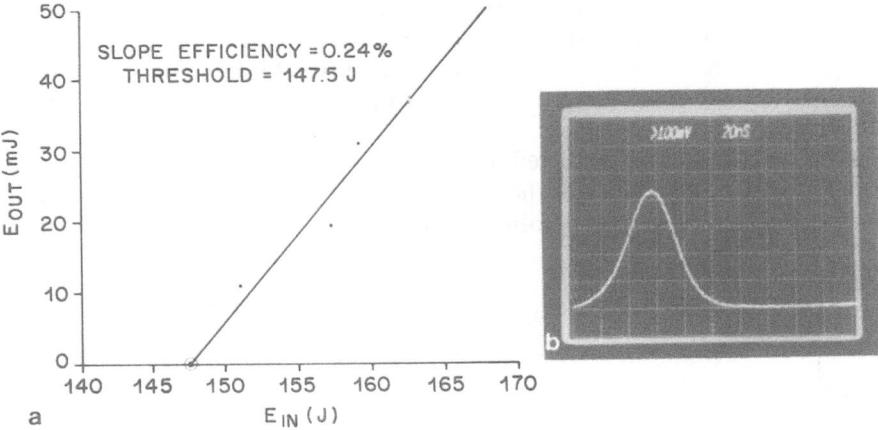

Fig. 9.39a,b. Single-mode performance at 748 nm of alexandrite in a SILO configuration. (a) Output vs flashlamp energy. (b) Temporal profile. (After *Harter* et al. [9.89])

hole burning is not nearly as significant a problem in alexandrite as it is in dye lasers because of the former's high energy storage. Nevertheless, performance is improved by eliminating this mechanism. Other factors are beneficial as well – the fact that component surface reflections do not feed back into the resonator, for example. Traveling wave, ring resonators applied to alexandrite have obtained a 150 mJ Q-switched output in a single longitudinal, single transverse mode [9.91].

f) Alexandrite Oscillator-Amplifiers

As difficult as it has been to develop a good alexandrite oscillator, alexandrite amplifiers work exceedingly well [9.13]. In fact, the amplifier is one of the major success stories in alexandrite laser development. The reason for this is that optical damage mechanisms, operative at high power, appear very much to be functions of the resonator kinetics. Amplifiers do not present the same feedback environment to these mechanisms and optical damage in amplifiers has been limited to that caused by surface defects. To date, no alexandrite amplifier rod has suffered bulk damage [9.92].

With the beam characteristics defined by the oscillator and the power density thereby controlled, it has proven possible to use a fluence in the amplifier that is twice the saturation fluence. A single-pass, oscillator-amplifier system under these conditions, operating at 10 Hz, has provided about 100 MW of peak power in a $7 \times$ diffraction-limited beam and 20 MW in a $1.2 \times$ diffraction-limited beam (adequate for efficient nonlinear frequency conversion by Raman or nonlinear mixing processes). The specific energy extraction achieved from the amplifier rod in this system illustrates the advantage of a low cross section material. Here, 3.5 J output was obtained at

10 Hz at the peak wavelength of 755 nm. Of this, 2.5 J was derived from the 4.5 mm apertured, standard rod amplifier, which corresponds to a *specific energy extraction of better than 1.4 J/cm³*. Figure 9.40 shows the performance of this alexandrite single-pass amplifier operating at its peak wavelength, under constant pump power, as a function of injected input energy. The curved line is model-predicted performance, and its deviation from the straight dashed line indicates the minor effect of saturation. Figure 9.41 gives the normalized small-signal gain as a function of wavelength from a similar

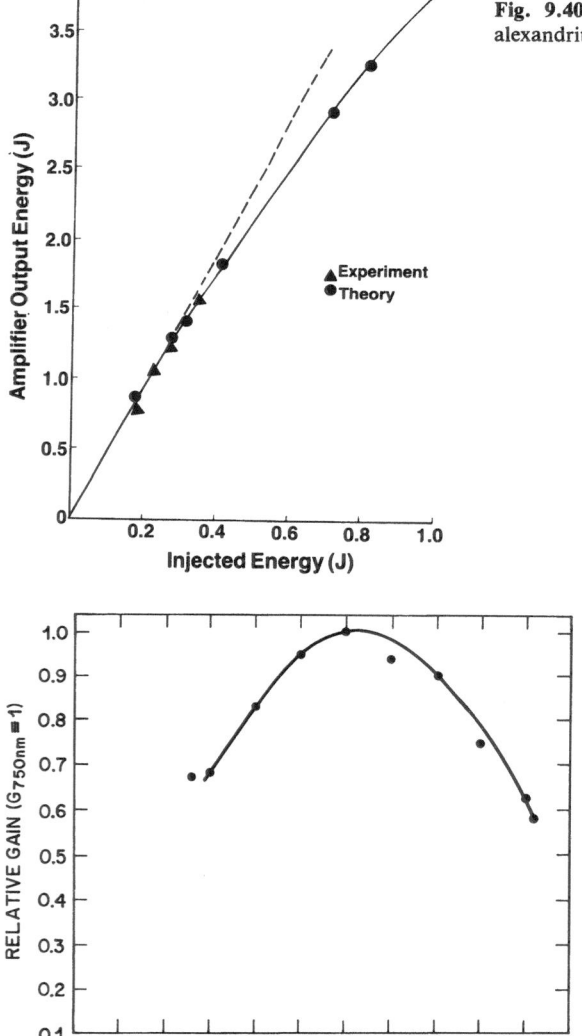

Fig. 9.40. Output vs input energy of an alexandrite amplifier [9.13]

Fig. 9.41. Relative gain vs wavelength in an alexandrite amplifier rod [9.13]. Rod is 6.35 mm × 10.8 cm with 0.12 at.% Cr. Input flash lamp energy is 650 J/pulse applied at 10 Hz. Rod temperature 93 °C

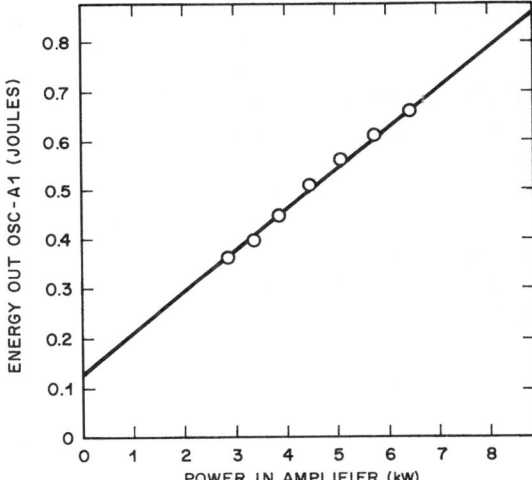

Fig. 9.42. Alexandrite oscillator-amplifier performance. Rods are 5 mm in diameter, 11 cm long, with Cr concentration of 0.23 at.%; operation is at 10 Hz, 95°C; pulse duration is 34 ns. (Figure supplied courtesy of D. F. Heller, Allied Corporation)

amplifier stage. Figure 9.42 shows the output energy from another 10 Hz, oscillator – single-stage-amplifier, as a function of time-averaged pump power to the amplifier. This latter system utilized 5 mm diameter rods, having a Cr concentration of about 0.23 at.%.

g) Zigzag Slab Alexandrite Lasers

The zigzag slab geometry illustrated in Fig. 9.43 is receiving increased attention because of its ability to circumvent the severe thermal lensing problems that plague high average power solid-state rod lasers. The lasing of the first alexandrite zigzag slab [9.93] was the first reported application of this technology to a material having natural birefringence; heretofore, it had been used primarily with isotropic and cubic materials (e.g., glass and Nd: YAG).

Alexandrite's laser properties are ideally suited for high average power operation in the slab geometry. Specific among these are (1) its high thermal loading strength that permits pumping at high average power, (2) its high specific energy extraction that leads to high energy pulses, (3) its low loss due to parasitic oscillations (ASE) that otherwise limit energy storage, (4) its natural birefringence that negates beam depolarization and consequential loss at polarizing elements, and (5) its high hardness that lends well to the fabrication of large, flat and parallel surfaces.

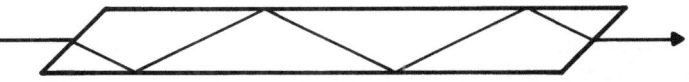

Fig. 9.43. Schematic of zigzag slab geometry

The performance obtained from the first alexandrite slab was modest as the experiment was an exploratory effort. Using a simple resonator with flat mirrors in long pulse operation, 1 J pulses were obtained at low repetition rates with about 1.5% slope efficiency. This performance was predictable, and no unexpected artifacts such as beam breakup or abnormal insertion loss were encountered. The beam fill was about 50% of the full slab aperture under these conditions, owing to finite wave front distortion (Fig. 9.44b). The filled region corresponded precisely to the region where the optical distortion constituted a local positive lens, stabilizing the resonator locally. Moiré deflectometry [9.94] has been shown to be an effective means to study thermal lensing in laser rods [9.95]. Figure 9.44a gives the Moiré diffraction pattern of the slab. A direct correlation can be seen between the local stability regions where the slope of the Moiré fringes cant to the right and regions where the slab is filled by the beam.

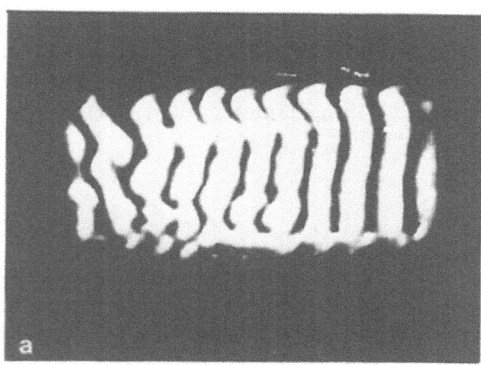

Moiré Deflectogram
of Slab
End Face

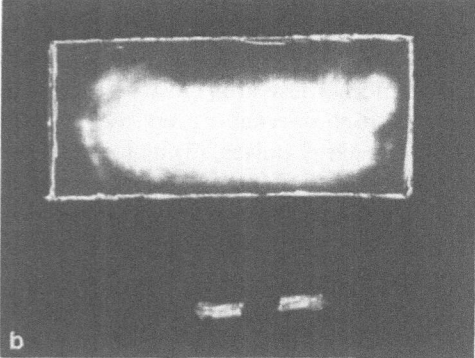

Filling of the
Slab with
Laser Radiation
Resonator:
Flat-Flat

← Burn Patterns

Fig. 9.44a, b. Ray deflection through alexandrite zigzag slab. (**a**) Moiré deflectogram of slab end face. (**b**) Filling of slab by laser beam in a resonator with flat mirrors, and burn patterns as indicated. The distortion corresponds to a 16 m focal length

9.5.2 Arc Lamp Pumped (cw) Alexandrite Lasers

Development of an arc lamp pumped cw alexandrite laser was encouraged by the success of earlier experiments [9.96] to pump ruby continuously with mercury arc lamps, which achieved nearly 2 W output power. The first experiments with cw alexandrite lasers were conducted along similar lines using both single and dual, silver coated, elliptical pump chambers and Hg capillary arc lamps. Early results were promising [9.97], but significant problems existed in the chemical stability of the pump chamber reflector, and in achieving adequate cooling. In these early experiments, the Hg lamps were driven from 60 Hz ac transformers. As a result, lasing was achieved only on the peaks of the power curve. With improvements in design, performance increased; ultimately 60 W was achieved using the ac driven Hg lamps, from a rod only 3 mm in diameter [9.98].

The Hg lamp spectrum overlaps well the alexandrite absorption bands as shown in Fig. 9.45. Because the Hg emission is in the regions of strong absorption, it was possible to achieve good capture efficiency with the smaller (3 mm) diameter rods. Even though Cr concentration can be varied, most of the development effort centered around material best suited for pulsed pumping with Xe lamps − a concentration of about 0.14 at.%. As it happens, 3 mm is about the optimum diameter for a rod of this concentration pumped with Hg.

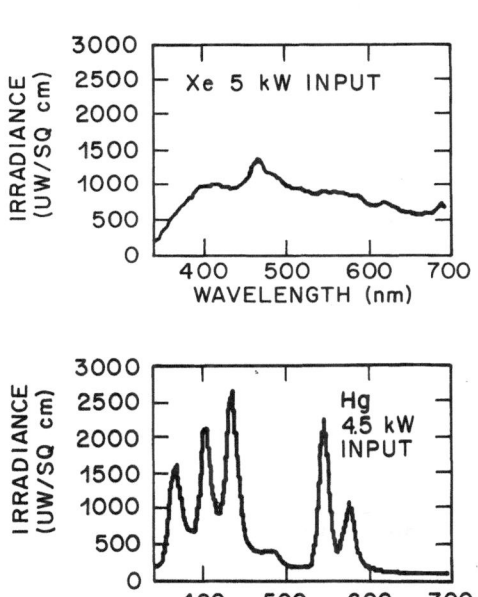

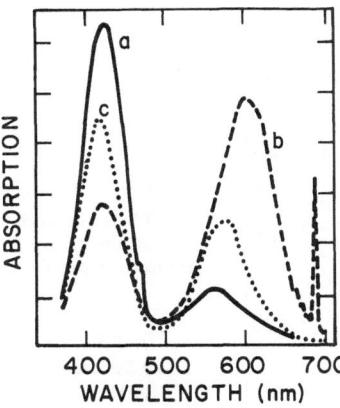

Fig. 9.45. Comparison of Xe and Hg arc lamp spectra with alexandrite absorption bands. (Figure supplied courtesy of H. Samelson, Allied-Signal Corporation)

The dc driven Xe arc lamps work best for achieving relatively low power operation with good mode control – 2 W in TEM_{00} for example. As shown in Fig. 9.44, the Xe emission spectrum is more uniform, not strongly peaked in the absorption bands of alexandrite. Larger diameter rods (5 mm) are used with Xe to give optimum excitation and thermal load uniformity from available concentration material.

An application of arc lamp pumping that is less sensitive to crystal quality is high repetition rate Q-switching. This is particularly true where the Q-switch is operated in synchronism with an ac driven arc lamp at the point of maximum gain. For a repetition rate of 120 Hz, 1.3 mJ pulses with a peak power of 2.3 kW and a pulse duration of 1 μs have been achieved [9.99].

The low gain of the cw alexandrite laser presents a significant challenge in line-narrowing applications, because birefringent tuners and etalons necessarily introduce further loss and can significantly reduce performance. On the other hand, the high Q of the cw alexandrite laser cavity facilitates the tuning and line-narrowing functions of these elements. There are, therefore, two compensating effects. Using a one-element birefringent tuner and two etalons it was possible to achieve a linewidth of $0.01\ \text{cm}^{-1}$ [9.88].

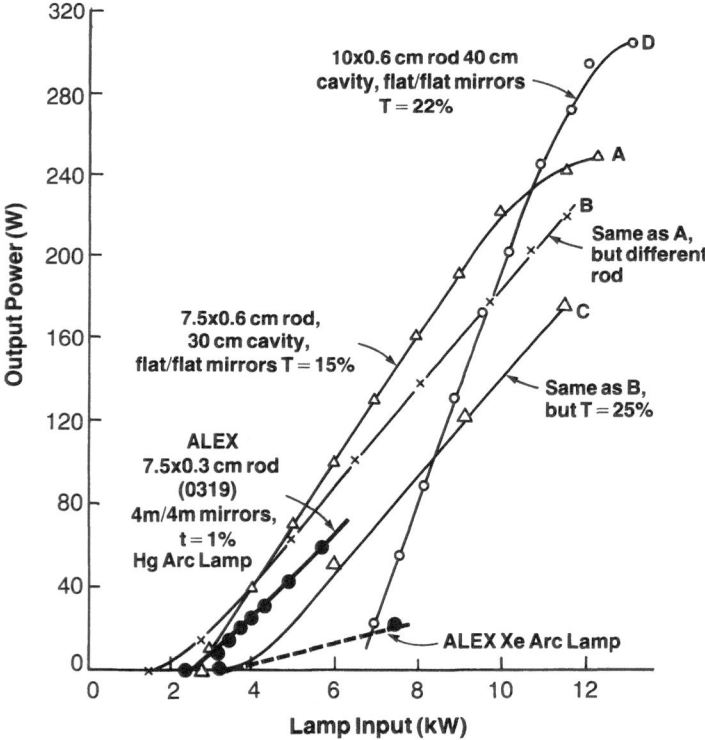

Fig. 9.46. Continuous-wave alexandrite performance compared with cw Nd:YAG [9.13]; Nd:YAG data from [9.12]

In spite of difficulties, progress has been reasonably good. The performance from cw alexandrite lasers, both in power and efficiency, can be compared with that of Nd:YAG, the industry standard. In Fig. 9.46, the multimode performance of both Xe and Hg pumped cw alexandrite lasers is compared to cw pumped Nd:YAG lasers. The Nd:YAG data represents systems of significantly larger size than in the alexandrite cases. Accounting for size, the performance is comparable, and suggests that with further development alexandrite may become important commercially as a cw device.

9.5.3 Optical Damage

At the onset of alexandrite laser development it was anticipated that optical damage would be of paramount importance because of the comparatively high saturation fluence. In fact, the fluences required for optimum alexandrite performance were higher than those then utilized in common industrial practice and considered safe by conventional industry standards. While the experience with the alexandrite laser as reviewed above has demonstrated that optical damage is a tractable problem in this material, high performance levels have not been accomplished without significant difficulty. The problem has not been limited by any means to the laser rod. Other components in the resonator are equally susceptible to optical damage. Fortunately, over the years of alexandrite development significant improvements in tolerance to high flux and fluence levels have been made in almost every optical component of the resonator.

The problem in reaching for higher fluence levels in optical resonators is not strictly an issue of materials quality. Even with material of the highest quality, there still remain formidable problems caused by the high fluences generated and the associated nonlinear instabilities. Lasers that operate in the regime of hundreds of megawatts per square centimeter experience a variety of power-related instabilities. Small scale self-focusing that breaks the beam into filaments is accompanied by other nonlinear mechanisms such as Brillouin and Raman scattering and by other less well understood phenomena. For example, during the development of a Q-switched pulse, any tendency toward filamentation can be enhanced by the lensing action of dissipated energy, and such dissipated energy comes from both the excited state absorption and the phonon release that always accompanies four-level laser amplification. Because of the large energy storage, a filament that starts with a large diameter will still have plenty of stored energy to draw upon as it contracts. As a consequence, any mechanism that causes filamentation will be enhanced by this effect [9.100]. (The time frame is too short for a strain response to occur during a Q-switched pulse, and the lensing parameters for the transient state are not known.)

Surface damage under conditions of very high power has a distinctive morphology. Elliptically shaped pits are formed, approximately $10-50$ µm across, with the major axis aligned with the a-axis in c-axis rods, Fig. 9.47.

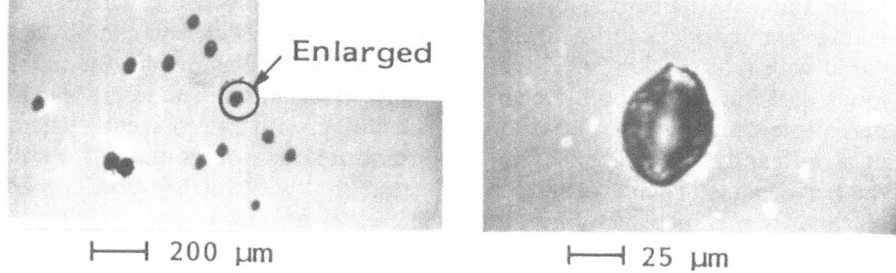

Fig. 9.47. Microphotographs of optical damage on surface of alexandrite laser rods produced under conditions of high flux; major axis of elliptical pit is aligned along the a-axis in c-axis rods

The causal mechanism is not known, but because these pits only occur at high power it seems likely that the pits are produced by power density filamentation. A filament of very high power density that experiences gain would deposit energy and thereby produce axial compression stress all along its length, but fracture might first occur at the surface where surface strain intensifies certain shear stress components locally. The elliptical shape could be explained by an anisotropic shear strength or by an anisotropic self-focusing mechanism which might produce filaments of elliptical cross section. Contamination or surface defects cannot be ruled out, but the pits occur on very well cleaned and carefully mounted rods, occur irrespective of an anti-reflection coating, and only occur at maximum power density (near 1.5 GW/cm^2).

The most serious problem, however, from a practical standpoint has been the impact of optical surface contamination. The ground state absorption coefficient of alexandrite depends exponentially on temperature and poses a potential thermal runaway condition if initiated by a partially absorbing surface contaminant. When the surface contaminant is heated locally by the beam, the underlying alexandrite begins to heat and become more absorbant itself. The ground state absorption first removes phonons and actually cools the material, but the excitations created soon relax liberating the phonon energy absorbed, and no net energy is absorbed or liberated. An increased excited state population is nevertheless produced, and the excited state absorption can lead to thermal runaway. Further heating by excited absorption produces a yet higher excited state population and more excited state absorption. As the temperature rises above 600 K, nonradiative relaxation takes over; at this point, all of the locally absorbed beam energy is liberated as heat. The molten appearance of damage sites when examined under the microscope is evidence for such thermal runaway. Pools of obviously solidified melt with a smooth surface that contain a crack, induced by cooling, have been seen at the bottom of damage pits. Finally, strongly absorbing contamination may be ablated from the surface by the beam and cause no appreciable damage.

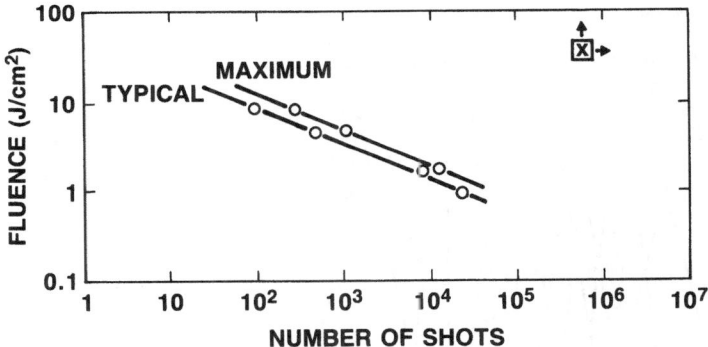

Fig. 9.48. Bulk damage resistance of alexandrite and ruby amplifiers compared. Ruby data (○ ○ ○) is from [9.12]. The alexandrite datum point (×) defines a lower limit, as no alexandrite amplifier rods displayed bulk damage in the controlled experiments refered to here. (Figure supplied courtesy of D. Heller, Allied-Signal Corporation)

Bulk damage occurred in early crystals of alexandrite because of iridium particles that became imbeded during crystal growth. In more recent material, bulk damage still appears related to bulk defects, but on a much finer scale and to a much reduced extent. Oscillator rods generally are able to withstand over 10^6 shots at an intraresonator power density of 10 J/cm^2; damage takes the form of bubbles that apparently develop randomly throughout the rod. Alexandrite amplifier rods have been able to tolerate much higher fluence. Figure 9.48 illustrates the lower bound to bulk optical damage in alexandrite amplifiers (since no amplifier bulk damage has yet occurred) as compared to ruby.

Damage in laser operation is extremely complex and difficult to quantify because of the number of mechanisms and parameters involved. The laser beam itself can even have a curative effect on the involved defects. Laser operating procedures and rules are very important. Practical experience ultimately determines safe operating limits.

9.6 Frequency Conversion

The principal (most widely used) nonlinear frequency conversion schemes are stimulated Raman scattering (SRS; usually performed in gases), frequency mixing and harmonic generation in nonlinear crystals, and the use of optical parametric oscillators (OPOs). As OPOs tend to be less convenient than the other methods, they are relied upon primarily to give tunable output from a fixed frequency pump. (New nonlinear crystals under development for OPOs may alter this situation however.) The other techniques mentioned are fixed frequency shifting methods and provide tunable output only when a tunable pump is used. Figure 9.49 illustrates how SRS and harmonic generation may be combined to greatly extend the pulsed alexandrite laser's frequency range.

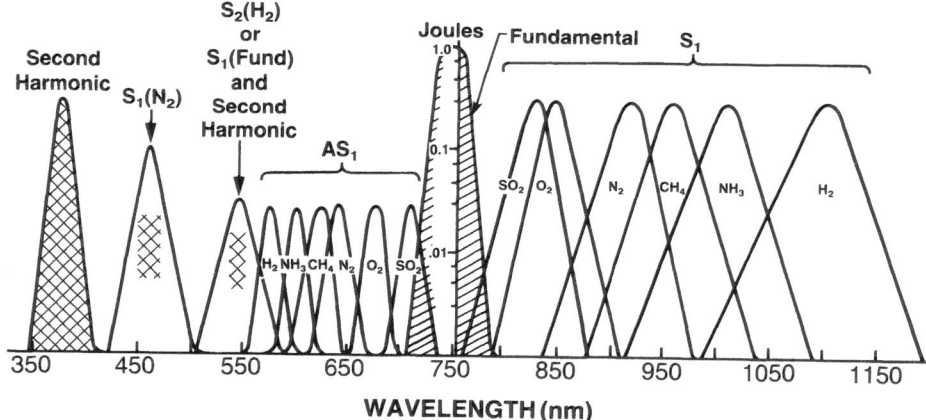

Fig. 9.49. Predicted performance of Raman shifting and/or second harmonic generation applied to the pulsed alexandrite laser: S_1 = 1st Stokes; S_2 = 2nd Stokes; AS_1 = 1st anti-Stokes. The second harmonic of S_1 in H_2 is shown. The second harmonic of the S_1 component of the other gases will span the gaps on the short wavelength side to provide complete coverage. (Courtesy of H. Samelson, Allied-Signal Corporation)

The high brightness achieved by use of the alexandrite oscillator-amplifier approach has direct application in nonlinear frequency conversion [9.13]. With it, 200 mJ tunable pulses at 380 nm have been generated in KDP at a pulse repetition rate of 10 Hz. Also, in Raman shifting experiments, 30% energy conversion into the first Stokes has been achieved from a single pass H_2 gas Raman cell, using an alexandrite laser source.

The efficiency of nonlinear conversion techniques depends to a large extent upon the quality and power of the fundamental beam. The shifted output is, however, often superior in quality because the nonlinear interaction tends to select the highest quality components of the primary beam for conversion. In the Raman case, a substantial improvement in beam quality is also derived from the low optical distortion of the gaseous medium in which the shifted beam develops. Consequently, some additional advantages are gained by use of a nonlinear frequency conversion step.

Most applications do not require a device of the broadest tuning range, but generally do require an efficient, simple and reliable device in the right wavelength range. As discussed earlier, broad bandwidths generally trade off against cross section and/or storage time, and maximum performance is potentially to be obtained from a laser with less than the greatest possible tuning range, but which is "inband" to be application's requirements. Tunability may be required, but often it is only necessary for achieving some specified wavelength.

One strategy for extended frequency coverage using vibronic lasers is to select for laser pumping a series of superior materials that together cover an extended frequency range. Nonlinear frequency conversion may still be applied to such a system provided adequate peak power is obtained. Adding a pump laser is generally a significant expense and complication, and flash lamp pumping, if possible, is usually preferred. For flash lamp pumping to be efficient, however, there must be broad pump bands in the visible. This places a limit on how short in wavelength the fundamental range may extend for a lamp-pumped device. For short wavelength coverage without conversion, a laser pump must be used, as exemplified by the excimer-pumped Ce^{3+} laser discussed in Sect. 9.3.

9.7 Applications

At this point the alexandrite laser is finding some applications that are basically new and which require its unique properties. The most evident of these are in photochemistry, lidar, and isotope separation. In addition, its unique combination of tunability, high energy storage (for Q-switching), and high average power is expected to open new research areas, particularly in nonlinear optics and photochemistry.

9.7.1 Photochemistry

The fundamental output of alexandrite is valuable for spectroscopic studies of numerous atomic and molecular species that have narrow line absorption or broad overtone bands in this range [9.101]. Alexandrite's first real application was in photochemical research [9.102], where experimenters sought to determine the internal relaxation time of supercooled tetramethyldioxetane. The fifth overtone vibrational state of this material was directly excited from the ground state, and the fluorescence linewidth observed provided an estimate of the state's lifetime. The results indicated that the time frame for performing vibrationally selective photochemistry is exceedingly short, on the order of tens of femtoseconds. The alexandrite laser was used because these measurements required a high fluence, narrow line, tunable source. High fluence was required in order to achieve an adequate signal-to-noise ratio, given the very low ($10^{-24} cm^2$) cross sections for both absorption and emission.

9.7.2 Lidar

Lidar (light based radar), in its different forms, provides a vast opportunity for remote atmospheric and geological diagnostics, performed from ground, aircraft, and space platforms. Of special interest is the monitoring of key weather parameters (temperature, pressure, water vapor content) and other important atmospheric constituents and contaminants. A space borne laser system of this type could also monitor the amplitude of waves on the sea, deducing therefrom the surface winds over much of the globe.

Table 9.7. Performance requirements of an alexandrite DIAL transmitter

Property	Requirement
Operating wavelength [nm]	727, 768, 769, 760, (759 ref.)
Tuning	Fine tuning is required about target wavelength
Output energy [J]	>0.15
Pulse duration [ns]	40 at 727 nm
	200 ns at other wavelengths
Bandwidth [cm^{-1}]	<0.007 at 727 nm
	0.02 at other wavelengths
Spectral stability [cm^{-1}/30 min]	±0.01
Beam quality (×diffraction limit)	<4
Beam divergence [mrad]	1
Overall efficiency [%]	0.075
Energy required for rated output [J]	<200

To perform these tasks, an efficient, compact, line-narrowed, short pulse duration, tunable source is required. In the past, dye lasers have been used for such applications, and in particular for differential absorption lidar (DIAL) [9.103]. For this application, however, dye lasers have serious drawbacks in size, convenience and efficiency (particularly important when eventual space operation is contemplated), but they also suffer from excessive background emission. To be sensitive, DIAL measurements require that the laser have a very quiet background emission around the narrowed laser line. However, this emission is directly dependent upon the emission cross section, which is large in dye lasers, but much lower in alexandrite. For this reason and its economy in size and efficiency, the alexandrite laser is strongly preferred.

Alexandrite lasers are being developed for various lidar applications [9.104] in connection with NASA programs [9.105]. One specific application is differential absorption lidar (DIAL) for the remote profiling of atmospheric characteristics, including pressure, temperature, and water vapor content. Oxygen is targeted for both pressure (at 769 and 760 nm) and temperature (at 768 nm); H_2O is targeted at 727 nm. The intensity and line shape of absorption due to these materials are monitored by means of backscatter of the laser beam from atmospheric aerosols, using a time-gated detector. A reference beam, which does not absorb but is equally attenuated by scatter, is used for calibration. The specifications for a multifunctional alexandrite laser DIAL transmitter are given in Table 9.7.

9.7.3 Isotope Separation

Isotopes have great value in medicine, science, and nuclear energy, but their separation by chemical means is generally costly and time consuming. Laser-based methods can reduce the cost of important isotopes. Many laser methods utilize the narrow band of the laser to selectively excite atoms or molecules of a specific isotopic species, which then are immediately ionized by a second

Table 9.8. Realized performance of 50 W alexandrite laser

Property	Performance
Output energy [mJ]	400
Pulse repetition rate [Hz]	120
Pulse width (doubly Q-switched) [µs]	1.6
Wavelength [nm]	790 – 793
Linewidth [nm]	0.005 ± 0.001
Beam profile	Smooth super-Gaussian
Full width beam divergence − aperture [mm − mrad]	15 ± 2

beam of less selective wavelength. The amount of isotope produced is therefore proportional to the photons available in the selective beam, and the cost of generating those photons largely determines the price of the isotope. It is important in this application that the laser be efficient and operable on a selected narrow line. For mass production, however, it is often required that the laser be also of high average power and be repetitively Q-switched. Based on these requirements, isotope separation is a potential application area for tunable transition metal ion lasers. At present, alexandrite is the only vibronic laser well enough developed for practical application in this area. In particular, a 100 W average power, tunable, line-narrowed, Q-switched alexandrite laser has been developed for such applications [9.106,107]. This system employs two independent line-narrowed oscillators each providing 50 W average power at 120 Hz, which are temporally multiplexed to provide 100 W, 240 Hz operation. Each oscillator utilizes two high average power pump chambers each equipped with a standard alexandrite rod. The realized performance of each 50 W unit is given in Table 9.8.

References

9.1 Y. Tanabe, S. Sugano: J. Phys. Soc. Jpn. **9**, 766 (1984)
9.2 M. H. L. Pryce: "Interaction of Lattice Vibrations with Electrons at Point Defects", in *Phonons in Perfect Lattices and Lattices with Point Imperfections* ed. by R. W. H. Stevenson (Plenum, New York 1966)
9.3 W. A. Harrison: *Solid State Theory* (McGraw-Hill, New York 1970)
9.4 D. E. McCumber: J. Math. Phys. N.Y. **5**, 221, 508 (1964); Phys. Rev. **135**, A1676 (1964)
9.5 D. E. McCumber: Phys. Rev. **133**, A163 (1964)
9.6 R. Englman: *Non-Radiative Decay of Ions and Molecules in Solids* (North-Holland, Amsterdam 1979)
9.7 P. F. Moulton: "Tunable Paramagnetic-Ion Lasers", in *Laser Handbook*, Vol. 4, ed. by M. Bass and M. Stitch (North-Holland, Amsterdam 1986)
9.8 D. E. McCumber: Phys. Rev. **134**, A299 (1964); ibid. **136**, A954 (1964)
9.9 M. L. Shand, J. C. Walling, H. P. Jenssen: IEEE J. QE-**18**, 167 (1982)
9.10 J. C. Walling, O. G. Peterson, H. P. Jenssen, R. C. Morris, E. W. O'Dell: IEEE J. QE-**16**, 1302 (1980)

9.11 M. L. Shand: J. Appl. Phys. **54**, 2602 (1983)
9.12 W. Koechner: Solid-State Laser Engineering, Springer Ser. Opt. Sci., Vol. 1 (Springer, New York 1976)
9.13 J. C. Walling, D. F. Heller, H. Samelson, D. J. Harter, J. A. Pete, R. C. Morris: IEEE J. QE-21, 1568 (1985)
9.14 M. L. Shand, H. P. Jenssen: IEEE J. QE-19, 480 (1983)
9.15 H. Samelson, J. C. Walling, D. Heller: Proc. Soc. Photo-Opt. Instrum. Eng. **335**, 85 (1982)
9.16 S. Guch, Jr., C. E. Jones: Opt. Lett. **7**, 608 (1982)
9.17 S. T. Lai, M. L. Shand: J. Appl. Phys. **54**, 5642 (1983); In Proc. Int. Conf. Lasers 83, ed. by R. C. Powell (STS Press, McLean, Va 1984) p. 165
9.18 S. T. Lai, M. L. Shand: In SPIE O – E LASE 86, 19 – 24 Jan. 1986, Conf. 622, Session 3 (1986)
9.19 J. Buchert, A. Katz, R. R. Alfano: In Proc. Int. Conf. Lasers 82, ed. by R. C. Powell (STS Press, McLean, Va 1983) pp. 791 – 798
9.20 M. L. Shand, S. T. Lai: IEEE J. QE-20, 105 (1984)
9.21 W. Kolbe, K. Petermann, G. Huber: IEEE J. QE-21, 1596 (1985)
9.22 J. Drube, B. Struve, G. Huber: Opt. Commun. **50**, 45 (1984)
9.23 G. Huber, K. Petermann: In *Tunable Solid State Lasers,* ed. by P. Hammerling, A. B. Budgor, A. Pinto, Springer Ser. Opt. Sci., Vol. 47 (Springer, Berlin, Heidelberg 1985) pp. 11 – 19
9.24 B. Struve, G. Huber: J. Appl. Phys. **57**, 45 (1985)
9.25 U. Dürr, U. Brauch, W. Knierim, C. Schiller: In *Tunable Solid State Lasers,* ed. by P. Hammerling, A. B. Budgor, A. Pinto, Springer Ser. Opt. Sci., Vol. 47 (Springer, Berlin, Heidelberg 1985) pp. 20 – 27
9.26 E. V. Zharikov, N. N. Il'ichev, S. P. Kalitin, V. V. Laptev, A. A. Malyutin, V. V. Osiko, V. G. Ostroumaov, P. P. Pashinin, A. M. Prokhorov, V. A. Smimov, A. F. Umyskov, I. A. Shcherbakov: Sov. J. Quantum Electron. **13**, 1274 (1983)
9.27 U. Brauch, U. Durr: Opt. Lett. **9**, 441 (1984)
9.28 U. Brauch, U. Durr: Opt. Commun. **49**, 61 (1984)
9.29 U. Brauch, U. Durr: Opt. Commun. **55**, 35 (1985)
9.30 P. Moulton: In Proc. Int. Conf. Lasers 81, pg. 359, ed. by C. B. Collins (STS Press, McLean, Va 1981)
9.31 P. F. Moulton: In SPIE O – E LASE 86, 19 – 24 Jan. 1986, Conf. 622, Session 3 (1986)
9.32 W. F. Krupke: In Proc. Int. Conf. Lasers 80, pp. 511 – 518, ed. by C. B. Collins (STS Press, McLean, Va 1981)
9.33 J. A. Caird, W. F. Krupke, M. D. Shinn, P. R. Staver, H. J. Guggenheim: Bull. Am. Phys. Soc. **30**, 1857 (1985)
9.34 S. Lai, H. P. Jenssen: In IEEE and OSA, Topical Meeting on Tunable Solid-State Lasers, Arlington VA, May 16 – 17, 1985, Tech. Dig., Talk FA71
9.35 C. E. Byvik, A. M. Buoncristiani: IEEE J. QE-21, 1619 (1985)
9.36 P. F. Moulton: In Proc. Conf. Lasers Electro-Opt., Anaheim CA, June 19 – 22, 1984, paper WA2
9.37 P. F. Moulton: IEEE J. QE-21, 1582 (1985)
9.38 W. Kunzel, W. Knierim, U. Durr: Opt. Commun. **36**, 383 (1981); Appl. Phys. B **28**, 233 (1982)
9.39 K. R. German, U. Durr, W. Kunzel: Opt. Lett. **11**, 12 (1986)
9.40 L. F. Johnson, R. E. Dietz, H. J. Guggenheim: Appl. Phys. Lett. **5**, 21 (1964)
9.41 L. F. Johnson, H. J. Guggenheim, R. A. Thomas: Phys. Rev. **149**, 179 (1966)
9.42 P. F. Moulton: IEEE J. QE-18, 1185 (1982)
9.43 L. F. Johnson, H. J. Guggenheim, D. Bahnck, A. M. Johnson: Opt. Lett. **8**, 371 (1983)
9.44 P. F. Moulton, A. Mooradian: Appl. Phys. Lett. **35**, 838 (1979)
9.45 D. J. Ehrlich, P. F. Moulton, R. M. Osgood, Jr.: Opt. Lett. **4**, 184 (1979)
9.46 D. J. Ehrlich, P. F. Moulton, R. M. Osgood, Jr.: Opt. Lett. **5**, 339 (1980)
9.47 Yu. S. Vagin, V. M. Marchenko, A. M. Prokhorov: Sov. Phys. – JETP **28**, 902 (1969)
9.48 R. Beck, K. Gurs: J. Appl. Phys. **46**, 5224 (1975)

9.49   H. P. Jenssen: unpublished data
9.50   A. Erbil, H. P. Jenssen: Appl. Opt. **19**, 1729 (1980)
9.51   L. F. Johnson, H. J. Guggenheim: IEEE J. QE-**10**, 442 (1974)
9.52   T. Maiman: Nature (London) **187**, 493 (1960)
9.53   W. H. Fonger, C. W. Struck: Phys. Rev. B **11**, 3251 (1975)
9.54   P. Kisliuk, C. A. Moore: Phys. Rev. **160**, 307 (1967)
9.55   B. K. Sevast'yanov, Yu. I. Remigailo, V. P. Orekhova, V. P. Matrosov, E. G. Tsvetkov,
       G. V. Bukin: Sov. Phys. – Dokl. **26**, 62 (1981)
9.56   R. C. Morris, C. F. Cline: U.S. Patent 3.997.853, Dec. 14, 1976
9.57   G. V. Bukin, S. Yu. Volkov, V. N. Matrosov, B. K. Sevast'yanov, M. I. Timoshechkin:
       Sov. J. Quantum Electron. **8**, 671 (1978)
9.58   M. L. Shand, J. C. Walling: IEEE J. QE-**18**, 1829 (1982)
9.59   M. L. Shand: In Proc. Int. Conf. Lasers 82, ed. by R. C. Powell (STS Press, McLean, Va
       1982) pp. 799 – 802
9.60   E. V. Zharikov, V. V. Laptev, E. I. Sidorova, Yu. P. Timofeev, I. A. Shcherbakov: Sov.
       J. Quantum Electron. **12**, 1124 (1982)
9.61   M. J. P. Payne, H. W. Evans: In IEEE and OSA, Topical Meeting on Tunable Solid-State
       Lasers, Arlington VA, May 16 – 17, 1985, Tech. Dig., Talk FA 4-2
9.62   L. F. Johnson, H. J. Guggenheim: J. Appl. Phys. **38**, 4837 (1967)
9.63   W. M. Fairbank, Jr., G. K. Klauminzer, A. L. Schawlow: Phys. Rev. B. **11**, 60 (1975)
9.64   M. O. Henry, J. P. Larking, G. F. Imbusch: Proc. R. Ir. Acad. **75**, 97 (1975)
9.65   M. L. Shand, J. C. Walling, R. C. Morris: J. Appl. Phys. **52**, 953 (1981)
9.66   L. Esterowitz, R. Allen, C. P. Khattak: In *Tunable Solid State Lasers,* ed. by P. Hammer-
       ling, A. B. Budgor, A. Pinto, Springer Ser. Opt. Sci., Vol. 47 (Springer, Berlin, Heidelberg
       1985) pp. 73 – 75
9.67   L. F. Johnson, R. E. Dietz, H. J. Guggenheim: Phys. Rev. Lett. **11**, 318 (1963)
9.68   P. F. Moulton, A. Mooradian, T. B. Reed: In Digest of Technical Papers, 10th Int. Quan-
       tum Electronics Conf., (Optical Society of America, Washington, D.C. 1978) Paper C.2,
       p 630
9.69   P. F. Moulton, A. Mooradian, T. B. Reed: Opt. Lett. **3**, 164 (1978)
9.70   S. Løvold, P. F. Moulton, D. K. Killinger, N. Menyuk: IEEE J. QE-**21**, 202 (1985)
9.71   B. C. Johnson, P. F. Moulton, A. Mooradian: Opt. Lett. **10**, 116 (1984)
9.72   P. P. Sorokin, M. J. Stevenson: IBM J. Res. Dev. **5**, 56 (1961)
9.73   J. C. Walling, H. P. Jenssen, R. C. Morris, E. W. O'Dell, O. G. Peterson: In Annual
       Meeting of the Optical Society of America, San Francisco CA, Oct. 31 – Nov. 3, 1978
       (Optical Society of America, Washington, DC)
9.74   C. F. Cline, R. C. Morris, M. Dutoit, J. Harget: J. Mater. Sci. **14**, 941 (1979)
9.75   *Internationale Tabellen zur Bestimmung von Kristallstrukturen* 1 (Bornträger, Berlin 1937)
9.76   E. F. Farrell, J. H. Fang, R. E. Newnham: Am. Mineral. **48**, 804 (1963)
9.77   C. E. Forbes: J. Chem. Phys. **79**, 2590 (1983)
9.78   R. C. Powell, Lin Xi, Xu Gang, G. J. Quarles, J. C. Walling: Phys. Rev. B **32**, 2788 (1985)
9.79   S. T. Lai: unpublished
9.80   B. K. Sevast'yanov, Kl. S. Bagdasarov, L. B. Pasternak, S. Yu. Volkov, V. P. Ore Khova:
       JETP Lett. **17**, 47 (1973)
9.81   M. L. Shand, S. T. Lai: In *Tunable Solid State Lasers,* ed. by P. Hammerling, A. B.
       Budgor, A. Pinto, Springer Ser. Opt. Sci., Vol. 47 (Springer, Berlin, Heidelberg 1985)
       pp. 76 – 79
9.82   A. M. Ghazzawi, J. K. Tyminski, R. C. Powell, J. C. Walling: Phys. Rev. B **30**, 7182
       (1984)
9.83   C. L. Sam, J. C. Walling, H. P. Jenssen, R. C. Morris, E. W. O'Dell: Proc. Soc. Photo-
       Opt. Instrum. Eng. **247**, 130 (1980)
9.84   D. J. Harter, H. Rainnee, A. J. Heiney: U.S. National Bureau of Standards Special
       Publication, *Laser Induced Damage in Optical Materials,* to be published
9.85   J. C. Walling, O. G. Peterson: IEEE J. QE-**16**, 119 (1980)
9.86   G. Boyd, H. Kogelnik: Br. Sci. Tech. J. **41**, 1347 (1962)

9.87   J. M. Eggleston, G. Givliani, R. L. Byer: J. Opt. Soc. Am. **71**, 1264 (1981);
        G. Giuliani, Y. K. Park, R. L. Byer: Opt. Lett. **5**, 491 (1980)
9.88   D. Harter: private communication
9.89   D. J. Harter, J. J. Yeh, A. J. Heiney, D. R. Siebert: IEEE and OSA, Topical Meeting on
        Tunable Solid-State Lasers, Arlington VA, May 16 – 17, 1985, Session FA2
9.90   R. Rapoport, J. J. Yeh, R. Sam: "High Efficiency Injection Locking of Q-Switched
        Alexandrite Lasers", in Conf. Lasers and Electro-optics (CLEO 83), Baltimore MD, May
        17 – 20, 1983, Session ThR4
9.91   J. Krasinski, P. Papanestor, D. F. Heller: In Conf. Lasers and Electro-optics (CLEO),
        Baltimore MD, May 21 – 24, 1985, Session WT2
9.92   D. F. Heller: In SPIE O – E LASE 86, 19 – 24 Jan. 1986, Conf. 622, Session 3 (1986)
9.93   J. C. Walling, H. Samelson: In Annual Meeting of the Optical Society of America, San
        Diego CA, Oct. 29 – Nov. 2, 1984, Session TuP5
9.94   O. Kafri: Opt. Lett. **5**, 555 (1980)
9.95   T. Chin, O. Kafri, J. Krasinski, D. F. Heller: In Conf. Lasers and Electro-optics (CLEO)
        Baltimore MD, May 21 – 24, 1985, Session ThR6
9.96   V. Evtuhov, J. K. Neeland: J. Appl. Phys. **38**, 4051 (1967)
9.97   J. C. Walling, O. G. Peterson, R. C. Morris: IEEE J. QE-**16**, 120 (1980)
9.98   H. Samelson, D. J. Harter: Conf. Lasers and Electro-optics (CLEO), Anaheim CA, June
        19 – 22, 1984, Session W14
9.99   H. Samelson: private communication
9.100 J. J. Barrett: interactive private communication
9.101 A. V. Nowak, B. J. Krohn: IEEE J. QE-**21**, 1607 (1985)
9.102 G. A. West, R. P. Mariella, Jr., J. A. Pete, W. B. Hammond, D. Heller: J. Chem. Phys.
        75, 2006 (1981)
9.103 C. L. Korb, J. E. Kalshoven, Jr., C. Y. Weng: Trans., Am. Geophys. Union **60**, 333 (1979)
9.104 F. P. Roulard, III: In *Tunable Solid State Lasers,* ed. by P. Hammerling, A. B. Budgor,
        A. Pinto, Springer Ser. Opt. Sci., Vol. 47 (Springer, Berlin, Heidelberg 1985) pp. 53 – 58;
        R. C. Sam, F. P. Roullard, III: Electro Optic System Design (EOSD) Magazine 37 (1982)
9.105 F. Allario, B. A. Conway: In *Tunable Solid State Lasers,* ed. by P. Hammerling, A. B.
        Budgor, A. Pinto, Springer Ser. Opt. Sci., Vol. 47 (Springer, Berlin, Heidelberg 1985)
        pp. 42 – 52
9.106 R. C. Sam, R. W. Rapoport, M. L. Shand: In Proc. Southwest Conf. Optics, Albuquerque
        N.M., March 4 – 8, 1985, ed. by R. S. McDowell, SPIE **540**, 264 (1985)
9.107 R. C. Sam, R. Rapoport, S. Matthews: In *Tunable Solid State Lasers,* ed. by P. Hammer-
        ling, A. B. Budgor, A. Pinto, Springer Ser. Opt. Sci., Vol. 47 (Springer, Berlin, Heidelberg
        1985) pp. 28 – 37

10. Recent Progress in Tunable Lasers

Clifford R. Pollock

With 4 Figures

This chapter briefly reviews the advances that have occurred since 1987 in the development of tunable laser sources. The last five years have seen much activity in certain fields, such as tunable solid state lasers and optical parametric oscillators. Other fields, like stimulated Raman scattering, have matured to the point where little additional progress has been made, or direct generation using new lasers has replaced the nonlinear sources. Extensive use is made of references to the current literature for the reader interested in more detail. Reference numbers follow on from the original chapters. The descriptions are by necessity rather brief, and build directly on the material presented in the preceding chapters.

10.1 General Principles and Some Common Features

The underlying operation and technology of tunable sources is basically the same as was described five years ago in Chap. 1. This reflects the fact that laser technology is maturing, so much effort is now directed into improving performance of existing technology. Two new topics deserve further elaboration: semiconductor diode pumping of solid state lasers, and the discovery of new schemes for ultrashort pulse generation.

10.1.1 Diode-Pumped Tunable Lasers

Several advances in the technology of diode-pumped tunable lasers have occurred in the last five years. Perhaps the most dramatic development has been the introduction of semiconductor diode lasers to pump tunable sources. The high efficiency and compact size of the semiconductor diode allow many new applications of tunable systems, which are otherwise restricted to laboratory benches with large-frame pump lasers. The state-of-the-art of semiconductor diodes has advanced to the point where cw powers exceeding 10 W are routinely available for prices of approximately US $ 1000/W. When possible, using semiconductor diodes for optical pumping is advantageous because it reduces the size of a total laser system, and it provides for a major increase in electrical conversion efficiency.

Most work to date has been directed at pumping nontunable solid state lasers, such as Nd: YAG. However, in the last two years the application to tunable sources has been explored. *Scheps* et al. [1.21] demonstrated the first operation of direct diode pumping of a Cr-doped solid state laser, alexandrite. Alexandrite has broad pump absorption bands in the visible region (see Fig. 9.21). Using recently commercialized red laser diodes operating in the 670–680 nm region, the laser was able to reach threshold at 12 mW of incident pump power, and had a slope efficiency of 25% at an output wavelength of 750 nm. The diodes replaced a cw argon-ion/dye laser pump source, making the overall laser much simpler and more compact. *Scheps* recently [1.22] demonstrated diode pumped Cr: LiCaF, producing 0.3 mW at 795 nm, with an incident pump power of approximately 18 mW obtained from two 10 mW 673 nm laser diodes. There is certain to be more active work in replacing optical pump sources with semiconductor diodes as the power and wavelength range of commercial diodes increases.

A second aspect of diode pumping of solid-state lasers is that the achievable spectral linewidth can be extremely narrow under proper cavity design configurations. *Day* et al. [1.23] characterized the linewidth of all-solid state diode-pumped Nd: YAG lasers, and found the linewidths to be in the kilohertz region. This narrow linewidth follows naturally from the use of an entirely solid state resonator (there are no moving parts to vibrate). Also, the semiconductor diode laser can be configured to deliver power only in the spatial mode of the laser, with little excess noise. This is in contrast to flashlamp or cw arc lamp pumped solid state lasers. In addition, index variations due to temperature fluctuations or gradients (dn/dT) are minimal, so the spatial mode and spectral linewidth are naturally well behaved. These characteristics have yet to be demonstrated with tunable laser hosts, but it is likely that similar performance can be achieved.

10.1.2 New Mode-Locking Techniques for Tunable Lasers

Several new mode-locking techniques have been developed for tunable solid state lasers. The precursor to all this work was the soliton laser, described in Sect. 6.4.2. In 1987, researchers studying the soliton laser observed several new effects that were inconsistent with solitons. The chief problem was that short pulse generation could occur with positive-dispersion optical fiber, a result totally inconsistent with the soliton theory, which requires negative dispersion. *Ippen* and co-workers [1.24] proposed a new theory based on the constructive interference of a pulse with a chirped counterpart, called Additive Pulse Mode locking (APM). Figure 10.1 illustrates the principle of additive pulse mode locking. The output pulse from a mode-locked laser is divided into two beams, one of which is coupled into a single-mode fiber. In the fiber, the pulse undergoes self-phase modulation. The pulse is retro-reflected back into the mode-locked laser. If the path lengths are adjusted to make the returning pulse coincident with a circulating pulse inside the laser cavity, the two pulses will

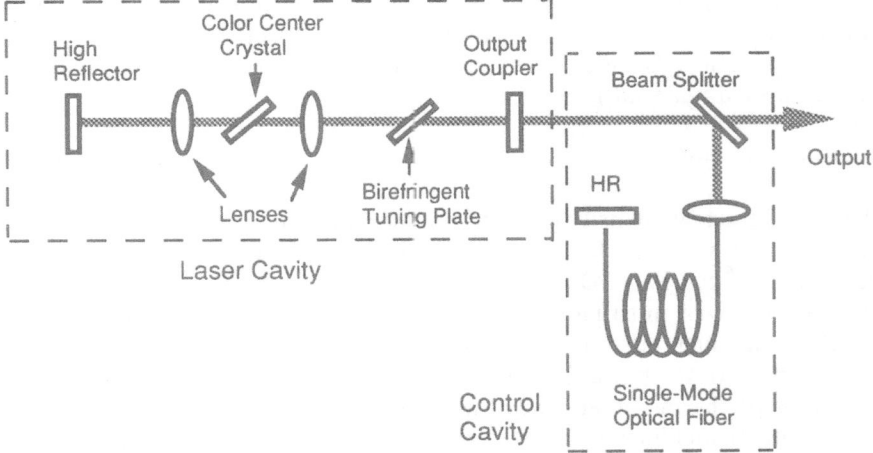

Fig. 10.1. The additive pulse mode-locked laser consists of a laser cavity, in this case a color center laser, coupled to a control cavity that contains a single-mode fiber. The length of the two cavities is adjusted such that the circulating optical pulses in each cavity strike the output coupler simultaneously. Interference between the pulses leads to short pulse generation

interferometrically combine on the output coupler. Since the returning pulse is chirped due to self-phase modulation, the interference between the pulses will vary between constructive and destructive across the length of the pulse. Optimum operation occurs when the pulse carries an excess phase chirp of about π radians [1.25]. Under this condition, the peaks of the two pulses will be in-phase and interfere constructively, while the tails of the pulses will be out of phase and interfere destructively. The result is that a temporally shortened pulse will be reinjected back into the laser cavity. The process of pulse shortening continues until dispersive effects in the optical system ultimately apply an equal pulse-broadening pressure. Typical pulse durations are 100 fs from such lasers.

The additive pulse mode-locking technique has been applied to several tunable lasers, such as the NaCl color center laser, which generates 100 fs pulses over the $1.48 - 1.75\ \mu$m region [1.26, 27], the Ti: sapphire laser, which generates 80 fs pulses in the $700 - 900$ nm region [1.28, 29], and fixed wavelength lasers such as the Nd: YAG at $1.06\ \mu$m [1.30] and the Nd: glass laser at $1.054\ \mu$m [1.31].

The virtue of the APM laser is that a tunable laser can be made to generate tunable femtosecond pulses. Unlike the fixed wavelength colliding pulse mode-locked dye laser, where a coincidence is required between the emission of the gain medium and the absorption profile of the saturable absorber, the APM laser is not critically dependent on spectral features of the gain. Thus the femtosecond laser is truly a tunable source.

In 1990, mode-locking techniques based on the Kerr effect were demonstrated, and are now commercially available. The Kerr effect causes an intensity-induced change in index in an optical medium. Specifically, the index of refraction of a material goes as

$$n(I) = \eta_0 + n_2 I \; , \tag{10.1}$$

where n_0 is the low intensity index of refraction for the material, and n_2 is the Kerr coefficient, which for most dielectrics away from resonance is on the order of $10^{-16} \, \text{cm}^2 \, \text{W}^{-1}$. Due to the small size of n_2, the Kerr effect is usually not noticed except in situations with extremely high intensity. This is exactly the situation which arises in mode-locked lasers with femtosecond pulses: the peak intensity of the pulses is many orders of magnitude greater than the average power of the mode, and weak nonlinear effects can become significant. The first mode locking attributed to the Kerr effect was discovered by *Sibbett*'s group in St. Andrews, Scotland in a Ti:sapphire laser [1.32], where 60 fs pulses generation was observed from a self-mode-locked laser, but only when the output mirror was slightly misaligned in order to create a mixture of TEM_{00} and TEM_{0n} modes. There were no modulators or coupled cavities involved in the modulation and phase locking of the modes of the laser. Several theories have been advanced to describe this effect, all using the Kerr effect as the effector of the coupling between modes. *Kafka* and *Baer* [1.33] propose that the addition of a χ_3 nonlinearity in the gain medium causes enhanced coupling between the modes [1.34]. If the gain medium is long enough (several Rayleigh lengths), and if the cavity design ensures that the off-axis modes are degenerate in frequency, then the coupling is sufficient to overcome normal phase wander, and the modes lock together, forming a short pulse.

A second scheme demonstrated by *Spinelli* et al. [1.35] utilizes the Kerr effect in Ti:sapphire to spatially alter the laser beam. In this design, the Kerr effect in the gain medium adds an intensity-dependent lens to the Ti:sapphire rod. Under low intensity cw operation, the Ti:sapphire rod acts like a passive element. When short pulses are formed, the high intensity of the pulse causes the index of refraction of the Ti:sapphire to increase slightly, forming a lens in the rod. The subsequent laser beam comes to a tighter focus on the output coupler than the cw beam. A simple aperture placed near the output coupler can be used to force operation in the lower-loss pulsed mode. With such a scheme, 100 fs pulses have been generated in the range $750-1000$ nm from the Ti:sapphire laser. This new mode-locking scheme should be universal for all solid-state lasers, and opens yet another door to the application of tunable lasers to ultrashort pulse generation, and secondarily, nonlinear frequency conversion via high peak powers.

Further details of ultrashort pulse generation for specific systems are described in Sects. 10.5, 10.6, and 10.9.

10.1.3 New References for Chapter 1

1.21 R. Scheps, B. M. Gately, J. Myers, J. Krasinski, D. F. Heller: Appl. Phys. Lett. **56**, 2288 (1990)
1.22 R. Scheps: Diode pumped Cr^{3+} : $LiCaAlF_6$ laser, in Conf. on Lasers and Electrooptics '91, Baltimore, MD (1991)
1.23 T. Day, E. K. Gustafson, R. L. Byers: Opt. Lett. **15**, 221 (1990)
1.24 J. Mark, L. Y. Liu, K. L. Hall, H. A. Haus, E. P. Ippen: Opt. Lett. **14**, 48 (1989)
1.25 B. J. Zook: Ph.D. Thesis, Cornell University (1990)
1.26 J. F. Pinto, C. P. Yakymyshyn, C. R. Pollock: Opt. Lett. **13**, 383 (1988)
1.27 C. P. Yakymyshyn, J. F. Pinto, C. R. Pollock: Opt. Lett. **14**, 621 (1989)
1.28 P. W. French, J. A. R. Williams, J. R. Taylor: Opt. Lett. **14**, 686 (1989)
1.29 J. Goodberlet, J. Wang, J. G. Fujimoto, P. A. Schultz: Opt. Lett. **14**, 1125 (1989)
1.30 J. Goodberlet, J. Jacobson, J. G. Fujimoto, P. A. Schultz, T. Y. Fan: Opt. Lett. **15**, 504 (1990)
1.31 F. Kraus, C. Spielmann, T. Brabec, E. Winter, A. J. Schmidt: Opt. Lett. **15**, 1082 (1990)
1.32 D. Spence, P. N. Kean, W. Sibbett: Opt. Lett. **16**, 42 (1991)
1.33 J. D. Kafka, T. Baer: Multimode mode-locking of solid state lasers, in Conf. on Lasers and Electrooptics '91, JMA 2, Baltimore, MD (1991)
1.34 C. L. Tang, H. Statz: J. Appl. Phys. **38**, 2963 (1967)
1.35 L. Spinelli, B. Coullaud, N. Goldblat, D. Negus: Conf. on Lasers and Electrooptics '91, postdeadline talk CPDP 7, Baltimore, MD (1991)

10.2 Excimer Lasers

Excimer lasers continue to play a major role in the generation of short-wavelength radiation. Recent progress has come in two directions: improved efficiency from broadly tunable excimer transitions in the visible, and the generation of ultrashort pulse continua based on amplified femtosecond pulses.

10.2.1 Broadly Tunable Excimer Lasers

There has been great activity in developing the $XeF(C \rightarrow A)$ excimer laser into a broadly tunable, efficient laser. Using a coaxially pumped flashlamp dye laser as an injection source, *Hamada* et al. [2.163] found that the amplified output of an electron-beam pumped $XeF(C \rightarrow A)$ laser was tunable from 470 nm to 500 nm with a spectral width of 0.6 nm. The tuning range was limited by mirror coatings and not by the gain medium. They found that electron-beam-induced transient absorptions could be saturated with sufficient injection energy, effectively reducing the valleys that appear in a free-running $C \rightarrow A$ laser. Improving the electron-beam energy deposition [2.164] led to improved energy efficiency, leading to output energy densities of approximately 1 J/l, with an intrinsic efficiency of 1% throughout the tuning range. In a similar experiment, *Voges* and *Marowsky* [2.165] demonstrated injection control over the tuning range 450−520 nm using a discharge excitation. Peak energies up to 4 mJ per pulse were obtained. These levels of performance are beginning to approach those of the more powerful $XeF(B \rightarrow X)$ laser. An analytical model was developed which closely describes the characteristics of the injection-controlled unstable resonator excimer laser [2.166]. *Nighan* and *Fowler* [2.167] analyzed the kinetic processes affecting the performance of the $XeF(C \to A)$ laser under conditions of short pulse excitation (10 ns), high pressure (6 atm), and using multicomponent gas mixtures composed of Ar-Kr-Xe-NF_3-F_2. These studies identified the primary transient species that lead to absorption in the blue-green region of operation.

To boost the power of the $C \to A$ laser, several scaling experiments have been performed. The first was reported by *Hirst* et al. [2.168] in 1989, where 0.7 J pulses were extracted from a gain volume of approximately 0.5 liter. The energy deposited in the gas was 120 J/l, yielding an intrinsic efficiency of approximately 1.2%. The energy improvement arose from using a larger electron-beam machine to pump a larger volume, along with a higher magnification unstable resonator. Free-running, the laser only produced 125 mJ per pulse, but when injection controlled, the output rose to 700 mJ. The system was operated at a 1 Hz repetition rate to achieve high average power. Under these conditions, the average pulse energy decreased, because the laser gas cannot thermally relax between shots within the small laser volume. More recent studies of the scaled system have investigated the effect of the intense electron-beam pumping on the beam quality within the resonator and the use of a gas flow system to rapidly exchange the laser gas between shots [2.169, 170]. In the latter case, energy extraction of 1.1% was achieved, with 1.2 J pulses being generated at a 1 Hz repetition rate. Injection control by a line narrowed dye laser resulted in operation of the $XeF(C \to A)$ laser with a linewidth of 0.001 nm [2.171]. Longer excitation pulse lengths (175 ns) have been investigated in a coaxial electron beam [2.172]. Without injection control, the system produced an output power of 215 mJ, corresponding to 1.4 J/l of stored energy. Output power can be expected to rise under injection control due to the faster cavity buildup time. Long pulse operation using 700 ns electron-beam pumping into a 1.6 atm gas mixture developed a 400 ns laser pulse, with output energy of 1 J [2.173].

10.2.2 Ultrahigh Peak Power Excimer Lasers

Significant work using XeCl and KrF excimer lasers to amplify femtosecond pulses generated by dye lasers has been underway. *Glownia* et al. [2.174] reported 350 fs pulses from a XeCl amplified mode-locked dye laser. The output of the dye laser (approximately 6 ps pulses) was pulse compressed using a fiber and grating. The compressed pulse was further amplified in a pulse dye amplifier operating at 10 Hz. The amplified pulses were frequency doubled, and then sent through an excimer gain module, with small signal gain of approximately 710. The output pulses had energies of 1.5 mJ at 308 nm. When focused into air, the pulses were sufficiently intense for self-phase modulation to increase their spectral bandwidth to approximately 1000 cm^{-1}. *Schwartzenbach* et al. [2.175] demonstrated a 20 mJ, 150 fs duration pulse using a KrF amplifier to boost the output of a frequency doubled, cavity dumped mode-locked dye laser. The spectral width of the 248 nm pulses was approximately 0.42 nm, corresponding to the transform limit of the pulse duration. This output has been improved to the point where it can be focused to small areas, leading to the generation of ultrahigh intensities. Average intensities exceeding 2×10^{19} W/cm^{-2} are reported [2.176]. Using a similar route, several other groups such as *Watanabe* et al. [2.177] and *Szatmari* et al. [2.178] report peak powers exceeding 1 TW. *Taylor* et al. [2.179] measured the energy extraction efficiencies of the XeCl discharge, and found that the saturating fluence depends upon the pulse duration, ranging from 1 mJ/cm^{-2} for 0.16 ps duration to 2.5 mJ/cm^{-2} for 600 ps duration. The increase in saturation fluence arises when the optical pulse duration exceeds the gain recovery time of 40 ps.

This work has been extended using the XeF$(C \rightarrow A)$ transition, which, due to its broader bandwidth, has greater application possibilities for short pulse amplification. *Hofmann* et al. [2.180] measured a saturation fluence of 50 mJ/cm^{-2} for 250 fs pulses in the XeF$(C \rightarrow A)$ amplifier. The higher saturating fluence allows for potentially smaller apertures, and thus easier construction, for generating high power pulses. Experimental results with the XeF$(C \rightarrow A)$ excimer system [2.180, 181] using subpicosecond pulses have demonstrated the improved energy storage capacity of the transition with respect to other excimer systems.

10.2.3 New References for Chapter 2

2.163 N. Hamada, R. Sauerbrey, W.L. Wilson, F.K. Tittel, W.L. Nighan: Proc. SPIE **894**, 50 (1988)
2.164 N. Hamada, R. Sauerbrey, W.L. Wilson, F.K. Tittel, W.L. Nighan: IEEE J. QE-**24**, 1571 (1988)
2.165 H. Voges, G. Marowsky: IEEE J. QE-**24**, 827 (1988)
2.166 C.B. Dane, Th. Hofmann, R. Sauerbrey, F.K. Tittel: IEEE J. QE-**27**, 2465 (1991)
2.167 W.L. Nighan, M.C. Fowler: IEEE J QE-**25**, 791 (1989)
2.168 G.J. Hirst, C.B. Dane, R. Sauerbrey, W.L. Wilson, K.F. Tittel, W.L. Nighan: Appl. Phys. Lett. **54**, 1851 (1989)
2.169 C.B. Dane, G.J. Hirst, S. Yamaguchi, Th. Hofmann, W.L. Wilson, R. Sauerbrey, F.K. Tittel, W.L. Nighan, M.C. Fowler: IEEE J. QE-**26**, 1559 (1990)
2.170 S. Yamaguchi, Th. Hofmann, C.B. Dane, R. Sauerbrey, W.L. Wilson, F.K. Tittel: IEEE J. QE-**27**, 259 (1991)
2.171 C.B. Dane, S. Yamaguchi, Th. Hofmann, R. Sauerbrey, W.L. Wilson, F.K. Tittel: Appl. Phys. Lett. **56**, 2604 (1990)
2.172 P. Peters, H.M.J. Bastiaens, W.J. Witteman, R. Sauerbrey, C.B. Dane, F.K. Tittel: IEEE J. QU-**26**, 1569 (1990)
2.173 A. Mandl, L.N. Litzenberger: Appl. Phys. Lett. **53**, 1690 (1988)
2.174 J.H. Glownia, G. Arjavalingam, P.P. Sorokin, J.E. Rothenberg: Opt. Lett. **11**, 79 (1986)
2.175 A.P. Schwartzenbach, T.S. Luk, I.A. McIntyre, U. Johann, A. McPherson, K. Boyer, C.K. Rhodes: Opt. Lett. **11**, 499 (1986)
2.176 T.S. Luk, A. McPherson, G. Gibson, K. Boyer, C.K. Rhodes: Opt. Lett. **14**, 1113 (1989)
2.177 S. Watanabe, A. Endoh, M. Watanabe, N. Surakura: J. Opt. Soc. Am. B**4**, 1870 (1989)
2.178 S. Szatmari, F.P. Schäfer, E. Müller-Horsche, W. Mückenheim: Opt. Commun. **63**, 305 (1987)
2.179 A.J. Taylor, T.R. Gosnell, J.P. Robens: Opt. Lett. **15**, 118 (1990)

2.180 Th. Hofmann, T.E. Sharp, C.B. Dane, P.J. Wisoff, W.L. Wilson, F.K. Tittel, G. Szabo: To appear in IEEE J. QE (1992)
2.181 T.E. Sharp, Th. Hofmann, C.B. Dane, W.L. Wilson, Jr., F.K. Tittel, P.J. Wisoff, G. Szabo: Opt. Lett. **15**, 1461 (1990)

10.3 Four-Wave Frequency Mixing in Gases

In the last five years progress has been made on several fronts in four-wave mixing, including advances in cw power and efficiency, new technology that allows improved performance, modified or new systems for generating atomic vapors, and femtosecond pulse formation. Two general review articles have appeared, one by *Gladushchack* et al. [3.265] and a second by *Hilbig* et al. [3.266]. Notable recent advances are summarized below.

Nolting et al. [3.267, 268] successfully demonstrated cw sources of VUV-radiation. The output intensities of the latter sources are still rather small, ranging from 5×10^{-13} to 6×10^{-8} W, but have definitely been improved compared to the $10^{-13} - 10^{-11}$ W of the first cw systems [3.33–35]. To obtain even higher conversion efficiencies, *Arkhipkin* et al. [3.269] have proposed using a gas-filled waveguide arrangement. The waveguide is expected to provide a conversion efficiency which is higher by about a factor of $10^2 - 10^4$ as calculated theoretically. This has not yet been demonstrated experimentally, due to the small dimensions required for the waveguide at visible wavelengths. The advantage of the waveguide would be a longer interaction pass length which is not limited by the condition for Gaussian beams that the length of the interaction region L be less than or equal to the confocal parameter b (see Sect. 3.4.3) of the beam. It is expected that the best conversion efficiencies will eventually be obtained with "quasi-cw" systems using mode-locked systems. The linewidth of the femtosecond pulses will, of course, be rather large compared to cw systems. The genuine cw systems (see e.g. [3.267]) have already come very close to the technical limits of existing cw lasers.

10.3.1 Mixing in Atomic and Molecular Vapors

Table 3.1 (Chap. 3) summarizes the nonlinear systems which had been investigated at the time of publication of the first edition of this book. Table 10.3.1 summarizes recent progress on modified or new systems for generating atomic vapors. As before, the nonlinear media are arranged according to the elements, and they are presented together with the wavelength region of the sum frequency wave and the method applied.

As well as in Ba, Ca and Zn, *second harmonic* generation has now also been seen in magnesium [3.276]. Although it has been stated again that a "symmetry breaking process" must be responsible, it has not yet been demonstrated what kind of process this might be. Hence one is still left waiting for

Table 10.3.1. Summary of recent four-wave mixing experiments

Nonlinear medium	Wavelength range [nm]	Method	References
Ar, Kr, Xe, CO, N_2	72; 90.4 – 102.5	3×1	[3.270]
Kr	72.5 – 83.5; 127 – 180	$1 + 1 \pm 2$	[3.271]
Kr	121 – 200	$1 + 1 - 2$	[3.272]
Xe, Kr	71 – 92	$1 + 1 + 2$	[3.273]
Hg	132 – 185	$1 + 1 + 1, 2$	[3.274]
Cd	138.1 – 140.3	$1 + 1 + 2$	[3.275]
Sr, Ca, Mg, Zn	133 – 277 (cw)	$1 + 2 + 3$	[3.267]
Sr	190 (cw)	$1 + 1 + 2$	[3.268]

"further detailed experiments". *Dinev* et al. have investigated the second harmonic generation (SHG) in various vapors rather extensively [3.277 – 280] and
measured the dependence on intensity, density and polarization involving s, p
and d states. They conclude that the most likely explanation for the second
harmonic generation is collisional l-mixing, and not removal of the spherical
symmetry by electric fields (see also [3.281]). Similar conclusions are drawn by
Tewari [3.282]. Unfortunately, detailed theoretical calculations are not yet
available. Also, due to low sensitivity, experiments in beams have not been able
to rule out the possibility of collisional effects. Hence, all conclusions are still
qualitative. This leaves open the door of inquiry for further research.

Similar to the experiments mentioned at the end of Sect. 3.5.3 *four-wave
mixing processes* have also been seen again in molecular gases such as CO
[3.283 – 285] and NO [3.286]. In the case of CO, the two-photon resonant intermediate electronic state was either the $C^1 \Sigma^+$ state [3.283] or the $A^1 \Pi$ state
[3.284]. The final electronic state in the first experiment was one of the
Rydberg states [3.283], whereas in the second experiment the $B^1 \Sigma^+$ state was
reached [3.284]. Using a pulsed free jet as the nonlinear medium [3.287] these
sources were applied for the first time to the spectroscopy of N_2, O_2 and CO_2
and the XUV spectral region.

Another interesting method for generating VUV in H_2 molecules has
recently been demonstrated by *Czarnetzki* and *Döbele* [3.288]. They use a
nonlinear process in which a spectrally narrow ArF excimer laser at 193 nm excites H_2 from the ground state to the $E, F^1 \Sigma_g^+$ state. This causes an amplified
stimulated emission (ASE) process to the lower $B^1 \Sigma_u^+$ state in the infrared,
which results in a population inversion of the B state with respect to higher
ro-vibronic levels of the electronic ground state. This radiation lies in the spectral range $\lambda = 130 – 160$ nm and is modified by a Stimulated Electronic Hyper-
Raman Scattering (SEHRS) process. The exact wavelength of the VUV step is
then determined by the corresponding phase-matching condition and can
therefore be tuned by about $20 \, \text{cm}^{-1}$ by the addition of other gases such as
N_2, D_2, and the noble gases. This process is similar to the tuning through
phase matching in parametric oscillators, but clearly different from the high

order anti-Stokes Raman scattering processes [3.289] which have previously been used for generating VUV radiation [3.30].

10.3.2 Application of VUV Sources

Advances have been made in the application of VUV sources to particular spectroscopic problems as covered in [3.5]. *Kung* and co-workers [3.390] obtained VUV sources of very high spectral brightness by starting with an amplified single-mode ring dye laser which can be calibrated, generating VUV, and sending the final VUV through a Czerny-Turner monochromator. By using the second mirror of the Czerny-Turner mount, the VUV signal is well focused and separated from the fundamental waves. With this technique a number of spectroscopic experiments have been performed such as the sensitive detection of H_2 molecules and the isotope shift of Kr. Other techniques for separating the visible from the VUV signal use dichroic mirrors [3.287] or chromatic aberration from a LiF lens, which is positioned off-axis with respect to the incident beam [3.291]. Another method uses two oppositely located prisms in series with an intermediate stop [3.292].

10.3.3 New Techniques for Creating Atomic and Molecular Densities

A recent development with the *heat-pipe oven* was achieved by *Milosevic* et al. [3.293]; they modified the oven described in Sect. 3.5.2 by adding an internal heater which superheats the vapor. The vapor is no longer saturated. As can be seen from [3.294] a thermal vapor of a metal contains atoms as well as molecules of various densities. Superheating the vapor allows the number density of absorbing molecules to be greatly reduced with respect to the atomic species wherever the broad molecular absorption is undesirable because of the associated optical depth (Sect. 3.4.2). On the other hand, the homogeneity is rather difficult to maintain over the length of the beam, which is very important for phase-matching as shown in Sect. 3.6.

The discussion on *inert gas nozzles* of Sect. 3.5.3 must be modified and extended. On top of page 81 one reads "A disadvantage is that because of the large density gradients in the vicinity of the orifice, pulsed jets are difficult to phase match and only small column densities can be used." Since the publication of the first edition, *Bethune* and *Rettner* [3.295] have carried out careful calculations assuming different analytical density profiles inside the jet. Since the optical path length is rather small, it turns out that the density profile is not very critical for the overall conversion efficiency. They obtained very good agreement with measured conversion efficiencies, demonstrating that, for small column densities, phase matching is not as important as originally suspected.

10.3.4 Femtosecond Pulse Generation

Work is in progress to convert suitable femtosecond pulses into the VUV range for applications in the photochemistry of molecules and certain transient phenomena. In order to efficiently transfer these pulses into the VUV, one must consider the available bandwidth. It is clear from the discussion of the review that two-photon resonant sum frequency mixing provides only a finite bandwidth, which is limited by the homogeneous linewidth of the two-photon resonance. As shown by *Scheingraber* and *Vidal* [3.26] some two-photon resonances are noticeably wider. It is expected that, due to power broadening of the high power femtosecond pulses, the homogeneous linewidth may become large enough to transfer these very short pulses without extending the duration of the pulse because of Fourier transform limitations. Work is underway in Berlin (*Vidal*), Hannover (*Wellegehausen*) and elsewhere on the amplification of femtosecond pulses. It should be noted that the work of *Rhodes* [3.29−31] described in the earlier edition uses only one pulse from a mode-locked femtosecond pulse train per second for amplification. New sources will eventually use the entire mode-locked pulse train with a repetition rate depending on the cavity round trip time, and will generate the "quasi-cw" source mentioned above.

10.3.5 New References for Chapter 3

3.265 V. I. Gladushchak, S. A. Moshkalev, G. T., Razdobarin, E. Ya. Shreider: Sov. Tech. Phys. **31**, 855 (1986)

3.266 R. Hilbig, G. Hilber, A. Lago, B. Wolff, R. Wallenstein: Comments At. Mol. Phys. **18**, 157 (1986)

3.267 J. Nolting, H. J. Kunze, I. Schütz, R. Wallenstein: Appl. Phys. B **50**, 331 (1990)

3.268 J. Nolting, R. Wallenstein: Opt. Commun. **79**, 437 (1990)

3.269 V. G. Arkhipkin, Yu. I. Heller, K. Popov, A. S. Provorov: Appl. Phys. B **37**, 93 (1985)

3.270 R. H. Page, R. J. Larkin, A. H. King, Y. R. Shen, Y. T. Lee: Rev. Sci. Instrum. **58**, 1616 (1987)

3.271 G. Hilber, A. Lago, R. Wallenstein: J. Opt. Soc. Am. B **4**, 1753 (1987)

3.272 J. P. Marangos, N. Shen, H. Ma, M. H. R. Hutchinson, J. P. Connerade: J. Opt. Soc. Am. B **7**, 1254 (1990)

3.273 K. Miyazaki, H. Sakai, T. Sato: Appl. Opt. **28**, 699 (1989)

3.274 R. Hilbig, G. Hilber, R. Wallenstein: Appl. Phys. B **41**, 225 (1986)

3.275 A. M. Schnitzer, W. Behmenburg: Z. Phys. D **8**, 141 (1988)

3.276 V. A. Kiyashko, A. K. Popov, P. Timofeev, P. Makarov, V. Sh. Epstein: Appl. Phys. B **36**, 53 (1985)

3.277 S. G. Dinev: J. Phys. B **21**, 1111 (1988)

3.278 S. G. Dinev: J. Phys. B **21**, 1681 (1988)

3.279 S. G. Dinev, G. B. Hadjichristov: J. Phys. B **24**, 307 (1991)

3.280 S. G. Dinev, G. B. Hadjichristov, I. L. Stefanov: J. Opt. Soc. Am. B **8**, 1846 (1991)

3.281 A. Guzman de Garcia: Ph.D. Thesis, Max-Planck-Institut für Quantenoptik, Garching (1984)

3.282 S. P. Tewari: A theory for second harmonic generation in dense atomic vapour, in *Coherence and Quantum Optics VI*, ed. by J. H. Eberly (Plenum, New York 1990)

3.283 F. Merkt, T. P. Softley: Chem. Phys. Lett. **165**, 477 (1990)

3.284 F. Merkt, T. P. Softley: J. Chem. Phys. **93**, 1540 (1990)

3.285 K. Tsukiyama, M. Tsukakoshi, T. Kasuya: Appl. Phys. B **50**, 23 (1990)

3.286 K. Tsukiyama, M. Tsukakoshi, T. Kasuya: J. Chem. Phys. **92**, 6426 (1990)
3.287 T. P. Softley, W. E. Ernst, L. M. Tashiro, R. N. Zare: J. Chem. Phys. **116**, 299 (1987)
3.288 U. Czarnetzki, H. F. Döbele: Phys. Rev. A **44**, 7530 (1991)
3.289 H. F. Döbele, M. Hörl, M. Röwekamp: Appl. Phys. B **42**, 67 (1987)
3.290 E. Cromwell, T. Trickl, Y. T. Lee, A. H. Kung: Rev. Sci. Instrum. **60**, 2888 (1989)
3.291 W. A. VonDrasek, S. Okajima, J. P. Hessler: Appl. Opt. **27**, 4057 (1988)
3.292 S. S. Dimov, C. R. Vidal: Chem. Phys. to be published
3.293 S. Milosevic, R. Beuc, B. Pichler: Apply. Phys. B **41**, 135 (1986)
3.294 A. N. Nesmeyanov: *Vapor pressure of the Chemical Elements* (Elsevier, New York 1963)
3.295 D. S. Bethune, C. T. Rettner: IEEE J. QE-**23**, 1348 (1987)

10.4 Stimulated Raman Scattering

Progress in Stimulated Raman Scattering (SRS) for the generation of tunable laser light has concentrated on new media, including optical fibers, and on the use of waveguides to improve the conversion efficiency.

10.4.1 SRS Using Lead

Several groups report the generation of SRS using lead vapor. *Zhang* et al. [4.301] used the third harmonic of a Nd:YAG laser to photodissociate PbI_2. The third harmonic then pumped the atomic lead vapor, generating Stokes lines at 394.3 nm and 443.9 nm, and an anti-Stokes line at 322.3 nm. The PbI_2 was contained in a heat pipe operated between 450 and 650 °C. Peak conversion efficiency was 0.4%. *Rieger* [4.302] generated SRS in a lead vapor heat pipe using a XeCl excimer laser. The XeCl laser was tunable over a 0.8 nm range near 308 nm with 310 mJ/pulse output energy. In single pass conversion, the excimer laser radiation was shifted to 459 nm with 80% conversion efficiency. Efficiency was found to increase as the linewidth of the source was reduced.

10.4.2 SRS Using Silica Fibers

Several groups have reported on SRS in silica fiber. *Mizunami* and *Takagi* [4.303] generated short wavelength (285 nm) radiation using a XeBr excimer laser operating at 281.8 nm. The output beam of the XeBr laser was focused onto the input face of an 80 µm core, fused silica fiber. The input intensity was limited by the damage threshold of the glass, which was found to be 350 MW/cm^2. The conversion efficiency was low (approximately 0.4%) due to strong two-photon absorption within the glass. The two-photon absorption cross section increases with decreasing wavelength. The authors point out that for wavelengths below 270 nm it will be impossible to generate SRS in a fused silica fiber due to this absorption.

To avoid relying on a spontaneous noise photon to extract the Raman gain established in pumped systems, *Selker* and *Lawandy* [4.304] demonstrated the use of a nonlinear self phase modulator continuum generator for seeding a SRS amplifier. They used a mode-locked, frequency-doubled Nd: YAG laser to pump a 15 m single-mode optical fiber. The output of the continuum generator was used to seed a CH_4 cell pumped by the same source. Threshold power for SRS in the amplifier was reduced by a factor of four using the seeding.

Foley et al. [4.305] investigated the role of Stimulated Brillouin Scattering (SBS) in low power fiber Raman amplifiers that are intended for optical communication systems. The Brillouin gain coefficient is three orders of magnitude larger than the Raman gain coefficient in silica fibers, so significant SBS can occur at pump powers useful for SRS. This problem can be reduced by using a larger bandwidth pump source, such as a multimode laser diode, however, this tends to limit the Raman gain as well. A second suggestion is to implement backward traveling pump pulses with temporal duration less than the inverse Brillouin linewidth. This is, in effect, the same as increasing the linewidth of the pump source.

Several new sources have been developed based on Raman scattering in silica fiber. Pask and Piper [4.306] observed strong SRS on three Stokes lines at 1.608, 1.744, and 1.896 μm when pumping a 12 km single mode fused silica fiber at 1.5 μm with a pulsed barium vapor laser. Peak input powers up to 25 W were used, however, conversion saturated for peak powers much above 10 W. Incomplete conversion was attributed to the formation of higher Stokes lines before the primary wave was depleted. In a similar work, Guasti et al. [4.307] reported on SRS in fused silica fiber using a copper vapor laser as the pump source. They observed up to nine Stokes lines in the 511 − 649 nm region, with a total conversion efficiency approaching 90%. To accommodate the nominally multimode spatial output beam of the Cu vapor laser, multimode fibers were used (100 and 200 μm core diameters).

10.4.3 SRS Using Molecular Gases

Broadly tunable SRS has been generated in a high pressure multipass cell, and in a high pressure hollow waveguide. MacPherson et al. [4.308] investigated the use of a multipass cell filled with hydrogen for SRS with a 532 nm single-mode frequency-doubled Nd: YAG laser. They found that second Stokes, backward Stokes, and anti-Stokes emission were negligible so long as the overall gain was kept low. By distributing the SRS gain over many passes, this low gain condition can be satisfied. Widely tunable infrared light has been generated by Brink et al. [4.309] using a hollow waveguide Raman hydrogen cell. The hollow waveguide has dimensions on the order of 1 mm, and was made by stacking specially coated plates of glass. The glass was coated with a semiconductor dielectric coating to reduce absorption loss for the dye laser pump source. The higher-order Stokes output from the cell was tunable from 5.5 μm to over 17 μm, and was only limited by the cut-off point of the ZnSe windows used to enclosed the pressure vessel.

Short pulse amplification and formation were studied by *Hooker* et al. [4.310]. Using a KrF laser at 249 nm, they generated gain in a methane cell at 268 nm. Input pulses at 268 nm of duration 4, 8, and 40 ps were used to measure the transient gain in the Raman cell. The T_2 time of the methane was estimated to be 30 ps. The measured transient gain agreed well with theory.

At the other extreme for temporal operation, cw SRS was reported by *Irrera* et al. [4.311] using a cw Nd:YAG laser coupled to a single-mode fiber. The laser had a linewidth of 40 GHz, and a coupled power of 5.7 W. Continuous wave output powers of 5 W and 0.2 W are reported at 1.112 μm and 1.18 μm. *Bryant* and *Golombok* [4.312] report cw SRS in a bulk liquid. A cw argon ion laser operating at 488 nm was focused to an intensity of approximately 100 MW cm^{-2} in a bulk organic liquid. They observed enhanced backscatter, and in addition, the backscattered light was a phase conjugate of the input wave.

10.4.4 New References for Chapter 4

4.301 J. Zhang, L. Zhao, B. Cheng, D. Zhang, Y. Zhao, T. Wang: Opt. Commun. **68**, 442 (1988)
4.302 H. Rieger: IEEE J. QE-**25**, 913 (1989)
4.303 T. Mizunami, K. Takagi: IEEE J. QE-**25**, 1917 (1989)
4.304 M.D. Selker, N.M. Lawandy: Electron. Lett. **26**, 409 (1990)
4.305 B. Foley, M. Dakss, R. Davies, P. Melman: J. Lightwave Tech. **7**, 2024 (1989)
4.306 H.M. Pask, J.A. Piper: Opt. Quantum Electron. **23**, S563 (1991)
4.307 A. Guasti, R. Pini, R. Salimbeni: Opt. Quantum Electron. **23**, S555 (1991)
4.308 D.C. MacPherson, R.C. Swanson, J.L. Carlsten: IEEE J. QE-**25**, 1741 (1989)
4.309 D.J. Brink, M. Budzinski: Meas. Sci. Technol. **1**, 754 (1990)
4.310 C.J. Hooker, J.M.D. Lister, P.A. Rodgers: Opt. Commun. **82**, 497 (1991)
4.311 F. Irerra, L. Mattiuzzo, D. Pozza: J. Appl. Phys. **63**, 2882 (1988)
4.312 C.H. Bryant, M. Golombok: Opt. Lett. **16**, 602 (1991)

10.5 Recent Advances in Optical Parametric Oscillators

The Optical Parametric Oscillator (OPO) continues to be a powerful solid-state source for broadly tunable emission in the visible and near infrared region. In the past five years, major advances have been made on several fronts. First, new and better nonlinear crystals have been developed which allow greater tuning range, resistance to optical damage, and higher efficiency. Second, the temporal operation of OPOs has been extended into the femtosecond domain. Finally, the use of optical waveguides to confine the optical intensity over substantial lengths has lowered the threshold power for OPOs, leading to the possibility of cw OPOs.

10.5.1 New Nonlinear Crystals

With a suitable nonlinear crystal, virtually any wavelength ranging from the UV to the IR can be reached with OPOs. Urea, which was highlighted in

Chap. 5, has been surpassed in performance by several new crystals. The recent availability of high quality BBO, LBO, MgO: LiNbO$_3$, and KTP crystals has opened the door to many new applications and abilities. For the $3-10\,\mu$m range, AgGaS$_2$ and AgGaSe$_2$ show great promise.

The first BBO (β-BaB$_2$O$_4$) OPO was reported by *Fan* et al. in 1986 [5.11] using a 9 mm crystal grown at the Fujian Institute in the People's Republic of China. It was a doubly resonant, plano-plano cavity pumped by the second harmonic of a Nd: YAG laser at 532 nm, and had a limited tuning range from 0.94 to 1.22 μm. The conversion efficiency was a modest 10%, however, this work served as a precursor to an abundance of subsequent work.

BBO has the advantage of being only slightly hygroscopic, mechanically hard, and chemically stable. It has a high damage threshold, and can be grown in usefully large sizes. It is transparent down to 191 nm, and has a relatively large birefringence. It can therefore be tuned, or phase-matched over a large spectral range. The relative d coefficient is not large compared to other crystals, but is adequate for reasonably long crystal lengths and pump powers. The properties of BBO are summarized in [5.12].

The BBO OPOs are pumped either by the harmonics of a Nd: YAG laser [5.13], or by a XeCl excimer laser [5.14]. The performance from either pump source is comparable for similar power and diffraction conditions. In the Nd: YAG case, the most practical source is the third harmonic at 355 nm. With such pumping, the entire spectral range from 415 nm to 2.5 μm can be covered with a single set of mirrors [5.15]. This is truly a versatile source. Using two crystals to compensate for walk-off loss, conversion efficiencies well over 30% have been achieved. The efficiencies are presently pump-length limited. The finite time the pump exists (typically $7-10$ ns) is much shorter than the time required for the parametric fluorescence to build up and saturate the parametric gain. With longer pulses, conversion efficiencies approaching 60% are possible.

Using excimer pumping, remarkable efficiency has been reported with a urea OPO. Using a 308 nm XeCl laser, conversion efficiencies approaching 66% were achieved at the noncritical phase matching condition [5.14]. Using BBO, tuning from 354 nm to 2.37 μm with approximately 10% conversion was achieved. A major problem with excimer pumping is that the output linewidth is rather large. With no line narrowing schemes in the OPO, the reported linewidth in a Type I OPO can vary from approximately 1 nm at 480 nm to 11 nm at 600 nm. By comparison, the corresponding linewidths with Nd: YAG harmonic pumping are nearly an order of magnitude narrower [5.16]. This could be an issue in spectroscopic applications of OPOs.

To narrow the linewidth, two methods have been demonstrated. The first is to use intracavity dispersing elements such as gratings [5.16]. Without beam expansion optics, such schemes have produced output with 0.1 nm linewidth. An alternative approach is to use Type II phase matching, where the signal is also an extraordinary wave, $k_p^{(e)} = k_1^{(e)} + k_2^{(o)}$ or $e \rightarrow e+o$ [5.17]. It is known that this scheme naturally leads to narrower linewidths. The disadvantage of

Type II phase matching is that the effective *d* coefficient is smaller, so longer crystals are needed. Using such schemes, *Bosenberg* and *Tang* [5.17] have demonstrated tuning over the 480 – 630 nm range with linewidths on the order of 0.05 – 0.3 nm.

10.5.2 Lithium Triborate

Lithium triborate (LBO) is a newly developed crystal which has good UV transparency, moderate birefringence and nonlinear coefficients, and a relatively high optical damage threshold [5.18]. Although the smaller birefringence in LBO compared to BBO tends to limit the tuning range, it also leads to the possibility of noncritical phase matching and larger acceptance angles for frequency conversion applications.

Using third harmonic Nd:YAG pumping, LBO OPO operation has been reported over the 540 nm to 1.03 µm range, with linewidths on the order of 6 nm [5.19]. Using XeCl excimer pumping at 308 nm, OPO output over the 372 nm to 1.8 µm range has been obtained using Type II phase matching with a 16 mm long crystal [5.20]. The linewidth was reduced compared to the Type I phase matching down to 0.15 nm in the 375 nm region.

Determining which crystal to use, LBO or BBO, is not a well-defined process at this time. There are many possibilities yet to be explored, but it is likely that each crystal will find complementary applications with particular design considerations.

10.5.3 Potassium Titanyl Phosphate

Recent advances in the growth of potassium titanyl phosphate (KTP) have made it suitable for development in OPO devices. KTP is chemically stable and mechanically robust. It can be well polished and coated if desired.

Using a diode-pumped pulsed Nd:YAG laser to pump a KTP OPO in a confocal cavity, *Marshall* et al. [5.21] reported conversion from 1.06 to 1.61 µm with 35% efficiency. *Vanherzeele* [5.22] reported generation of broadly tunable (600 nm to 4.5 µm) light using millijoule level picosecond pulses from a Nd:YLF laser (1.053 µm).

KTP has been actively used for synchronously pumped OPOs in order to generate picosecond [5.23] and femtosecond pulses [5.24]. In a synchronously pumped OPO, the pump beam consists of a train of short pulses, each with a spatial extent much shorter than the typical OPO cavity length. If the OPO cavity length is adjusted so that the cavity round trip frequency is identical to the pulse train frequency, the signal and pump pulse will meet and travel through the nonlinear crystal at the same time. Because the pulse has a much higher peak intensity than the average power of the pump, more efficient pumping can be achieved.

Bromley et al. [5.25] produced 70 ps pulses between 1.04 and 1.09 µm using a single resonant KTP OPO pumped with a mode-locked Nd:YAG laser.

10.5.4 Tunable Femtosecond OPOs

One of the more exciting results in the past five years has been the demonstration of tunable femtosecond pulse generation using a synchronously pumped OPO. While new femtosecond pulse sources are appearing every day, at a variety of wavelengths, such sources tend to operate over a limited tuning range, e.g., $1.45 - 1.75$ μm for a NaCl laser (Sect. 10.1), or $700 - 1000$ nm for Ti:sapphire (Sect. 10.9). The femtosecond OPO allows broad tuning (over octaves) using a single source. Due to its relatively large nonlinear coefficient, KTP is the preferred crystal for femtosecond OPOs. A serious constraint when working with femtosecond pulses is material dispersion. The large bandwidth of the femtosecond pulse is rapidly dispersed in most materials, and the femtosecond pulse is rapidly spread in time to picosecond or longer duration. It is essential to minimize the optical path length through any material. This raises a dilemma in the design of a femtosecond OPO: a long crystal provides more gain, but it also introduces deleterious dispersion. To overcome this, *Edelstein* et al. [5.24] incorporated a small crystal of KTP (50 μm thick) into a CPM dye laser cavity, and constructed a resonant OPO cavity about the synchronously pumped crystal. The OPO produced 100 fs pulses over the $820 - 920$ nm region, and over the $1.92 - 2.54$ μm region. Average output powers were on the milliwatt level. The limitations on the tuning range are simply due to mirrors: with three mirror sets, it should be possible to generate femtosecond pulses over the 700 nm to 4.5 μm range in this device.

10.5.5 New References for Chapter 5

5.11 Y. X. Fan, R. C. Echkardt, R. L. Byer, C. Chen, A. Jiang, CLEO '86, Postdeadline paper ThT4;
Y. X. Fan, R. C. Echkardt, R. L. Byer, J. Nolting, R. Wallenstein: Appl. Phys. Lett. **53**, 2014 (1988)
5.12 C. L. Tang, W. R. Bosenberg, T. Ukachi, R. J. Lane: IEEE Proc. Special Issue on Quantum Electronics, to appear 1992
5.13 H. Komine: Opt. Lett. **13**, 643 (1988)
5.14 M. Ebrahimzadeh, M. H. Dun, F. Akerboom: Opt. Lett. **14**, 560 (1989);
M. Ebrahimzadeh, A. J. Henderson, M. H. Dunn: IEEE J. QE-**26**, 1241 (1990)
5.15 L. K. Cheng, W. R. Bosenberg, C. L. Tang: Appl. Phys. Lett. **53**, 175 (1988);
W. R. Rosenberg, L. K. Cheng, C. L. Tang: Appl. Phys. Lett. **54**, 13 (1989)
5.16 W. R. Rosenberg, W. S. Pelouch, C. L. Tang: Appl. Phys. Lett. **55**, 1952 (1989)
5.17 W. R. Rosenberg, C. L. Tang: Appl. Phys. Lett. **56**, 1819 (1990)
5.18 C. Chen, Y. Wu, G. You, R. Li, S. Lin: J. Opt. Soc. Am. B **6**, 616 (1989);
S. Lin, Z. Sun, B. Wu, C. Chen: J. Appl. Phys. **67**, 634 (1990);
S. Zhao, C. Huang, H. Zhang: J. Crystal Growth **99**, 805 (1990)
5.19 Z. Xu, D. Deng, Y. Wang, B. Wu, C. Chen: CLEO '90, paper CWE6, Anaheim, CA (1990)
5.20 M. Ebrahimzadeh, G. Robertson, M. Dunn, A. Henderson: CLEO '90, paper CPDP26, Anaheim, CA (1990)
5.21 L. R. Marshall, A. D. Hay, R. Burnham: CLEO '90, Postdeadline talk CPDP-35, Anaheim, CA (1990)
5.22 H. Vanherzeele: Appl. Opt. **29**, 2246 (1990)
5.23 L. J. Bromley, A. Guy, D. C. Hanna: Opt. Commun. **67**, 316 (1988);

S. Burdulis, R. Grigonis, A. Piskarsas, G. Sinkevicus, V. Sirutkaistis, A. Fix, J. Nolting, R. Wallenstein: Opt. Commun. **74**, 398 (1990)

5.24 D. C. Edelstein, E. S. Wachman, C. L. Tang: Appl. Phys. Lett. **54**, 1728 (1989);
E. S. Wachman, D. C. Edelstein, C. L. Tang: Opt. Lett. **15**, 136 (1990)

5.25 L. J. Bromley, A. Guy, D. C. Hanna: Opt. Commun. **67**, 316 (1988)

10.6 Color Center Lasers

Over the past five years, there have been three significant developments in color center lasers. First, the structure of the powerful center responsible for the NaCl:OH$^-$ laser was described, and this led to the discovery of a new class of lasers in the 1.5 − 2 µm range. Second, a new class of color center laser based on the N center was demonstrated. Finally, several new schemes for ultrashort pulse generation were developed, using color center lasers. As was discussed in Sect. 10.1.2, these ultrafast techniques have been extended to other laser systems.

10.6.1 Oxygen-Stabilized Color Center Lasers

The structure of the color center responsible for the NaCl:OH$^-$ laser described in Sect. 6.2.10a has been identified by several groups. All groups agree that the laser-active center consists of an F_2^+ center located beside a substitutional O^{2-} ion [6.95, 96]. There is still some controversy concerning the kinetics of the formation of the center [6.97, 98], however, the spectroscopy of the center and its excited states are well established. Further support for the model has been derived from a perturbation analysis of the energy levels of an F_2^+ center in the proximity of a negative charge [6.99]. The excited state spectroscopy helped to explain why an auxiliary light is required for this system. In the presence of strong pumping, the laser-active F_2^+ center can absorb more than one photon and be placed in an excited state. Upon relaxation, the center reorients, effectively removing its dipole from the interaction with the optical fields. Auxiliary energy of the proper wavelength can directly excite these states, causing an immediate reorientation of the center. When applied by a mercury lamp transverse to the crystal [6.95], the light interacts with all orientations, and effectively keeps the entire population randomly distributed over the six possible orientations. *German* and *Pollock* [6.100] noted that the second harmonic of the Nd:YAG pump laser could be used as an auxiliary light, which would be collinear with the pump beam, and would require only milliwatts of power to be effective. A recent study by *Carrig* and *Pollock* [6.101] shows that the performance of the NaCl:OH$^-$ laser is basically independent of the choice of auxiliary light. The optimal choice is simply a matter of convenience.

Pulsed lasing from the NaCl:OH$^-$ laser was observed at cryogenic and room temperatures [6.102]. The pulsed laser was pumped by a Q-switched

Nd:YAG laser at 1.06 µm, with output energies up to 8 mJ for 100 mJ pumping. The laser used an unstable cavity for maximum gain volume, and displayed a slow fading of output power, decreasing by 3 dB after 10^6 shots. No fading was observed at cryogenic operation, where powers up to 25 mJ were obtained [6.103].

The oxygen-stabilized center has been demonstrated in several other lattices, and this result has extended the available tuning range of color center lasers. The center has been made to lase in electron-beam colored $KCl:O^{2-}$ over the 1.65 – 1.98 µm range with cw output powers up to 100 mW [6.96] and in $KBr:O^{2-}$ with power up to 30 mW. Since KBr requires pumping at 1.6 µm, the low power is due to the lack of a powerful pump laser. For reasons not well understood, the KCl system is not stable under long-term pumping, and slowly fades.

A defect similar to the $F_2^+:O^{2-}$ center has been recently developed by associating the F_2^+ center beside an O^{2-} anion and a foreign alkali metal (e.g. Na) cation. These so-called $(F_2^+)_{AH}$ centers have been produced in $KCl:Na:O^{2-}$ and $KBr:Na:O^{2-}$ [6.104]. Both centers can be created using additive coloration, so the centers are stable in storage. Using an argon laser for auxiliary light and a 1.3 µm Nd:YAG pump laser, the KCl system has produced over 600 mW cw from 1.7 to 2.0 µm. The KBr system requires pumping at 1.5 µm, and has delivered over 5 mW in the 1.98 – 2.17 µm region [6.105]. Both of these crystals open up new tuning ranges to the cw color center laser.

10.6.2 The N-Center Laser

A new class of laser center was demonstrated with the N center laser [6.106]. The N center has been known through its characteristic absorption band for over 40 years, however, the exact structure of the N center has been an unsettled controversy. Using a pulsed Nd:YAG laser, laser operation tunable over the 1.3 µm region was obtained. Figure 10.2 shows the average power as a function of wavelength for the N center laser.

The N center is the first stable neutral center laser. It operates in a pulsed mode when pumped by a Q-switched Nd:YAG laser at 1.06 µm, with average powers up to 40 mW when operated at 10 kHz. Like other color center lasers, the N center laser is temperature sensitive, and hence excess pumping can actually lead to lower output power.

Recent work shows evidence that the center is in fact a trigonal F center [6.107]. It is not yet clear why the laser operates only in a pulsed mode. Possibilities include excited state absorption, but the most likely reason is the formation of multiplets which both reduce the active population and potentially lead to absorption losses in the gain region.

10.6.3 Ultrashort Pulse Generation

Extremely short pulse generation using color center lasers was demonstrated using two techniques. The first, additive pulse mode locking, was demonstrat-

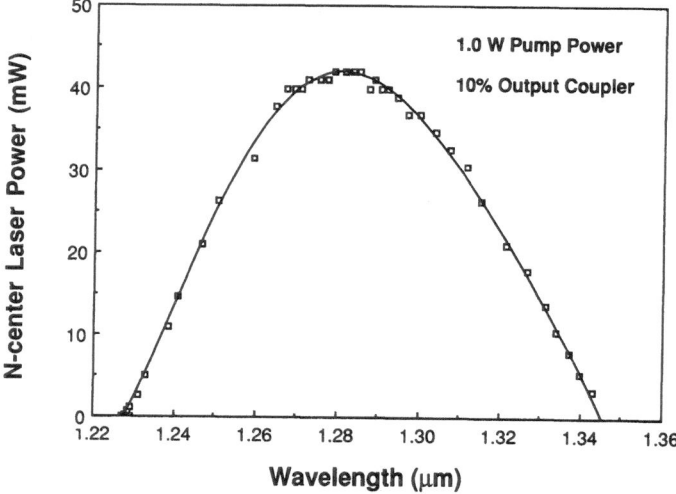

Fig. 2. The average power tuning curve for the N-center laser. The laser was pumped with a Q-switched Nd:YAG laser operating at a 10 kHz repetition rate

ed with KCl:Tl, and NaCl:OH, as described in Sect. 10.1.2. *Islam* and co-workers [6.108–111] have demonstrated passive mode locking with $F_A(II)$ center lasers in KCl:Li and RbCl:Li, and with the $F_2^+:O^{2-}$ center in NaCl using saturable absorbers based on semiconductors. A Multiple Quantum Well (MQW) saturable absorber based on InGaAs/InAlAs was placed in a tight focus within the color center laser cavity. The well dimensions of the MQW were adjusted to create an exciton absorption in the 1.6 µm region. Pulses of 275 fs duration were produced in the NaCl system, with peak powers of 3.7 kW, tunable over the 1.59–1.7 µm range by changing the quantum well sample. Extension of this technique to the 2.8 µm region was accomplished using HgCdTe multiple quantum wells. The $F_A(II)$ laser based on KCl:Li produced pulses as short as 120 fs, while the same type of center in RbCl:Li produced 190 fs pulses. These pulses are approximately 50 times shorter than can be achieved through synchronous pumping methods of mode locking. The pulse durations in all the reported systems correspond to the thermalization time of the excited carriers in the MQWs.

10.6.4 New References for Chapter 6

6.95 J.F. Pinto, E. Georgiou, C.R. Pollock: Opt. Lett. **11**, 519 (1986)
6.96 D. Wandt, W. Gellermann, F. Luty, H. Welling: J. Appl. Phys. **61**, 864 (1987)
6.97 E. Georgiou, J.F. Pinto, C.R. Pollock: Phys. Rev. B **35**, 7636 (1987)
6.98 W. Gellermann: J. Phys. Chem. Solids **52**, 249 (1991)
6.99 A. Sennaroglu, C.R. Pollock: J. Lumin. **47**, 217 (1991)
6.100 K.R. German, C.R. Pollock: Opt. Lett. **12**, 474 (1987)

6.101 T. J. Carrig, C. R. Pollock: J. Appl. Phys. **69**, 3796 (1991)

6.102 C. Culpepper, T. J. Carrig, J. F. Pinto, E. Georgiou, C. R. Pollock: Opt. Lett. **12**, 882 (1987)

6.103 C. R. Pollock: Unpublished result

6.104 D. Wandt, W. Gellermann: Opt. Commun. **61**, 405 (1987)

6.105 W. Gellermann: J. Phys. Chem. Solids **52**, 249 (1991)

6.106 E. Georgiou, T. J. Carrig, C. R. Pollock: Opt. Lett. **13**, 978 (1988)

6.107 E. Georgiou, C. R. Pollock: Phys. Rev. B **44**, 6608 (1991)

6.108 M. N. Islam, E. R. Sunderman, C. E. Soccolich, I. Bar-Joseph, N. Sauer, T. Y. Chang, B. I. Miller: IEEE J. **QE-25**, 2454 (1989)

6.109 M. N. Islam, E. R. Sunderman, I. Bar-Joseph, N. Sauer, T. Y. Chang: Appl. Phys. Lett. **54**, 1203 (1989)

6.110 C. E. Soccolith, M. N. Islam, M. G. Yopung, B. I. Miller: Appl. Phys. Lett. **56**, 2177 (1990)

6.111 C. L. Cesar, M. N. Islam, C. E. Soccolith, R. D. Feldman, R. F. Austin, K. R. German: Opt. Lett. **15**, 1147 (1990)

10.7 Fiber Raman Lasers

There have been several significant advances in fiber Raman lasers, mostly in the application of Raman gain to generate or maintain optical solitons. However, overall, it appears that research in this area has fallen off, possibly due to the increased interest in competing systems such as rare-earth doped fiber lasers. Here we summarize the results of several significant results since 1985.

Nakazawa et al. [7.61] have developed a complete theory of synchronously pumped fiber Raman lasers assuming parabolic gain shaping in both the frequency and time domain. A master equation for the Stokes pulse consisting of the group velocity dispersion, gain dispersion of the SRS, gain shaping due to pump curvature, and walk-off effect was derived. It is shown that a Stokes pulse shortening ratio of 3 – 5 should be attained for a pump pulse width of 200 ps. Saturation acts as a broadening effect that counteracts the pulse shortening mechanism. This paper has a fairly complete list of references up to 1986.

Several groups have developed femtosecond pulse sources based on fiber Raman lasers. *Kafka* et al. [7.62] reported a subpicosecond fiber Raman laser that operated at 1100 nm and was pumped by a 1.06 μm Nd: YAG laser. To compensate for the positive GVD in the fiber at that wavelength, a grating pair dispersive delay line was included in the fiber loop. Pulses as short as 800 fs were generated. *Dianov* et al. [7.63] developed a similar source that produced 400 fs pulses and was tunable over the 1.076 – 1.12 μm range. They used a single grating combined with a retroreflecting corner-cube prism to compensate for dispersion. The output wavelength could be tuned by translating a slit in front of the prism.

By moving to longer wavelengths, it has become possible to generate optical solitons through SRS. *Zysset* et al. [7.64] used a compressed cw mode-locked Nd: YAG laser to pump a dye laser that produced 1 ps pulses at 1325 nm. These pulses pumped a dispersion shifted fiber oscillator. They observed both Raman and four-wave mixing gain, and generated soliton-like pulses as short as 80 fs. *Kafka* and *Baer* [7.65] extended their previous results by pumping in single mode fiber with a mode-locked Nd : YAG laser operating at 1.319 μm. Their oscillator consisted of a totally integrated fiber system constructed with fusion splices and wavelength dependent couplers. Working in the negative GVD region eliminated the need for discrete dispersion compensation optics. They generated pulses as short as 160 fs, using nothing more than a cw mode-locked Nd : YAG laser as a pump. *Gouveia-Neto* et al. [7.66] demonstrated cascade Raman soliton generation at 1.6 μm by pumping a single mode fiber laser very hard, generating intense Stokes shifted light at 1.41 μm, which in turn served as a pump source for the second Stokes light at 1.5 μm. This process was extended to the third Stokes line at 1.6 μm, producing soliton-like pulses of 230 fs duration. *Islam* et al. [7.67] reported a Fiber Raman Amplification Soliton Laser (FRASL) using a mode-locked color center laser to pump the fiber loop. Using 10 ps duration pulses, output pulses as short as 240 fs were produced near 1.55 μm. A synchronously pumped dispersion compensated fiber Raman laser was reported by *Gouveia-Neto* et al. [7.68] that operated near 1.4 μm with nearly transform limited pulses. The pulse duration was 2 ps, with a 100 W peak power.

The theory of synchronously pumped fiber Raman lasers has been neatly summarized in a recent paper by *Golovchenko* et al. [7.69]. They investigated the dynamics of laser generation from spontaneous noise, and found numerical solutions describing stability regimes. They demonstrated smoothly tunable femtosecond pulses in the 1.07 μm region, and predicted that output powers as high as 1 W could be generated with sufficiently large core fiber. The dynamics of pulse formation were also modelled by *Band* et al. [7.70] with comparable results. The noise characteristics of femtosecond fiber Raman soliton lasers were measured by *Keller* et al. [7.71]. They measured a white noise floor 50 dB above the shot noise limit of their experiment. They found that timing jitter increases with fiber length, and is on the order of 5 ps for a 300 m fiber length. They concluded that fiber Raman soliton lasers are unsuited for applications where signal-to-noise issues are critical.

10.7.1 New References for Chapter 7

7.61 M. Nakazawa, M. Kuznetsov, E. Ippen: IEEE J. QE-**22**, 1953 (1986)
7.62 J. D. Kafka, D. F. Head, T. Baer: In *Ultrafast Phenomena V*, ed. by G. R. Fleming, A. E. Siegman, Springer Ser. Chem. Phys., Vol. 46 (Springer, Berlin, Heidelberg 1986) p. 51
7.63 E. M. Dianov, P. V. Mamyshev, A. M. Prokhorov, D. G. Fursa: Pis'ma Zh. Eksp. Teor. Fiz. **45**, 469–471 (1987)
7.64 B. Zysset, P. Beaud, W. Hodel, H. P. Weber: In *Ultrafast Phenomena V*, ed. by G. R. Fleming, A. E. Siegman, Springer Ser. Chem. Phys., Vol. 46 (Springer, Berlin, Heidelberg 1986) p. 54
7.65 J. D. Kafka, T. Baer: Opt. Lett. **12**, 181 (1987)
7.66 A. S. Gouveia-Neto, A. S. L. Gomes, J. R. Taylor, B. J. Ainslie, S. P. Craig: Opt. Lett. **12**, 927 (1987)
7.67 M. N. Islam, L. F. Mollenauer, R. H. Stolen: In *Ultrafast Phenomena V*, ed. by G. R. Fleming, A. E. Siegman, Springer Ser. Chem. Phys., Vol. 46 (Springer, Berlin, Heidelberg 1986) p. 46

7.68 A. S. Gouveia-Neto, P. G. J. Wigley, J. R. Taylor: Opt. Commun. **70**, 128 (1989)
7.69 E. A. Golovchenko, E. M. Dianov, P. V. Mamyshev, A. M. Prokorov, D. G. Fursa: J. Opt. Soc. Am. B **7**, 172 (1990)
7.70 Y. B. Band, J. R. Ackerhalt, D. F. Heller: IEEE J. QE-**26**, 1259 (1990)
7.71 U. Keller, K. D. Li, M. Rodwell, D. M. Bloom: IEEE J. QE-**25**, 280 (1989)

10.8 Tunable High Pressure Infrared Lasers

There appears to have been very limited work in this area over the last five years. The papers described below represent work in CO_2 lasers, and far-infrared lasers.

Marchetti and *Simili* [8.87] report on a high pressure, rf pumped CO_2 laser that operates at pressures up to 1.5 bar, with peak powers on the order of 600 W. They discuss the problems associated with coupling the rf power into the discharge, and the effects of the transient gas temperature on the laser frequency and linewidth. Pulse repetition rates are limited by thermal effects in their design, leading to optimum pulse repetition rates of approximately 100 Hz. He-free mixtures showed less output power, but could be operated much less expensively. They demonstrated continuous tuning over 506 MHz, with pulses of 15 µs.

The last two references refer to far-infrared laser research. *Lang* et al. [8.88] report on using a high pressure CO_2 laser to optically pump a D_2O laser. For pressure below 2 mbar, laser action on 58 lines was observed in the $26 \, \text{cm}^{-1}$ to $236 \, \text{cm}^{-1}$ region. For D_2O pressures above 2 mbar, far-infrared generation was found to occur by stimulated Raman scattering. The peak gain due to SRS had a bandwidth on the order of 10 GHz. Taking advantage of these broad gain profiles, the authors generated subnanosecond far-infrared pulses.

Finally, a thorough review article on optically pumped far-infrared lasers has been written by *Jacobsson* [8.89]. His paper summarizes the history of FIR sources and reviews the principles of operation, design and characteristics from an engineer's point of view.

10.8.1 New References for Chapter 8

8.87 S. Marchetti, R. Simili: Il Nuovo Cimento **13D**, 959 (1991)
8.88 P. T. Lang, W. Schatz, K. F. Renk: Opt. Commun. **84**, 29 (1991)
8.89 S. Jacobsson: Infrared Phys. **29**, 853 (1989)

10.9 Tunable Solid-State Lasers

The past five years have produced three dramatic advances in tunable solid-state lasers. First, Ti:sapphire lasers have been produced which provide watts of cw power under room temperature operation. The cw lasers have been tuned over the 670−1150 nm range with powers exceeding 1 W. The Ti : sapphire laser has essentially replaced all near infrared dyes as the choice for cw and mode-locked radiation in this region. Performance enhancement has come from advances in the control of crystal impurities during growth [9.108]. The second major development is the development of Cr^{4+} ions in new lattices such as YAG or forsterite. These are in contrast to the many Cr^{3+} that are described in Chap. 9. The Cr^{4+} lasers promise to extend the tuning range of optically pumped room temperature lasers into the 1−1.6 μm region. The last advance concerns the development of efficient new rare earth lasers that have tunability in the mid-infrared region. A good review of solid-state laser technology is provided in *Koechner*'s classic book, *Solid-State Laser Engineering* [9.12], now available in an updated third edition. The interested reader is directed to this new book for a complete description of the field. A general review of solid-state laser techniques as of 1988 can be found in the special issue on solid-state lasers of the *IEEE Journal of Quantum Electronics* [9.109].

10.9.1 Progress in Ti : Sapphire Lasers

Titanium-doped sapphire has emerged as an attractive solid-state laser material for laser and laser amplification applications. The crystal features high thermal conductivity, which allows for intense cw pumping. The gain cross section is relatively small ($\approx 3 \times 10^{-19}$ cm^2), which allows for high energy storage, although the radiative lifetime is relatively short (3.3 μs). This requires fast pumping to achieve large stored excited state populations [9.110].

The cw Ti : sapphire laser has undergone extensive commercial development, and is now available from many vendors. Using an Ar^+ ion pump laser, the tuning range of a cw laser has been extended from 670 nm to over 1100 nm. Figure 10.3 shows the tuning curves available from one such commercial system. Due to the broad tuning range of the crystal, several sets of mirrors are required to achieve the optimum power over the entire tuning range [9.111]. The optimal pump source for this system is a 15−20 W cw argon ion laser. As mentioned above, the breakthrough in achieving these high powers is a result of improved crystal quality. The theoretical modeling and design of longitudinally pumped lasers has been developed by *Alfrey* [9.112].

Significant progress has been made in the frequency control of Ti : sapphire lasers. *Schultz* [9.113] demonstrated a single-mode ring laser that produced over 500 mW when pumped with 7 W of power from an argon ion laser. The residual linewidth of the laser was 2 MHz averaged over a 10 s period. *Harrison* et al. [9.114] demonstrated an all-solid-state Ti-sapphire laser which was pumped by a frequency-doubled Nd : YAG laser, which itself was pumped by a semiconductor diode laser. This "diode-pumped" laser operated with a low threshold ring cavity, and produced an output power of 4.2 mW for a pump power of 173 mW. Operating in a traveling wave mode, the laser produced single-frequency output. Such a source should be useful for seeding, and for spectroscopy.

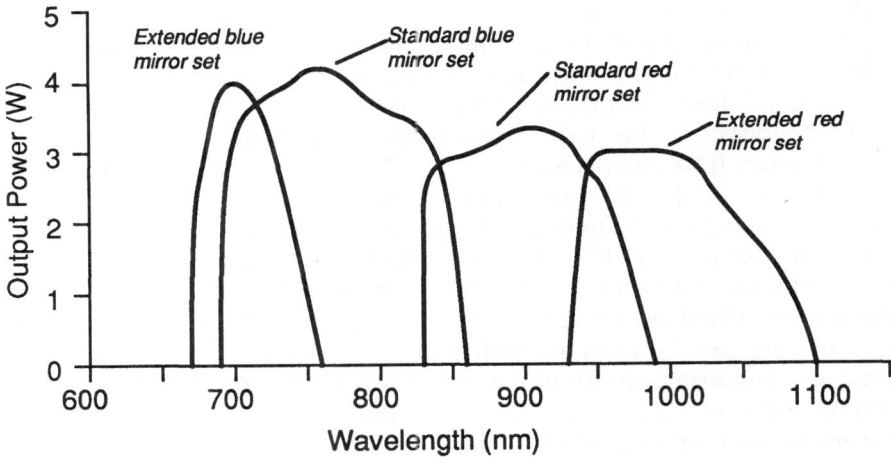

Fig. 10.3. Power tuning curve for a Ti:sapphire laser pumped by a 20 W argon ion laser. (Figure courtesy of Spectra-Physics)

High power pulsed Ti:sapphire lasers have been developed with both stable and unstable resonators, with and without injection control. *Kangas* et al. [9.115] demonstrated tunable single-mode pulse operation of a Ti:sapphire laser using a Littman configuration cavity. They were able to generate 2 mJ pulses with 2 ns duration, tunable from 746 to 918 nm. To increase the active volume of the pulsed laser, *Rines* and *Moulton* [9.116] explored the use of unstable resonators for high energy amplification. They explored both standing and traveling wave unstable resonator configurations, with a variety of output coupling schemes. In a standing wave cavity, they were able to generate 430 mJ with 1000 mJ pumping (532 nm). The power was limited by crystal damage. Using three prisms in a ring configuration, linewidths of 0.7 nm were obtained. However, an ultranarrow linewidth (< 100 MHz) required injection seeding from a cw single-mode laser.

Major advances have come in the formation of ultrashort femtosecond pulses using Ti:sapphire lasers. The principal results are summarized in Sect. 10.1.3, see [1.32, 33, 35].

10.9.2 New Chromium Lasers

The ion Cr^{3+} has been incorporated into a number of new host lattices. The interested reader is referred to the conference digests of annual solid state laser conferences for an overview on all the new systems [9.117−119]. Notable new crystals include $Cr^{3+}:LiCaAlF_6$ (Cr:LiCAF) and $Cr^{4+}:$forsterite. Cr:LiCAF lases in the 720−850 nm region, and has broad pump absorption bands in the 500−700 nm region [9.120]. In the LiCAF crystal, excited state absorptions are minimized, leading to efficient operation of the laser. With a

424   C. R. Pollock

radiative lifetime of approximately 200 μs, it is an ideal candidate for flashlamp pumping and energy storage. Recent results have generated output powers of several joules, with slope efficiencies approaching 50% [9.121].

Cr-doped forsterite (Mg_2SiO_4) has been demonstrated to lase in pulsed and cw modes over the 1.17–1.34 μm range. The radiative lifetime of this system ranges from approximately 25 μs at 77 K, to 3 μs at room temperature. *Petricevic* et al. [9.122] first reported room temperature laser action in forsterite using pulsed pumping. They identified the laser-active ion as being Cr^{3+}. *Verdun* et al. [9.123] suggested that the laser-active ion was in fact Cr^{4+}. This has led to a spirited debate in the community [9.124–126], and the issue is still not clearly resolved. *Petricevic* et al. [9.127] reported cw operation of forsterite by pumping with a cw Nd:YAG laser at 1.06 μm. They generated approximately 100 mW of power using a chopped pump, and about 40 mW under true cw pumping. *Carrig* and *Pollock* [9.128] reported multiwatt cw operation of forsterite tunable from 1.2 to 1.32 μm by cooling a 20 mm long crystal to liquid nitrogen temperature. A power tuning curve for Cr:forsterite is shown in Fig. 10.4.

Cr-doped YAG has been reported to lase from 1300 to 1550 nm [9.129] in pulsed operation. The fluorescence lifetime of Cr:YAG is nearly identical to Cr:forsterite in terms of magnitude and temperature dependence. Several spectroscopic studies have tentatively established the identification of the Cr^{4+} ion as being the laser impurity [9.130, 131]. In concert with forsterite, the two Cr^{4+} systems described here could eventually provide smoothly tunable operation over the 1.15–1.6 μm range, if the crystal refinement proceeds in a fashion similar to that for Ti:sapphire.

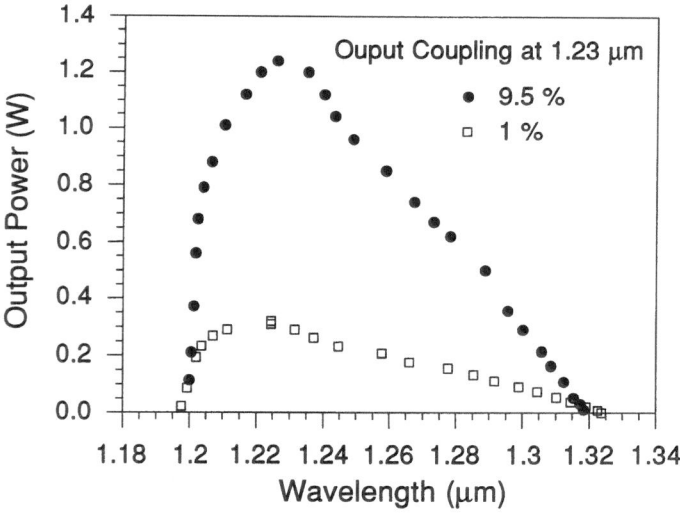

Fig. 10.4. Power tuning curve of a Cr:forsterite laser operating at 77 K with 6 W pump power from a 1.06 μm Nd:YAG laser. Room temperature power is approximately 35% of that shown here

10.9.3 Rare Earth Lasers

Notable progress has been made in the development of Ho lasers for medical and down-conversion application. Ho-doped crystals nominally lase at 2.09 μm, depending slightly on the lattice. The attractive feature of many Ho lasers is the use of sensitizers to enhance efficiency. Originally Er was used, but recent work with diode pumping [9.132, 133] has used Tm. The Tm ions form broad absorption bands which lie slightly above the upper laser level of the Ho ions, allowing for efficient energy transfer between the states. Detailed analysis of this process has been developed by *Antipenko* [9.134].

The thulium laser is emerging as a powerful mid-infrared laser. *Quarles* et al. demonstrated room temperature operation, tuning over the range from 1.94 to 2.09 μm in a flashlamp configuration [9.135]. This was later extended to cw operation using a Ti:sapphire pump source. The tuning range extended from 1.85 to 2.15 μm, with peak output powers of 170 mW [9.136]. Recent results using diode pumping have achieved cw powers exceeding 6.2 W using 20 W from a fiber-coupled GaAlAs diode laser array [9.137]. *Pinto* et al. have reported a mode-locked Tm laser that produces 35 ps pulses when pumped with a Ti:sapphire laser. The mode-locking was achieved by an intracavity acousto-optic modulator [9.138].

Finally, there has been much work on Er fiber lasers and amplifiers. The motivation for this work lies in the ability of Er-doped fibers to serve as optical amplifiers for 1.5 μm optical communication systems. Early results by *Desuvire* et al. [9.139] use argon ion laser pumping to create optical gains on the order of 22 dB in a single-mode fiber. More recent results include the use of 980 nm diode pumping [9.140] and diode or color center laser pumping at 1.48 μm [9.141]. Recent results in this field can be found in a review by *Laming* et al. [9.140] and in [9.141].

10.9.4 New References for Chapter 9

9.108 A. Sanchez, A.J. Strauss, R.L. Aggarwal, R.E. Fahey: IEEE J. **24**, 995 (1988)
9.109 Special Issue on Solid State Lasers, IEEE J. QE-**24**, 1988
9.110 P.F. Moulton: J. Opt. Soc. Am. B **3**, 125 (1986)
9.111 Power tuning curves from a Spectra-Physics model 3900 laser
9.112 A.J. Alfrey: IEEE J. QE-**25**, 760 (1989)
9.113 P.A. Schultz: IEEE J. QE-**24**, 1039 (1988)
9.114 J. Harrison, A. Finch, D.M. Rines, P.F. Moulton: Opt. Lett. **16**, 581 (1991)
9.115 K.W. Kangas, D.D. Lowenthal, C.H. Muller III: Opt. Lett. **14**, 21 (1989)
9.116 G.A. Rines, P.F. Moulton: Opt. Lett. **15**, 434 (1990)
9.117 M.L. Shand, H.P. Jenssen (eds): *Tunable Solid State Lasers*, Vol. 5, OSA Proc. Ser. (Optical Soc. of America, Washington, DC 1989)
9.118 H.P. Jenssen, G. Dubé (eds): *Advanced Solid State Lasers*, Vol. 6, OSA Proc. Ser. (Optical Soc. of America, Washington, DC 1990)
9.119 G. Dubé, L. Chase (eds): *Advanced Solid State Lasers*, Vol. 10, OSA Proc. Ser. (Optical Soc. of America, Washington, DC 1991)
9.120 S.A. Payne, L.L. Chase, H.W. Newkirk, L.K. Smith, W.F. Krupke: IEEE J. QE-**24**, 2243 (1988)

9.121 L.L. Chase, S.A. Payne, L.K. Smith, W.L. Kway, H.W. Newkirk: In Ref. [9.117]
9.122 V. Petricevic, S.K. Gayen, R.R. Alfano, K. Yamagishi, H. Anzai, Y. Yamaguchi: Appl. Phys. Lett. **52**, 1040 (1988)
9.123 H. Verdun, L.M. Thomas, D.M. Andrauskas, T. McColum, A. Pinto: Appl. Phys. Lett. **52**, 2593 (1988)
9.124 V. Petricevic, S.K. Gayen, R.R. Alfano: Appl. Phys. Lett. **53**, 2590 (1988)
9.125 K.R. Hoffman, S.M. Jacobsen, J. Casa-Gonzales, W.M. Yen: In Ref. [9.119], pp. 44–48
9.126 J. Casa-Gonzales, S.M. Jacobsen, K.R. Hoffman, W.M. Yen: In: Ref. [9.119], pp. 64–69
9.127 V. Petricevic, S.K. Gayen, R.R. Alfano: Opt. Lett. **14**, 612 (1989)
9.128 T.J. Carrig, C.R. Pollock: Opt. Lett. **16**, 1662 (1991)
9.129 N. Angert, N.I. Borodin, V.M. Garmash, V.A. Zhitnyuk, A.G. Okhrimchuk, O.G. Siyuchenko, A.V. Shestakov: Sov. J. Quantum Electron. **18**, 73 (1988)
9.130 W. Jia, M.M. Tissue, L. Lu, K.R. Hoffman, W.M. Yen: In Ref. [9.119], pp. 87–91
9.131 S. Kück, K. Petermann, G. Huber: In Ref. [9.119], pp. 92–94
9.132 G. Kintz, L. Esterowitz, R. Allen: Electron. Lett. **23**, 616 (1987)
9.133 T.Y. Fan, G. Huber, R.L. Byers, P. Melzcherlech: IEEE J. QE-**24**, 929 (1988)
9.134 B.M. Antipenko: In Ref. [9.119], pp. 207–209
9.135 G. Quarles, A. Rosenbaum, C.L. Marquardt, L. Esterowitz: Opt. Lett. **15**, 42 (1990)
9.136 R. Stoneman, L. Esterowitz: Opt. Lett. **15**, 486 (1990)
9.137 D. Shannon, D. Vecht, T. Kane, R. Wallace: IEEE LEOS '91 Conf. Digest, talk ELT6.4 (1991)
9.138 J. Pinto, L. Esterowitz, G. Rosenblatt: IEEE LEOS '91 Conf. Digest, talk ELT6.5 (1991)
9.139 E. Desuvire, J.R. Simpson, P.C. Becker: Opt. Lett. **12**, 888 (1987)
9.140 R.I. Laming, D.N. Payne, G.J. Cowle: Proc. SPIE **1314** (1990)
9.141 P.W. France (ed) *Optical Fibre Lasers and Amplifiers.* CRC, Cleveland, OH (1991)

Subject Index

Topics in Applied Physics Founded by Helmut K. V. Lotsch